PLANETARY SCIENCE
THE SCIENCE OF PLANETS AROUND STARS

PLANETARY SCIENCE
THE SCIENCE OF PLANETS AROUND STARS

George H A Cole
Department of Physics, University of Hull, UK

Michael M Woolfson
Department of Physics, University of York, UK

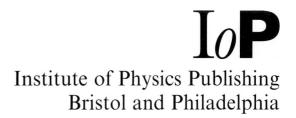

Institute of Physics Publishing
Bristol and Philadelphia

British Library Cataloguing-in-Publication Data

A catalogue record of this book is available from the British Library.

ISBN 0 7503 0815 X

Library of Congress Cataloging-in-Publication Data are available

Commissioning Editor: John Navas
Production Editor: Simon Laurenson
Production Control: Sarah Plenty
Cover Design: Frédérique Swist

Published by Institute of Physics Publishing, wholly owned by
The Institute of Physics, London

Institute of Physics Publishing, Dirac House, Temple Back,
Bristol BS1 6BE, UK

US Office: Institute of Physics Publishing, The Public Ledger Building,
Suite 1035, 150 South Independence Mall West, Philadelphia,
PA 19106, USA

Typeset by Academic + Technical, Bristol
Printed in the UK by Bookcraft, Midsomer Norton, Somerset

CONTENTS

INTRODUCTION

In choosing a title we had in mind that there are many planetary systems other than the Solar System. The book is concerned with the science associated with the planets, the stars that they orbit and the interactions between them. The relationships of several extra-solar planets to their parent stars differ from that of any Solar-System planet to the Sun and this can give clues either about the way that planets are formed or the way that they evolve after formation. For this reason we conclude with a chapter giving current ideas about the way that planetary systems come into being. There is general agreement that the formation of planets is intimately connected with the formation of stars—although there are important differences of view about the nature of the connection. To give a rounded and complete picture we include material on the formation, evolution and death of stars and those properties of the Sun that influence the planets of the Solar System.

The origin of the study of the Solar System, at a truly scientific level, occurred in the seventeenth century when Newton explained the motion of Solar-System bodies by the application of the laws of mechanics combined with the inverse-square law of gravitational attraction. With subsequent improvements in telescope technology and, more latterly, through the achievements of space science we now have detailed descriptions of many Solar-System bodies and have been able to analyse samples from some of them. The range of what constitutes stellar and planetary science has expanded in almost explosive fashion in the past few decades and includes aspects of many different conventional sciences—although physics and astronomy certainly predominate.

There are many excellent textbooks that describe stars and the Solar System in some detail and give qualitative explanations for some features and quantitative explanations where the underlying science is not too complicated. At the other extreme there are monographs and papers in learned journals that deal with aspects of stellar and planetary science in a rigorous and formal way that is suitable for the specialist and where, sometimes, jargon is used that is incomprehensible to the outsider. The readership we have in mind for the present work is the senior undergraduate student in physics or astronomy or the new graduate student working in planetary science who requires an overview of the whole subject before embarking on detailed study of one narrow aspect of it. Our analyses of aspects of stellar and planetary science are aimed to be accessible to such students—or, indeed, to any others meeting the field for the first time.

There are two main components of this text. The first of these is a general overview of the nature of stars and of the Solar System that can be read independently and quotes the important results that have been obtained by scientific analysis. For those unfamiliar with stellar properties or the overall structure of the Solar System we recommend that this part should be read before looking at the other material, to acquire a general picture of the system as a whole and the interrelationships of the bodies within it.

The second component is that which justifies the title of this work. It is a set of 41 topics in which the detailed science is described. The topics are very variable in length. Some, for example Topic A that deals with mineralogy, are as long as a normal chapter of a book. Others, for example Topic AG concerned with the mechanical interactions of radiation and matter, are one or two pages long. Together these topics provide a description of the great bulk of the underlying science required to explain the main features of the Solar System.

Problems are given at the end of chapters and most topics, designed to give the reader a quantitative feeling for stellar and Solar-System phenomena. Solving such problems clearly has some educational value but, even when the reader fails to solve a problem, reference to the provided solution may offer useful insights.

Our Earth and the other planets have undergone substantial changes in their states over many aeons by the action of natural forces. An understanding of the nature of the Solar System and of the influences that govern its behaviour may allow an appreciation to be developed of what can influence our planet in the future.

CHAPTER 1

THE UNITY OF THE UNIVERSE

Studies of the Universe, especially in recent times, have brought us to the realization that it is an entity and not a number of disconnected and unrelated units. Viewing stars, galaxies and gas clouds, either nearby or in deepest space (which is also equivalent to observing the Universe either in very recent times or long ago), shows the same formations involving broadly the same chemical compositions. The galaxies are made of essentially the same components with common mean chemical compositions. Gas clouds form a family of like objects. For this reason it is possible to consider a representative galaxy, a representative star or a representative cloud of material. There are, of course, variations of composition but the variations will be small deviations about some mean.

1.1. COSMIC ABUNDANCE OF THE CHEMICAL ELEMENTS

In most cases the composition of an object can only be studied from a distance. The radiation it emits is analysed in a spectrometer to determine the frequencies present and also the relative intensities of the spectral emission or absorption lines. Modern studies cover the whole range of the electromagnetic spectrum from the most energetic regions (gamma rays and X-rays) to the least energetic (radio waves). Radioactive elements emit particles during decay and these can also be detected. Each chemical element emits a characteristic range of frequencies when stimulated in different ways so that the component chemical elements can be determined. Molecules usually have vibration and electronic transitions, closely spaced in energy, that give characteristic spectral bands so enabling their presence to be determined if they are stable under the prevailing conditions.

This procedure has its limitations. Bodies cannot be examined in interior regions if the radiation or particles cannot escape so that, for example, the composition of the Sun can be found directly only in its surface regions. To infer the composition of the inner regions requires the exercise of theory, which may need to be modified as more information accumulates. Bearing this in mind, the cosmic abundance of the chemical elements is generally accepted as that given in table 1.1. The abundances are given relative to silicon as the unit.

Hydrogen and helium are overwhelmingly the most abundant elements. Helium is an inert gas; the second most abundant chemically active element is oxygen. The oxide of hydrogen is water, H_2O, so we will not be surprised to meet a high abundance of water in its various phases. The third and fourth most abundant elements, carbon and nitrogen, are also chemically active giving simple compounds such as CO, CO_2, NH_3, CH_4 and many others. The vast number of carbon-based organic compounds, many very complex and almost completely consisting of C, N, O and H, is the basis of life on Earth.

Table 1.1. *The cosmic abundance of the chemical elements relative to silicon.*

Element	Relative abundance (number of atoms with Si = 1)	Relative abundance by mass
Hydrogen, H	3.18×10^4	} 0.9800
Helium, He	2.21×10^3	
Oxygen, O	22.1	
Carbon, C	11.8	} 0.0133
Nitrogen, N	3.64	
Neon, Ne	3.44	0.0017
Magnesium, Mg	1.06	
Silicon, Si	1	
Aluminium, Al	0.85	
Iron, Fe	0.83	
Sulphur, S	0.50	} 0.00365
Calcium, Ca	0.072	
Sodium, Na	0.060	
Nickel, Ni	0.048	

Silicon is a relatively abundant element, as are magnesium, aluminium, calcium, sodium, potassium and iron. These elements, together with oxygen, form the great bulk of the silicate materials that constitute most of the Earth. Other non-silicate minerals, such as oxides and sulphides, also occur but they are much less common. Minerals form according to the local conditions of pressure and temperature and whether, in particular, cooling processes take place quickly or slowly. A systematic description of different kinds of minerals and the rocks that they form is given in Topic A.

For condensed matter, there are a few rules that control the general form. The controlling features are the affinities of different atoms to different types of crystalline bonds. In many cases the elements segregate into silicate, sulphide and metal according to the following general rules.

- Elements in Groups 1 and 2 of the Periodic Table have a tendency to combine with oxygen in oxides and silicates. Elements such as K, Ba, Na, Sr, Ca, Mg and Rb are called *lithophilic* elements (from the Greek for stone, *lithos*) because they tend to be found in stones.
- Elements in Groups 10 and 11 of the Periodic Table tend to appear as sulphides. Thus Cu, Zn, Pb, Sn and Ag are called *chalcophilic* elements (after the Greek *khalkos* for their leading member, copper).
- Some elements may combine in compounds as, for instance, silicates but may also appear in metallic form. Such elements are Fe, Ni, Co, As, Ir, Pt, Au and Ag. These are called *siderophilic* (after the Greek *sideros* for iron, their representative element).

A detailed analysis of the chemical affinities in different conditions is quite complicated but these simple rules are often useful in understanding particular situations.

1.2. SOME EXAMPLES

Before moving on it is useful to give some examples of the universality of chemical materials. The first involves the comparison between the mean compositions of the Solar System, the Orion nebula and a

Table 1.2. Log_{10} *(relative abundance), normalized to hydrogen as 12, for three different entities.*

Element	Solar system	Orion nebula	Planetary nebula
H	12.00	12.0	12.00
He	10.9	11.04	11.23
C	8.6	8.37	8.7
N	8.0	7.63	8.1
O	8.8	8.79	8.9
F	4.6		4.9
Ne	7.6	7.86	7.9
Na	6.3		6.6
S	7.2	7.47	7.9
Cl	5.5	4.94	6.9
Ar	6.0	5.95	7.0
K	5.5		5.7
Ca	6.4		6.4

planetary nebula. Data for 13 chemical elements are listed in table 1.2. The similarities between the three lists are very striking.

As expected, similar bodies frequently display similar mineral compositions. This is displayed in table 1.3 for a selection of minerals on the Earth and Venus, often regarded as sister planets because of their similar size and mass. As will be seen there are different compositions for different kinds of site on Earth and for the sites of the Russian Venera landers. The Venus data tend to be similar to the oceanic material on Earth. This suggests that Venus never underwent the processes forming crustal and continental material.

Meteorites form a rich source of information about minerals present in the Solar System. The ages of meteorites, as measured by radioactive dating (Topic B), are about 4.5×10^9 years—the accepted age for the Solar System as a whole. A list of the most common minerals found in meteorites is given in table 1.4. They are similar to those on Earth although they must originate from some condition or event in the very early Solar System.

These examples involving minerals exclude the most abundant elements, H and He, but these latter can be included by consideration of larger bodies. It is known that all normal stars are made

Table 1.3. *The relative abundance by percentage mass for regions of the Earth and Venus. The hostile conditions on Venus made rapid measurements essential; pressure, temperature and chemical attack destroyed each probe after about twenty minutes.*

Oxide	Earth (continental crust) (%)	Earth (oceanic crust) (%)	Venera 13 site (%)	Venera 14 site (%)
SiO_2	60.1	49.9	45	49
Al_2O_3	15.6	17.3	16	18
MgO	3.6	7.3	10	8
FeO	3.9	6.9	9	9
CaO	5.2	11.9	7	10
TiO_2	1.1	1.5	1.5	1.2
K_2O	3.2	0.2	4	0.2

Table 1.4. *The minerals most commonly found in meteorites.*

Mineral	Composition
Kamacite	(Fe, Ni) (<7% Ni)
Taenite	(Fe, Ni) (>13% Ni)
Troilite	FeS
Olivine	$(Mg, Fe)_2 SiO_4$
Orthopyroxene	$(Mg, Fe) SiO_3$
Pigeonite	$(Ca, Mg, Fe) SiO_3$
Diopside	$Ca(Mg, Fe) Si_2O_6$
Plagioclase	$(Na, Ca)(Al, Si)_4 O_8$

Table 1.5. *The main components of the visible regions of the Sun and the major planets by mass percentage.*

Molecule	Sun (%)	Jupiter (%)	Saturn (%)	Uranus (%)	Neptune (%)
H_2	85	89.9	96.3	82.5	80.0
He	15	10.2	3.3	15.2	19.0
H_2O	0.11	4×10^{-4}	—	—	—
CH_4	0.06	0.3	0.3	2.3	1.5
NH_3	0.016	0.026	0.012	—	—
H_2S	0.003	—	—	—	—

mostly of hydrogen with a substantial component of helium and a small admixture of other elements. This also applies to the major planets of the Solar System. A comparison between the main surface components of the Sun and the major planets is shown in table 1.5.

There is a general similarity between the different bodies but it must be stressed that these are surface components and do not refer directly to the interiors. There is, nevertheless, for Saturn and Jupiter the implicit assumption that these percentages would be very broadly the same if the full planetary inventory could be taken. The basis of this assumption is that these large and massive planets would have been formed from similar material that formed the Sun and that the escape velocity from them is so large that they would have retained all their original material. On this basis, the interior behaviour of Saturn, for example, is described in terms of a greater helium concentration than in surface regions. On the other hand the lesser proportion of hydrogen detected in Uranus and Neptune may be due to losses from these lower mass planets for which escape velocities are also lower.

Wherever we look in the cosmos we are seeing a very similar grouping of the chemical elements. It would seem safe to assume that the material composition of the Solar System and its neighbourhood is not untypical of such systems everywhere.

Problem 1

1.1   The density distribution of Saturn is modelled as

$$\rho = \rho_0 \left\{ 1 - \left(\frac{r}{R} \right)^{1/4} \right\}$$

where ρ_0 is the central density, r the distance from the centre and R the radius of the planet. The proportion of helium by mass is assumed to vary as

$$p = p_0(1 - \alpha r)$$

where p_0 is the central proportion and α is a constant. It is known that the surface proportion of helium is 0.033 and that the average for Saturn as a whole is 0.15.

(i)   What is the mass of Saturn in terms of ρ_0 and R?
(ii)   What is the mass of helium in terms of ρ_0, p_0 and R?
(iii) What is p_0?

CHAPTER 2

THE SUN AND OTHER STARS

2.1. THE INTERSTELLAR MEDIUM

The starting point for the formation of all objects in the galaxy is the interstellar medium (ISM). At first sight it appears a very unpromising source of material. The ISM is rather lumpy, less dense and hotter in some regions, denser and cooler in others. A characteristic state is with a density a few times 10^{-21} kg m^{-3} and temperature about 10 000 K. It is mostly hydrogen and helium but also contains 1–2% by mass of solid grains that may be ices, silicates or metal. Although it is so diffuse, the ISM actually contains a significant prtoportion of the mass of the galaxy. In the solar neighbourhood the density of stars is about 0.08 pc^{-3},[†] which is equivalent to 3×10^{-21} kg m^{-3}, assuming that the mean mass of a star is about $0.5 M_{\odot}$.[‡] Thus the mass of the ISM is comparable to the mass of the stars in the galaxy.

2.2. DENSE COOL CLOUDS

The first stage in the transformation of the diffuse ISM into dense objects like stars is the formation of dense cool clouds (DCCs). These can be detected because they obscure the light from stars behind them and appear as dark patches in the field of view of telescopes (figure 2.1). This raises the question of how a DCC can be formed from the ISM. A possibility that has to be considered is that of spontaneous collapse of the ISM under gravity and this will involve a consideration of the Virial Theorem (Topic C) and of the resultant Jeans critical mass (Topic D). A uniform gas sphere will experience gravitational forces tending to cause it to collapse and also thermal pressure forces that work in the opposite direction. A state of balance occurs when the mass of the sphere is given by

$$M_J = \left(\frac{375 k^3 T^3}{4\pi G^3 \mu^3 \rho} \right)^{1/2} \tag{2.1}$$

in which k is Boltzmann's constant, T the temperature, G the gravitational constant, μ the mean mass of a gas molecule and ρ the density of the gas. A general form is

$$M_J = Z \left(\frac{T^3}{\rho} \right)^{1/2} \tag{2.2}$$

[†] 1 parsec (pc) $= 3.083 \times 10^{16}$ m.
[‡] The symbol $\odot$ relates to the Sun so that $M_{\odot}$ is the mass of the Sun.

Figure 2.1. *The central dark feature is the Coal Sack, a dense cool cloud (David Malin, AAO).*

where $Z = 7.6 \times 10^{21} \, \mathrm{kg^{3/2} \, m^{-3/2} \, K^{-3/2}}$ for atomic hydrogen and $2.7 \times 10^{21} \, \mathrm{kg^{3/2} \, m^{-3/2} \, K^{-3/2}}$ for molecular hydrogen.

A typical mass of ISM material that could spontaneously collapse is found by putting $\rho = 10^{-21} \, \mathrm{kg \, m^{-3}}$ and $T = 10\,000 \, \mathrm{K}$ into (2.2) for, say, $Z = 5 \times 10^{21} \, \mathrm{kg^{3/2} \, m^{-3/2} \, K^{-3/2}}$ that allows for the presence of helium plus some molecular hydrogen. This mass is $1.6 \times 10^{38} \, \mathrm{kg}$ or $8 \times 10^{7} M_{\odot}$, much greater than the mass of a DCC. Actually, entities of this mass are unlikely to condense. The reason for this is that the timescale for the collapse would be so long that, well before any appreciable condensation of the ISM had happened in any particular region, the material in that region would have been stirred up and mixed with other material. The total time for a spherical body of initial density ρ_0 to collapse to high (theoretically infinite) density, the so-called *free-fall time*, is

$$t_{\mathrm{ff}} = \left(\frac{3\pi}{32\rho_0 G} \right)^{1/2}, \tag{2.3}$$

that is about 7×10^{7} years for the ISM. The form of collapse is such that at first it is very slow and then accelerates as shown in figure 2.2. Thus after 40% of the free-fall time the radius is still 90% of its original value. A theoretical derivation of the form of free-fall collapse is given in Topic E. It should be pointed out that for gaseous spherical bodies the effect of higher pressure and temperature as the body collapses slows down the collapse below the free-fall rate. However, for very transparent bodies, that quickly radiate away the heat energy produced by compression of the material, the form of the collapse is closely free-fall in the early stages.

As for the ISM, the conditions in DCCs are very variable. A typical condition is $\rho = 10^{-18} \, \mathrm{kg \, m^{-3}}$ and $T = 50 \, \mathrm{K}$. With the value of Z previously given this corresponds to a Jeans critical mass of $1.8 \times 10^{33} \, \mathrm{kg}$, or about $900 M_{\odot}$, and this is a not uncommon mass for a DCC. These DCCs are often referred to as molecular clouds because almost all the material within them is in molecular form—about 120 molecular species have been detected including H_2, CO and HCN.

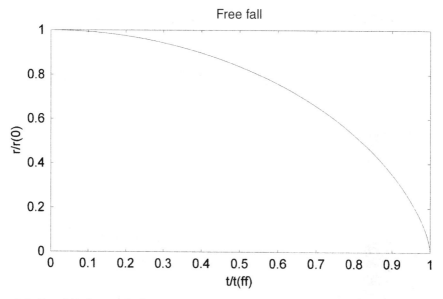

Figure 2.2. *Free fall, showing the fraction of the original radius against the fraction of the free-fall time.*

2.3. STELLAR CLUSTERS

The Sun is a *field star*, a star not closely associated with any other and moving independently through the galaxy. However, there are many stars that are seen in associations called *clusters* and in particular the formation of new stars, for example, as observed in the Orion nebula, seems to take place in a cluster mode.

There are two main types of cluster. The first of these is the *globular cluster* an example of which is shown in figure 2.3. It can be seen that there is a dense core in which individual stars cannot be resolved and the stellar density gradually reduces with distance from the centre. The total number of stars in a globular cluster is usually in the range 10^5 to 10^6 and the material of which they are formed contains very little ($\sim 0.01\%$) of any element heavier than helium, that is they have very low metallicity. These are old *Population II*, stars formed of hydrogen and helium—material very similar to that produced in the *big bang* that is believed to have initiated the creation of the universe. Globular clusters are found mainly in the diffuse halo that surrounds our galaxy. It is postulated that *Population III* stars may exist that consist of pure primordial big-bang material with absolutely no heavy element composition, but none have yet been found.

The second type of cluster is the *open* or *galactic cluster* (figure 2.4). Such clusters are *open* in the sense that the separation of the stars is large enough for them to be seen individually and *galactic* in the sense that they occur only in the galactic plane. Typical open clusters contain from 100 to 1000 stars and have a mean radius of between 2 and 20 pc, although they are usually quite irregular in shape. From their spectra it can be found that the constituent stars contain a 1–2% component of heavier elements by mass—that is they are *Population I* stars of similar composition to the Sun. The material in these stars has been through one or more cycles of star formation and evolution during which heavier elements have been produced. This material has been blown off the evolved star either in the form of a planetary nebula or in a supernova explosion and then became incorporated as part of the ISM. The ISM contains 1–2% of dust, representing a component of heavier elements, and stars produced from it will be Population I stars.

Figure 2.3. *The globular cluster M13 (Palomar Observatory/Caltech).*

Figure 2.4. *The Pleiades, a galactic cluster (Edinburgh Observatory/AAT).*

2.4.   A SCENARIO FOR FORMATION OF A GALACTIC CLUSTER

It is very likely that the formation of DCCs may be initiated by the effect of supernova explosions. To understand this we first need to consider the equilibrium of a DCC. Compared with the ISM, a DCC has a density larger by a factor of $\sim 10^3$ and a temperature lower by about the same factor. This means that the pressure in a DCC is similar to that in the ISM; it may be somewhat larger or somewhat smaller but it would not be too far from being in pressure equilibrium with the ISM.

The galaxy is traversed by energetic radiation and particles which we can identify as starlight and cosmic rays. These interact with matter, which can be in the form of dust grains, molecules, atoms or ions, and they are a source of heat. On the other hand there are also cooling processes taking place. The ISM and DCC material will be partially ionized and, because of the principle of equipartition of energy, low mass electrons will move quickly and frequently interact with other particles. As an example, an electron striking an atom or ion may promote one of its electrons into a higher energy state. Subsequently the electron will fall back into a lower energy state, emitting a photon. Because of the transparency of the medium this photon will leave the region and so carry away energy. The net effect in the vicinity of the interaction is that the original free electron loses energy, and this is equivalent to local cooling. If a DCC or a region of the ISM is to be in equilibrium then the heating and cooling processes taking place in it have to be in balance.

In regions well away from stars, cosmic rays will be the dominant source of heating and, to a crude approximation, cosmic ray heating in transparent regions is independent of the form of the material and can be expressed in units of $W\,kg^{-1}$. On the other hand all cooling processes are dependent on both density and temperature. We can regard the very low density ISM and DCCs as being perfect gases and so express the total cooling rate as a function of density and pressure. The form of this dependence is shown in figure 2.5 for a particular heating rate that we can take as the

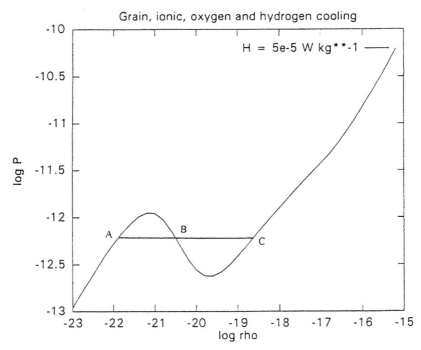

Figure 2.5. *The log P versus log ρ curve for typical ISM material and a cooling rate of $5 \times 10^{-5}\,W\,kg^{-1}$.*

cosmic-ray heating rate. We can see that the curve has a sinuous form and the horizontal line shows three points corresponding to the same pressure but three different densities (and hence temperatures) that give a cooling rate equal to the cosmic-ray heating rate. These points, A, B and C, are states of material that will be in pressure equilibrium with each other and in thermal equilibrium with the background radiation. Actually, point B represents an unstable state, corresponding to a part of the curve where increasing density leads to decreasing pressure. The other two represent a low-density, high-temperature state corresponding to the ISM (A) and a higher-density, low-temperature state corresponding to a DCC (C).

The action of a supernova explosion is both to inject heavier-element coolants into the ISM and also to compress it by a shock wave. Both these effects increase the cooling rate in the region. Cooling lowers the pressure in the region and the external ISM pressure then compresses it further, so increasing the cooling even more. It has been shown by numerical modelling (Golanski and Woolfson, 2001) that eventually the state changes from that at A, the ISM, to that at C, a DCC.

When the DCC is formed it will have a particular density and temperature, and if its mass exceeds the Jeans critical mass given by equation (2.1) then it will undergo a collapse. The collapsing DCC will be very turbulent with streams of gas occasionally colliding to give high-density, high-temperature regions. Because cooling processes are so efficient the high-density region quickly cools and may then satisfy the Jeans condition for further collapse to form a star. Computation has shown that sometimes colliding turbulent streams will give rise to binary or other multiple star systems (Whitworth *et al.*, 1995).

It has been shown theoretically (Woolfson, 1979), and also deduced from observations (Williams and Cremin, 1969), that the first stars produced are of mass about $1.35M_\odot$ and that later stars of decreasing and also increasing mass are formed (figure 2.6). The stream of decreasing mass is due to the increasing density of the DCC as it collapses. More, but smaller and more energetic, turbulent streams are formed and the smaller Jeans mass enables lesser mass stars to form. The stream of increasing mass stars in figure 2.6 is due to accretion by stars as they move through denser parts of the DCC. Stars produced directly by colliding streams spin fairly slowly, as is found from observation for lower mass stars. However, stars that become more massive by accretion also spin more quickly, again in line with observation.

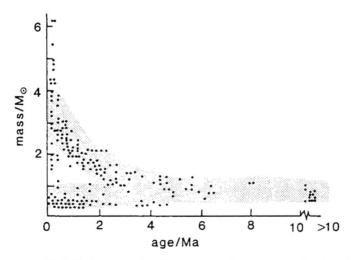

Figure 2.6. *The masses of individual stars produced in a young stellar cluster as a function of time. The origin represents 'now' (after Williams and Cremin, 1969).*

The scenario presented here for the formation of a galactic cluster is by no means certain but it does seem to explain in a rational way what is observed. In fact, all clusters eventually disperse as stars within them, exchanging energy in a many-body environment, occasionally attain escape speed. These escaping stars, or perhaps binary pair of stars, form the *field stars* of which the Sun is a typical member.

2.5. MAIN SEQUENCE STARS AND THEIR EVOLUTION

What we have described in section 2.4 is the formation of protostars—embryonic stars on the way to evolving into main sequence stars (Topic F). Main sequence stars are in a state of quasi-equilibrium (Topic G) where the energy emitted by radiation is balanced by energy generation through the conversion of hydrogen to helium in the stellar core (Topic H). The Sun, with an age of about 5×10^9 years, is half way through its main-sequence life and thereafter it will evolve through a red-giant stage until eventually it becomes a white dwarf, with almost its present mass but with a radius similar to that of the Earth. The form of the evolution of a star away from the main sequence is mass-dependent and is described in Topic I. The final state of the star—a white dwarf, a neutron star or possibly a black hole—is also mass-dependent. The theory of the structure of white dwarfs and neutron stars is given in Topic J.

2.6. BROWN DWARFS

The lowest mass for normal main-sequence stars, ones that emit in the visible part of the spectrum and so may be observed with optical telescopes, is about $0.075 M_\odot$. This is the lowest mass that gives rise to a sufficiently high temperature in the core to give hydrogen conversion to helium and hence a high enough effective black-body temperature to give visible light.

At somewhat below this mass the temperature in the core will still be high enough (a few million K) to give nuclear reactions involving deuterium, the stable heavy form of hydrogen. Deuterium forms a fraction about 2×10^{-5} of the hydrogen in the galaxy at large, an insufficient quantity to increase the core temperature to the point where reactions involving normal hydrogen can take place. Objects in this class are known as brown dwarfs and they have masses down to $0.013 M_\odot$, below which not even deuterium reactions can take place. It is usual to classify bodies with mass below this, corresponding to 13 times the mass of Jupiter ($13 M_J$), as planets. Planets are cold bodies for which gravity is a significant internal force. They range in mass from about 10^{19} kg upwards (Topic P). The radius increases with the mass up to a maximum R_{max} at a critical mass M_c ($\sim 2 M_J$ for hydrogen but other compositions are possible) when the strength of gravity inside begins to cause electron degeneracy (section J.2) in the central regions. The degree of degeneracy increases with the mass, progressively modifying the internal structure. The compressibility increases in consequence, the radius now tending to decrease with increasing mass, but the effect is small, especially for hydrogen up to the deuterium limit $13 M_J$ when the body ceases to be cold. The range of cold bodies, therefore, falls into the two categories $M < M_c$ and $M > M_c$, each with radii below R_{max}. M_J is the highest mass within the Solar System ($M < M_c$ there) and these bodies have been the model for planets in the past. The regime of cold bodies beyond M_c is yet to be explored in detail.

2.7. STELLAR COMPANIONS

Solar-type stars often, perhaps usually, have a gravitational link with one or more companions. A companion may have a mass comparable with that of the star, in which case the two form a binary

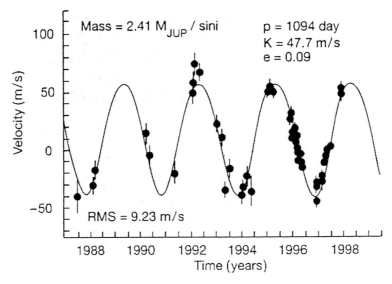

Figure 2.7. *The measured Doppler shifts from 47 Uma (after Butler and Marcy, 1996).*

star system. Alternatively, the mass may be less, and a star–brown-dwarf binary is formed, or the companion may even be a planetary body.

The first evidence of a planetary companion to a star came from observations of a pulsar. A pulsar is a rotating neutron star that emits a regular stream of radio pulses of period in the millisecond to a few seconds range. The period is constant to within about 1 part in 10^8, or better, over time periods of a year or so. If a star has a single companion then the two bodies move around the centre of mass with the speed of motion of the bodies inversely proportional to their masses. Thus if the plane of the orbit is not perpendicular to the line of sight then the star will have a component of motion along the line of sight, towards and away from the observer in a periodic way. This motion is with respect to the centre of mass of the star–companion system that may itself be moving with respect to the observer—but this just affects the mean observed period of the radio pulses. When the pulsar is moving away from the Earth then there is a cumulative delay in the arrival of pulses since they have farther to travel, and when it moves towards the Earth there is the opposite effect. If there is more than one companion then the motion of the star is more complicated but can still be interpreted. In 1992 it was reported that the pulsar PSRB1257+12 probably had three planetary companions and later these were confirmed to have masses $3.4M_\oplus$, $2.8M_\oplus$ and $\sim100M_\oplus$.[†]

The existence of planetary companions to normal stars was first detected in 1995 when a planet was detected orbiting Peg 51. The technique of detection depends on the continuous measurement of the Doppler shift of spectral lines in the light from the star (Topic K). Where the star has a single companion both bodies move around the centre of mass. If the plane of the orbit is in the line of sight then the radial motion of the star has a sinusoidal variation of velocity (figure 2.7), the measurement of which gives the mass and orbit of the planet. It is assumed that the mass of the star is known, which it will be because mass is closely linked to spectral class for main sequence stars.

[†] The symbol $\oplus$ represents the Earth so $M_\oplus$ is one Earth mass.

Table 2.1. *Planets detected around normal stars. The masses, in Jupiter units, are minimum masses.*

Star	Minimum planet mass (M_J)	Period (days)	Semi-major axis (AU)	Eccentricity
HD187123	0.52	3.097	0.042	0.03 ± 0.03
HD75289	0.42	3.51	0.046	0.054
τ-Bootis	3.87	3.313	0.0462	0.018 ± 0.016
51 Peg	0.47	4.229	0.05	0.0
υ-Andromedae	0.71	4.62	0.059	0.034 ± 0.015
	2.11	241.2	0.83	0.18 ± 0.11
	4.61	1266	2.50	0.41 ± 0.11
HD217107	1.28	7.11	0.07	0.14
55 Cnc	0.84	14.65	0.11	0.051 ± 0.013
	>5	>8 y	>4	
Gleise 86	4	15.78	0.11	0.046
HD195019	3.43	18.3	0.14	0.05
Gleise 876	2.1	60.85	0.21	0.27 ± 0.03
ρ-CrB	1.1	39.65	0.23	0.028 ± 0.04
HD168443	5.04	57.9	0.277	0.54
HD114762	11	84.03	0.3	0.334 ± 0.02
70 Vir	6.6	116.6	0.43	0.4
HD210277	1.28	437	1.097	0.45
16 CygB	1.5	804	1.70	0.67
47 Uma	2.41	3.0 y	2.10	0.096 ± 0.03
	0.75	7.11 y	3.73	<0.1
14 Her	3.3	1619	2.5	0.354 ± 0.088

However, in general the angle, i, between the normal to the orbital plane and the line of sight is unknown. The component of radial velocity of the star along the line of sight is still sinusoidal but the deduced mass of the planet is uncertain by a factor $1/\sin i$ so that only a minimum mass is known for the planet. Table 2.1 gives the characteristics of the first 18 detected planets around stars. They are listed in ascending order of semi-major axis.

The best conditions for detecting extra-solar planets are when the speed of the planet around the centre of mass is large, for then the Doppler shift is more easily measured, and when the period is short, for then measurements can be made in a relatively short time and several orbits can be followed. These conditions are met if the planet is massive and in a very close orbit. For this reason the extra-solar planets detected so far may not be a typical sample of the full range of masses of planets and orbits. The echelle spectrographs used for measuring the Doppler shifts can at present measure speeds with an accuracy of about $3\,\mathrm{m\,s}^{-1}$. The speed of the Sun around the Sun–Jupiter centre of mass (ignoring the other planets) is $13\,\mathrm{m\,s}^{-1}$ so Jupiter could be detected from afar although the measurements would not give great precision unless the time of the observations covered several years to include a significant fraction of a single orbit.

From the frequency of detection of planets around nearby stars it has been concluded that about 3–6% of stars like the Sun have at least one planetary companion. Not only does this influence the way that we think about the status of the Solar System in the galaxy and the universe but also about the possibility of the existence of intelligent life elsewhere than on Earth (Topic AM).

Problem 2

2.1   The proportion of stars with masses between M and $M + dM$ is given by

$$P(M)\,dM = CM^{-2.35}\,dM$$

where C is a scaling constant.
(i)   If the lowest mass main sequence stars have a mass of $0.075M_\odot$, then what is the mean mass of a main sequence star?
(ii)   If the same relationship extended into the brown-dwarf region, with a lower mass limit of $0.013M_\odot$, then what would be the ratio of the number of brown dwarfs to the number of main sequence stars?

CHAPTER 3

THE PLANETS

Although the Sun is the dominant body in the Solar System, containing some 99.86% of its total mass, it is just a star like others of its spectral type. The entities that make a solar *system* are the planets and other bodies bound together by the Sun's gravitational field. The most important of these other bodies, both on account of their masses and the fact that one of them is our home, are the planets.

3.1. AN OVERVIEW OF THE PLANETS

There are eight substantial planets in orbit around the sun, seemingly in two distinct families. Their relative sizes and mean orbital radii are shown in figure 3.1. The outer family, the major planets with widely spaced orbits, contains four relatively large bodies consisting mainly of gaseous material; the largest of them, Jupiter, has a diameter one-tenth that of the Sun. The four bodies of the inner family, the terrestrial planets with more closely spaced orbits, are much smaller and much less massive rocky bodies. The largest terrestrial planet, the third out from the Sun, is our own Earth. Its radius is less than one-tenth that of Jupiter and it supports a wide range of biological systems including humankind. It is of particular interest to understand the factors that enable it to do this. One of the important questions that scientists are addressing is whether planets in other systems can support life and, if that life is intelligent, whether communication can be established (Topic AM).

The two families of planets are separated by a gap as though a planet is missing, but this gap is actually occupied by a large number of very small bodies. These are the asteroids that will be discussed further in chapter 8. To complete the planetary inventory we must add a ninth body, although this planet, Pluto, the outermost one is extremely small—smaller even than the Moon. It also has an orbit that is more eccentric and more inclined than that of any of the others, which all adds to the general impression that this body is something of a misfit in the planetary family.

3.2. ORBITAL MOTIONS

Every planet moves around the Sun in an orbit that is influenced not just by the dominant effect of the Sun's mass but also, to a small extent, by the other bodies in the Solar System, in particular the other planets. The laws of planetary motion were first formulated by Johannes Kepler (1571–1630). These state that:

1. Planets have elliptical orbits with the Sun at one focus.

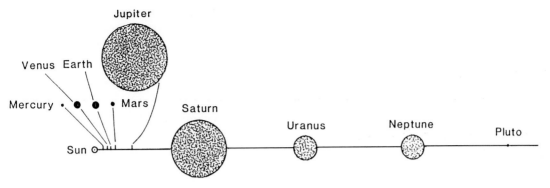

Figure 3.1. *The orbital radii and sizes of planets (different scales).*

2.   The line connecting the planet to the Sun sweeps out equal areas in equal time.
3.   The square of the orbital period is proportional to the cube of the semi-major axis of the orbit.

An ellipse is illustrated in figure 3.2. The point F is the focus occupied by the primary body, the Sun in our case, *a* is the semi-major axis and *b* the semi-minor axis of the ellipse. These quantities are related through the eccentricity, *e*, of the ellipse by

$$b^2 = a^2(1 - e^2). \tag{3.1}$$

For theoretical work connected with orbital motions an ellipse is best described in polar coordinates (r, θ) by

$$r = \frac{a(1 - e^2)}{1 + e\cos\theta}. \tag{3.2}$$

The semi-latus rectum of the ellipse, *p*, is shown in figure 3.2 and is the value of *r* when $\theta = \pm\pi/2$ or

$$p = a(1 - e^2). \tag{3.3}$$

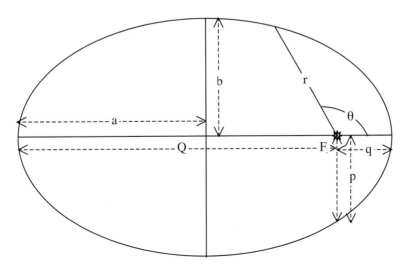

Figure 3.2. *The geometry of an ellipse.*

The closest distance of the planet to the Sun is when $\theta = 0$ and this distance, q, is known as the *perihelion distance* or sometimes just *the perihelion*, a term that is also used for the nearest point of the orbit. Similarly the *aphelion distance* (or *aphelion*) Q is the farthest distance corresponding to $\theta = \pi$. From equation (3.2)

$$q = a(1 - e) \quad \text{and} \quad Q = a(1 + e). \tag{3.4}$$

Kepler's deductions about the form of planetary motions were based purely on observations, mainly those of Tycho Brahe (1546–1601) to whom he was an assistant for many years. However, like Nicolaus Copernicus (1473–1543), who was the main protagonist for a heliocentric model of the Solar System in the Middle Ages, Kepler had no idea of the fundamental physics that led to these orbits. A brief historical account of the development of understanding and knowledge about the Solar System up to the middle of the seventeenth century is given in Topic L.

A complete understanding of orbital mechanics was not available until the basic work on gravitation by Isaac Newton (1642–1727), who contributed over a vast range of physics and mechanics and is arguably the greatest scientist who has ever lived. He started from the idea that the force that caused an apple to fall from a tree towards the Earth is the same force that holds the Moon in its orbit around the Earth and the Earth in its orbit around the Sun. He deduced that the magnitude of the force between two gravitationally attracting bodies was the same on each and was proportional to the product of the masses of the bodies and the inverse square of the distance between them, i.e.

$$F = G \frac{m_1 m_2}{r_{12}^2}. \tag{3.5}$$

The *gravitational constant*, G, has been determined by laboratory experiments and is $6.67 \times 10^{-11} \, \text{m}^3 \, \text{kg}^{-1} \, \text{s}^{-2}$. It is the least well determined of all the physical constants—although we can determine the product of G with mass quite accurately. This has implications for the estimates made of the masses of astronomical bodies.

It is possible to deduce the form of the law of gravitational attraction from Kepler's laws and, conversely, to deduce Kepler's laws from the law of gravitational attraction (Topic M).

While a and e completely define the size and shape of the orbit, in order to completely define the orbit in space it is necessary to add three orientation angles and a time fix, i.e. defining a time at which the orbiting body is at perihelion. Defining angles must be done in relation to a coordinate system and the *ecliptic*, the plane of the Earth's orbit around the Sun, is taken as the x–y plane in a rectangular Cartesian system. The positive z axis is taken towards the north so that all that remains is to fix an x direction in the ecliptic. Relative to the Earth the Sun moves in the ecliptic and twice a year it crosses the Earth's equatorial plane, at which times there are the *equinoxes* when all points on Earth have equal periods of day and night. The equinox when the Sun moves from north to south of the equator is called the *autumnal equinox* and the other, when the Sun moves from south to north is called the *vernal equinox*. The direction defined by the latter is called the *First point of Aries* and this is taken as the direction of the x axis. These planes and directions are illustrated in figure 3.3a.

The definition of the angles that fix a planetary orbit in space can be followed in figure 3.3b. The first angle that defines an orbit is i, the *inclination*, that is the angle made by the plane of the orbit with the ecliptic. The intersection of the orbital plane with the ecliptic defines the *line of nodes*; the point on this line when the orbiting body moves from south to north is the *ascending node* and when it moves from north to south the *descending node*. The other two angles that completely define the orbit are expressed relative to the ascending node. The first relates to the ecliptic plane and is the *longitude of*

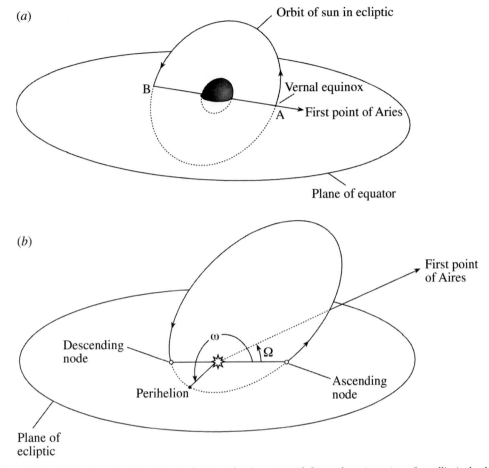

Figure 3.3. *(a) Defining the First Point of Aries. (b) Quantities defining the orientation of an elliptical orbit.*

the ascending node, Ω, which is the angle between the ascending node and the First point of Aries. The second angle relates to the orbital plane and is the *argument of the perihelion* ω, that is the angle in the direction of motion from the ascending node to the perihelion.

To define the position of the body at any time requires, in addition, some time-dependent information, usually the *time of perihelion passage*, T_{p}. With all six quantities (a, e, i, Ω, ω and T_{p}) given, the motion of the body is completely defined. Since the orbit must also be defined if both the position **r** and the velocity **v** are known, also requiring six items of information, then clearly it is possible to transform between the two sets of defining quantities.

3.3. ORBITS OF THE PLANETS

The quantities a, e and i for the planets are shown in table 3.1. The division into terrestrial and giant planetary families is just about evident in the distribution of the planetary orbital radii, although the gap between the two groups of planets is not too obvious. The orbital radii are given in *astronomical units* (AU), the mean Earth–Sun distance. The orbital characteristics reveal that most orbits are close to circular and close to being coplanar with the greatest deviations from this pattern being shown by the innermost and outermost bodies, Mercury and Pluto.

Table 3.1. *Characteristics of the planetary orbits.*

Planet	Mean distance (AU)[†]	Bode's law	Orbital eccentricity	Orbital inclination	Period (years)
Mercury	0.387	0.4	0.2056	$7°\,0'$	0.2409
Venus	0.723	0.7	0.0068	$3°\,24'$	0.6152
Earth	1.000	1.0	0.017	—	1.000
Mars	1.524	1.6	0.093	$1°\,51'$	1.8809
Asteroids	—	2.8	—	—	—
Jupiter	5.203	5.2	0.048	$1°\,18'$	11.8623
Saturn	9.539	10.0	0.056	$2°\,29'$	29.458
Uranus	19.19	(19.6)	0.047	$0°\,46'$	84.01
Neptune	30.07	(38.8)	0.0086	$1°\,47'$	164.79
Pluto	39.46	(77.2)	0.249	$17°\,19'$	248.54

[†] 1 astronomical unit (AU) $= 1.496 \times 10^{11}$ m.

The planets known to the ancients were those out to Saturn, and this state of affairs was unchanged until William Herschel (1738–1822) discovered Uranus in 1781. In 1772 Johann Bode (1747–1826) noticed that the planetary orbital radii out to Saturn, the outermost known planet, were close to a fairly simple mathematical progression, which became known as Bode's law. This is illustrated in the third column of table 3.1. The law is of the form

$$r_n = 0.4 + 0.3 \times 2^n \tag{3.6}$$

where the leading term is the radius of Mercury's orbit in AU and for the remaining planets $n = 0, 1 \ldots$ for Venus, Earth, etc. However, to fit the law $n = 3$ has to be assigned to the asteroids and, in fact, the mean of the orbital radii of the larger asteroids fit the law well. When Uranus was discovered and its orbital radius was found to fit the law, it was generally believed that the law had some fundamental validity, although without any physical basis being established for it.

In the early nineteenth century telescopes were quite refined and the principles of celestial mechanics were well established. When Uranus was discovered, a search of earlier records revealed that it had previously been observed but not recognized as a planet. Careful observations of Uranus over 20 years or so showed that it was departing from its expected path and it was suspected that some unknown perturber was the cause. In 1845 a Cambridge student, John Couch Adams (1819–1892), explained the residuals of the motion of Uranus in terms of an unknown planet and he sent the predicted position to Airy, the Astronomer Royal. Unfortunately, Airy was not inclined to waste telescope time on the predictions of a student and, in any case, he was sure that the residuals of Uranus were due to a breakdown of the inverse square law of gravitation at large distances. In 1846 a French astronomer, Urbain Leverrier (1811–1877), independently predicted the position of an unknown planet that agreed with that of Adams very closely. His prediction was sent to Galle at the Observatory in Berlin and the new planet, Neptune, was discovered straightaway.

The discovery of Neptune in 1846 gave the first major breakdown of Bode's law. In fact, even after the discovery of Neptune, observations of Uranus and Neptune suggested further unexplained residuals (now believed not to exist) and a search was mounted for Planet X, a planet even farther from the Sun. The first prediction of the mass of Planet X was six times that of the Earth—a substantial body—but nothing was found for a considerable time. The main searchers for the new planet were Percival Lowell (1855–1916) and William Pickering (1858–1938) of the Lowell Observatory with a telescope devoted to the search. Eventually the new planet, Pluto, was discovered in 1930 by Clyde Tombaugh, fairly close to the position predicted by Lowell and Pickering. Its orbital radius departed even farther from the Bode's law pattern. Most astronomers do not now believe that Bode's law has any fundamental significance although there are some theories of the origin of the Solar System that attempt to find it as an outcome.

Although Bode's law seems to have no physical basis for linking the orbital radii there are nevertheless some striking relationships between the *periods* of the major planets and Pluto. These relationships are in the form of orbital commensurabilities, that is where the ratio of the periods of a pair of planets is very close to the ratio of two small integers. Very often the ratio is not exact but close enough to suggest that some underlying physical process is trying to achieve the exact ratio. The most striking of these is the 'great' commensurability connecting the periods of Jupiter and Saturn. This is of the form

$$5n_S - 2n_J = 0.007\,127\,\text{year}^{-1}, \tag{3.7}$$

where n represents the *mean motion*, which is the mean angular speed in the orbit, in this case expressed in the units radians per year. If the right-hand side were zero then there would be an exact commensurability with the periods of Jupiter and Saturn in a 2 : 5 ratio.

Another striking commensurability is between Neptune and Pluto for which

$$2n_N - 3n_P = 0.000\,159\,\text{year}^{-1}. \tag{3.8}$$

Pluto's very eccentric orbit brings it at perihelion just within the orbit of Neptune and it might be thought that the gradual drift in the relative positions of the two planets implied by the finite value of the right-hand side of equation (3.8) would eventually bring them close together. Perturbations of Pluto produce the pattern that the conjunctions with Neptune (i.e. when the planets are aligned on the same side of the Sun) occur when Pluto is close to aphelion. Over many successive conjunctions Pluto librates (oscillates) about its aphelion with an amplitude of 80°. Computation has confirmed that this *gravitational evasion* could persist for much longer than the age of the Solar System.

Another, but much less striking, commensurability occurs for Uranus and Neptune where the ratio of periods is 1 : 1.962 although in this case there is no gravitational evasion. A possible origin of commensurable orbits is described in Topic N.

3.4. PLANETARY STRUCTURES—GENERAL CONSIDERATIONS

The details of the main physical characteristics of the planets are given in table 3.2. The size variations illustrated in figure 3.1 showed clearly the division into terrestrial and major categories, and this division is reinforced by the clear differences in masses and densities. The anomalous nature of Pluto, considered as a member of the planetary family, is also quite clear from its small size and mass as well as the extreme nature of its orbital eccentricity and inclination, as shown in table 3.1.

The main general components for the formation of planets, in descending order of density, are iron, silicates, ices and gases. From observations of the terrestrial planets and their overall densities they consist mainly of iron and silicates, together with atmospheres except for Mercury. The atmospheres are important in controlling the surface conditions on these bodies but contribute very little to the mass. The atmosphere of the Earth has a more complicated structure than that of any other terrestrial planet and is described in Topic O. It provides a basis for describing the atmospheres of other planets.

The major planets are sometimes called the *giants* or *gas giants* and the distribution of matter within them has been theoretically modelled as being an iron–silicate core, somewhat like a large

Table 3.2. *Characteristics of the planets.*

Planet	Mass $(10^{24}\,kg)$	Mass (Earth = 1)	Equatorial radius $(10^3\,km)$	Equatorial radius (Earth = 1)	Mean density $(10^3\,kg\,m^{-3})$	Surface gravity $(m\,s^{-2})$
Mercury	0.3302	0.0553	2.440	0.383	5.427	3.70
Venus	4.869	0.815	6.052	0.949	5.204	8.87
Earth	5.974	1.000	6.378	1.000	5.520	9.78
Mars	0.6419	0.107	3.393	0.532	3.933	3.69
Jupiter	1899	317.8	71.49	11.21	1.326	23.12
Saturn	568.5	95.16	60.27	9.449	0.687	8.96
Uranus	86.6	14.54	25.56	4.01	1.27	8.69
Neptune	102.4	17.15	24.76	3.88	1.64	11.0
Pluto	0.014	0.0025	1.44	0.226	2.03	0.45

terrestrial planet, with most of the mass in a surrounding ice and gas envelope. Thus Jupiter and Saturn may have iron–silicate cores of mass in the range $5-10M_\oplus$ surrounded by a mainly hydrogen + helium envelope. Uranus and Neptune may have an additional substantial component of fluid or solid molecular material, such as water, H_2O, methane, CH_4, and ammonia, NH_3, between the solid core and gaseous outer layers.

Despite the uncertainties of our knowledge concerning the internal structures of planets it is possible to develop a generalized approach to the structure of planets just based on the various energies associated with its component material treated as a collection of atoms (Topic P). It turns out from this approach that thermal energy is such a tiny part of the total energies involved that at this level, for all practical purposes, a planet may be regarded as a cold body. Nevertheless, when we consider the detailed behaviour of a planet then thermal energy cannot be neglected. Heated material sets up convection motions, often in the form of a pattern of cells, and these can be responsible for distortions in the overall shape of the planetary body, mass motions of its surface material, volcanism or earthquakes. The general equations of heat transfer are developed in Topic Q. Terrestrial seismology, and what we can learn from it, is described in Topic R.

Another difference between the two families of planets, but not such a distinct one, is seen in their spin periods, those of the giant group all being shorter than those of the terrestrial group. The spin characteristics of the planets together with information about their satellite families are given in tables 7.1 to 7.5. The effect of spin is to distort the planets away from spherical form and to a first approximation the distorted surface takes on the form of an *oblate spheroid*. A general ellipsoid (figure 3.4) is a surface described in rectangular Cartesian coordinates as

$$\frac{x^2}{a^2} + \frac{y^2}{b^2} + \frac{z^2}{c^2} = 1. \tag{3.9}$$

If $a = b$ then the z axis is a symmetry axis; $c > a$ gives a *prolate spheroid* (like a rugby ball) while $c < a$ gives an oblate spheroid, which is like the surface of a spinning planet flattened along the spin axis. If $a = b = c$ the surface is spherical.

The extent of the distortion of a planet can be expressed as the *flattening*

$$f = \frac{d_e - d_p}{d_e}, \tag{3.10}$$

where the equatorial diameter is d_e and the polar diameter is d_p. The values for the planets, except those for Mercury, Venus and Pluto that are too small to measure, are given in table 3.3. The general pattern

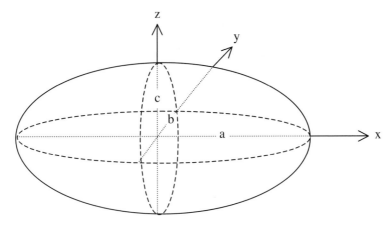

Figure 3.4. *A general ellipsoid.*

agrees with what would be expected. The giant planets, because they are more diffuse in structure and spin more rapidly, have the greatest flattening.

A matter of interest to planetary scientists is the distribution of matter within a planet. A non-spinning planet would have spherical symmetry and no information could be gained by external gravitational measurements since all the mass would act as though it was concentrated at the centre. A quantity that gives information about the distribution of matter is the moment of inertia (Topic S).

A uniform sphere of mass M and radius R has moment of inertia $0.4MR^2$ and where there is non-uniformity, but still spherical symmetry, the general form is αMR^2 where α is the *moment-of-inertia factor*. Thus for the Moon $\alpha = 0.392$ and this indicates that there is some central concentration of its mass—presumably in the form of an iron core. For a distorted spinning planet the moment of inertia about the polar spin axis (about which the planet actually spins), C, is different from that about a diametrical axis in the equatorial plane, A. Determination of both C and A can give important information about the internal distribution of matter within the body.

One consequence of a planet departing from spherical symmetry is that the gravitational potential in its vicinity no longer has a simple $1/r$ dependence and other terms appear in the expression for the gravitational potential. For an idealized distortion of a planet that maintains an axis of symmetry the extra terms involve higher powers of $1/r$ and Legendre polynomials that describe the θ dependence in a spherical polar coordinate system. Because of the r dependence, at great distances the extra terms become unimportant but at smaller distances their effect can be detected, for example by detailed measurements of the motions of satellites, both natural and

Table 3.3. *The flattening of planets.*

Planet	Flattening
Earth	0.0034
Mars	0.0065
Jupiter	0.0649
Saturn	0.0980
Uranus	0.0229
Neptune	0.0171

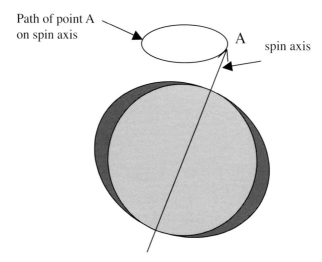

Figure 3.5. *Precession of the Earth's spin axis with a period of 26 000 years.*

artificial. Numerical coefficients of these extra terms, known as the *gravitational coefficients*, depend upon the distribution of matter in the planet and the most important of these coefficients, J_2, depends on the difference between the two moments of inertia, C and A. In addition to the effect of the planet's shape on the external gravitational field, the effective gravitational acceleration at a point on its surface will also be modified due to its spin. The gravitational effects of spin and distortion are described in Topic T.

Another consequence of departure from spherical symmetry is that the gravitation effect of an external body *on* the distorted planet is not equivalent to a simple force acting at its centre of mass. In addition, there is a net torque about the centre of mass. This situation occurs for the Earth–Moon system. The gravitational effect of the Moon acting on the Earth's equatorial bulge due to its spin causes precession of the Earth's spin axis with a period of 26 000 years (figure 3.5). The angle made by the polar axis stays at the same angle to the Earth's orbital plane but the intersection of the polar axis with the celestial sphere moves along a circular path. Thus Polaris, the pole star, has not always indicated north nor will it do so in the distant future. Similarly the intersection of the Earth's equatorial plane with the ecliptic changes with time, and hence so does the First point of Aries. An approximate derivation of the precession period, T_P, is given in Topic U. A characteristic of T_P is that it depends on the ratio of moments of inertia C/A.

If the mass M and mean radius R of a planet are known then a measurement of J_2 gives $C - A$ and an estimation of T_P gives C/A so that C and A can both be found. Knowledge of these two quantities then provides information about the distribution of matter within the planet or, at least, enables models to be tested. If a model based on some assumptions gives disagreement with the estimated values of C and A then that model can be ruled out. Unfortunately it is not always possible to find a precession rate, in which case a combination of the flattening, f, and J_2 can be used to estimate C and A but not as precisely as when T_P is available.

3.4.1. Planetary magnetic fields

Another important physical characteristic of most planets is an internally generated magnetic field. A description of the Earth's magnetic field is given in Topic V. It is not a simple matter to produce a picture of how such fields come about. The usual picture presented is that of a self-excited dynamo,

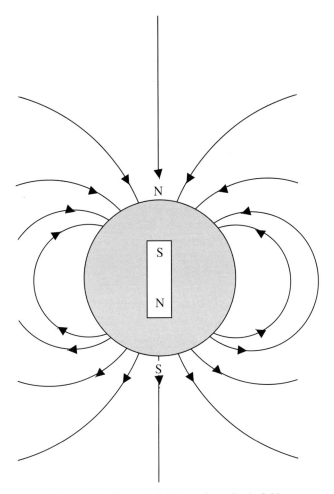

Figure 3.6. *Magnetic field lines for a dipole field.*

a device that can actually be constructed in the laboratory. A wire loop spun very quickly in a fairly weak magnetic field can generate a large alternating current that can be rectified and then be used to generate a magnetic field larger than the original one. How such a model can be applied to what goes on inside, say, the Earth is not clear. However, it is generally agreed that the Earth's magnetic field, corresponding to a dipole moment of $\sim 8 \times 10^{15}\,\mathrm{T\,m^{-3}}$, is generated by some kind of dynamo mechanism that converts mechanical energy into the energy of a magnetic field. Such a mechanism, whatever it is, must be capable of reversing the direction of the field from time to time in a fairly random way, as is indicated by observations (section V.6.3).

The planetary fields closely approximate a dipole form, which is the form of field produced by a conducting coil. This is illustrated in figure 3.6. However, the dipole form is only maintained for a small distance from the planet, beyond which the field is modified by interaction with the *solar wind*, a stream of energetic charged particles emanating from the Sun. The Sun itself produces a strong magnetic field with a quoted dipole moment $8 \times 10^{22}\,\mathrm{T\,m^3}$ but the structure of the field is very complex and can hardly be described in any simple way. It behaves like a large collection of different sources of dipole fields varying randomly in both polarity and strength. The Sun is also a source of ionized material that leaves the Sun spiralling around field lines until, at some distance, it becomes decoupled from the weakening field and then travels on to the edge of the Solar System and beyond. The solar wind is a plasma with equal numbers of protons and electrons; in the vicinity of the Earth it travels at about $400\,\mathrm{km\,s^{-1}}$ and it has a density that is very variable but is of the order of several million particles per cubic metre.

The way that the solar wind interacts with a planet or other body, e.g. the Moon, depends on the properties of that body. In the case of the Moon, a body with neither intrinsic magnetic field nor atmosphere, the solar wind simply bombards the surface and microscopic examination of lunar rocks reveals the damage caused by the impact of the charged particles. On the lee side of the Moon, and close to it, there will be a cavity without any charged particles but farther away particles will diffuse inwards. Of more interest is the interaction of the solar wind with a planet that has both an atmosphere and an intrinsic magnetic field and the Earth is such a planet. Figure W.8 illustrates the action of the solar wind on the Earth. The interaction of a fast-flowing plasma with a magnetic field is quite complex. The charged particles become coupled to the field and hence are influenced in the way that they move. Conversely, the magnetic field lines become *frozen in* to the plasma and are distorted by its motion. This complex interaction, giving rise to the pattern of field lines and particle distribution shown in figure W.7, is described more fully in Topic W.

Some particles break through the Earth's magnetic-field barrier and become trapped by field lines in such a way that they travel to-and-fro from pole to pole. They are concentrated in two doughnut-shaped regions known as the *van Allen belts*. When the motions of the charged particles bring them into the tenuous regions of the upper atmosphere they excite atmospheric atoms by collision. Subsequent de-excitation gives rise to spectacular displays in the form of streams and sheets of colour known as the *aurora borealis* at the northern pole (plate 1a) and the *aurora australis* at the southern pole (plate 1b).

Problems 3

3.1 Associating the planets from Mercury to Pluto with integers $n = 0$ to 8 then plot $\log(r_n)$ against n where r_n is the mean orbital radius of planet n. Draw a best straight line through the plotted points and so find the best relationship of type $r_n = a \times b^n$. Compare the values from your formula with the true values.

3.2 (a) The periods of Uranus and Neptune are 84.011 and 164.79 years respectively. Find a relationship, similar in form to equations (3.7) and (3.8), that links the mean motions of Uranus and Neptune.

(b) The orbital period of Venus is 0.723 years.

(i)   Find the mean motions of the Earth and Venus, n_E and n_V.
(ii) By considering integers, m, up to 10 find a relationship of form

$$m_E n_E - m_V n_V = c$$

with the minimum possible value of c.

CHAPTER 4

THE TERRESTRIAL PLANETS

So far we have considered the planets as of two major types, terrestrial and major, and also dealt with matters such as atmospheres, general structure and magnetic characteristics that are common to all or many planets. Now we consider the planets individually and highlight their characteristic features, starting with the terrestrial planets.

4.1. MERCURY

Mercury has the form of a triaxial ellipsoid, meaning that the values of a, b and c in equation (3.9) are different. The gravitational coefficient $J_2 = 8 \pm 6 \times 10^{-4}$ (Topic T). The mean moment-of-inertia factor is $\alpha_{\text{Merc}} \approx 0.33$ which is to be compared with $\alpha_{\text{Moon}} = 0.392$. This implies a greater concentration of material towards the centre of Mercury than for the Moon. There are indications of hemispherical asymmetry in the Mercury figure, the southern hemisphere being rather larger than the northern. Interestingly, this type of asymmetry also occurs with the Earth, the Moon and Mars.

Since it has neither a natural satellite nor an artificial one placed by man, its mass cannot be determined by the application of Kepler's third law. The best estimate available, given in table 3.2, comes from perturbation of Mariner 10 during its close passages to Mercury in 1974 and 1975. Its mean density, $5427 \, \text{kg} \, \text{m}^{-3}$, is 0.983 times that of the Earth, which is surprisingly high for such a small body. In fact, because of its larger mass, compression plays a significant role in the Earth and its intrinsic (i.e. pressure-free) density is closer to $4500 \, \text{kg} \, \text{m}^{-3}$—much less than that of Mercury.

The flybys established the *sidereal spin period* to be 58.65 days while the *sidereal orbital period* is 87.97 days. The term *sidereal* means that the frame of reference in which the motion is described is that of the fixed stars. Thus the sidereal orbital period is the period between which the Sun–planet line points towards the same fixed star and represents a true orbital rotation of 2π radians. By contrast the *synodic period* is the time between similar positions with respect to the Sun and the Earth. The *synodic orbital period* is the time between one conjunction, when the planet, Earth and Sun are collinear, and the next similar conjunction. Since, for Mercury, the sidereal spin period and the sidereal orbital period are in the ratio 2 : 3, the planet rotates three times on its axis during two revolutions of the Sun. This means that two regions, 180° apart in longitude, alternately face the Sun at successive perihelia. This forms two temperature 'hot' poles. The temperature extremes on the surface are wide—90 K and 740 K. The upper value is above the melting points of lead (601 K), zinc (693 K) and tin (505 K).

It is presumed that Mercury has the same age as the other members of the Solar System. The small size of the planet and the high temperature of its environment are consistent with it not having an atmosphere now. It may well have had one very early in its life but it will have evaporated very quickly (Topic O).

4.1.1. The surface of Mercury

The surface details were photographed during the Mariner 10 space vehicle missions of 1974 (29 March and 21 September) and 1975 (10 March) at distances of about 200 000 km. They were flyby missions and the vehicle passed the planet at high speed. The accuracy of the data is not high and imaging techniques have improved considerably since those early days. It was, as a consequence, not possible to get pictures of great detail but a mosaic formed from many pictures was obtained showing the surface details for the first time. This is shown in figure 4.1. Unfortunately, the vehicle had a 176-day orbit around the Sun, which is just twice Mercury's orbital period, so consequently

Figure 4.1. *A mosaic composition of part of Mercury's surface. The ring structures associated with the Caloris basin are on the left-hand side (NASA).*

at each bypass almost the same region, consisting of 45% of the surface, was photographed. The dangers of a partial photographic survey must be stressed. As an example, the first photographic views of Mars (the Mariner 4 encounter on 14 July 1965) showed only cratered terrain. This gave no indication of the very different types of terrain revealed by later photography. Although the Mercury data can be regarded as to a certain extent tentative, at present they are supplemented by restricted radar surveys from Earth ground-based apparatus and some firm conclusions about conditions on the Mercury surface can be inferred.

Superficially the surface of the planet appears very similar to that of the Moon. There will be some differences of course. Surface gravity is about three times that of the Moon and the very high temperatures can lead to plastic flows which tend to reduce the stability of initial crater heights. This will affect the trajectories of ejected material and the shape of elevations. The similarity between the surface appearances of Mercury and Moon enables some of the nomenclature devised for the Moon (for instance, using the word *basin* for a very large crater) to be useful for Mercury as well.

The surface area is $7.5 \times 10^{13} \, \text{m}^2$, which is slightly less than 15% that of the Earth and very nearly twice that of the Moon. The surface appears to be an entirely barren and densely cratered terrain with no present geological activity. Its surface seems to be covered by dust, formed from the underlying rocks by bombardment with high-energy radiation from the Sun, as has been found for the Moon. It is also similar to the Moon in the way that it polarizes scattered light and in its low visual albedo (Topic X), 0.106, compared with 0.07 for the Moon.

Like other solid bodies in the Solar System, Mercury shows the scars of heavy bombardment. The distributions of crater diameters in the highland regions of Moon and Mercury are shown in figure 4.2 with data for Mars included for comparison. The curves all have maximum values at around 100 km although that for Mercury is much sharper than that for the Moon. Both Moon and Mercury show a high proportion of larger (>400 km) diameter craters. These data suggest that the objects striking the surfaces were of comparable size and suffered impacts with comparable speeds, although it could be that the objects striking Mercury were rather more uniform in size than those striking the Moon. The frequency of impacts can be explored using the counts of crater density. The density of craters also shows some differences and it would seem that the intensity and scale of bombardment was greater on Mercury than on the Moon.

The surface regions of Mercury can be divided into five broad categories that can also be arranged in a general time sequence to portray early events. Four of the categories are very general

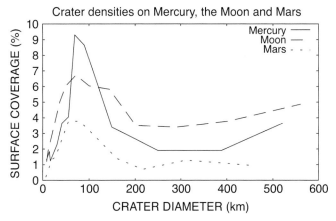

Figure 4.2. *A comparison of the density of craters on the Moon, Mercury and Mars related to the crater diameter.*

but the fifth is very specific. This last category is the major topological feature seen on the surface so far, the Caloris basin (figure 4.1). It lies near one of the 'hot poles' (hence the name Caloris) and spreads over a diameter of some 1500 km. An unusual hilly region at the antipodal region of the planet could be due to the focusing of seismic energy due to the collision with the substantial projectile that formed the basin. If this is so, it must have caused great structural damage to the interior of the planet.

The other four general features are: heavily-cratered terrain, inter-crater plains, smooth plains, and young craters. There are also *lineaments*, which appear to be *scarps* (or cliffs) and *trenches* (troughs). A representative scarp is up to 1 km high and several hundred kilometres long. The representative trench will be several hundred metres deep, about 120 km wide and will extend for several hundred kilometres. The global surface elevations are up to 3 or 4 km above and below the mean radius.

Several hundred linear features, such as scarps or troughs, have been identified and there are probably more to be discovered when a more detailed and complete survey of the surface is possible. They are not distributed at random but have a specific surface distribution. Except at the polar latitudes, the majority lie northwest–southeast and the remainder lie northeast–southwest. In the polar latitudes the orientation is more nearly north–south. The origin of these features is not understood. An overall contraction or expansion of the surface could well produce extensional cracks or compressional creases, similar to troughs and scarps, but it cannot be seen how this would lead to what is observed. Another possible explanation is the early de-spinning of the planet due to solar tidal effects that also gave rise to the 2 : 3 ratio of spin to orbital period. The time for de-spinning is difficult to estimate theoretically: values in the range between two million and two thousand million years have been proposed, depending upon whether the main body was taken as solid or liquid at the time. But either way, calculation suggests that the east–west stresses will be greater than the north–south stresses and this result could account, at least partially, for the observed distribution. There was, presumably, some change of mechanism in the high latitudes to provide the north–south alignments. These more symmetric forms could have resulted from contraction/compression of the regions as the newly formed planet approached a form of mechanical equilibrium.

The degree of overlapping of craters allows a relative chronology for the surface to be established. The relative chronology can be made absolute if (a) there is some other relevant body with an absolute chronology for comparisons to be made and (b) comparable processes can be assumed to have occurred on each surface. The dating of the lunar material returned to Earth by the Apollo missions has allowed an absolute chronology to be established for the Moon. The surface appearances of Mercury and the Moon are similar. A comparison between the lunar and Mercury craters may allow an absolute chronology to be established for Mercury.

Like the Moon, the Mercury surface shows every sign of being ancient. It is likely that it was once molten but cooled rapidly. It seems that the surface has been inactive over the last 3000 million years. Contraction faults would have arisen during the cooling of the surface and also from internal rearrangements. One is tempted to postulate a thick crust encasing a cold, solid interior because activity that accompanied the formation and early history of the planet was soon dissipated leaving a dead world. This may not be entirely true as we shall see.

The cratered terrain appears, as for the Moon, to be the result of the initial bombardment of large bodies that accompanied the formation of the Solar System. The impacts penetrated the solid crust quite extensively and resulted in a general volcanism that spread a layer of basalt across the solid surface. The structure of the intercrater plains is different from that shown by the Moon; there the basalt lava lies within each crater but for Mercury it lies between them. There are some impact craters on the plains, showing that the plains were formed before the later impacts.

Some smooth plains seem to have been the result of extensive volcanism over the surface at a still later date. The young craters in them are distinguished by the rays due to ejected material (similar to those on the Moon) still being evident. It seems that an occasional impact still occurs on the surface.

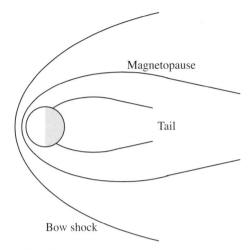

Magnetopause

Tail

Bow shock

Figure 4.3. *The Mercury bow-shock, magnetopause and tail.*

4.1.2. Mercury's magnetic field

Perhaps the most surprising result of the Mercury observations has been the recognition of an intrinsic magnetic field with a largely dipole form. This was recognized from observations of the interaction between Mercury and the solar wind. With no atmosphere the solar wind would be expected to strike the surface in the same way as it does with the Moon. There was much surprise when a bow shock structure was encountered characteristic of the effect of a magnetic field (Topic W). This is shown in figure 4.3. From the observed local speed and density of the solar wind it is possible to infer the strength of the opposing magnetic field. It appears that Mercury has a dipole moment of strength $4.9 \pm 0.2 \times 10^{12} \, \mathrm{T \, m^3}$. This would imply an equatorial surface field of $2 \times 10^{-7} \, \mathrm{T}$ (0.002 Gauss) or about one hundredth that of the Earth. As for the Earth, the direction of the field is anti-parallel to the rotation axis and the angle between the magnetic and rotation axes is less than $10°$ (section V.1). The origin of Mercury's field is not fully understood. A dynamo mechanism requires a rotating conducting fluid core and it is unlikely that the Mercury interior could provide these conditions although it is not ruled out completely. The magnetic field is too weak to sustain a system of trapped charged particles, such as the Earth's van Allen radiation belt, but it is strong enough to interact with the solar wind giving features observed near the Earth such as a bow shock and a distorted magnetosphere.

4.1.3. Mercury summary

Mercury presents an ancient surface and it is certain that heat loss is by simple conduction through the solid crust. There is evidence for substantial volcanism in the past so that early heat transfer from the interior would have been, to a large extent, by the advection[†] of lava materials.

In spite of appearing to be a dead body, the presence of a dipole field suggests that the interior may still be molten if the magnetic field has a dynamo origin.

[†] Advection is the mass motion of material, usually due to convection.

4.2. VENUS

Venus shows phases (figure L.9) to observers on the Earth but no surface details. It was evident to early observers that the planet's surface is covered by a thick cloud layer that reflects nearly all the solar radiation back into space—the visual albedo is 0.65. It is the nearest planet to the Earth and often the brightest object in the night sky. It is similar to the Earth in size, mass and density and, before modern exploration, it was referred to as a 'second Earth'. This view of Venus was changed in the 1950s when it first became possible to measure microwave radiation given off by the surface of a planet using Earth-based instruments.

These measurements indicated that the surface temperature must be in excess of 650 K, a conclusion confirmed by measurements from the American Mariner 2 flyby (27 August 1962) and by the early Russian Venera landers (starting with Venera 3 on 16 November 1965). Apart from confirming a surface temperature of some 730 K, the Venera landers measured a surface atmospheric pressure of about 93 bar—very nearly one hundred times that of the Earth. In spite of the dense atmosphere there was enough light to allow photographs to be taken (see figures 4.4 and 4.5 returned from Venera 13 and Venera 14). The environmental conditions were so harsh that each lander survived to function for no more than 30 min on the surface. This was, however, sufficient to allow soil samples to be tested automatically and the results are displayed in table 1.3. The composition and the density of the material, in the range 2700 to 2800 $kg\,m^{-3}$, are similar to those of terrestrial surface rocks.

Because of thick cloud cover, surface details cannot be found by photography with visible light. The first alternative method using Earth-based radar was attempted in the late 1950s using the Aricebo radio dish in Puerto Rico. This was successful within its limits but a global coverage of the surface required an orbiting vehicle.

4.2.1. The surface of Venus

Detailed radar imaging of the surface was first made by the Magellan orbiter and some 95% of the surface was surveyed. A general map of the surface derived this way is shown in figure 4.6. Like the

Figure 4.4. *The surface of Venus from Venera 13.*

Figure 4.5. *The surface of Venus from Venera 14.*

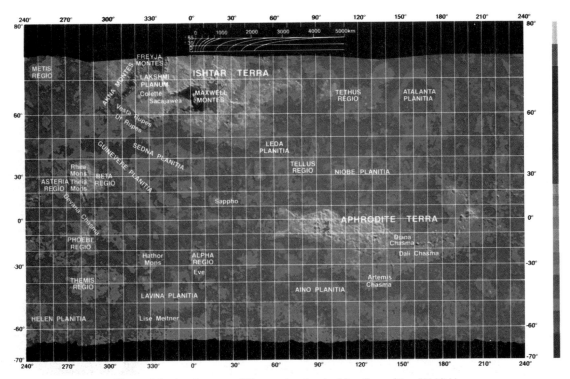

Figure 4.6. *A radar map of Venus taken by the Magellan orbiter (NASA).*

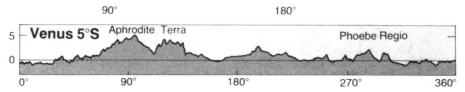

Figure 4.7. *A cross-section through a linear depression in Aphrodite Terra (NASA).*

Earth, Venus is a relatively smooth body. Some 60% of the surface is less than 500 m above the mean planetary radius and only 2% is more than 2 km above that. The range of elevations, however, is about 13 km and this is similar to that on Earth. The greatest depths are about 2 km below the mean radius (compared with 11 km on Earth). The majority of the surface is rolling plains that cover about 70% of the surface. Below them, lowland planes cover 20% of the surface and highland regions account for no more than 10%.

There are three substantial raised highland regions which could be compared with the continents on Earth. One is Aphrodite terra. This is a vast raised continent extending along the equator for some 10 000 km and with an area about half that of Africa. It is generally rough terrain and more complicated in structure than anywhere else. It is crossed by substantial linear depressions up to 300 km wide, 1000 km long and 3 km deep. The flanks of this massif appear to be moving away from the central ridge at the rate of a few centimetres per year. The cross-section from the western region is shown in figure 4.7, from which it is clear that there is symmetry about the centre very reminiscent of the mid-ocean ridge spreading structures on Earth. It is tempting to think of this as a region producing Venus surface rock.

A second highland region is Beta Regio, 60° longitude east of Aphrodite. This shows a large rift valley with interior faults. A volcano, Theia Mons, is set on the western boundary and has flooded the rift valley. This links with the highland rift valleys that seems to girdle the equatorial region. This region has the appearance of being relatively young.

The third highland region is Ishtar Terra and is very different from the other two highland regions. This region has been well explored by radar from the Russian orbiting spacecrafts Venera 15 and 16 and a picture is seen in figure 4.8. To begin with, it is at higher latitudes, between 50° and 80° north, and has an area similar to that of Australia. It has the form of an enormous plateau (Lakshmi Planum) some 2000 km in extent (about twice the size of Tibet). It is rimmed by mountain chains: to the east the Maxwell Montes (13 km high), to the north-north-east, Frejya Montes, to the west, Akna Montes, and to the south Dana Montes. There are two large craters (Colette and Sacajawea) on the plain that are probably calderas (volcanic craters) that provided much of the lava to form the plain. There is also an obscure circular feature (Cleopatra) that could be another caldera or an impact crater. This whole structure with its compression generated mountain belts is very similar to the folded uplifts that form along convergent plate boundaries on Earth. It is not clear whether this represents a simple subduction region. Indeed, it is by no means certain that Venus has, or has had, a plate tectonic system in any way comparable with that of Earth. More accurate data may clarify this question and show whether the system on the Venusian surface is an elementary form of that for the Earth or a specialized development unique to Venus.

The Russian Venera craft have explored the northern latitudes with a resolution of a few kilometres, covering 25% of the total surface. One feature of these data is that there are few craters. Certainly the thick atmosphere will affect the surface crater distribution but there should have been more if the surface had remained untouched since the earliest times. It seems that the surface is renewed and certainly more often than once every 1000 million years. This would be a rate some five times slower than that for the surface of the Earth.

Figure 4.8. *A view of part of Ishtar Terra from a Venera spacecraft.*

4.2.2. The atmosphere of Venus

A great deal of detailed knowledge of Venus's atmosphere was gained from the Pioneer Venus and Venera missions in the period from 1975 to 1982. The main constituents of the atmosphere of Venus are shown in table 4.1 together with those of the Earth, Mars and Titan. Apart from the constituents shown, Venus also contains 0.015% of sulphur dioxide (SO_2) and tiny amounts of other gases. The average vertical structure is shown in figure 4.9, which shows a troposphere with a characteristic adiabatic lapse rate and a stratosphere and thermosphere with much more gradual changes of temperature than are found for the Earth (Topic O). Because of the rather long days and nights on Venus, the thermal profiles have time to settle into an equilibrium state and so are somewhat different on the day side and night side. The absence of a mesosphere and the associated sinuous temperature profile is due to the lack of a strongly absorbing ozone layer. However, by comparison with the Earth, Venus has a very massive atmosphere with a surface pressure of 93 bar. With a surface acceleration of $8.89 \, \mathrm{m \, s^{-2}}$, this gives a column mass somewhat over $10^6 \, \mathrm{kg \, m^{-2}}$ (Topic O).

Table 4.1. *The main constituents of the atmospheres of three terrestrial planets and Titan.*

Component	Earth	Venus	Mars	Titan
Carbon dioxide	0.00033	0.96	0.95	—
Nitrogen	0.78	0.035	0.027	0.90
Oxygen	0.21	—	0.0013	—
Argon	0.009	0.00007	0.016	0.10
Water	0.001–0.028	0.0001	0.0003	—

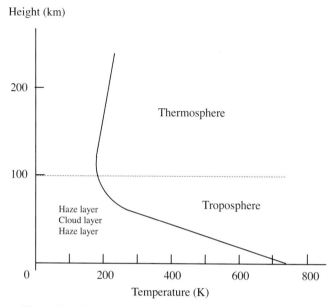

Figure 4.9. *The temperature profile of the atmosphere of Venus.*

It is the presence of such a high proportion of CO_2 plus the small amounts of H_2O in such a dense atmosphere which accounts for the high surface temperature through the *greenhouse effect*, which is also of great importance for the Earth. In figure 4.10 we show a somewhat simplified version of the absorption curve for radiation at different wavelengths by an atmosphere containing CO_2, H_2O and other greenhouse gases. Superimposed on this are the intensity versus wavelength characteristics (not at the same scale) of two black-body sources, one at the temperature of the Sun, 5800 K, and the other at 288 K, corresponding to the mean surface temperature of the Earth. It will be seen that the atmosphere is fairly transparent to the radiation from the Sun, at shorter wavelengths, so most of it penetrates to ground level and heats up the surface. The much lower temperature surface radiates at longer wavelengths and the atmosphere absorbs much of this and consequently heats up.

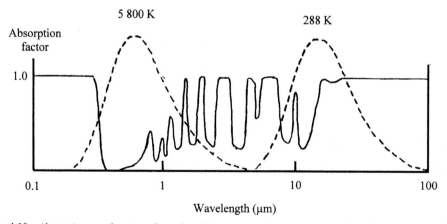

Figure 4.10. *Absorption as a function of wavelength due to greenhouse gases in the Earth's atmosphere. Black-body radiation curves corresponding to the temperatures of the Sun and the Earth are superimposed.*

Taking the Earth's albedo into account the surface temperature of the Earth without the greenhouse effect would be 255 K (Topic X), so that all water would be in the form of ice—making life at least difficult if not impossible. The greenhouse effect maintains a temperature some 33 K higher which is needed to give a balance of heating and cooling. In the case of Venus the greenhouse effect is amplified by the increased proportion of CO_2 and the very thick atmosphere to the extent that the surface temperature is 500 K higher than it would be otherwise.

The compositions of the original atmospheres of the terrestrial planets are not known but one possibility is that they were similar to those now found in the giant planets. If so then it is easily confirmed from equation (O.18b) that the hydrogen would be lost on astronomically short time scales ($<10^8$ years). At the distance of Venus from the Sun, with some greenhouse effect due to methane, its temperature would have ensured a heavy component of water vapour in the early atmosphere. Indeed, since Venus shows signs of volcanism in its early history the heating due to that activity alone could have led to an aqueous atmosphere. Since water is an effective greenhouse gas the surface temperature would have been raised thus producing more water vapour and also releasing carbon dioxide, another effective greenhouse gas, from carbonate rocks.

Water vapour in the upper atmosphere would have become dissociated by ultraviolet radiation from the Sun with a reaction described by

$$H_2O + \nu \rightarrow H + OH$$

$$2OH \rightarrow H_2O + O.$$

The hydrogen would have been lost and the oxygen would have reacted with any available material which was readily oxidized, including methane and carbon monoxide which would be available carbon sources to provide the now-dominant carbon dioxide. This runaway greenhouse effect would have continued until the present situation was reached. Most of the water on Venus has now disappeared and we shall comment further on this in section 12.12. A difficulty with this scenario, based on an original hydrogen-rich atmosphere, is that a considerable amount of helium should also have been present. The application of equation (O.18b) indicates that much of this helium should still be present, and it is not. Alternative histories of Venus, and of the Earth and possibly Mars, suggest violent episodes when most of the atmosphere, including the helium, was lost, with subsequent re-establishment of an atmosphere by degassing of the planet due to volcanism. Since the internal gases would include a good deal of water and methane the mechanism of water dissociation, loss of hydrogen and formation of free oxygen to oxidize methane would still apply.

The amount of CO_2 in the Earth's atmosphere has doubled since the start of the industrial revolution and the large-scale burning of fossil fuels. This has led to speculation that the increased greenhouse effect will lead to global warming (Topic AO) with dire consequences for the pattern of human life on Earth. Some have even speculated that we are taking the first steps on a path leading to a Venus-type atmosphere. However, the Earth's climate has fluctuated greatly in the past without human intervention and it is difficult to detect a gradual trend against a background of large fluctuations.

Venus has very heavy cloud cover at a very high level, some 50 km above the surface, with layers of haze or mist both above and below it. The main constituent of the clouds and the mists appears to be sulphuric acid, H_2SO_4, in sub-micron droplets in the mists but greater than micron size in the main cloud belt. Larger solid particles, probably chlorides, are thought to occur in the lower reaches of the cloud belt.

Due to the slow spin of Venus, once every 243 days in a retrograde direction, the Coriolis force is not very pronounced. The winds on Venus are mostly easterly varying from $100\,\mathrm{m\,s^{-1}}$ at high levels to just $1\,\mathrm{m\,s^{-1}}$ near the surface. The exact reason for this pattern of wind is not completely understood.

The very low ground wind speeds do help to explain why the surface rocks on Venus, photographed by landers from spacecraft, have sharp edges and show very little erosion. On the other hand it would be thought that a planet which rains sulphuric acid might give signs of chemical erosion.

4.2.3. *Venus and magnetism*

The Earth's 'twin' is different in yet another respect in that it appears not to be the seat of an intrinsic magnetic field. One cannot, of course, actually measure zero but the measurements that have been made imply that the magnetic dipole moment is certainly less than $10^{11}\,T\,m^3$, which is the limit of the resolution of the magnetometers that were used on the mission. Consequently, the surface field is less than $3 \times 10^{-8}\,T$. This has been linked with the slow rotation of the planet since it is believed that energy derived from the spin of a body is an important factor in the dynamo mechanism.

4.2.4. *Venus summary*

The study of the surface of Venus seems set to provide surprises in the future. The resolution of the radar data is not yet sufficient to allow the surface to be viewed with sufficient detail to reveal all the geological processes that might be involved. Some conclusions can be drawn but they may need modification later.

There is only limited evidence remaining of the early surface bombardment with large projectiles common to the terrestrial planets showing that the surface has renewed itself since then. There is evidence to suggest that that this occurs with a period rather less than 1000 million years. The mechanism for renewal is not certain. There is some evidence for tectonic activity on the surface but not the global distribution of volcanic structures familiar on Earth as the mid-ocean ridges. Data of higher resolution will indicate whether it is not there at all or whether it is actually present, but on a much smaller scale than on Earth. It is not clear, therefore, whether the passage of heat from the inside is by pure conduction through the crust or whether there are some, at least limited, processes of advection as well.

There is no detectable intrinsic magnetic activity on the planet. This has implications for the interior structure.

4.3   THE EARTH

The most obvious physical surface features of the Earth are the oceans and the continents. Of the total surface area of $5.1 \times 10^{14}\,m^2$ ($= 5.1 \times 10^8\,km^2$), 70.8% is covered by water ($3.61 \times 10^{14}\,m^2$ by the oceans, of which $2.7 \times 10^{13}\,m^2$ are marginal basins and $5.2 \times 10^{13}\,m^2$ continental shelves—the regions at the edge of continents which lie under the water to form a shallow edge to the oceans) and 29.2% is covered by the land ($1.49 \times 10^{14}\,m^2$ is continental). The mean height of the continents is 840 m while the mean depth of the oceans is 3800 m. This suggests, as a rough estimate, that the mass of the water of the oceans is some $1.4 \times 10^{21}\,kg$ while the mass of the continents above sea level is about $3.75 \times 10^{20}\,kg$. Water is clearly a significant feature of the surface structure of the Earth. We might notice that, for comparison, the mass of the atmosphere is $5.3 \times 10^{18}\,kg$. The highest point on Earth is the top of Mount Everest, 8.84 km above sea level (it is a dynamic feature which is still slowly rising) while the deepest point is 11.28 km below sea level (the bottom of the Mariana trench off the Philippines). This gives a total variation of about 19 km in the local radius of the Earth; with a mean equatorial radius of 6378 km, this variation is no more than 0.24%. The surface of the Earth is remarkably smooth! Indeed, the smoothest orange is far less smooth than the Earth.

The continents are covered by plains, mountain ranges, deserts, and volcanoes. Island arcs at sea show immediately the effects of oceanic volcanic activity and one example here is the Hawaiian islands. A more widespread volcanism, to be considered again later, is the mid-ocean ridge system which spans the entire globe. The main volcanic activity is, actually, found under the oceans. Some surface regions show signs of compression, others of extension.

4.3.1. The shape of the Earth

The mass of the Earth is $M_E = 5.976 \times 10^{24}$ kg; it is in steady rotation about the polar axis with an angular velocity $\Omega_E = 7.292 \times 10^{-5}$ radians s^{-1}. The effect of this slow rotation is to make the shape of the Earth (that is its *figure*) an oblate spheroid with flattening, $f = 3.35 \times 10^{-3}$. This is usually expressed as a fraction making $f = 1/298.187$. The effect of the shape of the Earth on the form of its gravitational field is described in Topic T.

4.3.2. Surface composition and age

The compositions of the continental and oceanic regions are different, indicating important structural differences below the surface. The continental material is generally less dense than the oceanic. The mean compositions of the oceanic and continental crusts are set down in table 1.3. The differences between continent and oceanic crust are not great for the major components but are more significant for the minor ones. In very general terms, the continental crust is silica rich and granitoid in composition while the oceanic crust is Mg,Fe-rich (mafic minerals) and basaltic in character (see table A.3).

The age of most of the surface material is found to be in excess of 100 million years and it seems that much of the surface is replaced about every 250 million years or so. Some rocks are older and the oldest are about 4000 million years old. The composition of the surface material can be found accurately for the past 200 million years and progressively less well before that, but for the first 800 million years of the Earth's history the global composition is not known directly. In this respect the Earth differs from the Moon where the earliest rocks are still present on its surface and the compositions can be measured or estimated. Estimates of the areas of continental basement (the underlying rocks) corresponding to various ages are listed in table 4.2. The division in the table at 2700 Ma marks the end of Archaean period of geology, a period that starts at the formation of the Earth. This embraces about 40% of the Earth's total history and only 1% of the present surface basement dates back to these times. The oldest solid material known so far is zircon crystals found in Western Australia which are estimated to date to 4.3×10^9 years ago (when the Earth was little more than 200 million years old). The crystals are held in a slightly younger meta-sedimentary rock formation and so are not preserved in their origin conditions, but they are a solid remnant of that time. Although virtually nothing is known of Archaean conditions it seems certain that some form

Table 4.2. *The approximate ages and areas of continental basement (after Hurley and Rand, 1969).*

Age (Ma)	Area (10^6 km^2)	Area (%)
0–450	38	29
450–900	41	32
900–1800	23	18
1800–2700	26	20
2700–3100	1–2	1
>3100	$\ll 1$	≈ 0

Time ago (Ma)	Period	Sub-period
Now	Quaternary	Pleistocene
1.64	Neogene	Pliocene
5.2		Miocene
		Oligocene
50	Palaeogene	Eocene
		Palaeocene
mammals 100	Cretaceous	Upper
		Lower
birds		Upper
150	Jurassic	Middle
		Lower
200		Upper
	Triassic	Middle
		Lower
250	Permian	Upper
land creatures		Lower
300	Carboniferous	Upper
		Lower
350		Upper
	Devonian	Middle
		Lower
400	Silurian	Upper
fish		Lower

Time ago (Ma)	Period	Sub-period
450	Ordovician	Upper
trilobites		Lower
500		Upper
	Cambrian	Middle
550		Lower
570		
	Vendian	
		Upper
1000	Riphean	Middle
green algae		Lower
2000		
	Apherian	
3000	Archaean	
first cells	*oldest known rocks*	
4000		
≈ 4550	*Formation of Earth*	

Figure. 4.11. *The Geological Column. These are the names most commonly in use; further sub-divisions are made but have not been included. The dating is by isotopic methods (Topic B) and is that most generally accepted. Further refinements are to be expected in due course.*

of solid continental structure was associated with the surface at that time. Dynamic resurfacing, always in progress on Earth, has destroyed virtually all details of the early surface. The geological history has been divided into a number of periods, of varying length, during which the surface conditions were broadly the same. The *geological column* describing the relation between geological periods is shown in outline in figure 4.11. Many sub-divisions can be added but are not of interest here. An indication of the most important events in the evolution of life on the Earth is also included.

There are indications that the surface has been produced continuously over the whole history of the Earth although the relative rates of production at different times are difficult to estimate. Present knowledge is not incompatible with the surface having been produced broadly at a constant rate throughout the life of the Earth and this is now usually assumed to have been the case. With a fixed total radius, continuous production of surface material is possible only if some form of recycling process is operative. The distribution of basement material over the surface is shown in figure 4.12. In very general terms, the older material is surrounded by the newer material that has formed around it. The old material then forms the core of each continent.

4.3.3.   Changing surface features

The surface of the Earth is a dynamic entity, constantly changing. Early thinking assumed that the forces required to bring about large changes were too great to be attainable in practice but there were many indications that substantial changes had, in fact, occurred over the years. Just one example of many is the discovery of fossils of marine creatures in the Alps, many hundreds of metres above the present sea level. Again, the Alpine, Himalayan and Carpathian mountain chains contain ophiolites, pieces of oceanic plate that have been pushed upwards and incorporated into continental material. It is now known that these changes are the result of the action of a variety of causes, some of them of a fundamental kind.

One cause is the simple erosion by wind and water that can be very effective in levelling the surface. The most fundamental changes, however, are due to the processes of making new oceanic surface material from material brought up from the interior of the Earth. This forms continental topography and changes the distribution of continents over the surface. Geological dating and magnetic studies have allowed the details of the changing surface and the chronology to be found and the sequence of the surface changes covering the past 220 million years are shown in figure 4.13. At this stage the continental land form a single large land mass named Pangaea. This then began to break into two parts; the northern part (comprising modern day Eurasia and North America) has been called Laurasia and the southern part (comprising modern South America, Africa, India, Australia and Antarctica) Gondwanaland. Between Eurasia and Gondwanaland was an ocean, rather wedge shaped, called the Tethys Sea. The later movements of Africa and India northwards to strike the Eurasian land mass largely eliminated the Tethys Sea presumably due to its subduction (see section 4.3.4) below the continents that rode over the top. The present day Mediterranean Sea, together with the Caspian and Black Seas, probably represents all that remains of this original ocean. As far as can be ascertained the land area has remained essentially constant during all these changes. The process of continental drift continues and can be monitored very accurately from spacecraft observations. The study of geography in another 100 million years, if there are any geographers about, will be very different from what it is today.

4.3.4.   Surface plate structure

These surface movements described in section 4.3.3 originate in a surface structure associated with a global ridge system of active volcanoes. This system is almost entirely under oceans (the so-called mid-ocean ridges), is some 40 000 km long and encompasses the globe as shown in figure 4.14. The

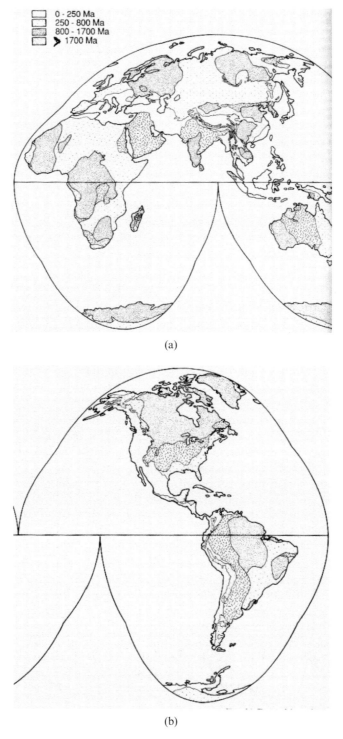

(a)

(b)

Figure 4.12. *The age of continental material on the Earth: (a) Europe, Asia, Africa and Australia. (b) The American continents.*

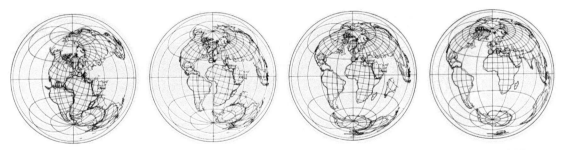

Figure 4.13. *The drift of continents showing (from left to right) the positions 220 Ma ago, 100 Ma ago, 60 Ma ago and the present (Moore & Hunt, 1983).*

ridges are at depths typically between 1 and 3 km but in Iceland and the East African Rift Valley the ridge is above water. The Icelandic ridge is very active and is continually creating new land. Ocean depths at different points on the same latitude give a height profile as shown in figure 4.15 for the North Atlantic Ocean: the constant latitude section through the African rift valley region is given for comparison.

Hot material issuing from the ridge system moves away horizontally, pushing the existing ocean crust outwards and cooling as it goes to form new ocean crust at a global rate estimated as some $17 \, km^3$ per year. The majority of the Earth's oceanic crust is believed to have been formed, and is still being formed, by this mechanism. Since there is evidence that the total surface area of the Earth remained constant during these changes, the continual appearance of new crust must imply the disappearance

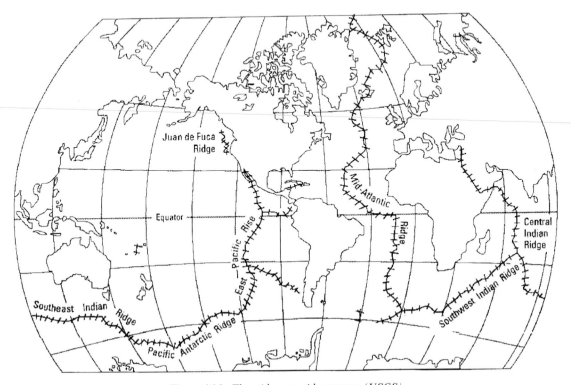

Figure 4.14. *The mid-ocean ridge system (USGS).*

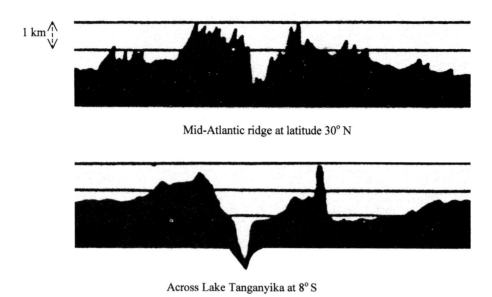

1 km

Mid-Atlantic ridge at latitude 30° N

Across Lake Tanganyika at 8° S

Figure 4.15. *A cross-section of the Mid-Atlantic Ridge compared to one through Lake Tanganyika that forms part of the African rift valley.*

of old crust. The full oceanic ridge system provides somewhat less than $2 \times 10^{10} \, \text{m}^3$ of crustal material per year or (assuming the rate of production is approximately constant) about $5 \times 10^{18} \, \text{m}^3$ in 250 million years. The surface area is about $5 \times 10^{14} \, \text{m}^3$ so that, to account for the volume of material, a 10 km thick layer of the near surface material must be involved in some form of recycling process. This mobile region of recycling is called the *asthenosphere*. It can be regarded roughly as starting at the depth below the surface where the temperature reaches about 1000 K. The near surface region is cool and solid: this we call the *lithosphere*.

The lithosphere is divided unevenly into a number of separate 'plates', most of which contain continental material but some (such as the Pacific plate) that do not (figure 4.16). The lithosphere being, on average, less dense floats on the higher-density asthenosphere. The horizontal motions of the lithosphere plates then reflect a general slow thermal overturning of the asthenosphere material downwards from the surface. The relative motion between the plates varies from place to place but is about 3 cm per year. The oceanic ridge system is where contiguous plates are forced slowly apart by the intrusion of basaltic magma from below (figure 4.17a). This molten region provides new lithosphere material. In other regions lithosphere material is lost by one plate sliding below another, a process known as *subduction* (figure 4.17b). Again, two plates may slide by each other without collision, for instance, the two plates forming the San Andreas fault in California (figure 4.18). If the two plates are colliding and have about the same density the result can be a buckling, forming a large mountain ridge (figure 4.19). Examples of this process are the Himalayas (the collision between the Indian plate and the Asian plate some 50 million years ago), the Alps (the collision between the African plate and the Asian–European plate at about the same time) and the Carpathian mountains. These structures are still gaining in height at the present time although the changes are, of course, very slow by human standards.

The continental mass rests on the plate below it and causes an elastic depression as a load on the plate. This can be detected by gravity measurements and by *bathymetric* (i.e. ocean depth) measurements across the plate. One example is the data referring to the north–south line passing

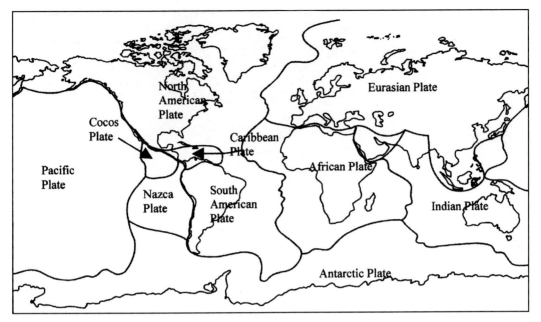

Figure 4.16. *The major tectonic plates.*

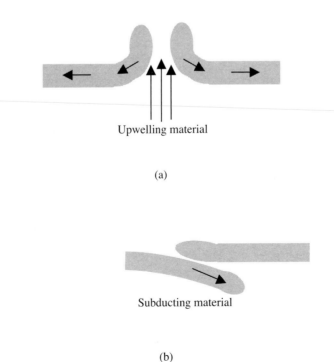

Upwelling material

(a)

Subducting material

(b)

Figure 4.17. *(a) Plate formation due to upwelling material at a ridge. (b) Plate subduction.*

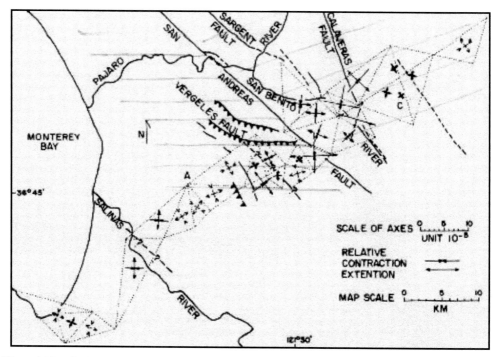

Figure 4.18. *The detailed structure of the San Andreas Fault. The two plates slide one with respect to the other (after Burford).*

through the island of Oahu, Hawaii, shown in figure 4.20. The island bends the plate downwards to form a trough. Simple modelling, presuming the plate to be homogeneous and using the normal elastic constants, explains the observations reasonably well.

4.3.5. *Heat flow through the surface*

There is a flow of heat through the surface showing that the Earth is slowly cooling. The measured near-surface temperature gradient is $30\,\text{K}\,\text{km}^{-1}$ and if this were maintained downwards, the temperature would reach $1000\,\text{K}$ at a depth of about $30\,\text{km}$. At this temperature surface rocks change their physical properties.

The total geothermal surface heat flux outward is estimated to be about $4.2 \times 10^{13}\,\text{W}$ implying a mean flux of about $8 \times 10^{-2}\,\text{W}\,\text{m}^{-2}$. Since the heat received by the surface from the Sun is nearly $700\,\text{W}\,\text{m}^{-2}$, some four orders of magnitude greater, the equilibrium temperature of the surface is

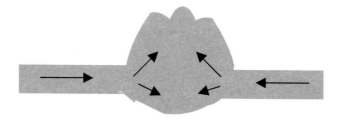

Figure 4.19. *The collision of two plates of equal strength and density giving rise to mountains.*

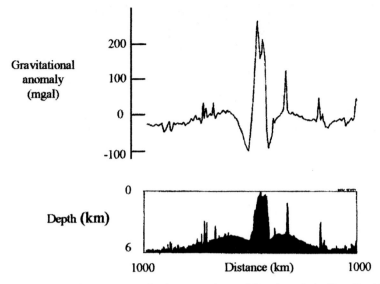

Figure 4.20. *Ocean depth and gravity anomaly along a north–south line through the Hawaiian island of Oahu. The weight of the islands bends the plate in their immediate vicinity. The gravity anomaly is in mgal units where 1 gal is $0.01\,m\,s^{-2}$.*

determined by solar radiation and not by the internal heat. Nevertheless, some heat is lost from inside the Earth and this heat loss has determined the nature of the surface structure involving the plates, discussed already. The mean heat loss from the interior is different for the oceanic and continental regions as is seen in table 4.3. This difference reflects different heat components—the oceanic crust involves a heat content brought up from below while the continental regions primarily involve internal radioactive heating. The continental structures are much thicker than the oceanic structures and continental material is richer in radioactive elements than the oceanic. These two circumstances combine to provide substantial amounts of radioactive material in continental crust. The main radioactive elements of geological interest are listed in table 4.4 together with their mean concentrations in surface rocks. The elements of importance for radioactive heating are isotopes of uranium, thorium and potassium. The particular isotopes involved and the rate at which they produce heat is set out in table 4.5. The data for heat generation are for the present epoch and will have been greater in the past by amounts that depend on the half life of the particular elements.

We have spoken of mean heat and energy flows so far but measurements vary greatly at different locations for local reasons. The mid-ocean ridge vents are exuding hot material from below. This is cooled through contact with ocean water giving material with a cool surface but hot interior. The

Table 4.3. *The mean heat flow from the surface regions.*

Region	% total surface area	Mean heat flow ($10^{-2}\,W\,m^{-2}$)	Heat loss ($10^{13}\,W$)
Oceans and oceanic basins	61	9.9	3.04
Continents and continental shelf	39	5.7	1.16
Total			4.2

Table 4.4. *The first three entries are the principal radioactive elements in various rocks. The last four are of importance in radioactive dating of rock samples.*

Element	Basalt (ppm)	Granitoid (ppm)	Shale (ppm)	Ultramafic (ppm)
Uranium (U)	0.5	4	4	0.02
Thorium (Th)	1	15	12	0.08
Potassium (K)	0.8	3.5	2.7	0.01
Neodymium (Nd)	40	44	50	2
Rubidium (Rb)	30	200	140	0.5
Samarium (Sm)	10	8	10	0.5
Strontium (Sr)	470	300	300	50

interior heat of the externally-cooled rock will move to the surface by conduction in a time depending on the thermal conductivity of the rock. Intuitively one might expect the temperature of the oceanic materials to decrease systematically from the mid-ocean vent reflecting its motion away from the source. Such a profile is well established in practice although the measurements are not easy to perform with accuracy, especially for the younger rock since the sea becomes turbulent when the rock temperature is high making the measurements difficult. Empirical formulae have been devised to describe this cooling process. For rock younger than 120 Ma the heat flow from the rock h (in units of $10^{-3}\,\mathrm{W\,m^{-2}}$) is related to the age t (in units of 10^6 years) after solidification by

$$h = 473t^{-1/2}. \tag{4.1a}$$

For rocks considerably older than this the relationship becomes instead

$$h = 33.5 + 67\exp(-t/62.8) \tag{4.1b}$$

showing a strongly decreasing heat flow with time.

Another important consequence of the cooling involves the material density. Cooler rock is denser than hotter and will lie lower in the asthenosphere. Consequently, in general, the depth of the ocean can be expected to be greater over old rock than younger rock, the ocean becoming deeper with distance from the ridge. This is, indeed, found to be the case. Depths of beds measured in the North Atlantic and the North Pacific increase with the age of the rock (and so with distance from the mid-ocean ridge) and empirical formulae have been devised to describe this effect. For rocks younger than 70 Ma, the observed bathymetric depth, d (km), is related to the age t (Ma) by

$$d = 2.5 + 0.35\sqrt{t}. \tag{4.2a}$$

Table 4.5. *The present rate of heat generation and half lives of the radioactive components in rocks.*

Isotope	Half-life (10^6 years)	Rate of heat generation ($10^{-5}\,\mathrm{W\,kg^{-1}}$)
^{235}U	4468	9.4
^{238}U	704	57
^{232}Th	14010	2.7
^{40}K	1250	2.8

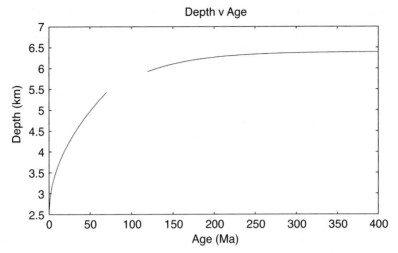

Figure 4.21. *The depth versus age relationship for the North Pacific Ocean.*

For rock older than 120 Ma the best relation involves becomes

$$d = 6.4 - 3.2\exp(-t/62.8). \tag{4.2b}$$

A plot of the mean depth with age for the North Pacific Ocean is shown in figure 4.21 with a gap in the region between 70 Ma and 120 Ma.

4.3.6. Earthquakes

Earthquake activity is the localized release of the energy of stresses within the main body of the Earth. The energy is dissipated throughout the volume and near the surface through a series of waves and the energy release can be very dramatic in its effects at the surface.

Almost any location on the surface feels the effects of a small earthquake occasionally but the substantial events are more confined to specific regions of the surface. The place within the Earth where the earthquake actually takes place is called the *focus*. The point on the surface immediately above the focus is called the *epicentre*. These are idealizations because the focus is generally not a point but a region with a typical dimension of several kilometres. There may be one large release of energy, lasting several seconds, or several short releases repeated during the period of an hour or even over a period of days. This makes it very difficult to assess the energy released during the whole disturbance. During the earthquake, the local surface shows both vertical and horizontal motions, corresponding to the presence of longitudinal and transverse waves. The resulting surface damage gives some indication of energy involved and varies from one earthquake to another. The strongest in populated areas give widespread devastation, often involving a large loss of life, while the smallest are hardly noticed at all.

The concept of the magnitude **M** of an earthquake was proposed by Richter (in 1935) on the basis of earthquakes observed in Southern California. This logarithmic Richter scale is well known although it is not the only scale. The magnitude is ultimately related to:

(i)   the maximum surface amplitude of the observed wave (measured in μm),
(ii)   the period of oscillation of the wave (measured in seconds),
(iii) the focal depth of the earthquake, a factor describing the decrease of amplitude with distance from the epicentre and
(iv)   the actual distance between the observer and the epicentre.

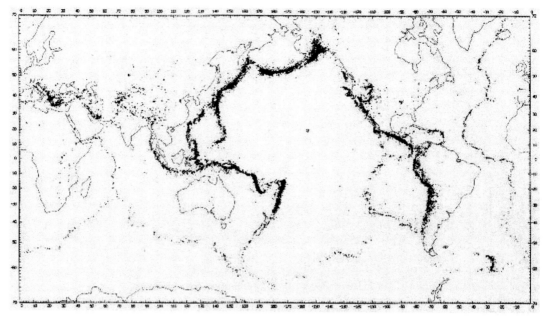

Figure 4.22. *Distribution of earthquakes at depths up to 100 km.*

The surface amplitude depends on the depth of the focus since the farther this is from the surface the smaller the proportion of energy entering the surface waves. Most scales represent the very strongest earthquakes of magnitude approaching 10 while the weakest have magnitudes about 1 or 2. The strongest earthquake recorded up to the end of the twentieth century was the 1933 Japanese

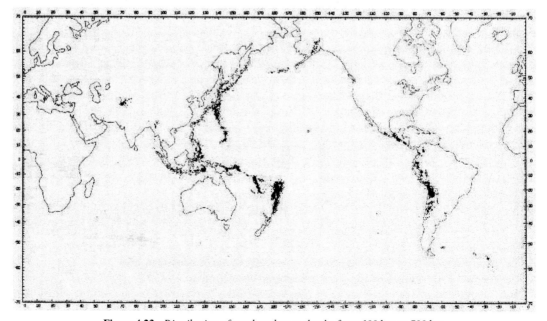

Figure 4.23. *Distribution of earthquakes at depths from 100 km to 700 km.*

Table 4.6. *The number of earthquakes against depth, 1918–1946 (after Gutenberg and Richter).*

Depth (km)	Number of earthquakes	Depth (km)	Number of earthquakes
<100	800	275–325	8
75–125	139	325–375	11
125–175	56	375–425	12
175–225	38	425–725	41
225–275	15		

earthquake with $\mathbf{M} = 8.9$. Generally, a shallow earthquake (focus near to the surface) with $\mathbf{M} \approx 6$ is extremely destructive; one with $\mathbf{M} \approx 5$ gives moderate damage while those with $\mathbf{M} < 3$ are hardly felt other than by instruments. An earthquake with $\mathbf{M} \approx 7$ releases energy of about 5×10^{15} J— rather more than the yield of a one megatonne nuclear weapon. The focus can be at any depth down to about 700 km but the most common are less deep than 100 km. About three-quarters of the total energy released in earthquakes comes from shallow events with depths below 100 km. The global distribution of earthquakes is shown in figures 4.22 and 4.23 and it will be seen that the main sites are associated with the Pacific plate.

The dependence of earthquake frequency on depth is shown in table 4.6, using data given by Gutenberg and Richter (1954) and covering the period 1918–1946. It is seen that the vast majority of all earthquakes have a focus less than 125 km deep. From table 4.7, which contains data given by Richter covering the period 1918–1955, it is also seen that the largest earthquakes have their source near the surface.

Although, in terms of human existence, earthquakes are destructive they do have an important scientific spin-off. Earthquake waves of various kinds travel through the interior of the Earth and by measuring their intensities at seismic observatories all over the world we have been able to build up a picture of the internal structure of the Earth. The model of the Earth that emerges is shown in figure 4.24 and basic structure of the different layers are now described. More detail about the layers together with a description of *seismology* and the way it gives us this information is given in Topic R.

4.3.6.1.   The crust

The crust consists of lower-density silicates, containing magnesium, aluminium, calcium, sodium and potassium. Oceanic crust is basalt with a high magnesium content and this more-or-less covers the whole Earth. In continental regions there is an overlay of lower-density silicates that tend to be aluminium-rich, such as granite. The thickness of the crust varies from about 10 km, or even less, below some oceans to 40 km in some continental areas.

Table 4.7. *Numbers of earthquakes of different magnitude and depth of focus.*

| Depth range | Magnitude | | |
	7.0–7.8	7.9–8.5	≥8.6
Shallow: ≤70 km	570	60	9
Intermediate: 70–300 km	214	8	1
Deep: ≥300 km	66	4	0

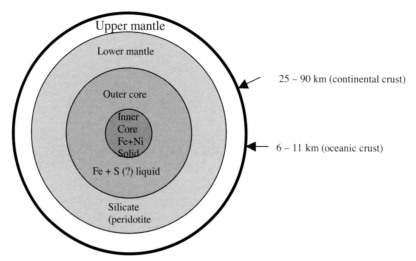

Figure 4.24. *The internal structure of the Earth as deduced from seismology.*

4.3.6.2. The mantle

The boundary between the crust and the mantle is marked by a seismic discontinuity known as the *Mohorovicic discontinuity* or, more popularly, as the *Moho*. This marks the top of the mantle which stretches down to a depth of 2890 km, the mantle–core boundary.

The mantle consists of denser silicates rich in iron and magnesium, mostly olivine. The upper part of the mantle is a rigid solid and, together with the crust, forms the lithosphere. Below a depth of some tens of kilometres the viscosity of the material, still technically a solid, is low enough for it to flow and this marks the upper bound of the asthenosphere, movements within which provide the dynamics of plate tectonics.

4.3.6.3. The core

Because of its very high density it is certain that the core consists of the ferrous elements (iron, cobalt and nickel—but mainly iron) since these are the only chemical elements with sufficient cosmic abundance to provide such density. Early seismic evidence suggested that the core was totally fluid, which is consistent with the very high temperature at the centre of the Earth, but later observations could be better explained by the presence of a solid inner core. This may be a combined effect of the very high pressures in this region plus some slight difference in chemical composition perhaps involving oxygen and sulphur.

4.3.7. The Earth's atmosphere

The structure of the Earth's atmosphere is described in detail in Topic O. It is the only atmosphere in the Solar System that can support advanced life in the form that we find it on Earth. There is the possibility that life could exist in other forms using the resources of other kinds of environment. Observations suggest that there may be salt-water oceans below the icy surface of Europa, one of Jupiter's Galilean satellites, and if so then some life form may be able to develop there. However, we know that changes being made to the atmosphere by the activities of humankind are giving noticeable effects on a time scale of a century and we also know that adaptive changes by evolutionary processes work on a much longer time scale. It is to be hoped that what we know influences what we do.

4.3.8.   *The Earth's magnetic field*

Chinese documents of the eleventh century speak of the Earth possessing a magnetic field at its surface; there is no suggestion that this is a new discovery so magnetism must have been familiar before this. It is detected at its crudest by suspending a small piece of lodestone on a string (an elementary compass) and finding that it aligns itself very roughly along the geographical north–south line. The geographical axis of the Earth and the magnetic axis are antiparallel; this arrangement is described in section V.1.

An early application of terrestrial magnetism was in navigation, especially navigation at sea. The age of western exploration, starting in the fifteenth century, required global magnetic measurements to be made and by the seventeenth century quite reliable world magnetic maps had been made, some overseen by Edmund Halley, showing the state of the magnetic field worldwide. Magnetic measurements of increasing accuracy were also made at certain magnetic observatories on land (especially at Kew outside London and at Potsdam near Berlin) so that accurate data for Europe are available from the sixteenth century onwards. Detailed data for the southern hemisphere are relatively recent. Developing technology has allowed the magnetic field at any location in the northern or southern hemisphere to be measured with ever-increasing precision, with the result that now very fine details of the field have been found which vary rapidly with time. The fine details, which vary on a short time scale, are associated with currents generated in the ionosphere due to Earth–Sun interactions while the gross features have their origin within the solid Earth. The spectrum of times associated with the Earth's field is shown in figure 4.25.

A full description of the Earth's magnetic field, taken as a prototype for planetary magnetic fields in general, is given in Topic V.

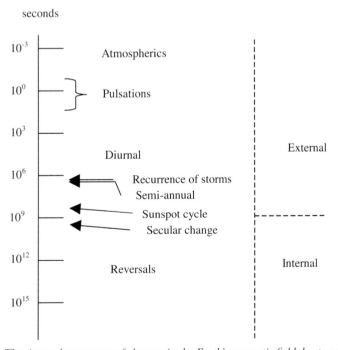

Figure 4.25. *The timescale spectrum of changes in the Earth's magnetic field due to various causes.*

4.3.9. Earth summary

The Earth's surface is dynamic and is divided into plates. These move relative to one another carrying the continents with them. The surface is recycled with a mean life of about 250 million years. This involves subduction of oceanic plates while new material is brought to the surface at the mid-ocean ridge system. The new material pushes back that already there to replace the existing oceanic bottoms. It appears that the mid-ocean ridge/tectonic plate system is the main way of transferring heat from the Earth's interior. It is supposed that the interior supports some kind of convective regime.

The temperature of the interior of the Earth will have been higher in the past, partly because there would have been more radioactivity and partly because more heat of formation of the Earth would still have been present. Presumably the plate system would then have moved more violently than now. It would follow that plate dynamics becomes weaker as cooling proceeds and ultimately will stop. Plate tectonics indicates loss of heat by convection of hot material (advection) from inside while a static surface gives heat loss by simple conduction. These conclusions are important in assessing the conditions of other terrestrial planets.

The Earth has a magnetic field with a major dipole component and a weak secular field. Both fields change with time, the dipole field changing more slowly. The direction of the axis of the dipole field has suffered complete reversals throughout geological history although the reversals are not periodic. The presence of the field acts as a constraint on theories of the interior.

4.4.   MARS

Perhaps the greatest change of attitude towards any one planet has resulted from increased knowledge of the planet Mars. Towards the end of the nineteenth century Schiaparelli claimed to have observed a series of 'canali' on the surface. The Italian word means 'channels' but was widely interpreted by English-speaking astronomers as 'canals'—implying artificial construction. These, it was claimed, changed appearance during the Martian seasons—the changes moving northwards from the equator as the summer season progressed. One interpretation was that this could be the advance of vegetation to higher latitudes with the summer warming. We now know that no vegetation exists and that there is no liquid water on the planet since the atmosphere is too thin to support it. However, water on the planet does pose interesting questions as will be seen later. Of all the planets Mars is probably the one that could most easily be made hospitable to mankind.

Its equatorial radius is 3392.2 km giving a surface area about the same as the continental area on Earth ($\sim 1.49 \times 10^{14}$ m^2). Its shape is a tri-axial ellipsoid with the radii in the ratios $1:0.9985:0.9934$. The mean density is 3930 kg m^{-3} giving an equatorial acceleration of gravity $g_{Mars} = 3.706$ m s^{-2}. The moment of inertia factor is $\alpha = 0.366$.

4.4.1.   The surface of Mars

The first direct contact with Mars was made early in the American space programme with a flyby by Mariner 4 in 1964 and photographs were returned to Earth of a portion of the southern surface. The pictures were not as expected. They showed a heavily cratered terrain reminiscent of the Moon. More information followed from an American long life orbiter (Mariner 9) which encountered Mars in November 1971. A Russian soft landing was made (by Mars 3) three weeks later. The American soft lander Viking I sent photographs from the northern surface during July 1976 and there were the first automatic searches (unsuccessful) for evidence of life.

The surface at the Viking lander site is shown in figure 4.26 and is very different from the cratered terrain shown in the first photograph by Mariner 4. There are no craters and the surface is barren and

Figure 4.26. *The view of the Martian surface from Viking I. The bar in the centre is part of the structure of the lander (NASA).*

smooth. This was the first indication that the surface of Mars has hemispherical asymmetry. The south is heavily cratered with an elevation of some 3 km above the mean radius (equivalent to sea level on Earth). The northern hemisphere is smooth and generally flat with a mean elevation up to 1 km below the mean radius. Data have been transmitted from the surface and from orbiting craft so considerable information has accumulated about the 'red planet' over many seasonal cycles. A general altitude map of the surface is shown in plate 2 and illustrates the hemispherical asymmetry of the planet.

An elementary chemical analysis of the soil of the Viking lander site was possible. It was found that the ratio Ca/Si $\approx$ 0–0.5 is broadly the same as for the Earth. On the other hand the ratio Fe/Si $\approx$ 0.7 is to be compared with the Earth value of about 0.3. The Martian 'soil' appears to contain about three times the iron content of Earth soil in the form of the oxide FeO. The surface of Mars is red due to the presence of rust! One might notice that the ratio Mg/Si $\approx$ 0.2 is similar to that for Earth basalts but the ratio Al/Si $\approx$ 0.2 is lower than for Earth basalts which have the ratio in the range 0.26–0.44. The Viking landing site seems deficient in aluminium in comparison with the Earth. It must be stressed that the surface 'soil' is not soil as known on Earth but a regolith (broken, older rocks) composed of micron spheres forming a dust. The polar caps are composed of water and CO_2 ices and they change during the year as the temperature changes, advancing and retreating with the seasons. The water component of each of the caps is more-or-less permanent but the CO_2 component evaporates at one pole and condenses at the other with the changing seasons. Mars shows essentially the same pattern of seasons as the Earth although they are extended in keeping with the longer Martian 'year'.

4.4.1.1. The highlands

These occupy all of the southern hemisphere and some of the northern. They are heavily cratered regions with elevations between 1 and 4 km above the mean radius. The very high density of craters suggests that the surface is very ancient, probably older than 3.8×10^9 years if lunar chronology is a guide. The lunar surface suffered heavy bombardment before this time and very little afterwards and the same may have been the case for Mars also.

The Martian highlands and craters have a somewhat different form from the lunar highland craters. Four differences are especially to be noted:

(i)   the Martian highlands are crossed by a number of valley complexes (see figure 4.27);
(ii) the Martian highlands have large areas of smooth terrain between craters;
(iii) the Martian craters often have ejecta surrounding them in the form of thin sheets (see figure 4.28), rather like a 'splash' in muddy water, while the lunar craters have coarse, well defined, ejecta often associated with lines of small secondary craters.

Figure 4.27. *Valley complexes in the Martian highlands (NASA).*

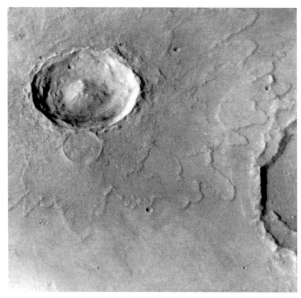

Figure 4.28. *The Martian impact crater Yuty. The ejected material forms layers as might be expected from muddy material that solidifies (NASA).*

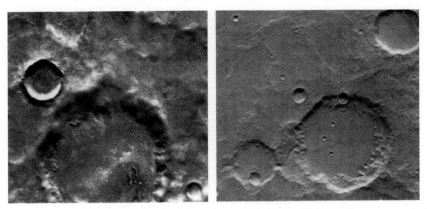

Figure 4.29. *The left-hand image shows sharply defined features of craters and also small craters near the Martian equator. The right-hand image shows that farther from the equator more erosion has taken place (NASA).*

(iv) the Martian craters away from the equator show a degree of degrading due to erosion while those nearer the equator do not show this (see figure 4.29). By comparison, the lunar craters give no evidence of degradation at all.

The form of the surrounding ejecta described in (iii) presents problems of interpretation. The most obvious is hidden water below the surface but the craters are often on the side of an elevated slope. Viewed in relation to (iv) it could be that there is considerable ice in the higher latitudes which produces a more complex rheology than for the simple rock alone. In figure 4.30 is seen a flow region (Protonilus Mensae, latitude $46°$ N) which could well be due to the presence of ice in the surface material. Such 'flows' are not seen near the equator where the temperature is higher and ice would not be expected. The solar heating will penetrate deeper into the crust in the equatorial regions giving the tendency for water to evaporate or polymerize at the surface.

4.4.1.2. The plains

The northern hemisphere is dominated by wide rolling plains with very few craters. This suggests a surface younger than 3.8×10^9 years and perhaps considerably younger. Certainly the intensity of

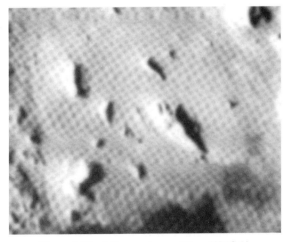

Figure 4.30. *A flow region on Mars (NASA).*

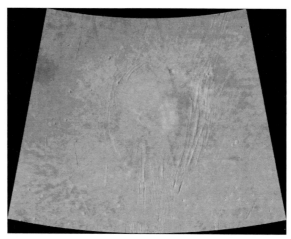

Figure 4.31. *Flows around Alba Patera, a large shield volcano. Some flows have travelled 225 km, much farther than any on Earth (NASA).*

the cratering in the region Lunae Planum suggests that it probably achieved its present form about 3.5×10^9 years ago. On the other hand, the sparsely cratered plains in Tharsis may be no more than 500 millions years old. The plains in the regions of volcanoes are probably formed of lava and ash. The flow along the very large shield volcano Alba Patera in the northern hemisphere is shown in figure 4.31. It is some 225 km long, which is far greater than anything known on the Earth or the Moon.

Other craters have different forms and give witness to a complicated environment linking the surface with the atmosphere.

4.4.1.3. *Volcanic regions*

Volcanic activity has been a central feature of the Martian surface over the great majority of its lifetime. Heavily degraded centres, now extinct, are to be found together with more recent structures of immense size. One of the ancient and now extinct volcanoes is Tyrrhena Patera in the southern highlands, shown in figure 4.32. The central crater is about 45 km across, showing the presence of substantial volcanoes early in the Martian development. The youngest volcanic features are in the Tharsis region on the equator (105° W) and Elysium at latitude 25° N (210° W). The Tharsis region, shown in figure 4.33, is a large bulge on the surface some 4000 km across and 10 km high at the centre. On the north-west boundary lie three large volcanoes, Arsia Mons, Pavonis Mons and Ascraeus Mons. Beyond to the north-west is Olympus Mons (see figure 4.34) the largest volcano in the Solar System. It rises 24 km above the surrounding plain and is about 600 km across at the base. It is one of the few features that were recognized on the surface by Earth-based telescopes before the space age, when it was called 'the snows of Olympus' (Nix Olympus). It has the form of a shield volcano like Mauna Loa of Hawaii (120 km base and 9 km above the ocean floor) so that it dwarfs the Earth structure as is seen in figure 4.35. Olympus Mons is not the only large structure: to the north of Tharsis lies Alba Patera which is 1500 km across although only a few kilometres high. Not all the volcanoes exuded lava. Tyrrhena Patera, north-east of the impact basin Hellas, most probably erupted ash rather in the nature of Mount St. Helens on Earth. The large size of Martian volcanoes is a characteristic feature of the planet. They are generally between 10 and 100 times the scale of comparable structures on Earth and give evidence for the presence in the past of large quantities of magma below the surface. Certainly the Tharsis bulge is very ancient. All the lava flows show a direction consistent with having travelled downhill from the present elevation.

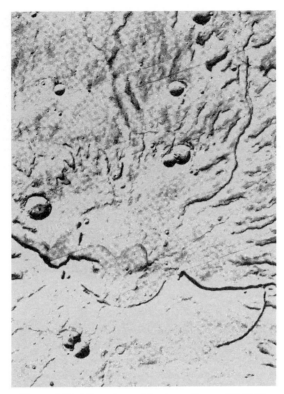

Figure 4.32. *The volcano, Tyrrhena Patera (NASA).*

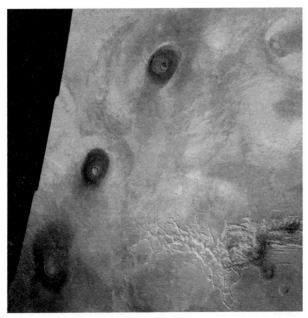

Figure 4.33. *Volcanoes in the Tharsis region. The edge of the Valles Marineris canyon can be seen on the right-hand side (NASA).*

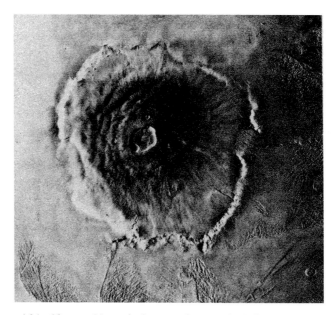

Figure 4.34. *Olympus Mons, the largest volcano in the Solar System (NASA).*

4.4.1.4.   *Channels and canyons*

To the east of the Tharsis Montes lies the Valles Marineris, a massive canyon complex extending 4000 km eastwards along the equator (figure 4.36). At the western end it starts with Noctis Labyrinthus at the summit of the Tharsis rise. On the east it merges with a region of strange topology known as *chaotic terrain*. The depth of the canyons varies from about 2 km at the ends to some 7 km in the middle, where it is 600 km wide. The canyon appears to have formed as *fault scarps* and shows many features common to such structures on Earth. There are, however, other features including subsidence and fluvial processes.

Fluvial activity is clearly indicated at several places, such as Candor Chasma (figure 4.37) with a dimension of some 200 km square. Here the geometry suggests seepage of groundwater at a past time. The darker floor seems to contain sediment of some kind (there has been speculation that it could be the floor of an ancient sea) and there is evidence of erosion on the walls of the canyon by water. Other features show evidence of water on the surface. For instance, Chryse Planitia involves a large flood plain scoured by water. The region has impact craters superimposed on it showing it is ancient,

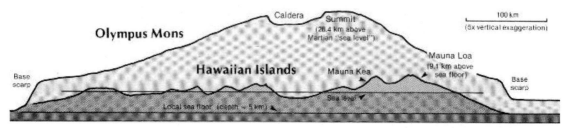

Figure 4.35. *A comparison of Olympus Mons and volcanic features of Hawaii.*

Figure 4.36. *The canyon, Valles Marineris. It runs along the equator for 4000 km (NASA).*

probably several thousand million years old. The many channels, which it seems could only have been made by the flow of water, are much more restricted than their counterparts on Earth. It would seem that they carried relatively small quantities of water and that this came from underground rather than from precipitation. The many channel forms remain isolated from each other showing that the channel systems nowhere controlled a water table on the planet. It would seem (from crater counts) that they were formed early in the planet's history and the water flows lasted for rather short geological periods of time. It may be that the chaotic terrain was formed by the movement of near-surface rock by water which allowed the surface to collapse.

Figure 4.37. *Candor Chasma containing dark deposits that may be volcanic ash or perhaps silt sediments on the bed of an ancient ocean (NASA).*

4.4.2. Consequences of early water

It would seem an unavoidable conclusion that water existed on the surface of Mars for at least a short period of time early in its history. This would imply a denser atmosphere than now and an increased surface temperature to allow the water to remain in the liquid state. The question then is what has happened to that atmosphere? There is evidence that there is considerable permafrost just under the surface and several authors have estimated the amount. It would possibly be equivalent to a uniform covering of a few hundred metres of water over the whole planet. There is also water trapped at the poles, particularly at the north pole, but the quantity is not easy to estimate.

One consequence of water early on is the possibility of the appearance of living entities at that time. There have been claims that a terrestrial meteorite sample that originated on Mars contains evidence of a very elementary life form but this conclusion is uncertain (section 10.2.2). Mars has provided many surprises so far and promises to give more in the future.

4.4.3. Later missions

Two American missions have provided more information about the surface conditions on Mars. The first, Mars Pathfinder, landed on the surface on 4 July 1997. The second, Global Surveyor, moved into orbit on September 1997. The first was designed to survey one location in detail while the other aims to make an accurate survey of the whole surface.

The chosen region of study for the Mars Pathfinder was the flood plane Chryse Planitia. This is an ancient flood plane containing rocks swept down from different initial locations upstream. The general view near the landing site is seen in figure 4.38. Several boulders are present—the closest to the Sojourner rover vehicle is nick-named Barnacle Bill, the chemical and mineral compositions of which are given in tables 4.8 and 4.9. An analysis of the oxides in samples of soil and rock were made using α-proton X-ray spectrometer methods. Some results are collected in table 4.8. The similarities between the rocks and the soils are noticeable but there is a discrepancy with the data from SNC meteorites that are believed to have originated on Mars on the basis of the atmospheric gases contained in small cavities in the meteorites (section 10.2.2). On the other hand, the rock and soil samples are each a mean over an area of a few centimetres and the material may not, in fact, be homogeneous to this extent. The data in table 4.8 for Mars can be compared with those of table 1.3

Figure 4.38. *The Sojourner vehicle from Mars Pathfinder close to the rock Barnacle Bill (NASA).*

Table 4.8. *The proportions by percentage mass of eight oxides in two rocks and soil samples at three close locations. The meteorites are believed to have come to Earth from Mars (NASA).*

Oxide	Rock Barnacle Bill (% mass)	Rock Yogi (% mass)	Soil (% mass) (three different locations)			SNC meteorites (% mass)
SiO_2	55.0	50.9	46.1	43.3	43.8	38.2–52.7
Al_2O_3	12.4	11.4	8.0	10.4	10.1	0.7–12.0
MgO	3.1	6.3	8.7	9.0	8.6	9.3–31.6
FeO	12.7	13.8	19.5	14.5	17.5	17.6–27.1
CaO	4.6	5.8	6.3	4.8	5.3	0.6–15.8
K_2O	1.4	1.1	0.6	0.7	0.7	0.02–0.19
TiO_2	0.7	0.8	1.1	1.1	0.7	0.1–1.8
MnO	0.9	0.5	0.5	0.5	0.6	0.4–0.6

for Venus and the Earth. The similarity between these three sets of data is striking although the larger iron content of the Mars surface is noteworthy. The analysis in terms of elements is especially useful since various ratios of abundance can be deduced from it and these can have a relatively higher accuracy for comparisons than the initial data itself. The analysis for some elements is contained in table 4.9 for the two rocks and for rock samples of table 4.8. The ratios Ca/Si and Fe/Si are shown in figure 4.39, which contains data for rocks from Earth and Mars. The relative Fe enrichment of the Martian rocks is clear. Ratios Mg/Si are plotted against the corresponding ratio Al/Si in figure 4.40. It seems that the Martian rocks have a lower Al content than Earth rocks. Nevertheless, the Martian rocks studied so far appear very Earth-like. For instance, the rock Barnacle Bill is classed geologically as an Andesite rock, similar to volcanic rocks on Earth which are named after the rocks first classified in the Andes mountains of South America.

The second American mission is the Mars Global Surveyor, the aim of which is to photograph the whole planet surface to an accuracy of a few metres. An especial interest is the accurate determination of the figure of the planet.

Table 4.9. *Element analysis of two rocks and soil at three locations on Mars (NASA).*

Element	Rock Barnacle Bill (mass %)	Rock Yogi (mass %)	Soil (mass %) Location A2	Location A4	Location A5
O	45.0	44.6	42.5	43.9	43.2
Si	25.7	23.8	21.6	20.2	20.5
Fe	9.9	10.7	15.2	11.2	13.6
Al	6.6	6.0	4.2	5.5	5.4
Ca	3.3	4.2	4.5	3.4	3.8
Na	3.1	1.9	3.2	3.8	2.6
Mg	1.9	3.8	5.3	5.5	5.2
K	1.2	0.9	0.5	0.6	0.6
P	0.9	0.9	—	1.5	1.0
S	0.9	1.7	1.7	2.5	2.2
Mn	0.7	0.4	0.4	0.4	0.5
Cl	0.5	0.6	—	0.6	0.6
Ti	0.4	0.5	0.6	0.7	0.4
Cr	0.1	0.0	0.2	0.3	0.3

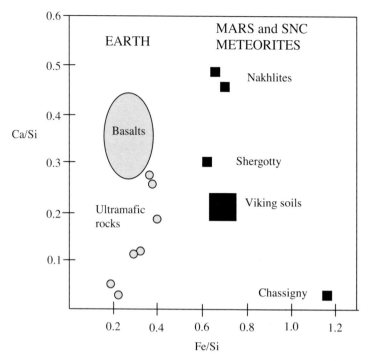

Figure 4.39. *The Ca/Si ratio plotted against the Fe/Si ratio for terrestrial and Mars samples. Shergottites, Nakhlites and Chassignites are the SNC meteorites that are presumed to have come from Mars.*

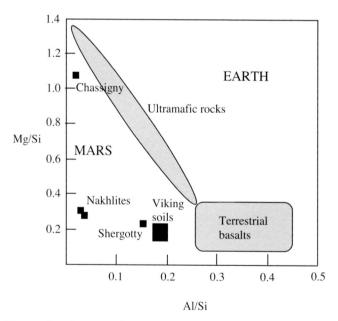

Figure 4.40. *Mg/Si ratio plotted against Al/Si ratio for various terrestrial and Mars samples, including SNC meteorites.*

4.4.4. *The atmosphere of Mars*

In 1976–1977 a Viking spacecraft studied the surface and atmosphere of Mars over a two-year period and most of our detailed knowledge of Mars' atmosphere comes from those observations. It will be seen in table 4.1 that in terms of the proportions of its components the composition of the atmosphere of Mars is very similar to that of Venus; apart from the components shown in the table, Mars also has a fraction 0.0007 of carbon monoxide, CO. However, there the similarity ends because the surface pressure of Mars is only 0.006 bar. With an acceleration due to gravity about one half of that on Earth, the column mass for Mars is just over 1% of the terrestrial value. Because the atmosphere is so thin there is virtually no greenhouse effect and the global average surface temperature of Mars, 223 K, is appropriate to its distance from the Sun. This temperature is an average over very wide variations; at the equator at mid-day the temperature may be as high as 290 K but at the same location at night it may plunge to 140 K. This is a consequence of the lack of substantial cloud layer or atmosphere to reflect back radiation from the surface.

The temperature profile with altitude is shown in figure 4.41. There is a region of almost constant temperature between heights of about 40 to 100 km, which plays the role of a mesosphere or stratosphere, and there is a layer of haze and dust at a height of about 32 km. Although the atmosphere is so thin there can be very high winds on Mars and this stirs up dust clouds that sometimes obscure large parts of the Martian surface for several days or even weeks.

Even before spacecraft visited the planet it was known that there are polar caps which advance and recede with the seasons. By analogy with the ice caps on Earth it was assumed that these were caps of water ice but now it is clear that, although the permanent part of the caps are water ice, the advance and retreat of the boundary is due to carbon dioxide. At the pole corresponding to the Martian winter there are snowfalls of solid CO_2 (a material in commercial use and called *dry ice*) which will occur down to the latitude at which solid CO_2 will form at a pressure of 6 mb. Then, as the summer returns the CO_2 will sublimate and return to the atmosphere while, at the same time, solid CO_2 is being deposited around the other pole.

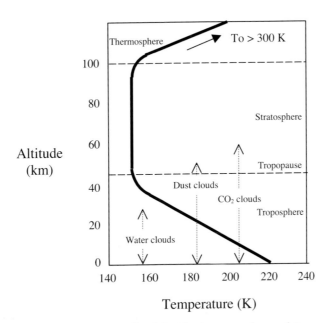

Figure 4.41. *The temperature profile of the Martian atmosphere and its composition.*

It is possible that there is more water on Mars than is indicated by the small atmospheric component. For one thing, at the low temperatures which prevail on Mars the vapour pressure of water is low. A great deal of water, in the form of ice but possibly in liquid form, could exist below the arid, dusty Martian surface. There is very clear evidence that a great deal of liquid water must once have existed on Mars, as can be seen from the sinuous channels which occur in many regions of the surface which resemble dried river beds. A possible scenario is that Mars once supported a denser atmosphere that provided a sufficient greenhouse effect to enable water to exist in liquid form in the equatorial regions. With an exosphere 400 km above the surface at a temperature of 2000 K, then the lifetime of a water vapour atmosphere would have been over 2×10^{10} years. Although Mars is more than twice as far from the Sun as Venus, the mechanism of dissociation of water vapour leading to the loss of hydrogen and available oxygen would still be effective. A paradox of Mars is that although it is less dense than the Earth or Venus it contains much more iron in its surface rocks, much of it in the form of iron oxide which gives the planet its characteristic red colour. Presumably in the process by which Mars was formed there was less segregation of iron into the core than in the more massive terrestrial planets so that more free iron was left on and near the surface. Since iron oxidizes very readily much of the oxygen produced by water dissociation would have been taken up in this way.

At higher latitudes water would have been stored in the form of ice and the loss of water in the equatorial region would have been slowly replenished from this source. Eventually too little ice remained to maintain this system and, in addition, the composition of the atmosphere could have changed by the loss of lighter components such as CO, O_2 and N_2. The greenhouse effect would have reduced in effectiveness until eventually the present situation was reached. This suggested scenario explains how water ice is still locked in the polar regions.

4.4.5. *Magnetism and Mars*

No intrinsic magnetization has been found so far to the limit of accuracy of the measuring equipment, either from surface measurements or from near-Mars observations of the solar wind. Any dipole moment is less than $1.4 \times 10^{12} \, T\,m^3$ and any surface field must have a magnitude less than $4 \times 10^{-8} \, T$ at the equator. It is perhaps surprising that Mars does not apparently possess a significant intrinsic field, whereas Mercury does. However, Mars is not without interest from the magnetic point of view. The Mars Global Surveyor measured the magnetism of the surface of Mars showing strip-like magnetization of alternating polarity, much as occurs on Earth (section V.7).

4.4.6. *Mars summary*

The surface of Mars has had a complicated history. The northern hemisphere is generally smooth while the southern hemisphere is cratered and mountainous. The join between the two hemispheres has a complicated form. It is almost as if two halves have been put together to form the single planet.

Volcanism is the major feature on the surface and this seems to have been the case from the earliest times. The major volcanoes are extremely large and, if as old as they appear to be, would require a substantial crust to have supported them over that length of time. It is possible, of course, that forces due to internal convection now support, or in the past have supported, these loads.

The present Martian atmosphere is thin (with a surface pressure about six thousandths that of the Earth). The mean surface temperature is about 210 K. Water could not exist on the surface in liquid form at the present time. There is, however, evidence of the effects of free water on the surface during the early history of the planet. This must have involved a thicker atmosphere than now, suggesting that atmospheric gases have either been lost or trapped underground. This is a problem to be solved later when manned exploration becomes possible. Certainly some crater regions appear to show a sludge surround that would support the view that water is now buried underground as a permafrost.

There is no evidence of intrinsic magnetic properties for the planet within the present accuracy of measurement. Any magnetic field will be small, certainly lower than one thousandth that of the Earth's field.

Problem 4

4.1   Calculate the synodic orbital period (sections 4.1 and 6.1), in years, for
  (i)   Venus, with a sidereal orbital period of 0.6152 years
  (ii)   Jupiter, with a sidereal orbital period of 11.862 years.

CHAPTER 5

THE MAJOR PLANETS AND PLUTO

5.1. JUPITER

Jupiter is by far the most massive of the planets, having about two-and-a-half times as much mass as all the other planets combined. It has a density similar to that of the Sun, with about one-tenth of the radius and one thousandth of the mass of the central body. Its visual albedo is 0.52, which makes it one of the brightest objects seen from Earth, except for Venus and Mars when they are closest to the Earth.

Views of Jupiter, as seen through telescopes from Earth, show a banded structure (figure 5.1). The bands are designated as *zones* if they are bright and *belts* if they are dark. They are seen to move around the planet at different speeds but with a near-equatorial spin period, known as the *System I* period, of 9 h 50 m 30 s. However, this is not the characteristic period of spin of the whole surface or of the planet as a whole. Regions more than 9° from the equator give a different period of 9 h 55 m 41 s, known as the *System II* period. A further spin period comes from the presence of relativistic charged particles trapped in radiation belts by Jupiter's intrinsic magnetic field (section 5.1.4) that generate radio waves with wavelengths of tens of metres (*decimetric radiation*) that can be detected on Earth. The tilt of the planet's magnetic axis with respect to the spin axis gives a periodic variation in the observed strength of the radio waves and the period is taken to be that of the bulk of Jupiter about the spin axis. This period is 9 h 55 m 30 s and this rotation, linked with the magnetic field, is known as the *System III* period.

5.1.1.   The internal structure of Jupiter

Because of its rapid spin, the most rapid of any Solar-System planet, and its gaseous nature, the flattening of Jupiter is quite large (table 3.3) and is easily detected visually (figure 5.1). The large distortion enables the values of both the gravitational coefficients J_2 and J_4 (Topic T) to be measured and this gives very tight constraints on possible models. A plausible structure for Jupiter is shown in figure 5.2. The central core is thought to have a mass in the range $10–20M_\oplus$. It could be like an enlarged version of a terrestrial planet with an iron central core, a surrounding shell of silicates and then perhaps another shell consisting of higher-molecular-weight materials containing hydrogen and carbon. The remainder is hydrogen and helium but at depths below the surface where the pressure exceeds a few million bars the hydrogen will be in the form of *metallic hydrogen*. In this state the hydrogen forms an ordered lattice, at least with local if not long-range order, and the electrons move freely through the proton lattice as they do in a conducting metal.

68

Figure 5.1. *Jupiter's Great Red Spot is lower centre. At the bottom, to the left, the satellite Io can be seen below the disk of the planet (NASA).*

The near-surface regions of the planet are gaseous and form an atmosphere, described in section 5.1.3.

5.1.2. Heat generation in Jupiter

From the Bond albedo (Topic X) of Jupiter, 0.343, it is possible to estimate how much energy it absorbs from the Sun and from an analysis of the spectrum of Jupiter it is possible to estimate its effective

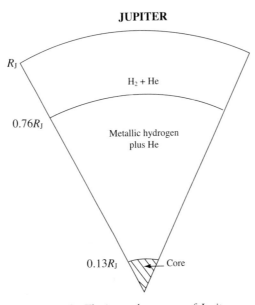

Figure 5.2. *The internal structure of Jupiter.*

temperature as a radiator of energy. In fact the radiated energy does not have perfect black-body characteristics but it is possible to ascribe temperatures to the surface, that vary from place to place, in the range 115–150 K with an average of 125 K. Taking this average temperature it can be shown that Jupiter emits about twice as much radiation as it receives from the Sun. The mechanism that give rise to this excess radiated energy is not known but two main suggestions have been advanced.

Gravitational contraction of Jupiter

This mechanism assumes that Jupiter is still settling down from its initial formation and so is slowly contracting. A contraction rate of the radius of less than 0.05 m per century (Problem 5.1a) is enough to release sufficient gravitational energy to explain the excess. Presumably the contraction rate would be slowing down but this rate of collapse over the lifetime of the planet (4.5×10^9 years) would have reduced its radius by only 3%, so the explanation is feasible.

Separation and settling of helium

If helium separates out from hydrogen and sinks towards the centre of the planet then this will release gravitational energy. The required rate of separation to explain the excess heat is very tiny compared with the total amount of helium in the planet and the change in the distribution of helium since the planet formed would be negligible (Problem 5.1b).

5.1.3.   The atmosphere of Jupiter

The first problem to be considered in describing the atmosphere of a major planet is to define what constitutes the atmosphere. The outermost region, which has similar densities to the atmospheres of terrestrial planets, would certainly be considered as part of the atmosphere but as one goes deeper into the planet the density eventually becomes more liquid-like than gas-like although, in fact, the material is still a very dense gas. The choice of the lower boundary of the atmosphere is thus quite arbitrary and the level where the pressure is 1 bar is often taken as the base of the atmosphere although atmospheric data are sometimes available from below this level.

Prior to the space age the only available method for studying the atmospheres of planets was spectroscopic analysis of the light from them. The different constituents of the atmospheres absorb at particular wavelengths—e.g. CO_2 absorbs at 4.3 and 15 μm and CH_4 at 3.3 and 7.7 μm. Thus analysis of the infrared spectra from distant planets enabled estimates of their compositions to be made. The advent of space probes that could orbit planets and take data over long periods has greatly improved what can be done.

Knowledge of the composition of Jupiter's atmosphere was greatly advanced by remote sensing from the Voyager 1 and 2 missions in 1979. The next important probe of the atmosphere was not of human origin but was provided by the spectacular impacts on Jupiter by fragments of the comet Shoemaker–Levy in July 1994. These fragments threw out material from deep within the atmosphere and gave information about composition at considerable depths. A more controlled probe was provided by the Galileo spacecraft in December 1995. This probe descended by parachute for 57 minutes, chemically sampling its environment as it went and measuring wind speeds. At its lowest point it was at a depth of 600 km where the pressure was about 24 bar. Some previous estimates of the composition had to be revised. The main component of the atmosphere is hydrogen, which was already well known, but there was 24.4% by mass of helium, which was much more than previous estimates that were in the range 11–19%.

Despite the many different ways that have been available for measuring Jupiter's atmosphere there are still many uncertainties. For example, the Voyager mission measurements suggested water

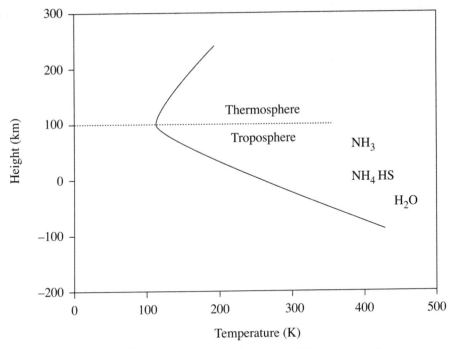

Figure 5.3. *The temperature profile and structure of Jupiter's atmosphere.*

abundance at twice the level the Sun would have if all its oxygen were contained in water. Analysis of the Shoemaker–Levy impact ejecta suggested a much higher abundance—perhaps up to ten times the solar level. By contrast the Galileo probe, the most direct measurement, suggested that the water abundance is at just about the solar level. It is quite possible that the components of Jupiter's atmosphere are not well mixed and that variations in the abundance estimates are indicative of variations from one site to another.

The vertical structure of the atmosphere is illustrated in figure 5.3. The thermal profile is a simple one corresponding to a troposphere with an adiabatic lapse rate and a thermosphere with a gradually rising temperature which goes to more than 1000 K (Topic O). The troposphere extends well below the arbitrary 1 bar level and has been explored down to just over 100 km below the 1 bar level, which corresponds to a pressure of about 10 bar. There are clouds at various levels, consisting of ammonia and probably of NH_4SH, ammonium hydrosulphide and water as shown figure 5.3.

Wind speeds measured by the Galileo probe varied from $150\,m\,s^{-1}$ at 0.5 bar, increasing to around $200\,m\,s^{-1}$ at a pressure of a few bar and then remained constant. The variation of wind speeds with latitude, as measured by the Voyager 1 spacecraft, is shown in figure 5.4. The speeds of the zones and belts correlate with the wind-speeds quite well so that neighbouring bands are in relative motion.

The atmosphere of Jupiter shows a great deal of structure (figure 5.1), the small features of which change on time scales varying from days to years but the largest features of which have remained more or less constant for 100 years or more. Ovals, swirls and wave-like features are seen on the surface, all of which corresponds to cloud patterns on Earth associated with different kinds of weather system. The most prominent feature in Jupiter's atmosphere is the *Great Red Spot* (GRS—actually more grey than red) which is about 35 000 km in length and 15 000 km in

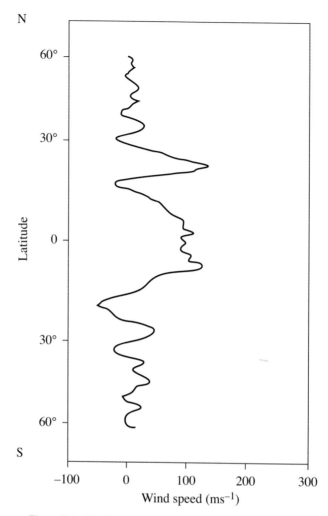

Figure 5.4. *Wind speed as a function of latitude for Jupiter.*

width. Three projections of the Earth would comfortably sit within the Red Spot. This has been seen for about 350 years, varying in intensity but never disappearing. It has the appearance of a high-pressure storm, an anticyclone, with the general motion within it counter-clockwise about the centre with a six-day period.

5.1.4. *Jupiter's magnetic field*

The dipole component of Jupiter's magnetic field has a dipole axis inclined at $10°$ to the spin axis with the centre of the field displaced from the centre of the planet by about one-tenth of Jupiter's radius (figure 5.5). The dipole moment, $1.5 \times 10^{20}\,\mathrm{T\,m^3}$, is about 19 000 times that of the Earth and the flux of the solar wind is only about 4% of that in the Earth's vicinity. This gives a magnetosphere much larger than that of the Earth and all other features connected with the interaction of the solar wind with the planetary field are similarly on a larger scale—for example the magnetotail may stretch to the orbit of Saturn.

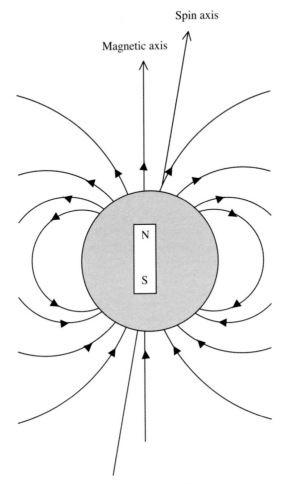

Figure 5.5. *The magnetic field of Jupiter.*

The closest of the Galilean satellites is Io (section 7.3.1) that has volcanoes and a tenuous atmosphere and ionosphere and it is a rich source of ions of heavier elements such as O, S and Na. The combination of a source of ions and a large magnetic field generates a current system that links Jupiter to Io's ionosphere.

The source of Jupiter's magnetic field is not known with certainty but could possibly be driven by electric currents in the metallic-hydrogen region. Currents within the metallic part of the core could also be responsible for part of the field.

5.1.5. Jupiter summary

Jupiter is by far the most massive of the planets and also the one that rotates most rapidly. Its surface shows a banded structure, with bands moving with slightly different angular speeds. Spots are also present, both bright and dark, representing weather systems, mostly short-lived. The very large GRS, an anticyclone system, has endured for more than 350 years.

The excess radiation emitted by Jupiter, over and above that it receives from the Sun, is probably due to some rearrangement of its mass—either a contraction of the total body or the settling of helium within it. Jupiter's large dipole moment, coupled with a lower flux of solar-wind particles, creates a magnetosphere and other magnetically related structures on a much larger scale than in the vicinity of the Earth.

5.2. SATURN

Saturn is smaller and farther from the Earth than Jupiter and so the main body of the planet shows less detail from Earth-bound observation. The main characteristic to be seen from Earth is its spectacular ring system, first seen in very low resolution by Galileo who thought it to be accompanying bodies of some kind. In 1659 Christian Huyghens (1629–1695) correctly interpreted it as a ring structure.

Spacecraft pictures show a banded structure, similar to that of Jupiter (plate 3), together with bright and dark oval features representing storm systems—although nothing with the scale and duration of the GRS is present. There is differential spin of the surface, as deduced from motions of the bands, varying from 10 h 15 m at the equator to 10 h 38 m closer to the poles. Radio emission detected by the Voyager spacecraft gave a System III period of 10 h 39.4 m, taken as the intrinsic period for the rotation of the planet as a whole.

5.2.1.   *The internal structure of Saturn*

Although Saturn is the second most massive planet it has by far the lowest mean density, this being about one-half that of Jupiter. Its lower total mass gives much less compression of interior material than is the case for Jupiter. In particular there is a much smaller region where the conditions are suitable for producing metallic hydrogen, a dense form of that element, as is seen in figure 5.6. The diffuse nature of Saturn and its rapid spin give it the largest flattening of any body in the Solar System (table 3.3) that is easily visible in telescope images taken from Earth. The moment-of-inertia factor of the planet, 0.210, requires a silicate + metal core with mass about $5M_\oplus$.

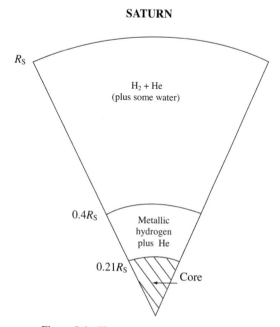

Figure 5.6. *The internal structure of Saturn.*

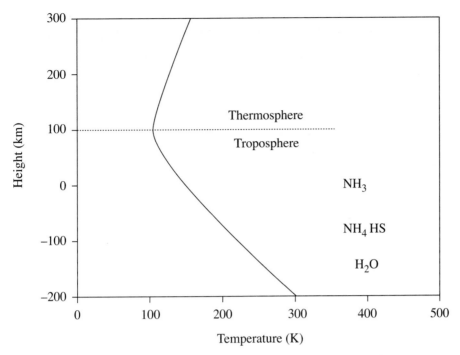

Figure 5.7. *The temperature profile and structure of Saturn's atmosphere.*

5.2.2. Heat generation in Saturn

From its temperature of 95 K it is found that Saturn radiates about twice as much radiation as it receives from the Sun so the heat generation within it is even greater per unit mass than is the case for Jupiter. Since Saturn would have cooled more rapidly after its formation it is generally believed that Saturn should by now have reached an equilibrium configuration and so should not be slowly collapsing, as has been suggested for Jupiter. This suggests that the separation and settling of helium could be the mechanism, a view supported by the depletion of helium in the atmosphere relative to hydrogen, assuming solar composition for the planet as a whole (section 1.2). Models of the planet, taking into account the values of J_2, J_4 and the flattening, are consistent with a concentration of higher-density helium towards the interior of the planet.

5.2.3. The atmosphere of Saturn

Like Jupiter, Saturn has an atmosphere dominated by hydrogen and helium although, as previously mentioned, the helium content estimated from remote sensing is much less than that of Jupiter. The vertical structure of the atmosphere is shown in figure 5.7 and like that of Jupiter is seen to be simple consisting just of a troposphere and thermosphere. The pattern of wind speeds, illustrated in figure 5.8, shows even higher speeds than found on Jupiter and this could be related to the internal energy generation.

5.2.4. Saturn's magnetic field

The magnetic dipole moment of Saturn, 4.2×10^{18} T m^3 is about 0.03 times that of Jupiter and, because it is twice as far from the Sun, the solar wind flux in its vicinity is one quarter as great. This gives a

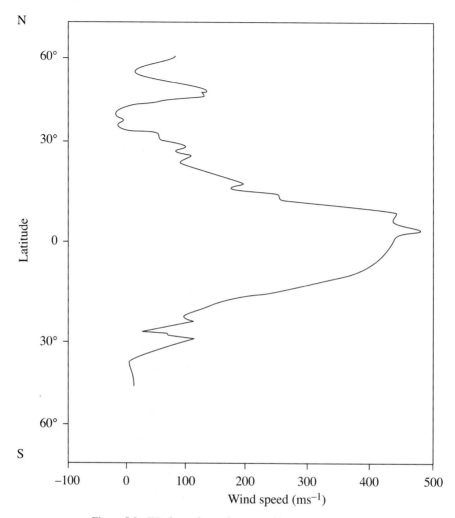

Figure 5.8. *Wind speed as a function of latitude for Saturn.*

magnetosphere and magnetic structure around the planet on a much smaller scale than around Jupiter. Just as for the Earth there are particle (van Allen) belts around Saturn and the accompanying aurorae have been photographed by the Hubble Space Telescope (figure **W**.14).

The magnetic dipole axis is inclined at less than $1°$ to the spin axis and the offset of the axis is about 0.04 of the planet's radius from the centre (figure 5.9). The comparatively small dipole moment suggest that the currents driving the magnetic field are much smaller than for Jupiter and this is consistent with the deduction that the amount of conducting metallic hydrogen was both smaller in volume and concentrated more towards the centre of the planet.

5.2.5. Saturn summary

In many ways Saturn appears to be a smaller version of Jupiter and shares many of its characteristics. The banded structure of the surface, oval spots representing storm systems and high speed winds are very similar, as are the internal generation of energy and the intrinsic dipole field. Such differences as

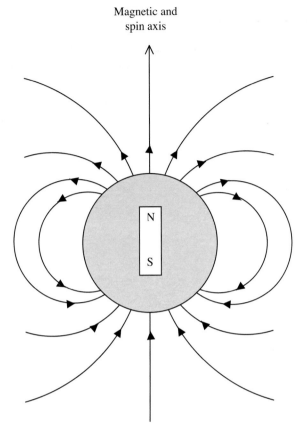

Magnetic and
spin axis

Figure 5.9. *The magnetic field of Saturn.*

do exist in the physical characteristics, e.g. a lower magnetic dipole moment, can be ascribed to the different masses of the two planets.

5.3. URANUS

Although Jupiter and Saturn are similar in many ways the much smaller body Uranus seems to be quite different. Its greenish blue colour comes from preferential absorption of red light that indicates the presence of methane in its atmosphere. The main atmospheric components are 83% hydrogen, 15% helium and 2% methane with small amounts of other substances, such as ammonia. Estimates of the spin period were very uncertain from Earth observation but Voyager 2 detected a periodically-varying magnetic field that gave a System III period of 17h 14m. The atmosphere is dominated by haze but the clouds below the haze appear to have a banded structure. Rather curiously, and in contrast to Jupiter and Saturn, the differential rotation of the bands gives slower motion at the equator and faster motion at higher latitudes, with periods varying in the range 17h to 15h.

A feature of Uranus that distinguishes it from all other planets is the extreme tilt of its spin axis. It shares with Venus a retrograde spin but, unlike Venus, its spin axis is almost in its orbital plane. This

URANUS

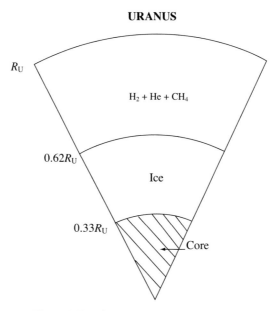

Figure 5.10. *The internal structure of Uranus.*

gives the curious pattern that if at some time a pole points almost towards the Sun then, half an orbit later, it points almost directly away from it.

5.3.1.   The internal structure of Uranus

From the motions of its inner satellites and one of its rings the value of J_2 has been determined and this, together with either the flattening or the spin period, can provide an estimate for C, the moment of inertia about the spin axis (section 3.4, table 3.3). Knowledge of the bulk composition of Uranus is not available so various models can be constructed on the basis of different assumptions—although they all involve H and He in solar abundance, various ices formed from H, N, C, O, plus silicates and iron. One model, consistent with a density of $1.29 \times 10^3 \, \mathrm{kg \, m^{-3}}$ and other known quantities, is shown in figure 5.10 but it is only one of many consistent models that can be produced.

5.3.2.   Heat generation in Uranus

The most interesting feature of heat generation in Uranus is the fact that, if it exists at all, it is only at the level of less than 20% of the oncoming solar radiation. This makes it an exception among the major planets and it is to be wondered why this is so. No convincing explanation is available for this difference.

5.3.3.   The atmosphere of Uranus

The structure of the atmosphere of Uranus is not well constrained by observations but a possible structure is shown in figure 5.11. It shows a simple troposphere–thermosphere structure. Assuming that a solar abundance of the elements should be present on Uranus then there seems to be far less nitrogen than expected, given that it should appear in the form of ammonia. The inferred ratio of C/N is 100 times greater than in the Sun.

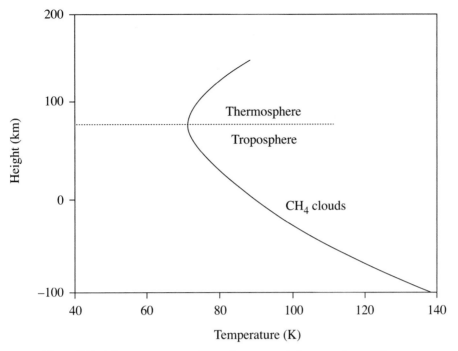

Figure 5.11. *The temperature profile and structure of the atmosphere of Uranus.*

The Voyager spacecraft measured winds in the upper atmosphere of Uranus at speeds up to $200\,\mathrm{m\,s^{-1}}$.

5.3.4. *The magnetic field of Uranus*

The magnetic dipole moment of Uranus is $3.8 \times 10^{17}\,\mathrm{T\,m^3}$ or about 0.0026 times that of Jupiter. The dipole axis is inclined at 59° to the spin axis, in sharp contrast to the small angles of inclination for the Earth, Jupiter and Saturn. In addition the dipole centre is offset from the centre of the planet by a distance 0.3 times the radius of Uranus along the spin axis (figure 5.12). Whatever electric currents are flowing in Uranus to give the magnetic field they must be very asymmetrically arranged relative to the equatorial plane. All the normal features are present in the magnetic environment of the planet, including van Allen belts with accompanying aurorae that have been detected. However, due to the tilt of the dipole axis the magnetic flux lines in the vicinity of the planet are somewhat distorted.

5.3.5. *Uranus summary*

Despite the Voyager flypast and associated measurements far less is known about Uranus than about the two largest major planets. In particular information about its interior structure is uncertain and there is no convincing explanation for why it differs from the other major planets in terms of its lack of excess radiation. Other observations for which no explanations are available are the large tilt of the magnetic dipole axis to the spin axis and the paucity of nitrogen observed in its atmosphere.

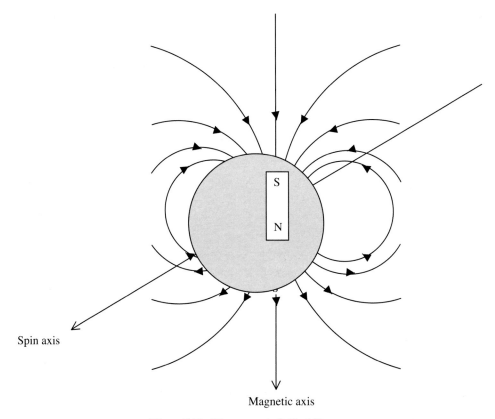

Figure 5.12. *The magnetic field of Uranus.*

5.4.   NEPTUNE

Neptune is often regarded as a twin of Uranus since its mass is comparable, although somewhat greater, and they are similar in size. As seen from Earth they are just rather bland featureless disks but when the Voyager spacecraft examined them both closely then important differences came to light. The most evident difference is that Neptune is much the more active body. It contains a feature, the *Great Dark Spot* (GDS; plate 4) that is a vast storm system similar to the GRS on Jupiter. Also seen in the figure are various white clouds, believed to be of condensed methane, which are driven along by high speed winds. The very rapid and variable easterly winds on Neptune causing large relative changes of position of atmospheric features. Thus the slightly darker spot seen in the lower right-hand part of the image in plate 4 is a feature that went around Neptune so fast that it overtook Neptune every six days.

Various ground-based measurements estimated Neptune's spin period as $17\,\mathrm{h}\ 47 \pm 4\,\mathrm{m}$ but Voyager detected periodically varying radio emission from which a System III period of $16.11\,\mathrm{h}$ was found.

5.4.1.   The internal structure of Neptune

The mass of Neptune is 18% greater than that of Uranus but its equatorial radius is about 3% less so its density is $1.638 \times 10^3\,\mathrm{kg\,m^{-3}}$, substantially greater than that of Uranus, $1.270 \times 10^3\,\mathrm{kg\,m^{-3}}$. It seems that compression effects cannot explain the difference so it is likely that the composition of the two bodies is different. In view of the uncertainty about the internal structure of Uranus, figure 5.10 will also serve as a tentative model for Neptune, with minor changes to make it consistent with the mass and overall density. A more massive silicate–iron core in Neptune may explain both the larger mass and the smaller radius.

5.4.2. Heat generation in Neptune

Taking into account its Bond albedo of 0.290 the equilibrium temperature of Neptune, with no energy production in that body, is 46.6 K. The actual black body temperature estimated from the radiation it emits is 59 K, indicating that it is emitting 2.6 times as much energy as it receives (Topic X). This contrasts with Uranus, which produces little if any heat of internal origin. Although the difference is unexplained the fact that Neptune produces this energy explains its much more active atmosphere.

5.4.3. The atmosphere of Neptune

The composition of the atmosphere of Neptune is very similar to that of Uranus although it seems to have a greater proportion of helium (19% against 15%). The vertical structure of the atmosphere is not well determined but a probable general structure is shown in figure 5.13. This is, of course, influenced by the excess radiation that it emits, giving extra heating from below.

5.4.4. Neptune's magnetic field

Neptune has a magnetic field with a dominant dipole characteristic with strength $2.2 \times 10^{17}\,\mathrm{T\,m^3}$, about one-half that of Uranus. As for Uranus there is a large tilt, $47°$, of the axis of the dipole to the spin axis. There is also a very large displacement of the centre of the dipole from the centre of

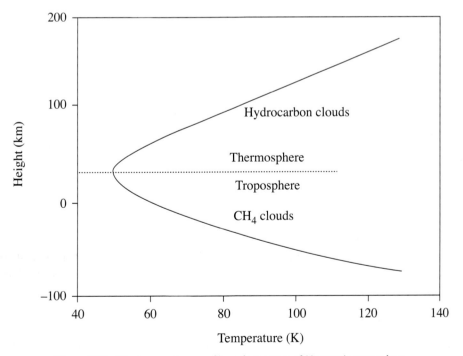

Figure 5.13. *The temperature profile and structure of Neptune's atmosphere.*

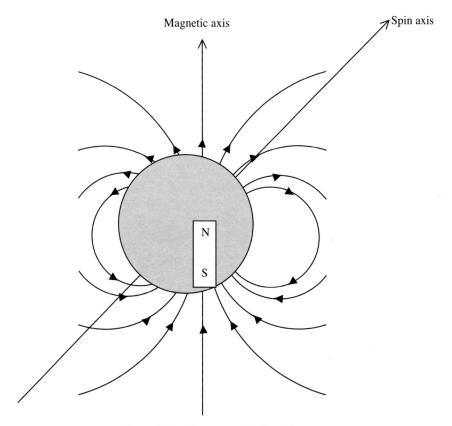

Figure 5.14. *The magnetic field of Neptune.*

the planet, by a distance 0.55 of the radius of Neptune in a direction making a small angle ($\sim$20°) with the equatorial plane (figure 5.14). The internal electric currents giving rise to this arrangement must be of a very complex and asymmetric form.

5.4.5. Neptune summary

Although in terms of mass and radius Neptune is somewhat similar to Uranus there are significant differences. The higher density is partly explained by greater compression effects but also suggests that either it has a larger silicate–iron core or a larger proportion of some other dense component, perhaps ices. The internal generation of heat suggests that internal processes are taking place that are not occurring in Uranus and the high asymmetry of the dipole field may reflect matching asymmetry of internal motions.

The differences in the properties of the two planets could be due to some difference in their origins or evolutionary processes—or perhaps even both. The unusual relationship of the spin axis of Uranus to its orbital plane may reflect some event in its early history that affected its subsequent development.

5.5. PLUTO

Pluto is not a member of the major-planet family and we are considering it here just on the grounds of its location. Its highly eccentric orbit ($e = 0.249$) brings it just within the orbit of Neptune at perihelion

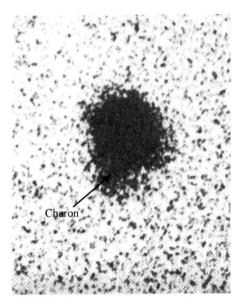

Figure 5.15. *The discovery of Charon, seen as a bump on the surface of the image of Pluto (US Naval Observatory).*

so that between 1979 and 1999 it was Neptune that was the outermost planet. It also has the highest inclination of any planet—17°. It may be linked with another family of bodies, forming the Kuiper-belt system (section 9.4) that exists in the region outside the orbit of Neptune.

5.5.1. *Physical characteristics of Pluto*

As described in section 3.3, the discovery of Pluto was based on a supposed unexplained residual of the orbit of Neptune and its mass was predicted to be six times that of the Earth. After its eventual discovery this estimate was gradually reduced and by about 1954 it was generally believed to have had about one-tenth of the mass of the Earth. The increased sensitivity of telescopes through the introduction of CCD (charged couple device) detectors led in 1978 to the discovery of a bump on the side of a telescope image of Pluto (figure 5.15) that turned out to be a satellite, Charon. Far higher resolution pictures from the Hubble Space Telescope (plate 5) have given a good estimate of the radius of Charon's orbit, 1.96×10^4 km, and this together with the orbital period, 6.39 d, gives an estimate for the combined mass Pluto + Charon. Detecting the position of the centre of mass then enables the individual masses of Pluto and Charon to be estimated. The mass of Pluto, 1.25×10^{22} kg or just under one-fifth that of the Moon, certainly suggests that it is not a true planet but more akin to the planetary satellites in nature.

The estimated radius of Pluto, 1.195×10^3 km, gives a density 2.1×10^3 kg m^{-3} that suggests that it is an icy body. A proposed structure for Pluto consists of a rocky core plus an ice mantle covered with a thin crust of methane ice, since methane and nitrogen are the main materials detected by spectroscopic investigation. The highlight in the image of Pluto in plate 5 indicates that it probably has a smooth reflective surface. The model also adds a tenuous atmosphere with surface pressure 3×10^{-6} bar.

5.5.2. *Relationship with Charon*

The radius of Charon, 593 km, and its mass, 1.9×10^{21} kg, make it much the largest satellite relative to its primary. It moves in a circular orbit and the two bodies are tidally locked so that they always present the same aspect to each other. This means that the spin periods of Pluto and Charon are both equal to the orbital period.

Problem 5

5.1   (a) Assume that the gravitational self-potential energy of Jupiter is

$$\Omega = -\frac{GM_J^2}{R}$$

where M_J is the mass and R is the radius of the planet. What rate of decrease of Jupiter's radius, in metres per century, would release gravitational energy at the rate that solar energy is received $(8 \times 10^{17} \text{ W})$?

(b) If Saturn contains 15% by mass of helium then what is the mass of helium it contains? How much helium would need to be transported from the surface region to 10 000 km from the centre to release gravitational energy equal to all that received by Saturn from the Sun in its lifetime $(2.4 \times 10^{34} \text{ J})$? What fraction of the total helium is this?

(For the purpose of this calculation assume that all the mass of Saturn is effectively concentrated at its centre.)

CHAPTER 6

THE MOON

Any discussion of the Moon is very often linked with that of the terrestrial planets. It is a rocky body in the terrestrial region and its general appearance is not very different from that of Mercury. It has been studied from the Earth with telescopes for about 400 years and it is the only body outside the Earth that has been visited by man. For this reason our knowledge of it is better than that of any other Solar-System body, except the Earth.

In this chapter we shall concentrate on observations and measurements and what has been learned about the Moon from them. Theories of the origin of the Moon and its relationship with the Earth will be deferred to chapter 12.

6.1. THE PHYSICAL CHARACTERISTICS OF THE MOON

By comparison with other satellites (except the tiny Charon) the Moon is large in relation to its primary body. The lunar radius, 1738 km, is more than a quarter of the radius of the Earth and its mass, 7.35×10^{22} kg, is approximately $1/81$ of that of the Earth. By virtue of its proximity and comparatively large mass the Moon exerts an important influence on the Earth so that precisely determining its physical characteristics is of some interest.

6.1.1. The distance, size and orbit of the Moon

Because the Moon is so close its distance can be estimated by using the phenomenon of parallax. If the Moon is observed at the same time from two different points on the Earth's surface then it will appear in different places against the background of the fixed stars. This principle was known to the ancients and was used by Ptolemy to obtain a reasonable estimate of the Moon's distance. With modern telescopes and communications, which ensures accurate measurements made at precise times, the distance can be estimated with high precision.

During the Second World War, when radar was developed for detecting approaching aircraft, it was noted that ghost echoes were appearing on the cathode-ray-tube screens corresponding to the time it would take a radar pulse, moving at the speed of light, to travel to the Moon and back. This was later developed into a technique for measuring the lunar distance although, because of the width of the radar pulses that were necessary, the distance estimates were no better than could be made by previous methods. For example, if the time of a return pulse can be measured to, say, 10^{-4} s then this corresponds to a possible error of 15 km in the estimate of distance. This may seem small compared

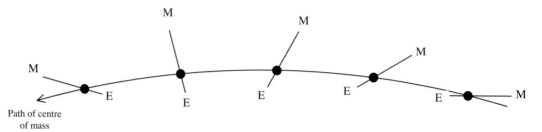

Figure 6.1. *The motion of the Moon and the Earth as the Earth orbits the Sun and the Moon orbits the Earth. The centre of mass of the Earth–Moon system moves on an elliptical path around the Sun.*

with an average Earth–Moon distance of 384 000 km but it is much larger than can be achieved by conventional surveying techniques.

In 1969 the Apollo 11 mission to the Moon left behind retro-reflectors, devices that reflect light back along its approach path, and further ones were left at the Apollo 14 and 15 landing sites. These enable laser-ranging experiments to be carried out, whereby a beam of light from a laser is timed on its journey to the Moon and back. Light from a laser source is not strictly parallel because of diffraction spread so that if, for example, the laser light, of wavelength λ, emerges from a circular aperture of diameter d then the central diffraction region is within a cone of semi-angle

$$\alpha = 1.22\lambda/d. \tag{6.1}$$

Although this angle is small it spreads the light into a large patch on the Moon. Since a telescope is a device in which the image has a larger angular size than the object, by firing the laser pulse *backwards* through a telescope the angular divergence may be much reduced. Laser pulses, of less than 1 ns duration containing several Joules of energy, can be used to measure distances to the retro-reflectors with an accuracy of a centimetre or so.

Once the distance of the Moon is determined accurately then its diameter can be found from the angular size. Variation in the angular size then gives its orbital characteristics—the semi-major axis, 384 403 km, and orbital eccentricity, 0.056—and the plane of the orbit, just over 5° from the ecliptic, can also be found. The Earth and Moon are both in orbit around the centre of mass of the Earth–Moon system and it is this centre of mass that is in elliptical orbit around the Sun (figure 6.1). At one time the parallax effect associated with the Earth's rocking motion about the elliptical orbit was used to estimate the ratio of the mass of the Moon to that of the Earth. With the advent of the space age and artificial satellites around the Moon it is possible to obtain more direct measurements. The Moon's mass can be found from

$$M_{\mathrm{m}} = \frac{4\pi^2 a^3}{GP^2}, \tag{6.2}$$

where the satellite orbit has semi-major axis a and period P.

The Moon always presents one face towards the Earth, which means that its spin period equals its orbital period. However, since its axial spin is at a steady rate while its orbital angular speed varies with its distance from the Earth, as seen from the Earth it appears to have a rocking motion, called *libration*, in going from one full Moon to the next (figure 6.2). Starting with the Moon at perigee, position 1, the shaded hemisphere, delineated in projection by AB, is seen from Earth. After one quarter of an orbital period, at position 2, the Moon's spin has rotated AB by $\pi/2$ but, because it is close to perigee, the Moon has gone more than $\pi/2$ around its orbit. The shaded surface that is now seen includes a

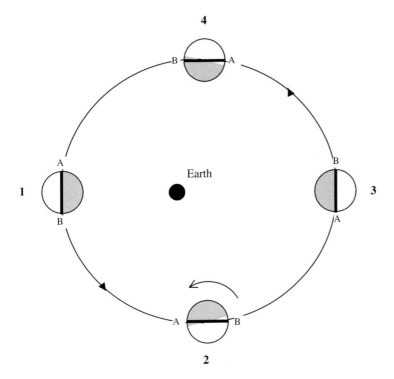

Figure 6.2. *Lunar libration, the apparent rocking motion due to the Moon's elliptical orbit around the Earth.*

wedge on the right-hand side that could not be seen at position 1. Similarly, at position 4 there is a wedge at the left hand side that comes into view. The effect of libration is that some 59% of the Moon's surface can be seen from Earth although the edges are seen in very oblique view.

There are two possible ways of defining the period of the Moon's orbit around the Earth. The *lunar sidereal period* is the time taken for the Earth–Moon direction to go through an angle 2π relative to the fixed stars (27.32 days) and the other, the *lunar synodic period*, is the time taken to go from one full Moon to the next (29.53 days). In figure 6.3 the Sun–Earth–Moon configurations are shown at the beginning and end of a sidereal period, t_{sid}. Since the Earth has moved around the Sun an angle θ it is clear that

$$t_{sid} = \frac{\theta}{2\pi} t_y \tag{6.3}$$

where t_y is a period of one year. The synodic period is longer than the sidereal period and

$$t_{syn} = \frac{2\pi}{2\pi - \theta} t_{sid}. \tag{6.4}$$

From (6.3) and (6.4) the following relationship is found

$$\frac{1}{t_{syn}} = \frac{1}{t_{sid}} - \frac{1}{t_y}, \tag{6.5}$$

which may be confirmed numerically, given $t_y = 365.256$ days.

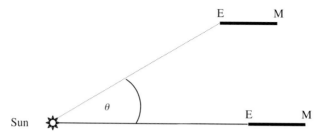

Figure 6.3. *The Sun, Earth and Moon at the beginning and end of a sidereal month.*

6.2. EARTH–MOON INTERACTIONS

Because of their closeness the Earth and Moon exert considerable gravitational effects on each other. Some of these are short term in nature, such as the production of diurnal tides on Earth, but others are much more long term, such as the precession of the spin axis of the Earth (section 3.4; Topic U). Another long-term effect, which has now stabilized, is that due to tidal force which constrains the Moon always to present the same face to the Earth. A theory of tidal action is given in Topic Y. Here we give general descriptions of the physical principles involved.

6.2.1. *The diurnal tides*

It is well known, particularly by those living close to open seas, that there are two tides per day and that these are due to the action of the Moon. The popular view is that the water is 'pulled up' towards the Moon but this simple description does not explain why there are high tides simultaneously in that part of the Earth facing the Moon and on the opposite side of the Earth. A more complete description is that the gravitational attraction (force per unit mass) exerted by the Moon on the nearside water is greater than that at the centre of the Earth and so the water experiences a force pulling it away from the Earth. Equally the attraction at the centre of the Earth is greater than that on the farside water so that the latter again experiences a force pulling it away from the Earth. This is illustrated in figure 6.4a.

Together with stretching forces along the Earth–Moon direction there are compression forces on the Earth, as seen in figure 6.4b. Expressed as force per unit mass, in the diametric plane perpendicular to the Earth–Moon line these have half the magnitude of the maximum stretching force.

It is shown in Topic Y that the tidal acceleration varies as the inverse cube of the distance. Since the eccentricity of the Moon's orbit is 0.056 the ratio of the maximum to minimum tidal acceleration is $\{(1+e)/(1-e)\}^3 = 1.40$. Another factor influencing the strength of tides is the Sun that, on average, gives about 46% of the Moon's effect. When the Moon and the Sun give reinforcing effects, which happens either at the time of the full Moon or new Moon (figure Y.5a) then *spring tides* occur. On the other hand when the Moon is in quadrature, which means that just one half of the hemisphere facing the Earth is illuminated, the solar-induced tide partially cancels out that due to the Moon giving *neap tides* (figure Y.5b).

In the theory above it was assumed that the Earth was not spinning. What rotation does is to pull out material, in particular the oceans and especially in the equatorial regions, thus giving the surface of the Earth the shape of an oblate spheroid. This distortion will be added to that of the tides but, since this effect is unchanging with time, there is no contribution to tides due to the Earth's spin.

In the open sea the difference between high and low tide is about 1 m or so but in an enclosed volume of water, for example, large lakes or inland seas, it will be much less and hardly detectable.

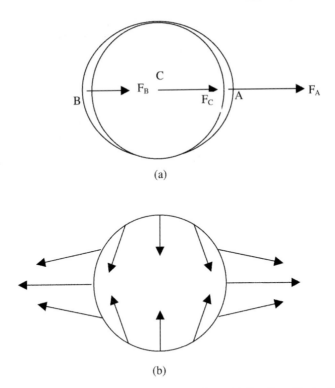

Figure 6.4. *(a) The relative forces per unit mass at points within a tidally affected body. (b) The distribution of force per unit mass, relative to the centre, at various points of the Earth's surface.*

Local coastal topography is also important in controlling the height and form of the tides. In some areas bordering the English Channel, tides of 12 m or more occur and backwash effects caused by the Isle of Wight give the port of Southampton four high tides per day instead of the usual two.

Because of the tilt of the Earth's spin axis to the plane of the Moon's orbit the tides on opposite sides of the Earth may have very different magnitudes. This also means that two consecutive tides at a particular location can also differ very much in height. In figure Y.4 the position of the Moon and of the Earth's spin axis are shown from which it is clear that after approximately 12 hours a particular spot on the Earth would go from position A to position A'. This *diurnal inequality* in the tides can be so large that there may be only one appreciable high tide per day.

6.2.2. The effects of tides on the Earth–Moon system

Because the day is shorter than the month the Earth's spin tends to drag the tides so that the high tide direction is ahead of the direction of the Moon (figure Y.6). The consequent form of the non-spherical symmetry of the Earth's mass exerts a non-radial force on the Moon that gives a net force in its direction of motion thus increasing the intrinsic angular momentum of its orbit. Similarly the Moon exerts a torque on the tidal bulges tending to reduce the angular momentum in the Earth's spin. Without the influence of the Sun the angular momentum of the Earth–Moon system would remain constant with the Earth spinning more slowly and the Moon gradually retreating from the Earth. Eventually a position would be reached in which the day and the month were equal in length (approximately 50 days) so that the Earth always presented one face to the Moon.

Following that stage there would be no drag on the tides and no advance of the tide ahead of the direction of the Moon.

Although these changes, which are going on today, conserve angular momentum they do imply a loss of energy. The Moon actually gains orbital energy as its orbital radius increases but this is more than offset by the loss of energy of the Earth in its spin. This is due to friction between the spinning Earth and the tidally raised material and has been likened to the dragging of brake pads on a spinning wheel. The loss of energy in the spin of the Earth is at the rate $\sim 2 \times 10^{12}$ W which causes the day to lengthen by about 1.6 ms per century. Although this is a small effect it can be detected by apparent time discrepancies in eclipses reported by ancient civilizations (section Y.3).

Another effect of the Earth–Moon interaction is to cause a precession of the spin axis of the Earth (Topic U; section 3.4). Since precession gradually changes the First point of Aires, or the Vernal equinox, this means that the seasons gradually drift with respect to the year. In the northern hemisphere the Vernal equinox (the beginning of spring in the northern hemisphere) will fall in the third week of September in the year 15 000 AD!

6.3.   LUNAR AND SOLAR ECLIPSES

A curious coincidence concerning the Earth–Moon–Sun system is that at the present time the angular diameters of the Moon and the Sun as seen from the Earth are so similar. Because of the eccentricities in the Earth's orbit around the Sun and the Moon's orbit around the Earth the angular diameter of the Sun varies between about 31.5′ and 32.5′ and that of the Moon between 29.5′ and 32.5′. The combination of this angular–size similarity and the relationship between the orbital plane of the Moon about the Earth and the ecliptic give the observed eclipses of the Sun and the Moon.

6.3.1.   *Solar eclipses*

A solar eclipse occurs whenever the Moon is between the Earth and the Sun and blots out all or part of the Sun's disk from view. If the plane of the Moon's orbit was coincident with the ecliptic then this would happen every new Moon but in practice there are about six solar eclipses per decade which can be seen either as total (figure 6.5) or annular somewhere or other on the Earth's surface. During a total eclipse the outer regions of the solar atmosphere, the light from which is usually swamped by total sunlight, can be seen and studied. Because of their scientific interest, expeditions are mounted to observe total eclipses, even when they occur in remote and hostile environments. Many other eclipses occur which are partial, where only a part of the Sun's disk is obscured, but they are of no particular scientific interest and, indeed, many would hardly be noticed by those on Earth if it was not brought to their attention.

The orbit of the Moon is inclined at 5° to the ecliptic and so crosses the ecliptic twice each month. There is a precession of the Moon's orbit because the Sun's pull on the Moon gives components of force on it not directed towards the Earth, so constituting a torque. Thus the two points at which the Moon crosses the ecliptic each month slowly change. The combination of this effect with the Earth's motion around the Sun creates a periodic pattern of eclipses with a period of just over 18 years, known as the *Saros*.

6.3.2.   *Eclipses of the Moon*

An eclipse of the Moon takes place when the Earth is directly between the Sun and the Moon so that the Moon is obscured. In fact there is always some light refracted by the Earth's atmosphere that falls on the Moon and this usually shows the Moon in a faint coppery glow.

Figure 6.5. *A total eclipse of the Sun.*

If the Moon's orbit was not inclined to the ecliptic then there would be a lunar eclipse at every full Moon. Because the Earth is larger than the Moon, lunar eclipses are far more common than solar eclipses and have a much greater duration—typically longer than one hour while the Earth's shadow passes over the Moon. Lunar eclipses are seen simultaneously everywhere on Earth from which the Moon can be seen but they are of not of great scientific interest.

6.4.   THE LUNAR SURFACE

Since the time of Galileo the Moon has been studied in detail by Earth-based telescopes. In general the Moon shows a melange of bright areas which are mountainous, and dark areas which are flat plains (figure 6.6). The latter are termed *maria* (seas) although most regions of the Moon contain no water, not even bound into minerals. It was long thought that the Moon contained no water at all but in March 1998 it was reported that water ice deposits had been discovered near both lunar poles by neutron spectrometers mounted on the orbiter *Lunar Prospector*. It is believed that this water was deposited by comets falling into deep craters near the poles. Since the Sun's rays cannot penetrate into these craters the water is protected from evaporation. Craters, which are seen all over the Moon but most densely in the highland regions, are caused by the impact of large projectiles. Other surface features are *rays*, which are splashes of brighter material thrown out of some of the newer craters, and *rills*, which are straight, curved or sinuous cracks or channels. Although the Moon appears bright at night it is actually a dark object with an average visual albedo of 0.07.

Most detailed knowledge of the Moon is now derived from space exploration and, in particular, from lunar material returned to Earth. One of the very early space missions to the Moon, the Luna 3

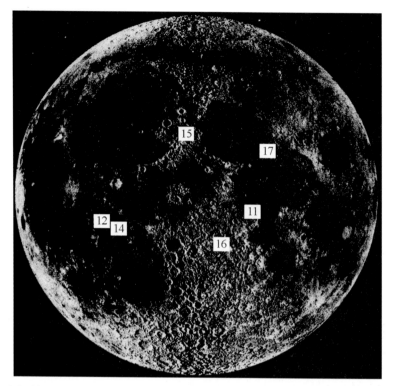

Figure 6.6. *The near face of the Moon showing the landing sites of the numbered Apollo missions.*

mission by the USSR in 1959, produced rather poor pictures of the far side of the Moon but they were good enough to show that the two hemispheres were substantially different. Whereas the side seen from Earth is dominated by maria the far side has few maria and those relatively small. The hemispherical asymmetry of the Moon is clearly a consequence of some important aspect of its origin and/or evolution.

The bulk of lunar material available for scientific examination on Earth has come from the American Apollo programme although some material was also returned by unmanned Soviet missions. The American missions and their sources of material are given in table 6.1 and their locations marked on figure 6.6.

6.4.1. The maria

Maria material is basalt with a composition similar to that from terrestrial volcanic eruptions. As is true in the terrestrial case, it can be assumed that the material is a sample of the interior composition at that location. The eruptions producing the mare deposits were spasmodic over a long period of time. Individual flow fronts can be seen in the Imbrium basin emanating from the vicinity of the crater Euler and a layered maria was reported by the Apollo 15 astronauts at Hadley Rille which had cut through the maria flow to reveal the layers forming it. The general appearance of the layers, plus the extreme flatness of the maria, suggests that the lava flows were much more fluid than on Earth and so spread farther and more evenly.

The time of eruption of maria material has been dated between 3.16 and 3.96×10^9 years ago which shows that volcanism occurred on the Moon at least during that period. Since older material

Table 6.1. *The Apollo landing sites and the types of returned material.*

Mission	Landing site	Material
Apollo 11	Mare Tranquillitatis	Simple maria
Apollo 12	Oceanus Procellarum	Simple maria
Apollo 14	Fra Mauro	Ejecta from Imbrium basin
Apollo 15	Hadley Rille	Eastern edge, Imbrium basin
Apollo 16	Descartes	Lunar highlands
Apollo 17	Taurus-Littrow region	Edge, Mare Serenetatis

tends to be covered up there may well have been volcanism even earlier, perhaps back to the origin of the Moon some 4.6×10^9 years ago.

The most common minerals in lunar maria are clinopyroxene and plagioclase which is usually in the range 80–90% anorthite (Topic A). Up to 20% olivine is present in some basalts although it is almost absent in most of them. Table 6.2 shows the compositions of basalts from the various Apollo missions plus one sample from the Soviet Luna returns. Although only about one hundred minerals have been identified on the Moon, compared with more than two thousand on Earth, three previously unknown minerals were discovered on the Moon. These were *pyroxferroite*, an iron-enriched pyroxene, *armalcolite* (subsequently discovered on Earth), an Fe-Ti-Mg silicate similar to ilmenite, and *tranquillityite*, an Fe-Ti silicate enriched with zirconium, yttrium and uranium.

A consequence of the deficiency of oxygen on the Moon, as compared to the Earth, is the presence of particulate iron in maria material and also iron only in the ferrous form, FeO, whereas the ferric form, Fe_2O_3 also occurs on Earth. Another difference is that the Moon is depleted in volatile elements compared with the Earth as illustrated in figure 6.7. As an example, lunar rocks contain about one-tenth the abundance of sodium and potassium compared with the Earth. If it is assumed that the basic raw material of which the two bodies formed was the same then it may be inferred that the more volatile elements were 'boiled off' the Moon at some stage, perhaps at the time of its formation. If so this would imply a temperature of at least 1700 K everywhere.

6.4.2. The highlands

Highland rocks are all igneous and the appearance of mineral crystals suggests that, unlike the maria, they cooled slowly. They are similar to maria material in containing particulate iron, in the absence of

Table 6.2. *Percentage compositions of basalts from Apollo missions.*

	11A	11B	12A	12B	14	15A	15B	16	17
Olivine	0–5	0–5	10–20	—	—	6–10	—	—	<5
Clinopyroxene	45–55	40–50	35–60	45–50	50	59–63	64–68	50	45–55
Plagioclase	20–40	30–40	10–25	30–35	40	21–27	24–32	40	25–30
Opaque	10–15	10–15	5–15	5–10	3	4–7	2–4	7	15–25
Silica	1–5	1–5	0–2	3–7	2	1–2	2–6	—	—

11A = Apollo 11, high K basalt; 11B = Apollo 11, low K basalt; 12A = Apollo 12, olivine basalt; 12B = Apollo 12, quartz basalt; 14 = Apollo 14, aluminous basalt; 15A = Apollo 15, olivine basalt; 15B = Apollo 15, quartz basalt; 16 = Luna 16, aluminous basalt; 17 = Apollo 17, high Ti basalt.
Opaque: minerals rich in ilmenite ($FeTiO_3$).
Silica: tridymite and cristobalite, high-temperature forms of silica.

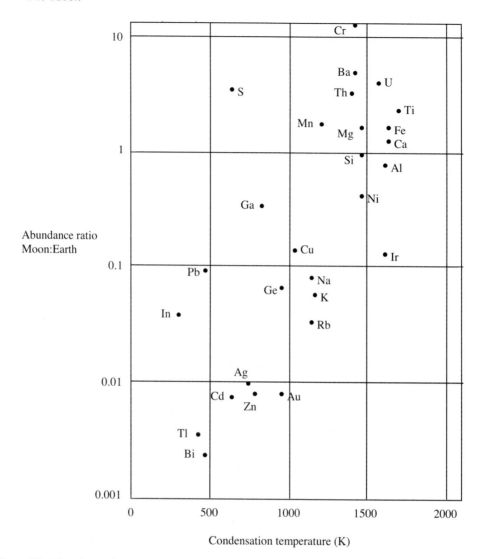

Figure 6.7. *The relative abundance of elements on the Moon and Earth related to the elemental condensation temperatures.*

water and in a deficiency of volatiles, but their chemistry and mineral compositions are quite different. Whereas lava is rich in iron, magnesium and titanium, the highland rocks are lighter in colour and are rich in aluminium and calcium; plagioclase, which accounts for over 50% of highland material, occurs with varying amounts of pyroxene, olivine and spinel (Topic A).

Highland rocks are similar to terrestrial rocks and so the terminology to describe them is similar. Most highland rocks are gabbro (table A.3) but some contain 90% or more of plagioclase and is called *anorthosite*. A few rock specimens are pure anorthosite. Unlike on the Earth there are only igneous rocks on the Moon, implying that the lunar surface was once completely molten and has never known oceanic conditions.

The measured ages of the highland rocks, that is from the time they became closed systems, are generally in the range 4.0–4.2×10^9 years. However, there is one sample from Apollo 17 which gives an age of 4.6×10^9 years, similar to that found for meteorites, and which is the generally assumed age of the Moon. There seems to be a 500 million year period following the formation of the Moon where either condensation of rocks did not take place over most of its surface or, alternatively, at the end of which previously condensed rocks were destroyed in some way. This is a problem that has not been resolved.

6.4.3. Breccias

Breccias are rocks formed from fragments of other rocks that have been broken up and then mixed together. On the Moon they are evidence of violent events and the individual fragments within them vary from a few millimetres to several metres in extent. They are found everywhere on the Moon but are most evident in the highlands. Some breccias from the Apollo 16 mission are an impact melt where many smaller rocks have been shock-welded together by the energy of the impact that formed some of the larger craters.

Breccias provide a unique probe for studying impact events. Apollo 14 breccias indicate that they were deposited at temperatures of at least 500 K and in the presence of a gas. The study of breccias leads to a better understanding of the thermal history of the Moon but the analysis involved is extremely difficult and complex.

6.4.4. Regolith: lunar soil

The Apollo missions found no solid bedrock on the Moon but everywhere loose pulverized material several metres thick known as *regolith*. It is also referred to as lunar soil although it has no relationship to terrestrial soil that is characterized by hydrous and organic components. A small glass spherule from the lunar soil is shown in figure 6.8.

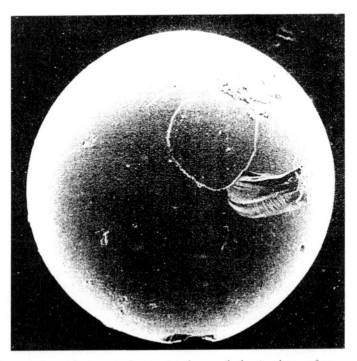

Figure 6.8. *A glass sphere (<0.2 mm in diameter) in lunar soil, showing damage features on its surface.*

Regolith has the composition of the surface on which it rests and is formed by bombardment of the surface by meteorites (some 1–2% of regolith consists of meteorite material) and by high-energy radiation from the Sun. It is slowly 'stirred' by incoming particles, a process called lunar gardening, which rearranges material near the surface. The timescale for this has been deduced from cores of lunar material brought back in the Apollo programme. The core returned by Apollo 15 from Hadley Rille was 2.42 m long and deposited as a result of a single impact 500 million years ago. Calculations suggest that the top 0.5 mm of the soil is turned over every 10 000 years and the top centimetre every 10 million years.

The measured radioactive age of the regolith is very uniform, in the range $4.4–4.6 \times 10^9$ years, which means that it seems to be older than the rocks from which it was formed! This paradox has not been completely resolved although a number of suggestions have been made which could disturb the relationship between radiogenic ^{87}Rb and its daughter product ^{87}Sr from which age estimates are made. One suggestion involves the presence of a component of the lunar soil called KREEP, on account of its high component of potassium (K), rare-earth elements (REE) and potassium (P). It is principally plagioclase and pyroxene and apart from the KREEP components it also contains more rubidium, thorium and uranium than is found in other lunar rocks. It forms a few percent of the lunar soil but cannot be that common within the Moon since its radioactive component would then make its presence known by melting of the lunar interior. The majority of KREEP material is found in the vicinity of Mare Imbrium and it could have originated at a depth of 25–50 km below the Mare Imbrium site and been brought to the surface by the collision which produced the mare basin. However, to explain the larger measured age of the rock requires a relative increase in the strontium component and another suggestion is that the heat generated by impacts may have preferentially removed rubidium which is more volatile than strontium.

6.5.   THE INTERIOR OF THE MOON

Knowledge of the lunar interior may be gained from many different kinds of measurement. The most obvious type is that from seismometers left on the surface of the Moon by the Apollo 11, 12, 14, 15 and 16 missions. Other information comes from external measurements of both the gravity and magnetic fields in the vicinity of the Moon and measurements of heat flow at the surface. We shall now review these measurements and indicate what is learned from them.

6.5.1.   *Gravity measurements*

We have already seen (section 3.4) that it is possible to estimate the moment of inertia factor of Solar-System bodies from a combination of measurements—for example the value of J_2 plus the rate of precession of the spin axis. In the case of the Moon the latter quantity cannot be measured easily but what is used instead are the lunar *physical librations*. In section 6.1.1 it was shown that the elliptical motion of the Moon around the Earth combined with the uniform spin of the Moon gives an apparent rocking motion, or libration. Another effect of the elliptical motion is that the tidal bulge on the Moon is not always precisely directed towards the Earth so that a torque, variable in strength and direction, is exerted on the Moon. This gives rise to a *physical libration* which is manifested as a very small periodic variation in the spin rate of the Moon. This can be measured and, combined with J_2, gives the estimate of 0.392 for the moment of inertia factor. This shows some central condensation of the Moon and is consistent with an iron core with radius up to 382 km.

When spacecraft move in orbit around the Moon they are found to experience accelerations and decelerations indicating gravitational anomalies. Negative anomalies are found where there are unfilled craters and are simply due to missing material. On the other hand larger craters and maria,

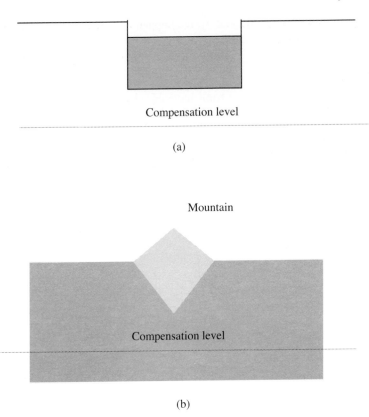

Figure 6.9. *Idealized representations of the principle of isostacy. (a) A crater filled with dense basalt such that the pressure is uniform at the compensation level. (b) A mountain of low density with a 'root' such that the pressure is uniform at the compensation level.*

with diameters more than about 200 km, have positive anomalies indicating concentrations of mass called *mascons*. It might be thought that this is due to the presence of dense basalt filling the mare basin, since basalt is much denser than the crust material. However, the material welling up through cracks to fill the basin would only rise to such a height that the pressure would be uniform at some lower *compensation level* (figure 6.9a). This is the principle of *isostacy* in geophysics that operates in a number of different ways. Thus in figure 6.9b the extra height of the mountain, which consists of low-density rock, is accompanied by a root of low-density rock so that the pressure at the compensation level is uniform. Ignoring the small variation of the acceleration due to gravity over the depth of the basin, if the isostatic principle was operating then the mass of infilling basalt would be the same as that of the removed crustal material so that no positive gravitational anomaly should be present. In fact it could be argued that since the dense material is lower, and hence farther from the measuring spacecraft, if there was a gravitational anomaly then it should be in the opposite sense. The anomaly is due to *overfilling* of the basins and craters and has been explained by Dormand and Woolfson (1989). When the Moon was young the cooling crust was shrinking and applying pressure on the material below. At the same time radioactive elements, particularly ^{238}U, ^{232}Th and ^{40}K, were more abundant than now and were heating up the interior, thus increasing the internal pressure even more. The net effect was that the magma was extruded under supra-hydrostatic pressure, rather like toothpaste being squeezed out of a tube. Each eruption would overfill the basin and subsequently slump due to the excess pressure but, as the surface regions cooled, the process would become more and more sluggish and the final outcome would have been a slightly overfilled basin, or a mascon.

6.5.2.   *Lunar seismicity*

Results from the seismic stations on the Moon have enabled deductions to be made concerning the lunar interior. Since, in seismic terms, the Moon is a much quieter body than the Earth it is possible to run the seismometers at maximum sensitivity, and vibrational amplitudes of the lunar surface as small as 10^{-8} m can be measured. Disturbances that have been recorded have been due to internal displacements of lunar material (moonquakes), the impacts of meteorites and also the impacts of parts of booster rockets and other spaceware deliberately made to collide with the Moon.

There are fewer than 3000 moonquakes per year and they all have strength less than 2 on the Richter scale. They are influenced by the Earth and are triggered close to apogee and perigee where tidal effects are changing from increasing to decreasing or vice versa. The total energy released by these events is 2×10^6 J per year or between 10^{-11} and 10^{-12} of that released by earthquakes. The moonquake energy is equivalent to less than 50 g of TNT per year; by contrast the collision of a Saturn third-stage booster rocket that crashed on to the Moon's surface was the equivalent of about 1 tonne of TNT. There are also impacts of meteorites greater than 1 kg mass at the rate of one per month within 200 km of each seismic station.

An earthquake disturbance on Earth decays in less than two minutes, due mainly to dissipation of energy in the fluid core and plastic regions of the Earth's interior. By contrast a moonquake may take up to an hour to decay, showing clearly that the bulk of the body is behaving like a near-perfect elastic solid. When the Saturn third-stage booster from the ill-fated Apollo 13 mission landed on the Moon, the Apollo 11 seismometer recorded the resulting disturbance for three hours and twenty minutes.

6.5.3.   *The interior structure of the Moon*

The essential structure of the Moon, as deduced from the seismic evidence, is shown in figure 6.10. The crust on the far side, with thickness 74 km, is thicker than that on the near side (48 km) and this provides an explanation for the difference of appearance of the two sides. Although the far side is deficient in maria it does have large impact basins, but these have not been filled by molten material from the interior. It seems that the molten material could not rise to the surface through the thicker crust. The asthenosphere is the region where material has greater plasticity and S-waves are attenuated there. The sources of most moonquakes are fairly close to the asthenosphere–lithosphere boundary.

The variation of P-wave and S-wave speeds in the Moon are not known with great precision and are approximately given in figure 6.11. There are some fine-scale variations close to the surface due to deposits of various materials but these are dependent on the location.

6.5.4.   *Heat flow and temperature measurements*

Surface heat flow measurements were made by the Apollo 15 and 17 missions. There was variation of measurements made at different places but the average surface heat flow is about 2×10^{-3} W m^{-2} with a near-surface temperature gradient of 1.8 K m^{-1}. The total surface heat flux, 1.9×10^{10} W, is some three orders of magnitude less than that of the Earth.

The surface temperature gradient, if maintained to greater depths, would imply an enormous internal temperature but clearly this is not so. One way of estimating internal temperatures is first to assess as well as possible the variation of electrical conductivity of the interior of the Moon, or

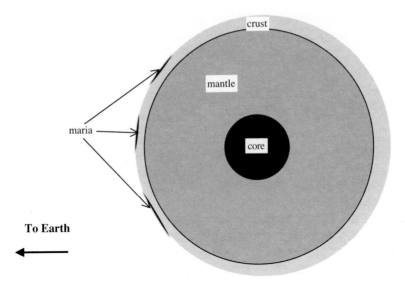

Figure 6.10. *A schematic cross-section of the Moon showing the difference of crust thickness on the two sides (exaggerated).*

any other Solar-System solid body. Silicate rocks are semiconductors, which is to say that their electrical conductivity increases with increasing temperature and, with assumptions about the type of rock present, conductivity can be translated into temperature. One way of inferring the interior conductivity is to record the electromagnetic response of the Moon to magnetic disturbances by means of magnetometers, some on the surface and others carried by orbiting spacecraft. The

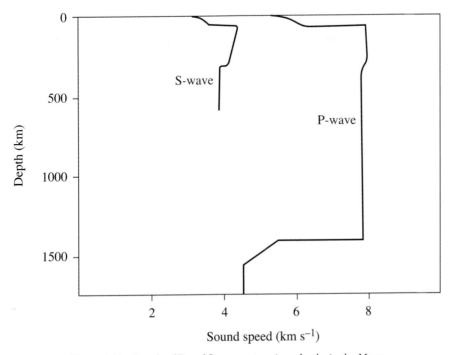

Figure 6.11. *Speeds of P and S waves at various depths in the Moon.*

magnetic disturbances are provided by variations of the solar wind that is coupled to the solar magnetic field. Analysis of such magnetic measurements indicates a temperature profile, similar to that shown as curve (e) in figure 6.14. The interior temperatures, if valid, would indicate partial melting of material in the core.

6.6. LUNAR MAGNETISM

In section 6.4.1 it was mentioned that particulate iron is present in the maria basalts and also elsewhere on the Moon. Iron is a ferromagnetic material, which means that it becomes permanently magnetized if placed in a magnetic field but it ceases to be ferromagnetic above a temperature known as the *Curie temperature*, which is 1043 K for iron. This may be simply interpreted as the temperature at which the elementary atomic magnets, of which the ferromagnetic material is formed, have so much thermal energy that they cannot remain aligned. When the mare basalt deposits were first formed they would be well above the Curie temperature for iron but they would quickly fall to below that temperature. The iron particles would become magnetized and would remain magnetized at the same strength even if the ambient magnetic field was subsequently reduced or even removed completely. The so-called natural remnant magnetism (NRM) enables an estimate to be made of the strength of the prevailing magnetic field when the magma solidified. This is done by subjecting returned samples of moon-rock to magnetic fields in the laboratory and then measuring the intensity of magnetization of the sample. Since the intensity of magnetization of any particular sample is approximately proportional to the magnetizing field this enables the original lunar magnetizing field to be estimated. From a measurement of the age of the magma, the variation of magnetic field strength with time may then be deduced.

Magnetometers orbiting the Moon measured a very small magnetic field and magnetometers on the lunar surface give measurements up to 38γ ($1\gamma = 10^{-9}$ T) in mare regions and up to 313γ in the highlands. However, the big surprise was the measurement of the NRM of samples returned to Earth. These indicated that the early Moon possessed a surface field greater than 10^{-4} T that then

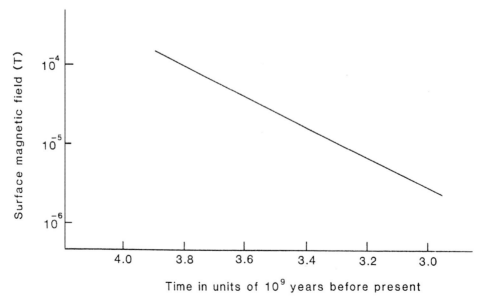

Figure 6.12. *The variation of the surface field with time on the Moon as deduced by NRM measurements.*

declined with time as shown in figure 6.12. To appreciate what this means the maximum terrestrial magnetic field, at the poles, is about 6×10^{-5} T.

The origin of the early magnetic field of the Moon has been the subject of much speculation. One obvious interpretation is that it was of internal dynamo origin so that the reduction of the field with time is then due to the weakening dynamo effect as the Moon became cooler. The problem with this is that the size of the Moon's core, as indicated by the moment-of-inertia factor, seems far too small to generate fields of the necessary strength.

Another explanation is that the Moon was magnetized by an external source but this raises the problem of finding an external source, transient in nature, which would have been able to give such large fields at the lunar surface.

6.7. SOME INDICATIONS OF LUNAR HISTORY

The radioactive dating of maria basalt indicates that soon after the formation of the Moon there was molten material just below the surface. On the near side the molten material was close enough to the surface to permeate through cracks produced by the collisions of the large projectiles which produced the mare basins although it must have gradually retreated away from the surface with time. On the far side the extra 20–30 km thickness of the crust was sufficient to reduce greatly the escape of magma so that few mare basins are evident there, and those that are present are small ones. A scenario which explains this pattern is that the Moon was formed by the very rapid accretion of inwardly falling material. In figure 6.13 the core of a forming Moon, with density ρ and radius x, is indicated with material raining in upon it. If the kinetic energy of the impacting material, assumed to be at free-fall speed, is completely transformed to thermal energy then the temperature of the deposited material, θ, will be given by

$$\tfrac{4}{3}\pi G \rho x^2 = c(\theta - \theta_0), \tag{6.6}$$

(Topic AE) where c is the thermal heat capacity and θ_0 the initial temperature of the material. This would give a temperature profile for the completely formed Moon similar to that shown as curve (a) in figure 6.14 which includes the effects of cooling at the surface, the transmission of energy

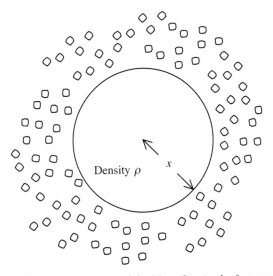

Figure 6.13. *A schematic view of the Moon forming by fast accretion.*

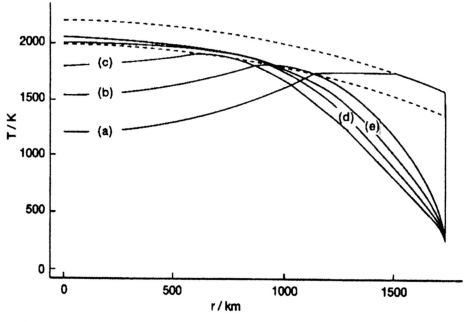

Figure 6.14. *Sequence of thermal profiles in the Moon: (a) at its formation; (b) after 10^9 years; (c) after 2×10^9 years; (d) after 3×10^9 years; and (e) at present (Mullis 1993).*

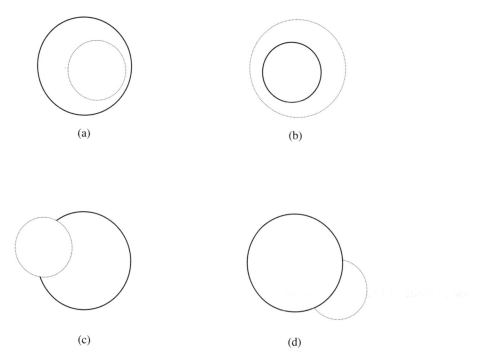

Figure 6.15. *Overlaps of an earlier crater (full line) and later crater (dotted line). In case (b) the later crater obliterates the earlier one. In the other three cases it is possible to detect both craters and to tell which is the earlier one.*

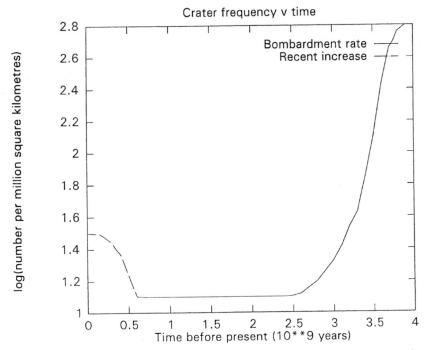

Figure 6.16. *The variation of the number of craters greater than 1 km in diameter per million km² with time. The recent increase is of unknown origin.*

inwards by shock waves and the latent heat of fusion when the material melts. Calculations show that, due to interior conduction and convection and surface cooling, the thermal profile would evolve in the way shown in the figure.

The highland regions of the Moon are saturated with craters but the number of them per unit area is much less in the maria as early craters have been erased by volcanism. Craters can be approximately time-ordered by their states of degradation, as they become less clear when covered by the dust of later collisions. Time sequences are also indicated when craters overlap. Some overlap situations are illustrated in figure 6.15. It is found that, whenever overlapping craters can be identified, in about 90% of the cases the later crater is smaller than the earlier one.

From the dating of mare basalt and the study of craters some idea of the variation of the rate of bombardment of the surface with time can be deduced. This is shown in figure 6.16. The situation farther back than 3.9×10^9 years ago is uncertain, but there does seem to have been a surge some 500 million years ago which probably indicates some major catastrophic event in the Solar System at about that time. From the relative sizes of overlapping craters and the dating of mare basins it also appears that the initial projectiles contained larger bodies but that the sizes of the bombarding bodies tended to reduce with time.

6.8.   MOON SUMMARY

The Moon is a large body compared with its primary body, the Earth. It always presents the same face towards the Earth although, due to libration, more than one hemisphere can be seen. Tidal effects give diurnal tides on Earth and also cause the Moon's orbit to slowly recede, the extra angular momentum required being provided by a slowing down of the Earth's spin. The present distance of the Moon can give total eclipses of the Sun, important events for the study of the Sun's outer regions. The forces of the Moon on the equatorial bulges caused by the spin of the Earth lead to a precession of the Earth's spin axis with a period of 26 000 years.

The surface of the Moon consists of highlands and mare basins. Other major features are craters, rays and rills. Highlands consist of igneous low-density rocks while mare material is higher-density magma produced by volcanism from the Moon's interior.

Seismic measurements and the electromagnetic response of the Moon to magnetic disturbances have enabled deductions to be made about the lunar interior. The crust is thinner on the near side than the far side, which explains the concentration of maria on the near side and the hemispherical asymmetry. Deduced interior temperatures suggest partial melting towards the centre. Although the Moon has little if any intrinsic magnetism, an examination of lunar rocks indicates that surface fields up to 10^{-4} T occurred early in its existence.

The bombardment history, as judged by the distribution of craters coupled with dating of lunar rocks, suggests an early bombardment by large numbers of large projectiles. Subsequently the intensity and size of the projectiles decreased although an upsurge in bombardment may have taken place 500 million years ago.

Problems 6

6.1 What is the orbital period of a spacecraft in orbit at a height of 50 km above the Moon's surface?

6.2 A laser of diameter 0.02 m fires a beam of light of energy 1 J and wavelength 560 nm on to the Moon. If all the energy falls in the central diffraction region then what is the average number of photons per m^2 in that region?

CHAPTER 7

SATELLITES AND RINGS

7.1. TYPES OF SATELLITES

The main satellites of the Solar System are listed in tables 7.1 to 7.5 and an examination of their orbital characteristics shows that they fall into two main types. One type is exemplified by the satellites of Jupiter out to Callisto. All these satellites have direct orbits very close to circular and almost in the planet's equatorial plane. Such satellites are often referred to as *regular satellites*; the satellites of Saturn out to Titan and all the satellites of Uranus given names in table 7.4 fall into this category, although Miranda's orbit is inclined by 3.4° to the equator. It is a common assumption that regular satellites are somehow connected with the formation of the parent planet. The general idea here is that the spin axis of the planet and the rotation axis of the satellites' orbits are both aligned with the net angular momentum vector of the original body of material that formed both the planet and its satellites. Although this assumption is probably true it must be treated with some caution. The three planets with regular satellites all spin rapidly and are therefore quite oblate. If the regular satellites were somehow placed in orbit around the planet not too long after its formation then the combination of the gravitational pull of the equatorial bulge plus energy dissipation due to a residual resisting medium, would pull the satellite orbit towards the equatorial plane.

The *irregular satellites* have as their largest and outstanding member the Moon, considered in some detail in chapter 6. There is no convincing theory for the coeval formation of the Earth and the Moon so it is generally believed that the Moon was acquired by the Earth by some process after the Earth had formed. The other large irregular satellite is Triton that has a circular but retrograde orbit well inclined to Neptune's equatorial plane. There must certainly be some unusual scenario associated with its relationship to Neptune.

We shall now describe the individual satellites of interest, moving outwards from the Sun, and comment on their probable structures and give ideas about their origin where appropriate. Table 7.1 gives the characteristics of planetary spins and of satellites for the terrestrial planets plus Pluto. Tables 7.2 to 7.5 give the same information for the individual major planets.

7.2. THE SATELLITES OF MARS

Mars has two small satellites, Phobos and Deimos, shown in figure 7.1. They are both small irregularly-shaped bodies, similar to typical asteroids in both appearance and size (cf. figure 8.4).

Table 7.1. *Spins and satellites of the terrestrial planets and Pluto.*

Planet or satellite	Spin period	Inclination[†]	Orbital radius (10^3 km)	Orbital eccentricity	Mass (10^{22} kg)	Mean diameter (km)	Mean density (10^3 kg m^{-3})
Mercury	58.65 d	28°					
Venus	243 d	177° (R)					
Earth	24 h	23° 27′					
Moon		23.4°	385	0.056	7.35	3476	3.34
Mars	24 h 37 m	23° 59′					
Phobos			9.4	0.021	9.5×10^{-7}	22	∼2
Deimos			23.5	0.003	1.9×10^{-7}	13	∼2
Pluto	6.39 d	122° (R)					
Charon		0?			∼0.12	∼1200	∼1.2

[†] For planets, inclinations are of spin axes with respect to the orbital plane. For satellites, inclinations are of orbits with respect to the planetary equator. (R) indicates retrograde spin.

Phobos, the larger and inner of the two satellites, has an orbital period of about 7 h 39.5 m that is less than the spin period of Mars so, seen from the surface of the planet, it would appear to rise in the west and set in the east. Phobos is roughly ellipsoidal in shape with axes $20 \times 23 \times 28$ km. It is covered with craters of a wide range of sizes, the largest being Stickney with a diameter of 10 km. The impact that produced Stickney must have come close to completely disrupting the satellite. Another characteristic of the surface is a series of parallel striations that are thought to be due to constant alternating tidal stresses on the satellite.

Deimos has dimensions $10 \times 12 \times 16$ km and an orbital period of 30 h 24 m. The largest crater on Deimos is about 3 km in diameter. Both satellites have a thin dust layer covering the surface,

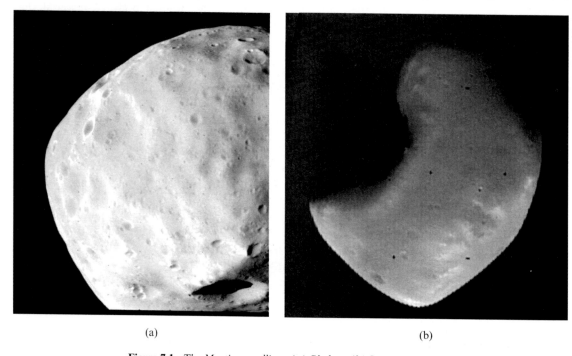

(a)   (b)

Figure 7.1. *The Martian satellites. (a) Phobos. (b) Deimos (NASA).*

representing fine material that happened to be retained when material was excavated to form the craters. Since the escape speed is low, most of the fine material would have escaped. However, from the somewhat smoother appearance of the surface of Deimos it is deduced that the layer of dust on it is thicker than on Phobos. This is surprising as it might be thought that the lower escape speed from Deimos would result in a thinner dust layer.

The visual albedoes of both satellites are low, between 0.05 and 0.07 and, combined with their low densities and their reflection spectra, suggest that they resemble C-type asteroids that are similar in composition to carbonaceous chondrites (section 10.2.1). An asteroid origin is likely and some scenario involving a collision between two asteroids close to Mars might explain their origin. Either the two asteroids could have been captured—if their speeds after the collision were both less than the escape speed—or, alternatively, only one asteroid could have been retained but in a disrupted form to give both the satellites.

7.3.   THE SATELLITES OF JUPITER

The four large satellites of Jupiter, first observed by Galileo in 1610 and now known as the Galilean satellites, were the first satellites, other than the Moon, to be seen by man but they were only seen as orbiting points of light. The era of space exploration has now enabled these bodies to be seen in great detail and they are found to be all very different and all with interesting and intriguing features.

7.3.1.   Io

Just before the Voyager 1 spacecraft made a rendezvous with Io in 1979 a paper by S J Peale, P Cassen and R T Reynolds appeared in the journal *Science* predicting that the satellite would show volcanic activity. To the astonishment of many this prediction was verified and figure 7.2 shows the plume

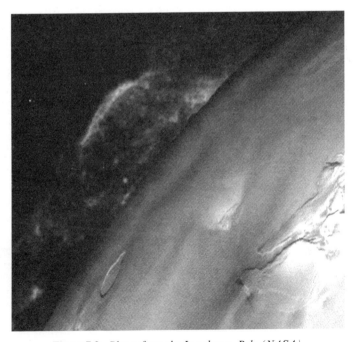

Figure 7.2. *Plume from the Io volcano, Pele (NASA).*

Table 7.2. *Satellite data for Jupiter.*

Satellite	Inclination of orbit to planetary equator ($^\circ$)	Orbital radius (10^3 km)	Orbital eccentricity	Mass (10^{21} kg)	Mean diameter (km)	Mean density (10^3 kg m^{-3})
Metis*	~0	128	~0	9×10^{-5}	~20	
Adrastea*	~0	129	~0	2×10^{-5}	~18	
Amalthea	0.5	181	0.003	7.2×10^{-3}	190	~2
Thebe*	~0	222	~0	8×10^{-4}	~100	
Io	0.03	422	0.000	89.2	3630	3.55
Europa	0.47	671	0.000	48.8	3126	3.04
Ganymede	0.18	1070	0.001	149.7	5276	1.93
Callisto	0.25	1880	0.007	106.8	4820	1.82
Group of four*	25–29	11 110–11 740	0.12–0.21	6×10^{-6} to 9×10^{-3}	~15 to ~185	
Group of four	147–164 retrograde	20 700–23 700	0.17–0.38	4×10^{-5} to 2×10^{-4}	~30 to ~50	

Jupiter's spin period is 9 h 55 m and its spin axis is inclined at $3^\circ\,5'$ to the ecliptic.
Satellites marked * and one satellite of the marked group were discovered from spacecraft.

from the volcano Pele that was first seen. Subsequently ten other volcanoes were discovered and there may be even more.

The basis of the prediction was that the satellites Io and Europa had commensurate periods in the ratio 1:2. Because of this Europa and Io are always at *inferior conjunction*, i.e. closest together, at the same point of Io's orbit. The perturbations to Io's orbit by Europa are thus enhanced by a resonance effect and the orbit becomes slightly eccentric. This eccentricity is very small and to three significant decimal figures is zero (table 7.2). However, given Jupiter's large mass and its proximity to Io, it is enough to give a periodic tidal flexing of Io that generates some 10^{13} W of internal heating (Topic AB).

The general appearance of the surface of Io is seen in plate 6—it has been likened to a pizza! The general coloration is orange and white, due to sulphur and sulphur compounds in the ejected volcanic material. The white areas are covered with a layer of solid sulphur dioxide, and sulphur in the form of gaseous S_2 has been detected in the plume of the volcano Pele. The volcanoes on Io are more violent than those on Earth with material ejected with speeds up to 1 km s^{-1} compared to a normal 100 m s^{-1} or so on Earth. The higher ejection speed combined with the lower gravity gives volcanic plumes up to 300 km in height, equivalent to one-sixth of the satellite's radius. At least some of the erupted material, at temperatures in the range 1500–1900 K, is silicate highly enriched with sulphur, sulphur compounds and other volatile materials. The volcanoes on Io are neither cone-shaped nor like the more laterally-extensive gently sloping types known as *shield volcanoes*. The sources of volcanic eruption are at a low level and flows from them spread for hundreds of kilometres around. More recent flows can be recognized from the higher local surface temperature—300 K or more, rather than the 130 K that is characteristic of the majority of the surface. Because of the active volcanism no craters are seen on the surface, as any that do form are covered in a geologically short time. The general age of the surface of Io is of the order of a few hundred thousand years.

The satellite possesses a very tenuous atmosphere with a pressure about 10^{-10} bar. This atmosphere is ionized and the resultant plasma interacts with the Jovian magnetic field. Jupiter

Plate 1. *(a) Aurora Borialis seen from the surface of the Earth. The red colour shows that it occurred at high altitudes (J Curtis, ACRC). (b) Aurora Australis photographed from Space Shuttle Endeavor. The display spreads from lower altitudes (green) to the higher ones (red) (NASA).*

Plate 2. *Martian topography with height indicated by a spectral sequence. The red highland regions in the south and the green-blue northern plains give an approximate hemispherical asymmetry. The Hellas basin, the deepest feature on Mars, is seen at the bottom left (NASA).*

Plate 3. *Saturn and its ring system (NASA).*

Plate 4. *Neptune showing the Great Dark Spot (NASA).*

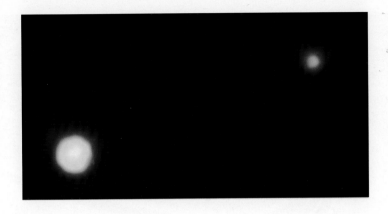

Plate 5. *Pluto and its satellite, Charon (HST).*

Plate 6. *The Galilean satellite Io (NASA).*

Plate 7. *A false-colour view of Saturn's rings (NASA).*

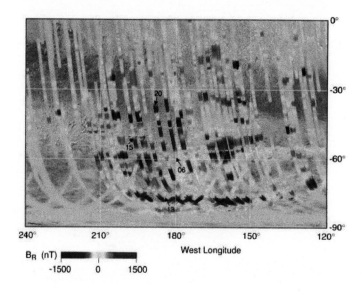

Plate 8. *Magnetic data for the surface of Mars obtained from the Mars Global Surveyor (NASA). The field measurements are in units of nT (gammas).*

possesses an equatorial plasma torus, co-rotating with the planet, within which Io is embedded. The plasma is rich in sodium that can be detected by the characteristic yellow glow that it emits, but the main components are sulphur and oxygen, derived from the volcanic emissions.

7.3.2. Europa

The general appearance of the surface of Europa is shown in figure 7.3. The surface is icy and the ice is cracked in a fairly complicated way. One interpretation of the appearance is that what is being seen is sea ice so that under the surface is liquid water. The cracking is caused by stresses on the ice due to tides induced by Jupiter. The validity of this interpretation cannot be checked at present but what is quite certain from the overall density of Europa is that the layer of water, be it in solid or liquid form, must be quite thin, probably considerably less than 100 km in thickness. If there is water below the frozen surface then what maintains it in that state is clearly of interest. Because of the R^{-6} dependence of the energy input due to tidal effects [equation (AB.6)] the tidal heating of Europa would only be about 6% that of Io. On the other hand the other parameters that appear in equation (AB.6) might be more favourable for heating than they are for Io, so tidal heating remains a plausible source of energy. The possibility of another body in the Solar System, other than the Earth, that might have large quantities of liquid water has suggested to some biologists that some form of life could have developed there.

There are only three observed craters on the surface of Europa, and they have diameters between 18 and 25 km. Since the scars of the early bombardment that affected the whole of the early Solar System are absent, this indicates that the surface was at one time slushy or watery and has frozen to

Figure 7.3. *The cracked icy surface of Europa (NASA).*

cover traces of early damage. If the eccentricity of Europa's orbit had once been much larger than at present then tidal heating could have been considerable and liquid, or near liquid, water a distinct possibility.

7.3.3. Ganymede

Ganymede has the distinction of being the largest and most massive satellite in the Solar System and its diameter is actually 10% greater than that of the planet Mercury. However, because of its low density, $1930\,kg\,m^{-3}$, its mass is only 45% that of Mercury. From its density and ice-covered surface it may be deduced that its bulk composition is approximately one-half ice and one-half silicates. It could have a small iron core which might then be the source of its magnetosphere, detected by the Galileo spacecraft.

The surface of Ganymede is very variable in appearance from one region to another. Some parts are heavily cratered (figure 7.4a) and have probably been undisturbed since the early heavy bombardment in the Solar System some 4000 million years ago. The background surface on which the craters are superimposed is dark and more recent craters show up as bright features with rays radiating from them. By contrast about one half of the surface looks as though it has been fashioned by tectonic forces of some kind. Complicated intersecting systems of almost parallel ridges can be seen where the ridges are tens of kilometres in lateral extent, hundreds of metres high and many thousands of kilometres long (figure 7.4b). In some places there are lateral displacements

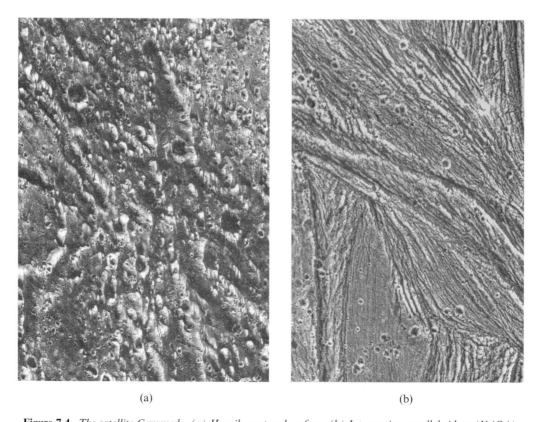

(a)   (b)

Figure 7.4. *The satellite Ganymede. (a) Heavily cratered surface. (b) Intersecting parallel ridges (NASA).*

of grooves along fault lines, similar to the slippage between tectonic plates on Earth. It is clear that there has been reformation and movement of Ganymede's surface in these grooved regions but the exact processes that have been involved are not known.

7.3.4. *Callisto*

Callisto is the second largest of the Galilean satellites and is only slightly less than Mercury in size. Its density indicates that, like Ganymede, it must consist of about one-half ice and one-half silicates.

The surface of Callisto must be extremely old since it is very heavily cratered. It must be understood that although its surface is mainly water ice, generally considered to be a frangible and rather fragile material, at the low temperature of Callisto (about 100 K) ice will behave like a strong rock. There are several multi-ringed basins on Callisto, similar to the Caloris basin on Mercury and the 'bullseye' feature, Mare Orientale, on the Moon. The largest of these multi-ringed basins is *Valhalla* (figure 7.5) the central basin of which is about 600 km in diameter. When the projectile that caused this feature struck Callisto the ice liquefied in its vicinity and concentric waves swept across the surface like circular ripples in a pond when a stone is tossed into it. These waves would have been created even in the solid surface far from the collision. The waves were gigantic in scale but in the extreme cold they quickly froze leaving the wavy profile permanently sculptured on the surface. There are at least ten frozen ripples although, depending on how surface features are interpreted, it is sometimes claimed that the number of ripples exceeds thirty. Valhalla must have formed after the early period of bombardment in the Solar System since there are comparatively few craters in the central region.

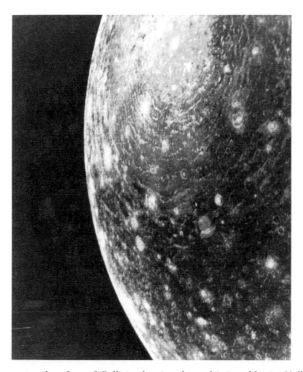

Figure 7.5. *The cratered surface of Callisto showing the multi-ringed basin, Valhalla (NASA).*

7.3.5.   *Commensurabilities of the Galilean satellites*

The periods of Io, Europa and Ganymede are close to being in the ratios $1:2:4$. Although these ratios are not precise what *is* precise is the relationship

$$n_1 - 3n_2 + 2n_3 = 0 \qquad (7.1)$$

where n indicates the *mean motion*, or average angular speed in the orbit, and suffices 1, 2 and 3 correspond to Io, Callisto and Ganymede respectively. In addition the three satellites can never line up on the same side of Jupiter; the allowed conjunctions and oppositions are shown in figure 7.6.

A general mechanism for establishing the approximate $1:2:4$ resonance system was suggested by C F Yoder in 1979. The effect of Jupiter on its satellites is similar to that of the Earth on the Moon, as described in Topic Y and illustrated in figure Y.6. For example, Io raises a tide on Jupiter but the planet's rapid spin drags the tidal bulge forward. The gravitational effect of this is to pull Io in a forward direction so increasing its angular momentum and moving it outward. The effect is strongest at perijove, the closest point on the orbit to Jupiter. If the effect *only* occurred on perijove then it is easily seen that the new orbit would have the same perijove distance but that the semi-major axis would increase—i.e. the eccentricity, e, would increase. However, it is clear from equation (AB.6) that the greater is the eccentricity the greater is the dissipation of energy within Io

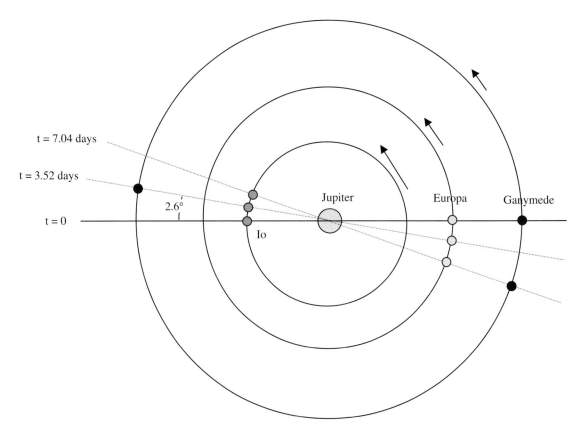

Figure 7.6. *Allowed alignments of the three inner Galilean satellites. The orbits and Jupiter are drawn to scale but angles are exaggerated.*

itself. A stable equilibrium is established where the eccentricity due to Jupiter and the other satellites gives a loss of energy that just balances the input of energy by Jupiter. If the period of Io, P_{Io}, was below $\frac{1}{2}P_{\text{Europa}}$ then Jupiter's tidal influence, acting most strongly on Io, would have pushed it outwards until its period was just below $\frac{1}{2}P_{\text{Europa}}$. The effect of Europa would then be to increase Io's eccentricity until the resultant loss of energy prevented any expansion of Io's orbit relative to Europa. At this stage the tide on Europa would drive out the coupled satellites, Io and Europa until P_{Europa} was just less than P_{Ganymede}. In principle the three coupled satellites could be driven out to link up with Callisto but in practice the Solar System will not last long enough for this to happen.

We shall see that several of Saturn's satellites have commensurable orbits and these can be explained in terms of the general principles described here for the Jupiter triplet of connected satellites.

7.3.6. The smaller satellites of Jupiter

Of the three small satellites within the orbit of Io the largest is Amalthea, with the distinction of being the last satellite to be discovered by telescopic observation (in 1892 by Barnard). It is ellipsoidal in shape with its major axis pointing towards Jupiter. Its temperature indicates that it may have some form of heating other than by radiation from the Sun and Jupiter. This could be by internal currents induced by moving in Jupiter's non-uniform magnetic field or some form of tidal heating due to a slight eccentricity of its orbit. Its surface is reddish in colour, that could be due to a sulphur layer derived from Io.

The two innermost and very small satellites, Metis and Adrastea, were discovered by the Voyager flypast in 1979. They are possibly of asteroid origin and their main interest is that they probably act as shepherd satellites to stabilize Jupiter's ring system (section 7.8.3).

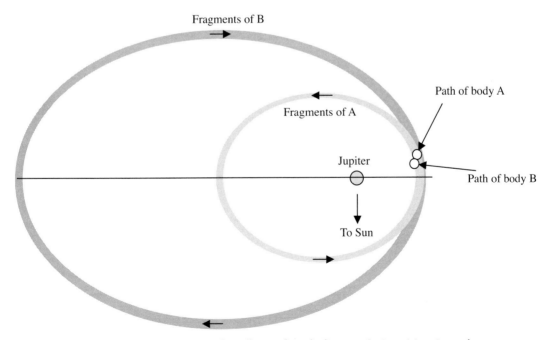

Figure 7.7. *A schematic representation of a collision of two bodies near Jupiter giving rise to the two outer families of satellites.*

From table 7.2 it will be seen that there are two groups of small satellites, each with four members, at much greater distances from Jupiter than the others. The innermost group, orbiting at distances between 11 and 12 million km, has quite eccentric orbits that have inclinations between 25 and 29° with Jupiter's equatorial plane. The outermost group, orbiting between 20 and 24 million km from Jupiter, has retrograde orbits that are even more eccentric. These orbits are substantially perturbed by the Sun, although they are well inside the sphere of influence of Jupiter (section AA.4) so that the satellites are securely retained. They are almost certainly the result of some capture process, and various speculations have been made from time to time about the mechanism for their capture. It may be significant that the apojove distance, 14.18 million km, of Elara, a member of the inner group, is just greater than the perijove distance, 13.75 million km, of Pasphaë, a member of the outer group. A collision between two large asteroids in the vicinity of Jupiter could have led to fragments left behind in the two sets of orbits now observed (figure 7.7).

7.4.   THE SATELLITES OF SATURN

The known satellites of Saturn are listed in table 7.3. All those larger than Hyperion, plus Pheobe, were known from telescope observations prior to the space era; the remainder were discovered by spacecraft observations. They will be discussed in groups within which individual satellites have some relationship linking them.

7.4.1.   *Titan and Hyperion*

As a satellite Titan (figure 7.8) is second only to Ganymede in both diameter and mass and is the only other satellite actually larger than Mercury. However, on account of its much smaller density, $1900 \, \mathrm{kg\,m^{-3}}$, it has only just over 40% of the mass of that planet. The density suggests that it must have a roughly equal mix of silicate and ice.

Table 7.3. *Satellite data for Saturn.*

Satellite	Inclination of orbit to planetary equator (°)	Orbital radius (10^3 km)	Orbital eccentricity	Mass (10^{21} kg)	Mean diameter (km)	Mean density ($10^3 \mathrm{kg\,m^{-3}}$)
Group of six*	<0.3	133–151	<0.007		~20 to ~200	
Mimas	1.5	186	0.02	0.045	390	1.4
Enceladus	0.0	238	0.005	0.074	510	1.2
Tethys	1.1	295	0.000	0.626	1050	1.2
Telesto*		295			~30	
Calypso*		295			~26	
Dione	0.0	378	0.002	1.05	1120	1.4
Dione B*		378				
Rhea	0.3	527	0.001	2.28	1530	1.3
Titan	0.3	1222	0.029	136	5150	1.9
Hyperion	0.6	1483	0.104	0.017	280	
Iapetus	14.7	3560	0.028	1.88	1460	1.2
Phoebe	150 (R)	12950	0.163		~200	

Saturn's spin period is 10 h 39 m and its spin axis is inclined at 26° 45′ to the ecliptic.
The satellites marked *, including all the group of six, were discovered from spacecraft.

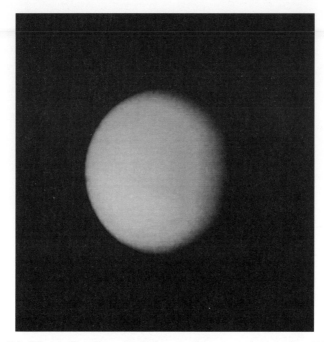

Figure 7.8. *The satellite Titan. Its thick atmosphere conceals its surface (NASA).*

Titan is exceptional in having a very dense atmosphere, comparable with those found for the terrestrial planets. The atmosphere consists of nitrogen with about 1% methane, some argon and traces of other gases, many of them hydrocarbons. The surface pressure is about 1.6 bar which, together with a surface acceleration due to gravity of $1.4\,\mathrm{m\,s}^{-2}$, gives a column mass (section O.1) of $1.1 \times 10^{5}\,\mathrm{kg\,m}^{-2}$, more than 10 times the Earth value. Although it is a smaller body than the Earth it has a greater total mass of atmosphere. The application of equation (O.18b) to Titan with an exosphere radius of 2800 km and temperature 200 K, which is on the high side, gives the lifetime of a nitrogen atmosphere to be effectively infinite.

Because of its extensive atmosphere Titan's surface cannot be seen. It has been suggested that it might be covered by shallow oceans of liquid nitrogen and liquid hydrocarbons, corresponding to its atmospheric content.

Hyperion is an irregularly shaped object with roughly the appearance of a very flat oblate spheroid. Its large eccentricity suggests that it is not a regular satellite but, on the other hand, its closeness to Titan must have a substantial effect on its orbit. This is also suggested by its close-to 4 : 3 orbital resonance with Titan.

7.4.2. Mimas, Enceladus, Tethys, Dione and co-orbiting satellites

As their densities indicate, all the satellites of this group are icy. At the low prevailing temperatures water ice is as hard as rocks on Earth and the surfaces of these bodies all show the results of bombardment in the past. In particular, Mimas shows an impact feature that is huge in relation to its size (figure 7.9); the event that produced this feature must have gone close to completely disrupting the satellite.

Enceladus (figure 7.10) shows some smooth regions that suggest surface activity subsequent to an early heavy bombardment. It is interesting that the non-adjacent pairs of satellites Mimas–Tethys and

Figure 7.9. *The surface of Mimas showing a huge impact feature (NASA).*

Figure 7.10. *The surface of Enceladus showing cracks and flow features (NASA).*

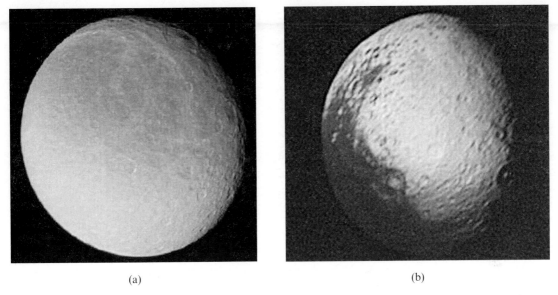

(a)   (b)

Figure 7.11. *(a) The well-cratered surface of Rhea, the second largest satellite of Saturn (NASA). (b) The edge of the dark hemisphere of Iapetus is at the lower left (NASA).*

Enceladus–Dione have a close-to 2 : 1 orbital resonance and the orbital eccentricities of Mimas and Enceladus are appreciably different from zero. When equation (AB.6) is applied to Mimas and Enceladus it is found that the energy generation by tidal flexing within them is now of order 10^8 to 10^9 W. Given that each of them has less than 0.001 of the mass of Io, it is possible that tidal heating could have been important in the past.

One member of the inner group of six satellites given in table 7.3 may be co-orbital with Mimas although in view of its small estimated diameter, 10 km, its existence is not certain. However Tethys certainly does have two co-orbiting satellites, one 60° ahead of it and the other 60° behind, as is found for the relationship of the Trojan asteroids to Jupiter (section 8.2; Topic AD). Dione B co-orbits with Dione in much the same way.

7.4.3.   Rhea and Iapetus

These are both relatively big satellites with Rhea showing the characteristics of a regular satellite but with Iapetus almost certainly irregular on account of its large inclination.

Rhea seems to have an ancient surface that is covered by a large number of small craters (figure 7.11a). The hemisphere facing the direction of motion around Saturn is much lighter in colour than the trailing hemisphere although the reason for this is unknown. The same phenomenon is found for Iapetus (figure 7.11b) where the albedoes of the leading and trailing hemispheres have been estimated as 0.5 (bright) and 0.05 (dark).

7.4.4.   Phoebe

The main interest in this satellite, which is small and almost certainly a captured body, is its retrograde orbit. It was not closely observed by the Voyager spacecraft so little is known of it, other than that it is approximately spherical in shape.

Figure 7.12. *The interchanging orbits of the satellites 1980 S1 and 1980 S3.*

7.4.5. Other small satellites

Some of the smaller satellites that have been discovered play an important role in stabilizing Saturn's rings and their action will be described in section 7.8.1. Two satellites, 1980 S1 and 1980 S3 orbiting within the ring system, have an interesting dynamical relationship illustrated in figure 7.12. The orbits are almost identical with a difference of radii not very different from the sum of average radii of the satellites. The mean diameter of 1980 S1 is about 100 km and about 70 km for 1980 S3 so that they exert only small gravitational effects. However, whichever of them is the inner satellite at any particular time slowly catches up the other and when they approach closely their mutual gravitational interaction causes the inner one to be swung into an outer orbit and the outer one to swing into an inner orbit. This behaviour then repeats itself and seems to be stable over long time periods.

7.5. THE SATELLITES OF URANUS

The five largest satellites of Uranus—Miranda, Ariel, Umbriel, Titania and Oberon—were known from telescope observations and all fall into the category of regular satellites. The Voyager 2 spacecraft, that visited the planet in 1986, revealed the presence of 10 smaller satellites, all within the orbit of Miranda, and also took detailed photographs of the surfaces of the larger ones. The densities of the larger satellites suggest that they consist of mixtures of silicate and ice. The surface of Miranda (figure 7.13a) shows signs of considerable disturbance with large-scale features consisting of regions of parallel ridges and troughs. It may possibly have been disturbed by tidal action from Uranus. The surface of Ariel (figure 7.13b) shows evidence of bombardment and also extended valley systems that may indicate tension in the crust at some stage in its development.

Table 7.4. *Satellite data for Uranus.*

Satellite	Inclination of orbit to planetary equator (°)	Orbital radius (10^3 km)	Orbital eccentricity	Mass (10^{21} kg)	Mean diameter (km)	Mean density (10^3 kg m^{-3})
Group of nine*		50–75			40–80	
Puck*		86			170	
Miranda	3.4	130	0.017	0.075	484	1.3
Ariel	0.0	191	0.003	1.4	1160	1.7
Umbriel	0.0	266	0.004	1.3	1190	1.4
Titania	0.0	436	0.002	3.5	1610	1.6
Oberon	0.0	583	0.007	2.9	1550	1.5

The spin period of Uranus is 17 h 14 m and its spin axis is inclined at 97° 52′ to the ecliptic. The satellites marked * were discovered from spacecraft.

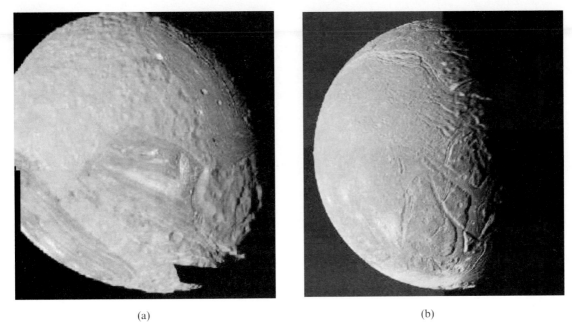

| (a) | (b) |

Figure 7.13. *(a) The surface of Miranda showing parallel ridges and troughs. (b) The bombarded surface of Ariel showing fault scarps (NASA).*

7.6. THE SATELLITES OF NEPTUNE

Only two satellites were known for Neptune prior to the flypast of Voyager 2 in 1989, but these were both remarkable and certainly not regular. The smaller satellite, Nereid, is distinguished by having the greatest orbital eccentricity of any satellite, 0.749. Its large average distance from Neptune, 5.56 million km, made it readily visible by telescope observation although a larger satellite, 1989N1 discovered by Voyager 2, orbiting much closer to Neptune (table 7.5) was previously unobserved.

Before the visit of Voyager 2 it was suspected that Triton might be the most massive satellite in the Solar System, or at least closely rival Ganymede in that respect. Actually it turns out to be the seventh most massive satellite with about one-seventh of the mass of Ganymede. It is in a close orbit of small eccentricity but with an inclination of 160° that means that the orbit is retrograde

Table 7.5. *Satellite data for Neptune.*

Satellite	Inclination of orbit to planetary equator (°)	Orbital radius (10^3 km)	Orbital eccentricity	Mass (10^{21} kg)	Average diameter (km)	Density (10^3 kg m^{-3})
Group of five*		48–74			55–190	
1989N1*		118			400	
Triton	159.9 retrograde	355	0.000	21.4	2760	~2
Nereid	27.7	5560	0.749	0.021	340	~1

Neptune's spin period is 16 h 7 m and its spin axis is inclined at 29° 34′ to the ecliptic. The satellites marked * were discovered from spacecraft.

and substantially inclined to the planet's equatorial plane. Its relationship with Neptune is usually taken to imply that it is a captured body.

A very tenuous atmosphere of nitrogen with traces of methane has been detected with a surface pressure of 10^{-5} bar. The surface of Triton has few distinct features, suggesting that it is a comparatively young surface. There is some evidence of flow features and volcanic activity based on volatile materials such as nitrogen, methane and possibly water.

7.7. PLUTO'S SATELLITE

Charon was previously discussed in relation to Pluto in section 5.5. Ideas about the origins of Pluto and Charon and a possible relationship with Neptune and Triton are described in section 12.11.2.

7.8. RING SYSTEMS

The ring system associated with Saturn was detected by telescopes and has been studied for four hundred years. A ring system for Uranus was inferred from stellar occultation but its exact form was not known at the time of its discovery. Observations from the Voyager 1 and Voyager 2 spacecraft have shown these ring systems in more detail and also revealed that the other two major planets, Jupiter and Neptune, also possess rings. Thus a ring system accompanies each of the major planets and it is tempting to believe that there is something conducive to ring formation for large planets well removed from the Sun.

7.8.1. The rings of Saturn

When Galileo turned his telescope on to Saturn, the outermost planet known at that time, he recorded the presence of two lumps on either side of the planet looking like a pair of 'accompanying globes'. He was also surprised to find that this feature disappeared after a few years. We now know that what he saw with his crude instrument were Saturn's rings and that they disappeared because Saturn had moved in its orbit to a position where the rings were seen edge-on from Earth.

As seen with good modern telescopes Saturn's rings are quite beautiful and detailed in structure. There are two bright rings, an outer ring A and an inner ring B separated by the Cassini division. Closer in there is a fainter ring C much harder to see and usually referred to as the Crepe ring. It was shown by Clerk Maxwell in 1857, as an Adams Prize Essay, that the rings could only be stable if they consisted of small solid bodies in independent orbit around Saturn. Modern infrared data show that they are either icy bodies or perhaps ice-covered silicate bodies. Radar observations indicate that the size of the bodies varies from being small grains up to having diameters of several metres. The Cassini division has as its cause the same kind of perturbation that produces the Kirkwood gaps in the periods of asteroids (section 8.3). It turns out to correspond to one half the period of Mimas, one third that of Enceladus, a quarter that of Tethys and one-sixth that of Dione. Other fine divisions were located by careful observation and these are found to be commensurate either with the periods of Mimas or Enceladus (figure 7.14).

When photographs of the ring system were returned by Voyager 1 they were found to consist of hundreds of separate rings of varying thickness and brightness separated by gaps of different widths (plate 7). Outside the outer boundary of ring A there are three more faint rings, F, G and E, that cannot be seen from Earth. Within the C ring there is a D ring rather indeterminate in its inner edge. The precise nature of the mechanics that maintains the pattern of Saturn's rings is not completely understood although some aspects are known. For example, two of the innermost small

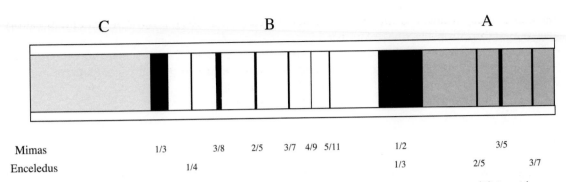

Figure 7.14. *A representation of the major divisions in Saturn's rings showing their commensurabilities with the periods of Mimas and Enceladus.*

satellites, Pandora and Prometheus, are called *shepherd satellites* and they straddle the F ring and prevent it from spreading either inwards or outwards. The action of the shepherd satellites is rather subtle but can be explained fairly simply. In figure 7.15 there is shown the motion of a particle in an orbit inside that of Pandora, the outer of the shepherds. The particle has a higher angular speed and during the part of the orbit when it is behind the particle, but catching up, it will gain energy from the satellite. At the same time it is deflected slightly towards Pandora. After the particle passes the satellite it now loses energy and to a first approximation the gain and loss of energy cancel each other. However, in the energy-losing part of the motion the particle is slightly nearer Pandora so that the energy lost is slightly larger than that gained. The loss of energy corresponds to an orbit farther in so the effect of Pandora is to repel the particle away from itself. By a similar argument it can be shown that if the particle's orbit is outside that of Prometheus then it is repelled by Prometheus so constraining it to have an orbit farther out. The net effect of the shepherds is that particles orbiting between them feel a pinch effect that prevents them from diffusing inwards or outwards.

The complete pattern of rings and gaps cannot be explained by a small number of shepherd satellites and there is still much to be understood about this complicated feature of the Solar System.

There is also uncertainty about the origin of the ring system. The rings are all within the Roche limit (Topic AA) and it has been suggested that the orbit of a small satellite of Saturn decayed until it strayed within the Roche limit and was then torn apart. This would not have happened if the satellite had been too small; a rocky or even a rock + ice satellite would have sufficient mechanical strength to resist tidal disruption if it had a diameter less than about 100 km. Even if the satellite had been large enough to be disrupted it would probably have broken up into bodies a few tens of kilometres in

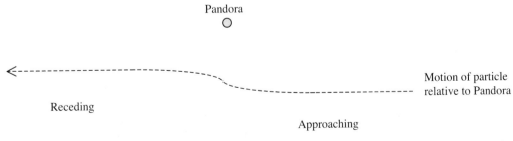

Figure 7.15. *The motion of a ring particle relative to Prometheus and within its orbit. On approach the particle gains energy and on receding it loses energy. The loss slightly exceeds the gain because it is a little closer.*

diameter that could have survived. Subsequent collisions between these might then have produced the finer material to form the rings. Another idea is that two bodies, originally in heliocentric orbits, collided close to Saturn and directly produced the small size bodies that now constitute the rings. A further possibility is that the ring particles are simply material that was not incorporated into the planet when it formed and that it was just left behind in orbit.

7.8.2.   *The rings of Uranus*

In 1977 the occultation of a star by Uranus was being observed. The observers were surprised that some time before the occultation was due to appear the star was obscured five times by objects close to the planet. After the main occultation by Uranus there was a symmetrical set of smaller occultations on the other side of the planet indicating that they were due to a ring system. Additional occultation measurements increased the number of inferred rings from five to nine. Later, when Uranus was visited by Voyager 2 the number of rings observed was 11 with the possibility that other very faint rings may be present but unobserved. All the rings are between 42 000 and 51 000 km in radius and seem to consist of metre-sized chunks of ice that are very dark in colour. The outermost ring is stabilized by two shepherd satellites, members of the inner group of nine (table 7.4).

7.8.3.   *The rings of Jupiter*

In 1979, after the rings of Uranus had been detected, the Voyager 1 spacecraft discovered that Jupiter too had a ring system, albeit much less substantial than that of Saturn. The rings are so thin that when viewed from most directions they are transparent and almost invisible. However, at a high angle of incidence they can be seen in scattered light. From the infrared spectrum of the scattered light and the form of the scattering it is deduced that the particles forming the ring system are small ($\sim$3 μm in diameter) silicate grains.

A picture of Jupiter's rings is shown in figure 7.16. There is some structure in the rings but this is not so sharply defined as for Saturn. As mentioned in section 7.3.6 the small satellites Metis and Andrastea may serve as shepherd satellites to stabilize these rings.

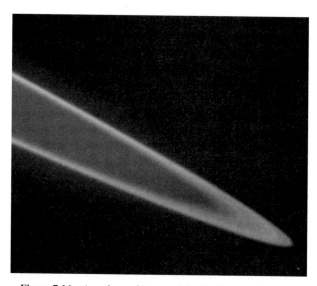

Figure 7.16. *An enhanced image of Jupiter's rings (NASA).*

7.8.4.   *The rings of Neptune*

With the three inner giant planets known to have ring systems there was considerable interest in what Voyager 2 would find when it reached Neptune in 1989. In the event the last of the giant planets also had a set of rings, five in number. Two of the rings are sharply defined while the others are rather faint and tenuous, like the Crepe ring of Saturn. The outer distinct ring is quite lumpy in character with three concentrations of density strung out around it like wide beads on a string.

7.9.   GENERAL OBSERVATIONS

Spacecraft observations have shown that all the giant planets have substantial families of satellites. As a general rule the largest satellites are for the planets closest to the Sun; Triton, that might seem to be an exception to this rule, is also clearly irregular and is probably a captured body.

Because of their large masses each of the giant planets exerts considerable tidal effects in its own vicinity and so a Roche-limit origin for ring systems may be feasible. However, these large bodies, well away from the Sun, also have large spheres of influence. The debris from bodies in energy-absorbing collisions in the vicinity of one of these planets is likely to be captured and may then form a ring system. Indeed the very small satellites that are close in and have been detected by Voyagers 1 and 2 for each of the planets might constitute the larger debris from collisions. They also indicate that small bodies can remain intact within the Roche limit, as indicated by section AA.3.

Problem 7

7.1   Find the ratio of the total masses of satellites to the mass of the parent planet for Jupiter, Saturn, Uranus and Neptune. Comment on the results.

CHAPTER 8

ASTEROIDS

8.1. GENERAL CHARACTERISTICS

By the end of the eighteenth century, especially after the discovery of Uranus, it was strongly believed that Bode's Law was fundamental in nature and that the gap between Mars and Jupiter should be occupied. Astronomers searched for the missing body and on 1 January 1801 such a body was found by Giussepe Piazzi (1746–1826) who called it Ceres, after the guardian god of his native Sicily. Although the body was obviously rather small to be considered as a planet, it filled the gap in Bode's law in a very satisfactory way. The discovery of other similar, if even smaller, bodies over the next few years was the prelude of many other discoveries of *asteroids* which continues to the present day. These bodies have a variety of characteristics in terms of orbit, shape, size and composition and these will now be discussed together with ideas that have been proposed concerning their origin.

Table 8.1 shows the characteristics of a number of asteroids, chosen because they span the period from the first discovery to comparatively recent times and also because they illustrate a wide range of characteristics. The orbits of most of them are shown in projection in figure 8.1 without regard to the relative directions of their perihelia. Gravitational forces in large bodies tend to force them into a shape of minimum energy, which is a sphere, but the mechanical strength of the material of small bodies can

Table 8.1. *Characteristics of some important asteroids. The ones given are: a, the semi-major axis, e, the eccentricity, i, the inclination and D, the diameter (u = unknown).*

Name	Year of discovery	a (AU)	e	i (°)	D (km)
Ceres	1801	2.75	0.079	10.6	1003
Pallas	1802	2.77	0.237	34.9	608
Juno	1804	2.67	0.257	13.0	250
Vesta	1807	2.58	0.089	7.1	538
Hygeia	1849	3.15	0.100	3.8	450
Undina	1867	3.20	0.072	9.9	250
Eros	1898	1.46	0.223	10.8	∼25
Hildago	1920	5.81	0.657	42.5	15
Apollo	1932	1.47	0.566	6.4	u
Icarus	1949	1.08	0.827	22.9	∼2
Chiron	1977	13.50	0.378	6.9	u

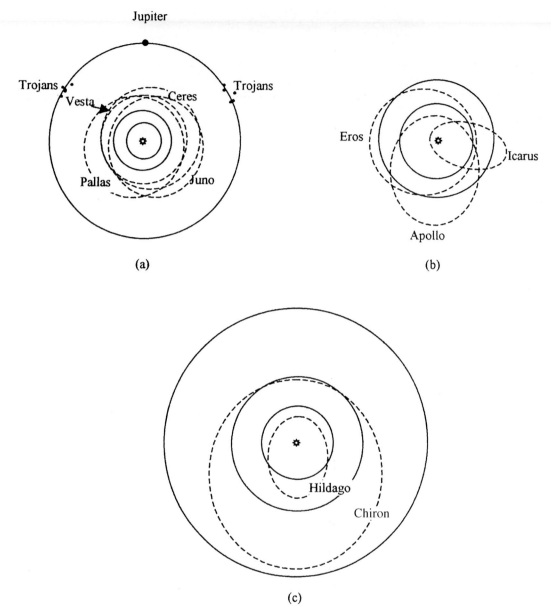

Figure 8.1. *(a) The orbits of the planets Earth, Mars and Jupiter (full lines), of Ceres, Pallas, Juno and Vesta (dashed lines) and the positions of the Trojan asteroids relative to Jupiter. (b) The orbits of the Earth and Mars (full lines) and of Icarus, Eros and Apollo (dashed lines). (c) The orbits of Jupiter, Saturn and Uranus (full lines) and of Chiron and Hildago (dashed lines).*

enable them to retain a non-spherical form. It can be shown that the largest asteroids, those with diameters more than about 300 km, are spherical or nearly so because the strength of their material is insufficient to support much departure from spherical symmetry (section P.7.1).

The usual way of finding asteroids is by photography of the night sky where the camera is rotated so that the stars remain stationary in the field of view. If the plate is exposed for a long period the stars

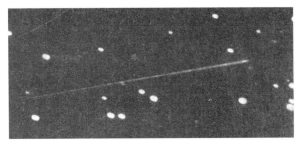

Figure 8.2. *The image of an asteroid seen as a streak against the fixed stars.*

will appear as points but any object moving within the Solar System will appear as a streak in the photograph, as shown in figure 8.2. The information this gives is the direction of the transverse component of motion and its angular velocity with respect to the Earth. The problem of determining the orbit of an asteroid from such observations is a difficult one and it is necessary to observe it at well separated points in its orbit. Unless there is some way of determining its distance the only information that is gained from an instantaneous observation is its direction. This problem, which is also the problem of finding planetary orbits, was first solved by Pierre Laplace (1749–1827) but a better form of solution was given later by Johann Gauss (1777–1855). From observations of three directions of an asteroid it is possible by some fairly complex calculations to work out its orbital elements. The orbit so determined will not be very precise but can be subsequently refined by further observations and the application of Newtonian mechanics. Only when an orbit has been determined is an asteroid given a name and there are many thousands that have been seen but not yet named.

The sizes of asteroids are difficult to determine in general. One of the most accurate ways to measure a dimension of an asteroid is if by chance it passes in front of a star. By measuring the duration of the occultation and knowing the orbit of the asteroid, it is possible to obtain quite a precise estimate of the distance across the asteroid along the line of the star's relative motion.

8.2. TYPES OF ASTEROID ORBITS

The great majority of asteroid orbits lie in the region between Mars and Jupiter—the 'gap' that eighteenth century astronomers were so keen to fill to complete the Bode-law progression. All known asteroid orbits are prograde, which is to say that they orbit the Sun in the same sense as do the planets, and most orbits have eccentricities less than 0.3 and inclinations less than 25°. There are some notable exceptions to these general rules, some of which are shown in table 8.1. For example, Hildago has an inclination of 42.5° but a few asteroids have even larger inclinations up to 64°. The eccentricity of Icarus, the largest for any known asteroid, combined with its small semi-major axis, gives a perihelion distance of 0.19 AU, the closest that any asteroid approaches the Sun. It is one of a number of asteroids that have Earth-crossing orbits—another is Apollo that was the first observed to have this characteristic. In the year of its discovery, 1932, Apollo came to within 3 million kilometres of the Earth—just seven to eight times the distance of the moon. Since the discovery of Apollo a few tens of other small asteroids, with diameters not more than a few kilometres, have been discovered with Earth-crossing orbits and these are known collectively as the *Apollo asteroids*. Another class of asteroids closely associated with the Earth are the so-called *Aten* group with orbits which lie mostly within that of the Earth. Very few of these bodies are known, and they are small, but the known ones could be representatives of a much larger population which stays well within the Earth's orbit. Some Aten and Apollo asteroids have a theoretical possibility of striking the Earth and it is possible that in its long history the Earth has undergone collisions from asteroids from time to time. It is generally believed that it was the consequences of an asteroid collision about 65 million years ago that led to the demise of the dinosaurs, which became extinct within a short period, having been the dominant living species on Earth for hundreds of millions of years.

There are a number of asteroids which have Mars-crossing orbits but with perihelia outside the Earth's orbit. The first such to be discovered was Eros, in 1898, but several more are now known. In a favourable conjunction Eros can approach the Earth to within 23 million km and in such an approach in 1975 it was studied by radar and found to have a rough surface. In common with many other small asteroids it is of irregular shape and it is somewhat elongated with a maximum dimension about 25 km.

Two other interesting groups of asteroids are the *Trojans* that move more-or-less in Jupiter's orbit, one group following Jupiter and 60° behind it and the other group leading Jupiter and 60° ahead of it. In general it is only possible to obtain an analytical solution for the motion of two bodies under mutual gravitational attraction. However, there are some special three-body problems that give a solution where there are two bodies of finite mass plus one body of negligible mass. One such problem is that of the Trojan asteroids and in Topic AD we demonstrate that the three bodies can be in equilibrium in circular orbits about the centre of mass of the system. The positions of the Trojan asteroids relative to Jupiter are shown in figure 8.1a.

The original belief about asteroids, once it was realized that there were many of them, was that they were all confined to the region between Mars and Jupiter. The discovery of the Apollo and Aten asteroid groups made it clear that this belief was not true. The 1977 discovery of Chiron, which moves mainly in the region between Saturn and Uranus, raised the question of whether there might be unknown families of asteroids well beyond Jupiter and perhaps too small to be observed from Earth. The diameter of Chiron is unknown but it must be at least 100 km for it to be observable and it may be much larger. At perihelion it crosses Saturn's orbit and in the seventeenth century it came within 16 million km of that planet—not far beyond the orbit of Phoebe, Saturn's outer retrograde satellite with an orbital radius of just under 13 million kilometres.

Even beyond the Chiron region many other bodies have been detected, with estimated diameters in the range 150 to 360 km, which are close to, or farther out than, the orbit of Neptune. These are known as *Kuiper-belt objects* and since they are usually considered as linked to comets rather than asteroids they will be dealt with in chapter 9. There is some uncertainty about the relationship between asteroids and comets—whether they are two manifestations of some common source of material or represent two different types of material with completely different origins. All the objects we describe as asteroids are in direct orbits and mostly have moderate inclinations and eccentricities. By contrast, as we shall see in chapter 9, comets are frequently in retrograde orbits, have a wide range of inclinations and eccentricities and also manifest a considerable content of volatile material. Most of them are associated with regions well outside that occupied by the planets. Ideas about the origin of various types of object in the Solar System are heavily linked with ideas of the origin of the system itself and the possible origins of asteroids and comets will form part of the discussion in chapter 12.

8.3. THE DISTRIBUTION OF ASTEROID ORBITS—KIRKWOOD GAPS

A diagram giving the frequency of asteroid periods, such as figure 8.3, indicates that the distribution has prominent gaps. These were first explained by the American astronomer Daniel Kirkwood in 1866. He pointed out that there are two very prominent gaps corresponding to one-third and one-half the

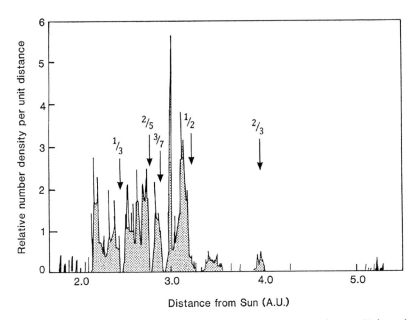

Figure 8.3. *The frequency distribution of asteroid orbit semi-major axes showing Kirkwood gaps.*

period of Jupiter, and that these gaps were a manifestation of some resonance phenomenon. For example, an asteroid with one-half the period of Jupiter will make two complete orbits while Jupiter is making one. Thus the two bodies will always be in conjunction in the same region of the asteroid's orbit so that the perturbation by Jupiter at closest approach will always be modifying the asteroid orbit in the same way. The asteroid's period will steadily change in one direction until the asteroid and Jupiter are sufficiently out of resonance for the nearest approaches, and hence maximum perturbations, to occur all around the asteroid's orbit with much diminished effect. The reader should be able to confirm that for the one-third resonance the asteroid is perturbed at two points on opposite sides of its orbit. With the determination of increasing numbers of asteroid orbits other *Kirkwood gaps* became more clearly seen and gaps at two-fifths and three-sevenths of Jupiter's period are also shown. To illustrate the complexity of this process we should note the small *concentration* of asteroid orbits that correspond to two-thirds of Jupiter's period. It is evident that the Kirkwood gap phenomenon is related to the formation of the gaps in Saturn's rings due to perturbation by the inner satellites Mimas and Enceladus.

8.4. THE COMPOSITIONS AND POSSIBLE ORIGINS OF ASTEROIDS

Recent observations of asteroids have greatly increased our knowledge about them in terms of their appearance and composition. A near passage of the asteroid *Gaspra* by the Galileo spacecraft gave the very detailed photograph shown in figure 8.4. This ellipsoidal object has dimensions $11 \times 12 \times 10$ km, shows craters due to collisions with objects much smaller than itself and appears to be covered with a layer of rocky dust. In common with many other asteroids it is in a tumbling motion with a period of about four hours. It much resembles in appearance the Martian satellites, Phobos and Deimos (figure 7.1), which have long been thought to be captured asteroids. The pock-marked appearance of Gaspra, and of the Martian satellites, suggests that collisions involving asteroids take place and such collisions are almost certainly the source of most of the material

Figure 8.4. *The asteroid Gaspra (NASA).*

which reaches the Earth in the form of meteorites. Meteorites are very important in aiding our understanding of the evolution and perhaps also the origin of the Solar System and will be discussed in greater detail in chapter 10. It suffices to say here that the main types of meteorite are *stones*, consisting mainly of various types of silicate, *irons* which are mostly iron with some nickel and *stony-irons* containing intimate mixtures of stone and iron regions. Within the stony classification is an important subclass, the *carbonaceous chondrites* that are very dark in appearance and contain volatile materials.

The main information about asteroid composition has come from visible and near infrared spectroscopy. Reflectance spectra have been measured for many hundreds of asteroids in visible light and in the infrared range up to 1.07 μm and it has been possible to match these spectra with those measured for meteorites in the laboratory. Figure 8.5 shows the match between spectra from four asteroids and four meteorites that indicate clearly the relationship between the two classes of object. On the basis of their spectral characteristics asteroids have been divided into six types. The two most common types, which together account for 80% of the spectrally observed asteroids, are designated as C, associated with carbonaceous chondrite material and S, mostly associated with stony irons. These asteroid types occur at all distances within the main belt of asteroids between Mars and Jupiter but there is a distinct tendency for the C type asteroids to have larger orbital radii—which is illustrated in figure 8.6.

There seems little doubt on the basis of the observational evidence that the study of meteorites is also tantamount to the study of asteroids. The question that then becomes of importance is the way in which asteroids are related to planets for, if they are intimately related, then the information from laboratory meteorite studies could be directly applied to the problem of the origin and evolution of the planets. However, there is no clear consensus on the form of this relationship. Some time ago it was thought that asteroids were the debris from a broken planet but this raised two difficulties. The first concerned the source of energy which could break up a planet and the second was that of disposing of the planetary material since the total mass of known asteroids is much less than a lunar mass. A second theory, which assumes planetary formation by an accumulation of asteroid-sized objects, asserts that the presence of Jupiter would exert considerable perturbations on objects in the asteroid-belt region and so prevent their accumulation by a continual stirring process. If there

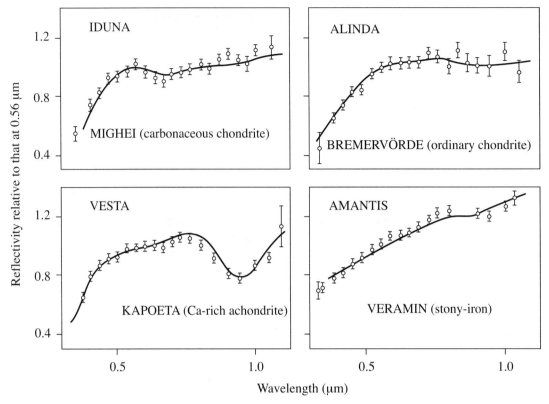

Figure 8.5. *Reflection spectra of asteroids, shown as points with error bars, and meteorites (full lines).*

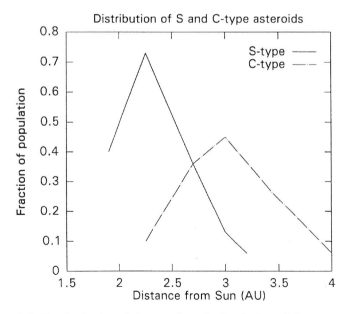

Figure 8.6. *The distributions of distances from the Sun for S- and C-type asteroids.*

was originally enough material in the asteroid region to produce a planet then this again raises the question of disposal. Certainly all observed solid bodies in the Solar System show signs of damage by large projectiles so many former asteroids can be accounted for in this way. Others may have been swept up by the major planets without leaving any visible evidence of their former existence.

If the relationship between asteroids and meteorites is accepted then studies of meteorites throw up many interesting problems. The first is that there is unequivocal evidence that some meteorite material has come from bodies that were molten and in which segregation of material by density had taken place. If an asteroid a few hundred kilometres in radius had formed by the accumulation of smaller bodies then there would not be enough gravitational energy available to melt silicates or metals (Topic AE).

For asteroids formed by accumulation to have melted, some other source of heating must have been available. There is evidence in meteorites for the one-time presence of a radioactive isotope of aluminium, ^{26}Al, which has a half-life of 720 000 years. If something like 2 parts in 10^5 of the aluminium of the minerals in asteroids had been ^{26}Al then this would have been enough to melt asteroids of diameter 10 km or so (see Problem 8.1). This has introduced the question of the isotopic composition of early Solar-System material, a matter that will be discussed in relation to meteorites in the chapter 10.

Problem 8

8.1   A newly-formed asteroid contains 1% by mass of aluminium of which a proportion 5×10^{-6} is ^{26}Al.

(i)   How many atoms of ^{26}Al are there per kilogram of asteroid?

(ii)   If each disintegration of an ^{26}Al atom yields 10 MeV of energy then what is the total energy produced in the asteroid in units J kg^{-1}?

(iii)   If the specific heat capacity of the asteroid material is 1000 J kg^{-1} K^{-1} then what is the rise of temperature of the asteroid due to ^{26}Al disintegration? You should ignore heat losses and any latent heat effects.

CHAPTER 9

COMETS

The general appearance of a comet is well known—a luminous ball with a long tail often, but mistakenly, thought of as trailing in the wake of its motion (figure 9.1). Some comets are the most spectacular astronomical objects that can be seen with the naked eye and so they have long held a fascination for mankind. In bygone days, when knowledge of astronomy was limited and superstition abounded, they were held to be the harbingers of an impending disaster. The apparition of what we now know as Halley's comet in 1066 was interpreted as an evil omen foretelling King Harold's defeat and death at the battle of Hastings—although William of Normandy would have given a more favourable interpretation. Such feelings about comets survived for a long time. When Shakespeare wrote Julius Caesar he had Calphurnia, Caesar's wife, say '*When beggars die, there are no comets seen; the heavens themselves blaze forth the death of princes.*' The play was set in Roman times but the superstition belonged to the Elizabethan age.

Figure 9.1. *A typical comet—Halley taken in 1986 (NASA).*

9.1. TYPES OF COMET ORBIT

It was Edmund Halley who first recognized that comets were bodies in orbits around the Sun and should therefore repeat their appearances in a periodic way. He postulated that the comet seen in

132

1682 was identical with comets in 1607, 1531 and possibly 1456 that had similar orbits, and he predicted the return of the comet in 1758. He did not live to see his prediction confirmed but this is the comet still seen in modern times and which bears his name. Halley had the benefit of Newton's analysis of planetary orbits, which applied also to comets, and he was thus able to deduce that the semi-major axis had to be of order 18 AU. That value, combined with the comet's close approach to the Sun, implied a high eccentricity, and this was the first deduction that bodies were moving around the Sun in very eccentric orbits. Measurements of the periods of comet orbits show a very wide variation and it is customary to divide comets into two categories—*short-period* for those with periods less than 200 years and *long-period* otherwise. It is well to note that this is a fairly arbitrary division but it roughly distinguishes those that move mainly within the region occupied by the planets and those that move well outside that region. There have been about 100 short-period comets observed. Those with periods of more than about 20 years have inclinations which are more-or-less random—for example Halley's comet, with a period of 76 years, is retrograde in its motion around the Sun and has an inclination of 162°. Comets are significantly perturbed by planets, especially by the major planets, and the period of Halley's comet can vary between 74 and 78 years due to this cause.

There are about 70 short-period comets with orbital periods, mostly between three and ten years, which have direct orbits and have fairly small inclinations, less than about 30°, and eccentricities mostly in the range 0.5 to 0.7. The orbital characteristics of a selection of such comets, together with those of Halley, are shown in table 9.1. These form the *Jupiter family* of comets and they are presumed to have originally been long-period comets which interacted with Jupiter either in a series of small perturbations on their occasional incursions into the inner Solar System or, possibly, in one massive perturbation (but see section 9.4). This is much more likely to happen for comets with small inclinations and also for those that are in direct orbits, for then their speeds relative to Jupiter during the interaction will be smaller and there will be more time to generate a strong perturbation. Another factor that would reinforce the interaction is if the comet is close to perihelion when it approaches Jupiter as then the motions of the two bodies are nearly parallel. Figure 9.2 illustrates an interaction with Jupiter that will remove energy from the comet's orbit and so make smaller its semi-major axis.

Table 9.1. *The characteristics of some members of the Jupiter family of comets and of Halley. The perihelion distance is q and the aphelion distance Q. q_{init} is the estimated original value of Q before perturbation by Jupiter.*

Comet	q (AU)	Q (AU)	q_{init} (AU)	e	i (°)
Grigg–Skjellerup	1.00	4.94	5.40	0.66	21.1
Tempel 2	1.36	4.68	5.83	0.55	12.5
Tempel 1	1.50	4.73	5.82	0.52	10.5
Wirtanen	1.26	5.16	5.35	0.61	12.3
Di Vico–Swift	1.62	5.21	5.47	0.52	3.6
Neujmin 2	1.34	4.84	5.67	0.57	10.6
Tuttle–Giacobini–Kresák	1.15	5.13	5.36	0.63	13.6
Tempel–Swift	1.15	5.22	5.23	0.64	5.4
Pons–Winnecke	1.25	5.61	4.99	0.64	22.3
D'Arrest	1.17	5.61	4.93	0.66	16.7
Forbes	1.53	5.36	5.33	0.56	4.6
Koppf	1.57	5.34	5.36	0.55	4.7
Giacobini–Zinner	0.99	5.98	4.51	0.71	31.7
Halley	0.59	35.0	0.99	0.967	162.3

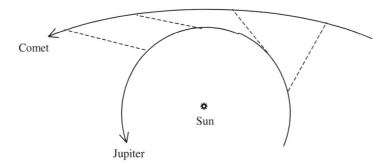

Figure 9.2. *The passage of a comet past Jupiter. The dashed lines join the comet and Jupiter at different times and it can be seen that when the effect of Jupiter is greatest it removes energy from the comet.*

When a comet has a close interaction with a planet with which its orbit is coplanar, or nearly so, then the values of a and e before and after the interaction are linked by the Tisserand criterion. This is

$$\frac{1}{2a} + \left\{ \frac{a(1 - e^2)}{a_{\mathrm{P}}^3} \right\}^{1/2} = T_{\mathrm{PC}} \qquad (9.1)$$

where T_{PC} is a constant, a and e are the orbital elements of the comet before or after the interaction and a_{P} is the radius of the planet's orbit, assumed circular. The value of T_{PC} can be calculated for the Jupiter family of comets from their present orbital elements, assuming that they have not changed since the original Jupiter perturbation—although in practice they must have changed a little through subsequent planetary perturbations. If it is assumed that the original orbit of the comet was one with a very large value of a, so that the first term in equation (9.1) is negligible, and had an eccentricity close to unity, then it is possible to calculate the perihelion of the original orbit. This comes from

$$q_{\mathrm{orig}} = a(1 - e) = \frac{T_{\mathrm{PC}}^2 a_{\mathrm{P}}^3}{1 + e} \approx \frac{1}{2} T_{\mathrm{PC}}^2 a_{\mathrm{P}}^3. \qquad (9.2)$$

The values of q_{orig} calculated for the members of Jupiter family of comets in table 9.1 are between 4.51 and 5.83 AU compared with the mean Jupiter orbital radius of 5.20 AU. For the very closest approaches a single interaction might have been sufficient to bring about the present orbit from one with a large original semi-major axis but for the others several interactions were probably necessary.

An outstanding recent example of the influence of Jupiter on a comet is the case of comet Shoemaker–Levy. This comet had passed very close to Jupiter and was torn apart by tidal forces to form a string of cometary fragments all moving close to the original orbit. Further perturbation by Jupiter led to a direct collision of the cometary fragments with that planet in July 1994 (section 5.1.3).

Since perturbations by Jupiter, or any other major planet, can add energy to a passing comet as well as remove it then the Jupiter family must represent the few comets, of very many which have interacted with Jupiter and which happened to have been suitably perturbed.

Another extreme class of comet orbits is where the periods extend from tens of thousands to millions of years. These comets have very large major axes and since comets can only be observed when they have small perihelia, usually less than 3 AU, then this implies that the orbital eccentricity must be close to unity. It is extremely difficult to measure the orbital characteristics of such comets accurately enough to confidently distinguish an extreme elliptical orbit from a parabolic or even marginally hyperbolic one. The characteristic of interest is that of the orbit before it approached the inner Solar System making it necessary to correct for planetary perturbation, which gives another possible source of error. A planet approaching the inner Solar System with an apparent near-parabolic but marginally hyperbolic orbit would need to have had almost zero velocity relative to the Sun at a large distance. This is extremely improbable and, taking the possible errors of measurement into account, it is safe to assume that all such comets approached the inner Solar System in extreme elliptical orbits. Comets with such orbits are called *new comets*. The category implies that the comet has never been so close to the Sun on a previous occasion and will never again be on an orbit with similar characteristics. Why this is so will be explained in section 9.3.

9.2. THE PHYSICAL STRUCTURE OF COMETS

Comets at more than about 5 AU from the Sun are difficult to see. At that distance a typical comet is a solid inert object, usually with low visual albedo and a diameter of a few kilometres. As it approaches the Sun its appearance changes dramatically. Vaporized material escapes, sometimes in jets from limited regions of the solid *nucleus*, and forms an approximately spherical gaseous *coma* that can be of large extent, from 10^5 to 10^6 km in radius. In the process of outgassing, some dust is also ejected and forms part of the coma. The coma becomes visible due to the action of sunlight on its constituents. Outside the coma, with ten times its dimension, there is a large cloud of hydrogen that emits no visible light but can be detected by its ultraviolet emission. Finally the comet develops a tail, or often two tails, which are acted on by the stream of particles from the Sun, the *solar wind*, so that the tails point approximately in an anti-solar direction. These features are shown in a schematic form in figure 9.3. This brief description of a comet and its behaviour as it approaches the Sun is the background for discussing its composition.

Current belief about the structure of the nucleus is that it is, as Fred Whipple once described it, a 'dirty snowball'. The best model is of an intimate mixture of silicate rocks and ices—perhaps similar to the composition of frozen swampy ground on the Earth although in the comet nucleus the ices would not all be water ice. The outer material of a comet that had made several perihelion passages would be relatively deficient in volatile components and would probably be in the form of a frangible rocky crust. As the nucleus approaches the Sun so it will absorb solar radiation and heat energy will eventually penetrate into the interior, causing sublimation of the volatile material—that is to say transforming it directly from the solid to the vapour state. The pressures so produced will eventually fracture the crust in weaker regions and jets of vapour will escape to form the coma. The molecules in the coma will *fluoresce*, that is they will be excited by ultraviolet radiation from the Sun and then emit visible radiation when they return to the ground state, and it is from this visible radiation that the composition of the coma can be found. The full range of atoms, molecules and free radicals detected in the nucleus is listed in table 9.2. Free radicals are fragments of molecules, produced by the disruptive effects of ultraviolet radiation, which readily react with other material to form stable molecules. In the coma the density is very low and free radicals will survive for a relatively long time before they come in contact with other material with which to react. It is for this reason that their presence can be detected spectroscopically. In figure 9.4 there is seen a spectrum for comet Bradfield taken with spectrometers on the IUE (International Ultraviolet Explorer) satellite covering the ultraviolet region. The composition of the halo is just what would be expected from a mixture of common Solar-System volatile materials—H_2O, CO_2, NH_3 and CH_4 plus traces of some less common volatile material and also silicate dust. Dissociation of the hydrogen-containing volatile substances will lead to the production of free hydrogen atoms. By virtue of their small mass they will be very mobile and move out to large distances, so producing the hydrogen cloud that can be detected from satellite or rocket-borne telescopes by a strong ultraviolet Lyman-series emission at 121.6 nm.

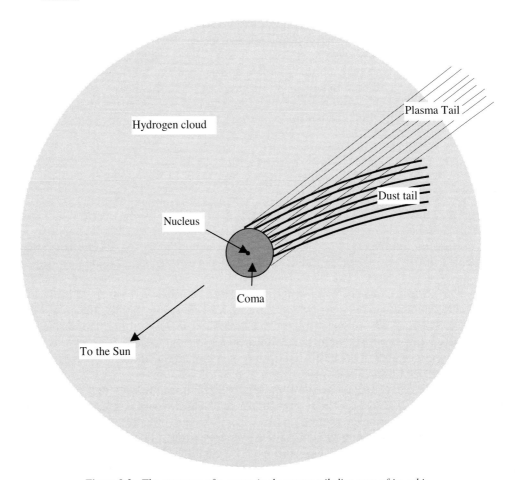

Figure 9.3. *The structure of a comet in the near perihelion part of its orbit.*

The largest feature of a comet is the tail or, more precisely, its tails. Molecules in the coma not only dissociate into atoms and free radicals but they and their by-products are also ionized to form a plasma, a mixture of positive and negative charged particles with no net overall charge. The plasma tail is swept along by the solar wind, itself a flux of charged particles that interact with the Sun's magnetic field. Since the solar wind moves radially outwards from the Sun so does the plasma tail, or very nearly so. The types of ions detected in the plasma tail are listed in table 9.3. The plasma tail has a blue colour due to strong emission from CO^+ at 420 nm. It may be about 1 AU in length; in the apparition of Halley's comet in 1910 the Earth passed through its tail and there were many who predicted dire consequences from this since the poisonous gas cyanogen was known to occur in the tail. However, in the event, the density of the tail was so low that there were no effects that could be detected.

Table 9.2. *The atoms, molecules and free-radicals detected in the coma of a comet.*

C	C_2	C_3	Ca	CH	CH_3CN	CN	Co	CO	CO_2
Cr	CS	Cu	Fe	H	H_2O	HCO	K	Mn	Na
NCN	NH	NH_2	NH_3	Ni	O	OH	S	S_2	V

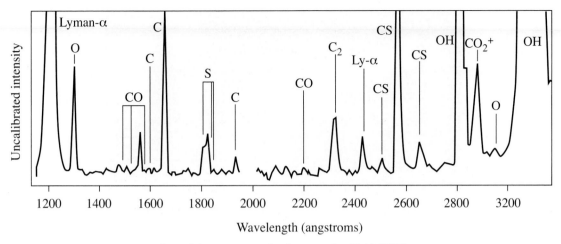

Figure 9.4. *A spectrum for the comet Bradfield (IUE).*

The dust tail consists of very fine silicate grains and is usually yellow in colour since most of the light from it is just reflected sunlight. Study of the polarization of the reflected light suggests that the silicate grains are typically of 1 μm dimension. The forces on the dust particles are, first, that due to the gravitational attraction of the Sun, which will tend to send it into Keplerian orbits, and second, the action of the solar wind, which will tend to move it radially outwards from the Sun. The solar wind gives the stronger forces so the net effect is a tail that is *almost* anti-solar in direction but usually distinct from the plasma tail. However, dust tails are generally much shorter than plasma tails and rarely exceed 0.1 AU in length. The two tails show up particularly well in figure 9.5, a photograph of comet Mrkos taken in 1957.

The Halley comet apparition of 1986 was a rather poor one from the point of view of Earth-bound observers since the perihelion passage was on the opposite side of the Sun from the Earth. However, it was the first to occur in the space age and there were several missions to observe it. The European Space Agency sent the spacecraft *Giotto* that passed within 600 km of the nucleus. Photographs were taken of the nucleus (figure 9.6) which showed that it had a peanut shape with a maximum dimension of about 15 km and the surface showed features that could be described as valleys and mountains. In addition to the photographs, measurements were made of the charged-particle density, magnetic fields and the compositions of the dust particles. Many of the particles were clearly silicates, similar in composition to a class of meteorites called *carbonaceous chondrites* (section 10.2.1), but others were rich in H, C, N and O and were presumed to be grains containing organic material. The rate of loss of dust was estimated at 3 tonnes s^{-1} but there was a much larger mass loss from large fragments of the nucleus being prised free and ejected through the impulse imparted by the escaping volatile material. The rate of loss of icy material is of the order of 50 tonnes s^{-1} during the perihelion passage and such a rate implies that the lifetime of Halley, and of other comets, as vapour-emitting bodies must be limited. The expected lifetime of comets is estimated to be in the range of hundreds to thousands of orbits, after which they will be small,

Table 9.3. *Ions detected in the plasma tails of comets.*

C^+	Ca^+	CH^+	CN^+	CO^+	CO_2^+	CS^+	CS_2^+	Fe^+
H_2O^+	H_3O^+	N_2^+	Na^+	O^+	OH^+	S^+	S_2^+	

Figure 9.5. *The comet Mkros photographed in 1957 showing the long plasma tail and the shorter but thicker dust tail (Palomar Observatory/Caltech).*

dark, inert objects very difficult to detect. The fact that short-period comets have such a short life compared with the age of the Solar System indicates that there must be a reservoir of comets somewhere that constantly replenishes their number.

Observations were made of Halley from other spacecraft although from much greater distances. Vega I and Vega II, launched by the USSR, had previously surveyed the planet Venus and then were

Figure 9.6. *A view of Halley from Giotto. The dark outline of the nucleus can be seen together with the emission of material at the left edge (ESA).*

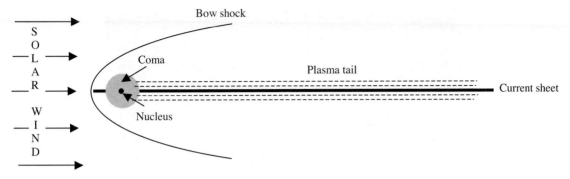

Figure 9.7. *The magnetic and charged-particle environment around a comet due to the solar wind.*

redirected towards Halley. They passed within 10 000 km and provided the first pictures of the nucleus. There were also two Japanese spacecraft, Sakigake and Suisei, which made measurements of the comet's environment at even greater distances. Sakigake made observations of the tail, at a distance of about 7×10^6 km, and found that there were considerable variations in the flow of gas and dust which depended on the orientation of active regions of the nucleus with respect to the Sun. Suisei, which went to within 1.5×10^5 km of the nucleus, recorded the bow shock due to the interaction of the stream of particles from the comet with the solar magnetic field. The data from the five probes has given the best information yet of the structure of a comet and its immediate environment. Giotto actually had more measurements to make after its Halley fly-past. The rather turbulent environment, close to the nucleus, through which Giotto passed damaged its cameras, which could no longer be used, but the other instrumentation was still functioning. The spacecraft was diverted after the Halley interaction to rendezvous with Giacobini–Zinner (table 9.1) and important charged particle and magnetic field measurements were made which complemented those from Halley. The local magnetic and charged-particle environment of a comet somewhat resembles that of the Earth and other planets (Topic V). The magnetic field is concentrated locally by the stream of charged particles forming the plasma tail. The bow shock is where the solar wind particles meet this concentrated field. On the side of the comet away from the Sun the arrangement of the magnetic field traps a thin layer of charged particles forming a current sheet, another feature of the terrestrial environment (figure 9.7).

Future plans for the exploration of comets include a mission to Wirtanen (table 9.1) in which there will be an orbiter and at least one lander that will have provision for taking samples from below the surface of the nucleus. If this mission is successful then it should provide answers to many unanswered questions about the structure of comets. However, comets would still pose interesting problems for the planetary scientist, as we shall see in the following section.

9.3. THE OORT CLOUD

In section 9.1 the existence of new comets was mentioned and we shall now explore the significance of these in more detail. Before we do so we shall consider how a comet's orbit may be described in terms of its intrinsic energy. Since intrinsic energy is proportional to the inverse of the semi-major axis, cometary scientists find it convenient to express it in units of $1/a$ or in AU^{-1}. Thus a comet with intrinsic energy $0.001 \, \mathrm{AU}^{-1}$ has a semi-major axis of 1000 AU and, since it must have a small perihelion if it has been observed, then its aphelion distance will be approximately $2a$, or 2000 AU. In figure 9.8 there is reproduced a histogram of values of $1/a$ given by Marsden, Sekanina and

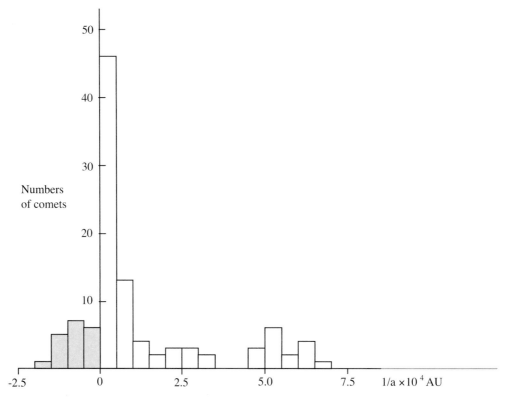

Figure 9.8. *Numbers of comets with values of 1/a of small magnitude. Positive values correspond to bound (elliptical) orbits. The small number indicated as having hyperbolic orbits (negative values of 1/a) are probably also in elliptical orbits (after Marsden, Sekanina and Everhart, 1978).*

Everhart (1978) for the pre-planetary-perturbed orbits of new comets which had been estimated up to that time. It will be seen that there is a concentration with very small negative energies corresponding to elliptical orbits with very large aphelia. There are also some comets with positive energies but, because of reasons given in section 9.1 it is certain that most of these, if not all, actually have elliptical, but near-parabolic orbits.

The concentration of orbits with $0 < 1/a < 5 \times 10^{-5}$ AU^{-1} had previously been noted by Jan Oort in 1948. The significance of the term *new comets* arises from their very small and negative intrinsic energies. Such a comet moving in the region of the giant planets gains or loses intrinsic total energy by planetary perturbations and the expected magnitude of the gain or loss, $\sim 10^{-3}$ AU^{-1}, is much greater than the initial magnitude of E, $\sim 10^{-5}$ AU^{-1}, when it is approaching. If it gains energy then the new E will almost certainly be positive so that the comet will leave the Solar System on a hyperbolic orbit and will never be seen again. On the other hand, if it loses energy then it will be more negative and it will move to an orbit with a much smaller value of a so that it moves out of the class of *new* comets.

Oort concluded that there exists a cloud of about 10^{11} comets in highly eccentric elliptical orbits, mostly at distances of several times 10^4 AU. The estimate of the total number of comets was based on the assumption that, at very large distances from the Sun, the comets would have been perturbed by stars to such an extent that all directions of motion and all speeds up to the escape speed from the Solar System are equally likely. Oort then calculated the proportion of comets with such motions that would pass within 3 AU of the Sun and so be seen in a year. From the observed frequency of new comets, somewhat less than one per year, he was thus able to make his estimate of the total number of comets in what we now call the *Oort cloud*. While Oort's assumptions that gave the number may not be completely certain, the existence of a cloud with a substantial number of comets cannot be denied. The inclinations of new comets are more-or-less random with direct and retrograde orbits equally likely. Detailed analysis of the orbits shows a slight tendency for clustering where groups of comets coming from some directions are found to have similar orbital elements.

The existence of the Oort cloud raises several interesting problems, the first being 'how did it get there?'. It seems very unlikely that new comets could have come from a source outside the Solar System. A body approaching the Solar System from a great distance with finite velocity will make a hyperbolic orbit around the Sun and will never be a bound member of the system. While it may be possible to invent scenarios where planetary perturbations could give capture, each individual capture event would be so unlikely that it would not be feasible to capture a vast number of bodies in this way—as is required for the Oort cloud. For this reason most planetary scientists prefer to have a scenario where the comets arose as part of the general process that led to the formation of the Solar System. We shall return to this topic when theories of the origin of the Solar System are discussed in chapter 12.

Another interesting problem about the Oort cloud is 'how does it stay there?' since the individual comets are exposed to many perturbing influences. These are treated fully in Topic AF but here we just indicate the general nature of the sources of perturbation.

Perturbations by stars

The farthest members of the Oort cloud stretch out almost half way to the nearest stars and during the course of its lifetime much closer passages of stars to the Solar System must have occurred. Closely passing stars will perturb cometary orbits significantly. The perturbation of the comet comes from the difference between the gravitational force exerted by the star on the Sun and that which it exerts on the comet.

Perturbation by Giant Molecular Clouds

Another possible type of perturber is the Giant Molecular Cloud (GMC). A typical GMC has a mass of $10^5 M_\odot$ and a diameter of order 20 pc (1 pc $\approx$ 206 000 AU). They are rather clumpy in structure and contain concentrations of mass distributed throughout their volume. Clube and Napier (1984) have suggested that during its lifetime the Solar System would have passed through four or five GMCs and that during these passages near interactions with massive clumps would have completely swept away the Oort cloud. They then propose that GMCs are themselves a source of new comet material and at the same time as the pre-exiting comets are removed so replacement comets are provided by the GMC. This seems an unlikely model; more convincing is a suggestion by Bailey (1983) that there exists an inner reservoir of comets, at distances of a few thousand AU, that are perturbed outwards to replenish the Oort cloud when the more loosely bound comets are removed.

Perturbation by the galactic tidal field

The Solar System lies within the disk of the Milky Way galaxy. To a first approximation the disk can be regarded as a slab of material of uniform density. A body in the mean plane of the disk experiences no force because it is pulled equally in two opposite directions normal to the disk. However, a body moving away from the mean plane, but still within the disk, experiences a force towards the mean plane proportional to its distance from that plane. Thus, in addition to their orbital motion about the centre of the galaxy, the Sun and other disk stars oscillate with approximate simple harmonic motion about the mean plane.

If the vector connecting the Sun to the comet has a component perpendicular to the disk then the forces on the Sun and the comet will be different. It is this differential force that constitutes the perturbation.

Perturbers within the Oort cloud

It has been suggested that there may be planetary-mass objects within the Oort cloud that perturb individual comets from time to time. Such perturbations would involve fairly close interactions and there would be a tendency for the aphelion of the perturbed comet to be at the distance of the perturbing object, as is the case for the Jupiter family. This would explain the slight tendency for the clustering of observed cometary orbits where the members of the cluster have similar orbital parameters.

9.4.   THE KUIPER BELT

In 1951 Gerard Kuiper suggested that outside the orbit of Neptune there exists a region in which comet-style bodies orbited close to the mean plane of the Solar System. His main argument was that it seemed unlikely that the material in the Solar System abruptly ends beyond the orbits of Neptune and Pluto. He proposed that in the outer reaches of the Solar System there could be material in the form of small bodies that is too widely separated ever to have aggregated into a larger body.

Kuiper's idea did not receive much attention for the next 20 years but then received reinforcement from another direction. In section 9.1 a mechanism was suggested for the origin of the Jupiter family of short-period comets through the perturbation of long-period comets by Jupiter. It is certainly true that such a mechanism *can* occur but what was being challenged in the 1970s was whether it was an efficient enough process to explain the large number of observed short-period comets. With approximately 100 short-period comets, each with a lifetime of, say, 10 000 years, one short-period comet per 100 years would need to be formed to maintain the population. With the availability of suitable computing power, calculations showed first that the mechanism of perturbing long-period comets by Jupiter was very inefficient. Second, they showed that, despite the qualitative arguments given in section 9.1, such short-term comets that were produced would have almost random inclinations rather than being close to the ecliptic as is actually observed. From these calculations it was suggested that the source of short-period comets contained bodies moving close to the ecliptic and, since Oort cloud inclinations are randomized by stellar perturbations, such bodies had to be from a region much closer to the Sun. This renewed interest in Kuiper's idea and a search began for bodies beyond Neptune in what became known as the *Kuiper belt*.

In 1992 the first such body was found, 1992 QB_1, with an estimated diameter of 200 km at about 40 AU from the Sun. Since that time there has been a steady stream of discoveries and there are now several hundred known *Kuiper belt objects* (KBOs); there were 125 discovered in 1999 alone. All the bodies have orbits close to the ecliptic and have semi-major axes larger than that of Neptune, although the eccentricities of some bring them within Neptune's orbit. In figure 9.9 the values of a and e for 32 of these objects, plus Pluto, are shown; it is clear that in this subset of KBOs there is a family of 12 KBOs which share Pluto's 3 : 2 orbital resonance with Neptune (section 3.3).

It has been suggested that the Kuiper belt could contain more than 70 000 bodies with a diameter greater than 100 km, and this would suggest a very much larger population of smaller bodies to form potential short-period comets of the usual size. Perturbations in the Kuiper-belt region are sufficiently

Eccentricity

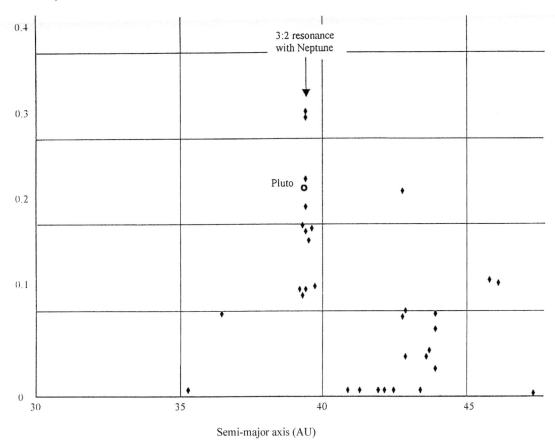

Semi-major axis (AU)

Figure 9.9. *The a and e values for the first 32 discovered Kuiper-belt objects.*

small to enable most objects there to have survived for the lifetime of the Solar System but the effect of Neptune is sufficient to cause a small inwards transfer of bodies. Once they have penetrated the inner Solar System they can be perturbed either to become short-period comets or to be thrown outwards to regions well beyond the Kuiper belt.

Problems 9

9.1   A comet is observed with perihelion 2.5 AU and aphelion 10.5 AU. What is its Tisserand constant with respect to (i) Jupiter and (ii) Saturn? If it had been an Oort cloud comet perturbed by one or other of these planets, then what would the original perihelion have been in each case?

9.2   A comet from the Oort cloud, with perihelion 6 AU, is perturbed by Saturn into an orbit with perihelion 4 AU. What is the semi-major axis and eccentricity of the new orbit?

CHAPTER 10

METEORITES

10.1. INTRODUCTION

It has been known for thousands of years that solid objects fall from the skies and early falls were recorded in China in 644 BC, in Greece in 466 BC and by the Romans. Meteoritic iron objects have been found in Egyptian pyramids and the hieroglyphic symbol used to describe them translates as *heavenly iron*. They were sometimes regarded as of religious significance and the holy black stone enshrined in the Caaba of the Islamic faith in Mecca is possibly of meteoritic origin. To the Eskimo people of the Arctic, meteoritic iron was an important material and was used in the making of tools and hunting weapons although it is somewhat soft for such a purpose.

The oldest collected meteorite is one that fell outside the town of Ensisheim in Alsace on 7 November 1492 and is kept in the local town hall. A woodcut illustrating the event describes the falling object as a 'donnerstein' or 'thunderstone' and that is how they were commonly described. The noises accompanying a fall at Hatford, near Oxford, on 9 April 1628 were described as like thunder or the noise of cannon. The fall consisted of a shower of objects; one recovered near Hatford was described as 'In colour outwardly blackish, somewhat like iron: crusted over with that blacknesse about the thicknesse of a shilling.' During a period of 14 years, beginning in 1789, nine well-recorded falls occurred, including one at Wold Cottage in Yorkshire which caused a considerable stir in the British scientific community. Even with all this observational evidence there was some reluctance to accept that these falls were truly of objects from space. Several alternative theories were put forward including that they were terrestrial stones thrown up by lightning strikes and then falling back to Earth or perhaps accumulations of volcanic dust, somehow concentrated in the atmosphere.

In addition to the witnessed falls there were many finds of objects which clearly were out of place where they were found—very large lumps of iron in South America, South Africa and Siberia, for example. In view of the size and solidity of these objects the alternative theories of meteorite production seemed increasingly far-fetched and by the very early years of the nineteenth century all doubts seem to have been resolved and the fact that meteorites were indeed objects from space became firmly established.

The amount of meteoritic material striking the Earth amounts to between 100 and 1000 tonnes per day. Although this may seem a large amount it would account for only one part in 10^7 of the Earth's mass over the lifetime of the Solar System. Although this would cover the Earth uniformly with a 20 cm thick layer of material, in view of constant tectonic activity it has left no visible trace. The material that falls covers the size range from fine dust to objects of kilometre size. Bodies must

Figure 10.1. *A 35 tonne iron meteorite from Cape York (Greenland) on display at the American Museum of Natural History, New York (McCall, 1973).*

enter the atmosphere at more than $11 \, \text{km s}^{-1}$ (the escape speed from the Earth). Larger objects are decelerated by the atmosphere but still strike the ground at high speed. Their passage through the atmosphere heats them, causing surface material to melt and a fusion crust to form. They are also subjected to very large forces by atmospheric resistance which may cause them to fragment and form a shower of smaller objects. Rather curiously very tiny objects may survive the passage to Earth almost intact. Because of their large surface-to-volume ratio they radiate heat very efficiently and they are quickly braked by the atmosphere and then gently drift down to Earth.

The largest meteorites known have masses up to about 70 tonnes, most of them being of iron (figure 10.1). The largest stone meteorite found is one that fell in Jilin, China, in 1976 and has a mass of 1.77 tonnes. There is evidence that much larger objects must have fallen on Earth from time to time. Observation of surface features of the Moon indicate that early in the history of the Solar System objects fell on it that were able to form the major mare basins, and these must have been asteroids from tens to perhaps more than 100 km in diameter. While both the frequency and size of colliding bodies seem to reduce with time there is a continuous bombardment pattern up to the present. There is no reason to suppose that what happened on the Moon did not also happen on Earth but features produced on Earth in the 10^9 or so years after its formation would have long been expunged by tectonic activity. There is some evidence that a very large object may have fallen to Earth at the end of the Cretaceous period, some 65 million years ago. Marine clays deposited at that time have a high iridium content and iridium is a much more common element in meteorites than it is on Earth.

More direct evidence for the fall of larger bodies can be seen in the craters that exist in various parts of the Earth. The largest of these is the Barringer crater in Arizona (figure 10.2) which is more than 1200 m in diameter and 170 m deep. Small amounts of meteoritic iron have been found in the vicinity and it is estimated that the crater was formed by the fall of an iron meteorite of mass approximately 50 000 tonnes—that is with a diameter about 25 m if it was a sphere. The time at which the crater formed has been estimated as 35 ± 15 thousand years ago. Attempts to discover the main body of the meteorite have been unsuccessful and it has been concluded that the material would have become widely dispersed by the explosion which produced the crater.

Figure 10.2. *The Barringer crater.*

In 1908 there was a huge explosion in the Tunguska River region of central Siberia. The noise was heard more than 1000 km away and some observers saw a fireball, brighter than the Sun, crossing the sky. The event was recorded on seismometers all over the world. However, no attempt was made to locate the source of the explosion until 1921 and it was not until 1927 that an expedition discovered a region of about 2000 km² of uprooted trees with their directions of fall radiating away from the centre of the region. However, no crater was found nor were any fragments discovered that could be identified as of meteoritic origin. In the 1950s fine fragments of meteoritic dust were found embedded in local soils and the current belief is that the event was caused by the impact of a small comet with the explosion centre 10 km above the Earth's surface.

Figure 10.3. *Mr Pettifor examining the stony meteorite that landed in his garden (Natural History Museum, London).*

Table 10.1. *Numbers of falls and finds up to the end of 1975.*

	Falls	Finds	Totals
Stones	791	593	1384
Irons	46	610	656
Stony-irons	11	67	78

As already indicated, meteorites which are recovered may be classed as either *falls* or *finds*. The former category indicates those where the object is seen to fall and is recovered shortly afterwards. A notable example of such a fall is that of a fist-sized stony meteorite which struck a tree in the garden of Mr Arthur Pettifor near Cambridge in May 1991 and finished up close to his feet (figure 10.3). This small meteorite is now in the British Museum.

Meteorites fall into three major general types—stones, irons and stony-irons—and table 10.1 shows the proportions of falls and finds for the different types. It will be seen that the proportions of the different kinds of meteorites differ for falls and finds, a notable feature being the much larger proportion of irons in the finds. We may take the proportion of falls as representing the relative numbers of the different types of meteorites that fall to Earth. This is just saying that the chance of spotting a falling meteorite of a particular type is simply proportional to the number of that type that fall on Earth. On the other hand a 'find' depends on recognizing that the object, which may have fallen tens or hundreds of thousands of years previously, is actually a meteorite. A stone meteorite which lands in Europe, say, will be exposed to weathering which will erode its surface and soon make it indistinguishable from rocks of terrestrial origin. If Mr Pettifor's meteorite had landed in the middle of a stony field it would not have looked too unusual to the untrained eye even before weathering processes. On the other hand we are not accustomed to finding large lumps of iron on Earth and if we found a very dense object of blackish appearance then we might suspect that it was indeed a meteorite. Again an iron object would not weather in the same way as one of stone and would maintain its integrity for a much longer period.

Factors which assist in the recognition of stony meteorites are, first, that it should as far as possible maintain its original appearance and, second, that it should stand out in its environment. One type of environment where these conditions prevail is in deserts where, because they are arid, weathering processes are greatly diminished and where large rocks may stand out in places where rocks are rare. In 1969 Japanese meteoriticists began a series of expeditions to Antarctica that have since yielded thousands of meteorites (figure 10.4); these are not included in table 10.1. Meteorites

Figure 10.4. *A meteorite just underneath Antarctic ice (Smithsonian Astrophysical Laboratory).*

landing in Antarctica are more-or-less preserved in cold storage and move around in great sheets of ice. Sometimes they arrive in regions where the ice is eroded by winds and they may end up either on or close to the surface. Since the Earth's rocky surface in these regions is buried a great distance, 3 km or so below the surface of the ice, the meteorites are readily recognized for what they are.

Meteorites are an abundant source of extra-terrestrial material and the ways in which they are similar to terrestrial materials and the ways in which they differ have much to tell us about the origin of the Earth and of the Solar System.

10.2. STONY METEORITES

Stony meteorites are of two general types, *chondrites* and *achondrites*, which differ chemically from each other. Most, but not all, chondrites contain *chondrules* from which they get their name. These are small glassy millimetre-size spheroids which are embedded in the fine-grain matrix which constitutes the main body of the meteorite. A section of a typical chondrite, within which is a clear chondrule, is shown in figure 10.5. The other type of stony meteorite, achondrites, contain no chondrules and virtually no metal or metallic sulphides. In some ways they are similar to terrestrial and lunar surface rocks. These various types of stony meteorite will now be described in greater detail.

10.2.1. *The systematics of chondritic meteorites*

Before embarking on a description of chondritic meteorites it is necessary to mention some of the important minerals which they contain because these minerals are important in defining the various sub-types of chondrites that occur. A fuller description of most of these minerals is given in Topic A. The most abundant mineral is olivine which accounts for ~45% by mass of the commonest type of chondrite. The second most abundant mineral is pyroxene, about 25% by mass of the commonest chondrites, and this too is a magnesium–iron silicate. Another fairly common mineral, at the 10% level, is plagioclase, sometimes referred to as plagioclase feldspar. Native metal in the form of iron with some nickel is also present together with the mineral troilite, FeS. The metal occurs in the form of two iron–nickel compounds—*taenite* which is nickel-rich with 13% nickel and *kamacite*, a nickel-poor compound with 5–6.5% nickel. Where nickel occurs in iron at levels between 6.5 and 13% then plates of kamacite form, surrounded by a taenite matrix.

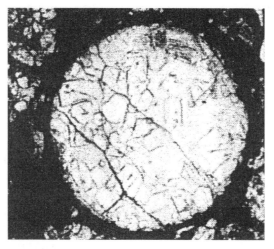

Figure 10.5. *A well-formed chondrule (about 1 mm in diameter).*

Table 10.2. *The table gives the percentage by mass of total iron and of iron occurring in the form of metal for the three divisions of the ordinary chondrites.*

	H	L	LL
Total iron (%)	27	23	20
Metallic iron (%)	12–20	5–10	2

Within the chondrite classification there are three designated sub-types, *ordinary*, *enstatite* and *carbonaceous*. The ordinary chondrites are so named because they are the commonest type and these are further subdivided into three further divisions—H (high iron), L (low iron) and LL (low iron–low metal)—distinguished by their content of iron and the form in which it occurs, as summarized in table 10.2.

The different total and metallic iron contents are also linked to differences in oxygen content where the amount of oxygen present increases from H to LL. About 30% of the non-metallic iron in the H group occurs as the sulphide, troilite, whereas only 15% of non-metallic iron in the LL group is troilite.

Within each ordinary chondrite group based on iron composition a further distinction can be made on the basis of texture and mineral content—the so-called petrological classification. This also applies to the enstatite and carbonaceous chondrites so table 10.3, which describes the petrological classification refers to these types of chondrite as well.

The petrographic type 3 ordinary chondrites show signs of having been very rapidly cooled from about 1700 K to 1100 K or less. Pyroxene crystals formed very quickly and in a poorly ordered form while olivine occurs in a form which is also indicative of very rapid cooling. Its iron content varies from 0 to 40% and the assemblages of minerals are non-equilibrated, which means they were quenched and atoms were trapped and made immobile before they could move to form mineral assemblages closer to thermodynamic equilibrium. Another sign of rapid cooling is that the chondrules are clear and glassy; with slower cooling, crystals would have formed in the chondrules making them more opaque. It is evident that the quick quenching to temperatures of 1000 K or less was never followed by heating to above that temperature.

Table 10.3. *Chondrite classification according to appearance and mineralogy.*

	Type 1	Type 2	Type 3	Type 4	Type 5	Type 6	Type 7
Chondrules	Absent	Sparse	Many and distinct		Visible	Indistinct	Absent
Chondrule glass	—	Clear and isotropic		Opaque	No glass present		
Matrix	Fine and opaque	Opaque		Transparent micro-crystals	Granular	Granular coarse-grained	
Silicate uniformity	—	>5% variation		Variation 0–5%	No variation		
Carbon (% mass)	3–5	0.8–2.6	0.2–1.0	<0.2	<0.2	<0.2	<0.2
Water (% mass)	18–22	2–16	0.3–3.0	<1.5	<1.5	<1.5	<1.5
			Carbonaceous chondrites (Mg/Si = 1.05)			—	—
	—	—	Ordinary chondrites (Mg/Si = 0.95)				
	—	—	—	Enstatite chondrites (Mg/Si < 0.85)			

The petrographic type 4 ordinary chondrites either cooled more slowly or were re-heated, so enabling crystals to grow within the chondrules and making them more opaque. Olivine has become fairly uniform in composition but pyroxene crystals still show signs of cooling too quickly to give crystals either well-formed or uniform in composition. By the same considerations petrographic types 5 and 6 ordinary chondrites show signs of very slow cooling or reheating to 1100 K and 1200 K respectively, while type 7 seem to have been heated over a long period to 1500 K causing chondrules to melt and disappear and metal also to melt and drain away.

We now consider enstatite chondrites of which there are only about 20 examples available. These meteorites are characterized by their very low content of oxygen, so much so that none of the iron present is combined with oxygen but either appears as metal or as sulphide. Another feature is a very low magnesium-to-silicon ratio (<0.85) which means that there is no olivine present since in a pure magnesium olivine there would be twice as many magnesium atoms as there are those of silicon. The common mineral in these meteorites, some 65% by mass, is the pure magnesium pyroxene, enstatite, from which they get their name. Plagioclase is also present together with many sulphides—even sulphides of sodium and potassium which occur extremely rarely. It is difficult to imagine how enstatite chondrites originated. They seem to require an oxygen-poor but metal-rich source and it has been postulated, but without much conviction, that this could be from some region close to the Sun—perhaps on Mercury or somewhere near that planet.

Finally we consider carbonaceous chondrites which probably provide the richest source of information from meteorites. They alone give us the petrographic groups 1 and 2 and from table 10.3 it will be seen that they are rich sources of carbon and water. The diagnostics which distinguish carbonaceous chondrites are that they are rich in oxygen and also have a high magnesium-to-silicon ratio—approximately 1.05:1. With one exception they contain very little free metal and they are dark in colour, almost black. An interesting characteristic is that they consist of mixtures of minerals that formed at very different temperatures. Some of the material contained in these meteorites consists of very high-temperature condensates while at the same time they contain a high inventory of volatile materials.

Carbonaceous chondrites fall into four groups, each designated by the name of a representative member. One group, with five members, is called CI (C = carbonaceous; I = Ivuna, the name of the representative member). They are all of petrographic type 1, containing no chondrules but are carbon rich with an abundance of minerals such as serpentine, $Mg_3Si_2O_5(OH)_4$, magnetite, Fe_3O_4, and epsomite, $MgSO_4$, which are either highly hydrated or, in the case of epsomite, actually dissolve in water. The meteorites contain about 20% of water in a bound form and, if very volatile materials are excluded, then their composition matches closely that observed for the Sun; in table 10.4 the chemical composition of the Sun is shown together with the compositions of typical stony meteorites.

The second group of carbonaceous chondrites, the largest with 14 members, is CM2 (M = Mighei with petrographic group 2). These also contain serpentine, with somewhat less magnetite, epsomite and water (10%) than the CIs. They contain chondrules, small grains of olivine and some small regions containing high-temperature minerals. Another group of carbonaceous chondrites are the four members of CV2 (V = Vigarano) and another four designated CV3. These contain very little carbon, which seems a little odd for carbonaceous chondrites but they are nevertheless quite dark (figure 10.6). They do contain chondrules and the high Mg:Si ratio which is what really characterizes carbonaceous chondrites. They also contain white inclusions of high-temperature minerals rich in calcium, aluminium and titanium (figure 10.6). The fourth group is CO3 (O = Ornans) which like the CVs are carbon-poor and contain high-temperature inclusions. They are mainly characterized by an abundance of small (0.2 mm) closely-packed chondrules. There is also one CO4 specimen.

Table 10.4. *The percentage composition by mass of the silicate and metal components of typical stony meteorites. The sums do not total 100% as oxygen and some other elements are not included. The relative proportions of elements for the sun are normalized to Si = 20.*

	Silicate component							Metal component			
	Si	Mg	Fe	Al	Ca	Na	H_2O	Fe	Ni	FeS	C
Ordinary chondrites	19	14	9	1.4	1.3	0.70	0.3	11.7	1.3	5.9	—
CI	11	9	18	0.9	0.9	0.56	20.5	0.11	0.02	16.7	3.8
CM	13	11	21	1.1	1.2	0.40	13.2	—	0.16	8.6	2.4
CV	17	14	20	1.3	1.8	0.40	1.0	2.3	1.1	6.1	0.5
CO	17	15	22	1.4	1.4	0.41	0.7	1.9	1.1	5.7	0.3
Sun	20	21	16	1.7	0.7	1.21	—	—	1.0	—	235
Enstatite chondrites	19	13	1	1.0	1.4	0.74	0.6	19.8	1.7	10.7	0.3
Eucrites	23	5	12	4.8	4.9	0.28	0.6	1.2	—	0.6	—
Howardites	24	7	12	4.7	4.8	0.25	0.3	0.4	0.1	0.6	—
Diogenites	26	16	12	0.8	1.0	0.03	0.1	0.8	0.03	1.1	—
Ureilites	19	21	10	0.2	0.6	0.13	1.1	8.1	0.15	—	0.7
Aubrites	27	22	9	0.3	0.7	0.09	1.1	2.3	0.2	1.3	—
Pallasite	8	12	5	0.2	0.2	0.05	0.2	49.0	4.7	0.5	—
Mesosiderite	10	4	4	2.2	2.1	0.13	0.7	46.0	4.4	2.8	—

There is a single carbonaceous chondrite that fell in Australia in 1972 which falls into the petrographic type 5. The chondrules are not very distinct and the olivine is of uniform composition, both factors indicating either slow cooling or some reheating after formation.

10.2.2. Achondrites

This type of meteorite contains no chondrules—hence their name—and are mostly silicate-rich igneous rocks although some of them were formed as 'soils' by the mixing of rocky fragments. They

Figure 10.6. *A CV3 carbonaceous chondrite. White high-temperature inclusions can be seen against the dark matrix material.*

contain very little native metal or sulphides. There are different kinds of achondrite and several different ways of classifying them have been suggested. For our present purpose we shall initially restrict ourselves to five groups of achondrites, *eucrites, howardites, diogenites, ureilites* and *aubrites,* and then later deal with a few achondrites that fall outside this grouping. The chemical compositions of the five groups, given separately for their silicate and metal components, are given in table 10.4.

We start with a description of diogenites, which have been described as being similar to the material of a planetary mantle. The mantle of the Earth is mostly olivine with some pyroxene but the diogenites are dominantly pyroxene, about 25% of which is the Fe-end member of that class of mineral. They have mostly been crushed and then reassembled as *breccias,* which means rocks formed by assemblages of small rocky fragments. All the fragments forming the diogenites are of the same kind of rock and such assemblages are called *monomict.*

The eucrites, most of which are a more-or-less equal mixture of plagioclase and a calcium-containing pyroxene, have been likened to lunar lava or terrestrial basalt produced by volcanism. When mantle material is melted calcium–aluminium–sodium rich plagioclase, which is of lower density, will tend to accumulate at the top of the solidifying melt. It also has a low melting point, which means that other minerals would solidify earlier and sink to the bottom of the cooling melt. Most eucrites are monomict although there are some that appear to be in an original uncrushed state.

Howardites, which consist of aggregated fragments of different kinds of rock (*polymict*) in the main contain rocks similar in nature to eucrites and diogenites. Their origin seems to be that of a 'soil' formed on some solid body of size sufficient to produce an appreciable gravitational field so as to give the required degree of compression (figure 10.7). They show radiation damage due to proton bombardment from the solar wind, similar to that found in lunar soils. It has been inferred that, on

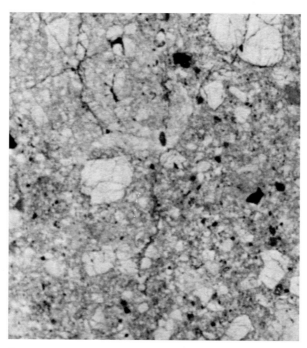

Figure 10.7. *Section through a howardite achondrite, consisting of compressed fragments of different kinds of rock (Dodd, 1982).*

whatever body they originated, the degree of radiation damage suggests that it must have been somewhere in the inner part of the Solar System.

Aubrites consist mainly of enstatite, the magnesium end-member of the pyroxene series, and might be thought of as related to enstatite chondrites except for the much smaller component of metal and sulphur. Almost all the aubrites are brecciated.

The ureilites are mostly interesting for their carbon content (up to 1%), much of which is in the form of micro-diamonds. The assumption is that the high pressures and temperatures required to produce diamonds came from a collision in space between the ureilite parent body and some other object.

There are ways of ascribing ages to meteorites that are dealt with in Topic B. By an age in this context we mean the time from which the meteorite became a closed system trapping all the materials contained within it. Thus, if radioactivity occurred in the meteorite then both the parent and daughter products would be retained and deductions can then be made about the time from which the system became closed. Most meteorites have ages which are around 4.5×10^9 years which is the accepted age of the Solar System. However there are some unusual achondrites with very different ages and we shall now consider these.

Some of the eucrites are cumulates (section A.3.1). They show a very coarse texture and a preferred orientation of crystallites that indicate that they formed by crystals falling to the bottom of a molten magma in a gravitational field. There are several achondrites, not belonging to the five classes already mentioned, which are also cumulates and have unusual characteristics. Some of these fall into three groups called the *shergottites, nakhlites* and *chassignites*—named after the three prototype achondrites Shergotty, Nakhla and Chassigny—and together they are referred to as the SNC meteorites. It is the age of the SNC achondrites that provides a puzzle, for the crystallization ages of their materials, when they would have achieved closure, seem to be 6.5×10^8, 1.4×10^9 and 1.4×10^9 years respectively. The question arises of their source, or sources, which had to contain molten material within the last 1.4×10^9 years. This would seem to rule out asteroids as sources because even the largest asteroid would have completely solidified long before then, and even postulating continuous heating from radioactivity does not give a plausible scenario. The suggested source of these achondrites is the planet Mars, which could have been volcanically active recently enough to explain the ages. Analysis of some of the gases trapped in these meteorites supports this idea as the gases are rich in CO_2, and resemble in composition the Martian atmosphere as measured by spacecraft. In 1996 a group of scientists at NASA reported that they had discovered microfossils within an SNC meteorite with associated hydrocarbons and other possible products of bacteria, such as magnetite. Great excitement was engendered about a second possible source of life in the Solar System but subsequently considerable doubt arose about the biological origin of what was seen.

10.3. STONY IRONS

Stony irons are meteorites which consist of roughly equal proportions of stone and iron component. They nearly all fall into one of two groups, the *pallasites* and the *mesosiderites*. The pallasites are named after a find in the latter half of the eighteenth century by Pallas, a German natural historian who was making a survey of Siberia on behalf of Catherine the Great of Russia. The pallasites are in the form of olivine crystals set in a continuous metal framework (figure 10.8). Quite a considerable amount of troilite is also present. The probable formation mechanism for these meteorites is that the metal was forced under pressure into the a region where olivine crystals had formed and were cooling, contracting and cracking. A probable scenario for this is within a cooling

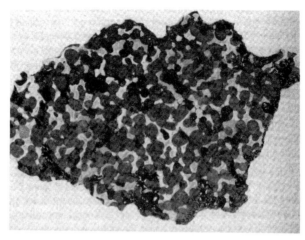

Figure 10.8. *A pallasite, showing light-coloured metal surrounding darker olivine regions (McCall, 1973).*

solid body in which there had been separation of dense metal and less dense stone in a gravitational field, with pallasites deriving from the interface region.

The appearance of mesosiderites is quite different (figure 10.9). The rock is in fragments, mainly of plagioclase and calcium-bearing pyroxene, together with smaller amount of olivine in spheroidal form. The minerals are somewhat similar to those in the howardites although differences in, for example, the Ca/Al ratios rule out a common origin. Mesosiderites contain minerals that are only stable at pressures below 3 kbar that suggests that they are not directly derived from deep within a large solid body. The metal is present both as globules and also as veins running though the meteorite.

Figure 10.9. *A mesosiderite. The highly fragmented rock intimately mixed with small metal regions suggests a highly chaotic environment when it formed (McCall, 1973).*

10.4. IRON METEORITES

Most iron meteorites consist of an iron–nickel mixture that formed as cumulates from an initially liquid state and are thought to originate in the cores of planetary bodies. However, there are some iron meteorites that look as though they have never been completely molten. The metal appears as mixtures of the two iron–nickel alloys, taenite and kamecite. In figure 10.10, the phase diagram for the formation of these two alloys, α represents the body-centred cubic kamacite and γ the face-centred cubic taenite (figure 10.11). For a meteorite with less than about 6.6% nickel the process of cooling can be followed along the line PT in figure 10.10. In the liquid phase, PQ, the material will be an intimate and structureless mixture of iron and nickel. When it becomes solid it quickly forms itself into taenite until it reaches a temperature below 1200 K, at point R, when kamacite begins to appear in the form of plates in equilibrium with the taenite. The mixture becomes richer in kamacite until, at point S, only kamacite is present. According to the phase-diagram, taenite should reappear at about 600 K but in practice this will not happen; the transformation from kamacite to taenite requires atoms to be able to diffuse through the solid, which becomes a progressively slower process as the metal cools. At a temperature of about 650 K whatever structure exists will be frozen-in and will not subsequently change.

There are about 50 iron meteorites with between 5 and 6.5% nickel which consist only of kamacite with some troilite. Surfaces of these meteorites examined under a microscope show a

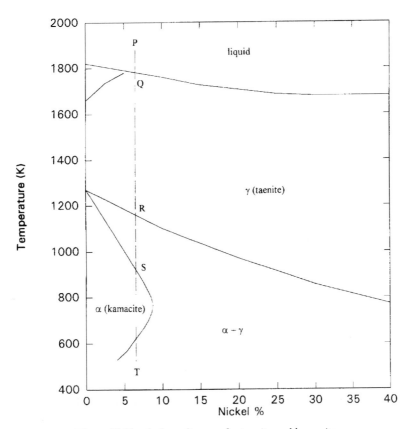

Figure 10.10. *A phase diagram for taenite and kamacite.*

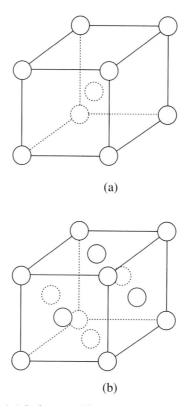

(a)

(b)

Figure 10.11. *(a) Body-centred kamacite. (b) Face-centred taenite.*

characteristic pattern due to the cubic structure of kamacite and so they are known as *hexahedrites* (a hexahedron is a regular solid with six faces, i.e. a cube). On the other hand for a very nickel-rich meteorite, say with 20% nickel, by the time a temperature was reached at which kamacite should appear the diffusion of atoms would be very slow, so that the material would be totally in the form of taenite. Such meteorites show no structure under the microscope and so are called *ataxites* (meaning 'without form' in Greek). If the nickel content is between 6.5 and 13% then the meteorite will contain a mixture of kamacite and taenite. This will be of the form of kamacite plates surrounded by taenite rims. Kamacite forms on the corners of the original cubic crystals of taenite so that the plates run parallel to the faces of an octahedron giving rise to an *octahedrite* structure (figure 10.12). At one time these textural terms—hexahedrite, ataxite and octahedrite—were the only way that iron meteorites were classified.

When polished and etched sections of octahedrites are examined they frequently show a characteristic *Witmanstätten pattern* (figure 10.13). These patterns contain information about the cooling rate of the iron; for a given nickel concentration the slower the cooling rate the longer would the kamacite plates have to form and the larger they would be. Cooling rates for different iron meteorites showing Witmanstätten patterns have been estimated as between 0.5 and 7.5 K per million years. This has important implications for the size of the bodies within which the cooling took place. The lowest cooling rates are consistent with the interiors of asteroids some hundreds of kilometres in diameter or being close to the surface of even larger bodies. Very few asteroids of such large size now exist. Witmanstätten figures also occur in the iron components of mesosiderites and for these even lower cooling rates of 0.1 K per million years are estimated which would imply a

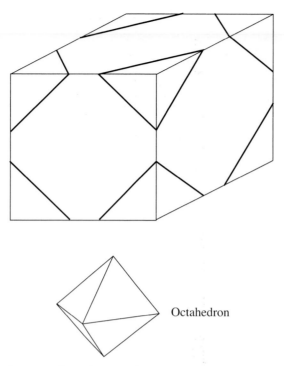

Octahedron

Figure 10.12. *Kamacite formation (heavy lines) at the corners of a taenite crystal give plates parallel to the faces of an octahedron.*

Figure 10.13. *Witmanstätten pattern on an etched section of an iron meteorite showing kamacite plates surrounded by darker taenite (McCall, 1973).*

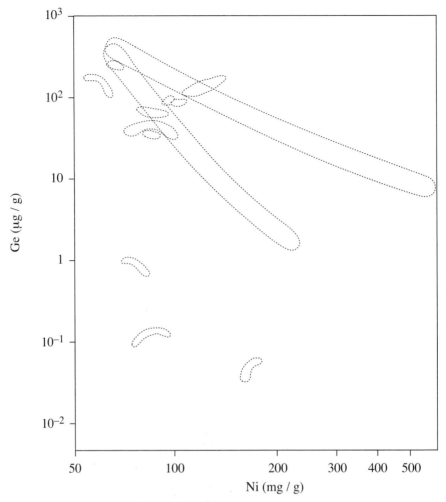

Figure 10.14. *The correlations between nickel and germanium concentrations in iron meteorites. The distinctive regions suggest 10 or more sources.*

parent body some 2000 km in diameter, which is twice that of Ceres. These conclusions pose some very difficult problems for those wishing to deduce how the Solar System evolved.

With new techniques available it has become possible to classify iron meteorites according to the concentrations of trace elements they contain relative to the amount of nickel. The diagram, figure 10.14, for the trace element germanium shows that the Ni–Ge concentrations are highly correlated and that observations fall into distinct connected regions. Interpretation of such diagrams for many trace elements taken together suggest that there are up to 16 distinct types of iron meteorite. Although the exact number is not universally accepted, what is clear is that there must have been at least 10 different sources to explain the variety of trace-element compositions. Taken in conjunction with the cooling-rate estimates, it now seems that something like 10 very large asteroids would need to have been broken up to provide the range of observed characteristics of iron meteorites. Since there are so few such bodies in existence at present, this is difficult to accept and some other explanation may well be necessary.

10.5. THE AGES OF METEORITES

Techniques for finding the ages of rocks from when they became closed systems are described in some detail in Topic B. Most estimated ages of chondritic meteorites fall in the range 4.3 to 4.5×10^9 years. Allowing for some period of time during which the material was settling down into closed systems, the top end of the range is the best estimate of the age of the Solar System—assuming that this was synchronous with the formation of meteorites.

There are some younger ages measured for achondrites, some between 3.2 and 3.7×10^9 years, and these are taken to indicate that there was some re-heating process which reset the radioactive clocks of these meteorites. We have already referred to the very young ages of the SNC meteorites that have been interpreted as indicating a Martian origin.

10.6. ISOTOPIC ANOMALIES IN METEORITES

Most elements have more than one stable isotope and all elements are capable of having a mutiplicity of isotopes, many of which may be radioactive. Thus oxygen has three stable isotopes, ^{16}O, ^{17}O and ^{18}O, and on Earth these occur in the proportion $0.9527 : 0.0071 : 0.0401$, a composition referred to as SMOW (Standard Mean Ocean Water). The isotopic compositions of some meteorites for some elements differ from the terrestrial standards; the differences, which can be very precisely measured, are usually very small and some convenient way of describing the variation is needed. This can be done by the δ notation that is described here in relation to oxygen. Let us assume that concentrations of ^{16}O and ^{17}O have been measured in a meteorite sample as $n(^{16}O)$ and $n(^{17}O)$. Then we write

$$\delta^{17}O(‰) = \frac{\{n(^{17}O)/n(^{16}O)\}_{\text{sample}} - \{n(^{17}O)/n(^{16}O)\}_{\text{SMOW}}}{\{n(^{17}O)/n(^{16}O)\}_{\text{SMOW}}} \times 1000 \qquad (10.1)$$

where the symbol ‰ indicates 'per mille' or 'parts per thousand'.

When terrestrial samples are analysed for their oxygen isotopic composition, they are found to vary, but in a very systematic way. A plot of $\delta^{17}O$ against $\delta^{18}O$ for different samples gives a straight line with slope 0.5 and this can be explained by mass-dependent fractionation. Various kinds of physical or chemical process, e.g. diffusion in a temperature gradient or the rate of a chemical reaction, may be linearly dependent on mass. For this reason the departure of the behaviour of ^{17}O from ^{16}O in a particular situation may be half as great as the departure of the behaviour of ^{18}O from ^{16}O. The factor of 0.5 then appears as the slope of the full line in figure 10.15, which is known as a *three-isotope plot*.

10.6.1. Oxygen in meteorites

Three-isotope plots for oxygen from the Earth and the Moon all lie on the expected fractionation line and samples of eucrites and some other achondrites also seem to give lines of similar slope but slightly displaced. In 1973 R N Clayton *et al.* found that samples taken from carbonaceous chondrite anhydrous materials gave an oxygen three-isotope plot with a slope of approximately unity; later work has refined the slope to 0.94 ± 0.01. This is shown in figure 10.15 together with the terrestrial line. Later it was found that samples from some ordinary chondrites gave a slope close to 1.0 although, as seen in figure 10.15, the line they give is well displaced from the carbonaceous chondrite line. A slope of unity can be explained as the result of mixing some standard pool of oxygen, which could be SMOW but could be something else, with various amounts of pure ^{16}O. If

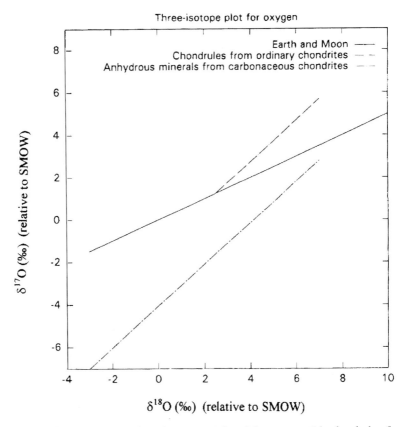

Figure 10.15. *Oxygen three-isotope plots for terrestrial and lunar material, chondrules from ordinary chondrites and anhydrous minerals from carbonaceous chondrites.*

the proportion of pure ^{16}O added changed the amount of ^{16}O in the mixture by a factor $1 + \alpha$ then, from equation (10.1)

$$\delta^{17}O = \frac{\dfrac{n(^{17}O)}{n(^{16}O)}\dfrac{1}{1+\alpha} - \dfrac{n(^{17}O)}{n(^{16}O)}}{\dfrac{n(^{17}O)}{n(^{16}O)}} \times 1000 = -\frac{1000\alpha}{1+\alpha}, \tag{10.2}$$

and the same result would be found for $\delta^{18}O$. This would automatically give the required unit slope.

Various ideas have been put forward as to how the pure ^{16}O could have been produced and how it entered the Solar System. The usual assumption is that the ^{16}O was produced by nuclear reactions in stars whereby the common isotope of carbon, ^{12}C, reacts with an alpha particle (^4He). A suggested scenario is that grains highly enriched in ^{16}O entered the Solar System and that normal oxygen in the Solar System mixed with it by diffusion processes in which most of the original oxygen left the grain and was replaced by normal oxygen.

10.6.2. Magnesium in meteorites

When the oxygen isotopic anomaly was discovered it was clearly of interest to see whether there were any similar effects for elements such as silicon and magnesium that are linked to oxygen in meteorites.

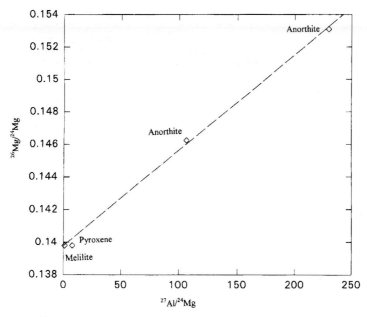

Figure 10.16. *Excess* ^{26}Mg *against total aluminium in CAI, high temperature inclusions in carbonaceous chondrites.*

For magnesium there are three stable isotopes, ^{24}Mg, ^{25}Mg and ^{26}Mg, which occur in the approximate ratios 0.790 : 0.100 : 0.110. For most samples the three-isotope plots of magnesium showed a slope of 0.5, characteristic of mass-dependent diffusion. However, in 1976 Lee, Papanastassiou and Wasserberg discovered that in some of the white high-temperature inclusions in carbonaceous chondrites (figure 10.6) there was an excess ^{26}Mg which was proportional to the amount of aluminium in the sample (figure 10.16). The interpretation of this result is that the ^{26}Mg was produced by the decay of radioactive ^{26}Al by

$$^{26}\text{Al} \rightarrow {}^{26}\text{Mg} + \beta^- + \nu, \tag{10.3}$$

in which a proton in the nucleus of ^{26}Al is transformed into a neutron and an electron, together with a neutrino. The assumption is that when the meteorite sample was formed the aluminium it contained was a mixture of ^{27}Al, the only stable isotope, with a small proportion of ^{26}Al. In all grains of any particular sample the proportion of ^{26}Al to ^{27}Al was the same so that the amount of ^{26}Mg produced, appearing as an excess in company with normal magnesium, would be proportional to the total aluminium content no matter what the mineral content of the grain.

The inferred proportions of ^{26}Al in the original aluminium of different meteorites vary from about 2×10^{-5} down to 10^{-8} or less. The onetime presence of ^{26}Al has some interesting possible implications for the formation and evolution of the Solar System. The half-life of ^{26}Al is 720 000 years and it may be assumed that, to observe the anomaly, the rocks within which it was included had formed closed systems within a few ^{26}Al half-lives from the radiosynthetic event which produced it—probably a supernova. This is a constraint, but not a very tight one, on the timescale for forming the Solar System. Another possible implication is that if ^{26}Al was widespread as a component of aluminium in the early Solar System then it would have formed an important source of heating and could have melted even small asteroids. Many features of meteorites suggest melting, and even re-melting, in the early period of the Solar System and the assumption of ^{26}Al heating makes it more plausible that the parent bodies of meteorites were of large asteroid, rather than of planetary, size (see Problem 8.1).

10.6.3. Neon in meteorites

Some meteorites contain a great deal of gas and have been widely studied in this respect. When meteorites are heated the atoms or molecules of gas trapped within them become sufficiently energetic to diffuse through the material and escape; they can then be collected and analysed. A normal procedure is to step-heat the meteorite whereby, after all the gas has been expelled at a particular temperature, the temperature is then raised to a new value and the next gases released are separately collected. This enables the composition of the trapped gases to be related to the strength with which they are bound to the rocky host material.

Normal neon has three stable isotopes, ^{20}Ne, ^{21}Ne and ^{22}Ne, in the approximate ratios $0.9051 : 0.0027 : 0.0922$. Neon collected from various stony meteorites show a wide range of compositions and figure 10.17 shows the outcome from three-isotope points plotted from many samples of gas-rich meteorites. This diagram differs from the one given for oxygen; since the compositions vary so much, it simply plots ratios of isotopic concentration rather than δ values and these ratios are given with respect to ^{22}Ne, which is not the commonest isotope. Most of the samples give points within a triangle marked ABC in the figure; the point C corresponds to normal cosmogenic neon and it is assumed that points A and B refer to some other naturally occurring sources. All the points within the triangle could then come about by taking various mixtures of the three components, A, B and C. However, there are samples giving points well outside the triangle, some of which are very close to the origin—corresponding to almost pure ^{22}Ne. An obvious interpretation of the presence of neon much enriched in ^{22}Ne, called neon-E, is that it was produced by the decay of radioactive sodium, ^{22}Na. There is only one stable isotope of sodium, ^{23}Na, but ^{22}Na is readily produced in the kinds of nuclear reactions which occur in a supernova. A sodium-

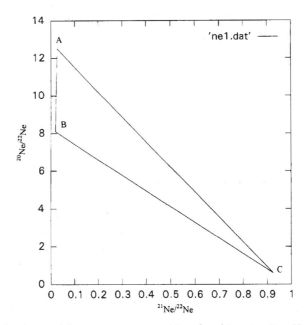

Figure 10.17. *The triangle containing most neon compositions found in meteorites. Neon E specimens occur near the origin.*

containing mineral produced from the debris of a supernova would certainly contain a small proportion of ^{22}Na. If there was no neon initially in the sample then the neon coming off when the meteorite was heated would be pure ^{22}Ne; if there was originally some neon of normal isotopic composition present then the neon would be highly enriched in ^{22}Ne.

There is a reluctance to accept ^{22}Na as the source of neon-E because the half-life of ^{22}Na is only 2.6 years and this would imply that the meteorite rock had become a closed system within a few half-lives of the production of the isotope. Other scenarios have been put forward whereby the neon-E was produced by particle irradiation in the meteorite after it formed or that it came from some unknown source outside the Solar System.

10.6.4. Other isotopic anomalies

A number of interesting isotopic anomalies have been found in grains of silicon carbide, SiC, which occur in some chondrites. Silicon has three stable isotopes, ^{28}Si, ^{29}Si and ^{30}Si, but three-isotope plots show linear correlations with a slope about 1.3—very different from either the mass-dependent diffusion value or that obtained by adding pure ^{28}Si to silicon of normal isotopic composition. The carbon in these grains shows a very variable ratio of $n(^{12}C)/n(^{13}C)$, sometimes below 20, whereas the terrestrial value is 89.9. This is termed 'heavy' carbon. There is also 'light' nitrogen for which the ratio of $n(^{14}N)/n(^{15}N)$ is much higher than the terrestrial figure and some 'heavy' nitrogen as well. Some SiC samples contain 'heavy' neon in which both ^{21}Ne and ^{22}Ne are enhanced and a linear relationship exists between $n(^{20}Ne)/n(^{22}Ne)$ and $n(^{21}Ne)/n(^{22}Ne)$ from different grains.

The ones described in this section are by no means the only isotopic anomalies in meteorites but they illustrate the range and complexity of those that occur. Clearly these anomalies contain a message about conditions in the early Solar System. In section 12.12 a theory to account for the ones which have been mentioned here will be briefly described.

Problems 10

10.1 Grains from a meteorite give the following ratios of concentrations of oxygen isotopes:

$n(^{17}O)/n(^{16}O)$	$n(^{18}O)/n(^{16}O)$
0.007419	0.04209
0.007453	0.04226
0.007484	0.04246
0.007459	0.04232

Plot the $\delta^{17}O$ and $\delta^{18}O$ values in a three-isotope plot and find the slope of the best straight line. Comment on your result.

10.2 A grain of almost pure SiC is formed within a meteorite. Trapped within it is some normal nitrogen gas that, in mass terms, forms one part in 10^5 of the grain. If 0.01% of the original carbon in the grain is ^{14}C that decays to ^{14}N, all of which is retained, then what is the final ratio of $n(^{14}N)/n(^{15}N)$ in the grain. The measured ratios for 'light' nitrogen are generally less than 2000. Is the value you find reasonable? The following information is required: Atomic mass of Si $= 28$ amu; atomic mass of C $= 12$ amu (1 amu $= 1.66 \times 10^{-27}$ kg). In normal nitrogen $n(^{14}N)/n(^{15}N) = 270$.

CHAPTER 11

DUST IN THE SOLAR SYSTEM

11.1. METEOR SHOWERS

On a clear moonless night brief streaks of light, popularly known as 'shooting stars', are occasionally seen crossing the sky. These are caused by tiny particles of dust, known as *meteors*, that enter the upper reaches of the atmosphere and become incandescent and vaporize at a height of about 120 km. Normally they are few and far between, referred to as *sporadic meteors*, but at predictable times of the year they can be very frequent and give what are known as known as *meteor showers*.

In chapter 9 it was mentioned that comets are rather frangible objects and when the volatile materials within them vaporize then parts of the comet's surface break away. Some of this material is in the form of fine debris and it moves with the comet in its orbit. Due to the effect of solar radiation and the solar wind, the paths of individual particles gradually depart from that of the comet and they spread out into a fairly thick band that at any time is distributed all around the comet's orbit (figure 11.1). Depending on the spasmodic nature of the comet's disintegration the density of debris might vary greatly around the orbit with bunching of material in some places and a comparative lack of material in others. Sometimes the debris may be present long after the comet itself has ceased to be visible although visible comets also have their debris trails.

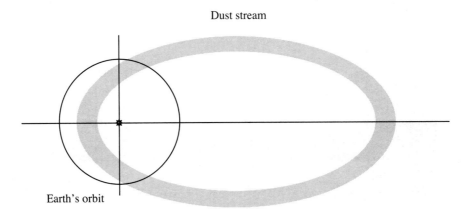

Figure 11.1. *The relationship between a meteor dust stream and the Earth's orbit. Because of orbital inclinations the orbits may only intersect at one point.*

Table 11.1. *The main meteor showers throughout the year.*

Shower	Time of maximum rate	Average hourly rate	Contributing comet
Quadrantids	3–4 January	110	Unknown
Lyrids	22 April	12	Thatcher 1861
Eta Aquarids	4–5 May	20	Halley
Delta Aquarids	27–28 July	35	Unknown
Perseids	12 August	68	Swift–Tuttle
Orionids	21 October	30	Halley
Taurids	4–8 November	12	Encke
Leonids	17 November	10	Tempel–Tuttle
Geminids	14 December	58	Unknown

When the Earth in its orbit moves through one of these debris trails a meteor shower is seen. Since the locus of the debris and the Earth's orbit are both fixed in space, a particular crossing will happen at the same time each year but the intensity of the meteor shower will depend on the concentration of debris in the crossing region. Table 11.1 gives the major meteor showers and their approximate timings throughout the year. The average hourly rate can be greatly exceeded; if the Earth happens to pass through a very dense part of the meteor trail then *meteor storms* giving tens of thousands of meteors per hour can be seen, although usually for a period less than one hour.

The meteors are all travelling on parallel paths as they enter the atmosphere and the effect of perspective makes it appear that they are radiating outwards from one point in the sky, called the *radiant*. It is the stellar constellation within which the radiant falls that gives the shower its name.

A typical meteor particle is about the size of a grain of sand, 1 mm in diameter, but occasional larger objects, a centimetre or so in diameter, can give a spectacularly brilliant trail referred to as

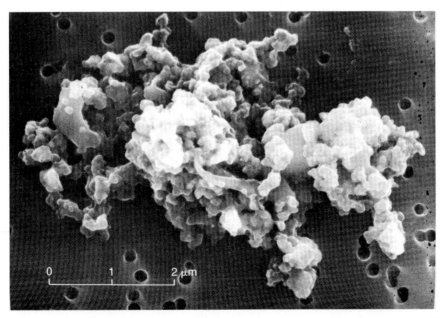

Figure 11.2. *A typical Brownlee particle consisting of large numbers of sub-micron particles cold-welded together (NASA).*

a *fireball*. Even larger objects are rare in meteor showers but when a larger body enters the atmosphere then some of it will survive to ground level and may be found as a meteorite. Very fine material, less than about 200 μm in dimension, has such a large ratio of surface area to volume that it cools efficiently and can survive the impact with the atmosphere. Such material remains in the atmosphere for some time and very slowly drifts down to Earth. Spacecraft have collected such material in the upper atmosphere by trailing sticky surfaces on to which the particles adhere. These *Brownlee particles* (figure 11.2), named after the scientist, D Brownlee, who suggested collecting them, are studied to learn something about the nature of comets, from which it is believed they are derived.

11.2. ZODIACAL LIGHT AND GEGENSCHEIN

After a considerable time the dust from comets that has not been swept up by the Earth or other large bodies will depart appreciably from the original orbit and will spread throughout the Solar System. In addition there will be other sources of dust. Most meteorites are derived from asteroid collisions and such collisions will produce debris over a wide size range. Thus the Solar System is occupied by a large amount of small-scale material varying upwards from microns in size. This moves around the Sun in a variety of orbits but tends to be concentrated near the plane of the ecliptic.

   Although individual particles are far too small to be seen, what can be observed is the sunlight that they scatter. On a dark clear night, well away from any artificial sources of light, a faint band of light can be seen, going from the direction of the Sun, which is below the horizon, and distributed about the ecliptic. This is the *zodiacal light* and it indicates the presence of small reflecting bodies between the Earth and the Sun. Observations by satellites indicate that the zodiacal light dust pervades the inner Solar System and has a density of about 10^{-8} particles m^{-3} in the vicinity of the Earth. Since these local particles are moving with the Earth in orbit around the Sun their motion relative to the Earth is small and they do not give rise to many sporadic meteors.

   Small particles are most effective in scattering in the forward direction but they also backscatter quite efficiently. In the antisolar direction there is a faint band of light in the sky called the *gegenschein* or sometimes *counterglow*. The zodiacal light and gegenschein are manifestations of a large amount of material, much in the form of very tiny particles upwards of microns in size, occupying the space between the planets. We have referred to sources of such material—from comets, that would have to be perturbed in from the Oort cloud in the first place, and by asteroid collisions. Unless there is some way of disposing of such material, its concentration in the Solar System would gradually build up. In fact there are disposal mechanisms so that the concentration of dust in the Solar System is controlled by the balance of the processes of formation and elimination. The total mass of the material between the Earth and the Sun is about 10^{16} kg, a significant mass but, to put it in perspective, well under 1% of that of the Earth's atmosphere.

11.3. RADIATION PRESSURE AND THE POYNTING–ROBERTSON EFFECT

Energy and matter pour out of the Sun in two main forms, as electromagnetic radiation and as the solar wind. These both have important effects on planets in affecting their temperatures and their magnetic and charged particle environments but they have no significant effect on their dynamical behaviour. The situation is quite different for very small orbiting bodies. We have seen that the plasma tail of a comet, consisting of ions, is pushed in an antisolar direction by the solar wind (section 9.2). The dust tail is also seen to point generally in an antisolar direction but not exactly so and it also has a rather curved appearance.

The main influence on the dust is the photon flux from solar radiation. This travels radially outwards and impinges on dust particles. Each photon has momentum associated with it, depending on its energy or wavelength and this momentum is absorbed by the dust particle. The total absorbed momentum per unit time gives the total force on the particle due to photon bombardment. This force acts in the opposite direction to the gravitational force due to the Sun. Since the flux of photons will vary as r^{-2}, where r is the distance from the Sun, and the gravitational force varies in the same way, for particles of a given size it is as though the mass of the Sun was effectively reduced. This argument is put into a mathematical form in Topic AG.

There is another way in which bodies are affected by radiation and that is through the Poynting–Robertson effect. The vector representing the momentum of the photon absorbed by a particle is in a radial direction. However, in a state of thermal equilibrium the particle will be re-radiating this energy, albeit with lower-energy photons corresponding to a lower temperature. This re-radiated energy is isotropic with respect to the particle but in the framework of the Sun it has a net momentum, tangential to the orbit, in the direction of travel of the particle. Because of momentum conservation the particle itself must lose momentum in the tangential direction, or in other words the intrinsic angular momentum of its orbit gradually declines. This causes the particles to spiral inwards towards the Sun. The effect is a large one on the timescale of events of the Solar System. Thus a 1 cm silicate sphere orbiting the Sun in the vicinity of the Earth will spiral into the Sun in about 20 million years. For such a large particle the effective reduction of the gravitational field due to radiation pressure would be negligible. Smaller particles, and those closer to the Sun, will be absorbed by the Sun more quickly but even a 2 m radius boulder in orbit at the distance of the Earth soon after the origin of the Solar System would be absorbed by now.

A mathematical treatment of the Poynting–Robertson effect is given in Topic AG.

Problem 11

11.1 The Earth moves through a comet dust train within which the density of particles capable of giving a visible meteor is 10^{-16} m^{-3}. The Earth moves at right angles to the direction of the stream, which is of thickness 10^9 m and the speeds of the Earth and the dust particles in their respective orbits are 30 km s^{-1} and 40 km s^{-1}. Assuming that any meteor track within a horizontal distance of 400 km from an observer can be seen then estimate the number of meteors per hour that will be seen from a given point on Earth and also the duration of the shower.

CHAPTER 12

THEORIES OF THE ORIGIN AND EVOLUTION OF THE SOLAR SYSTEM

Since we know comparatively little about extra-solar planetary systems our main emphasis in this chapter will be to explain what we know about the Solar System. However, although it is possible that the Solar System may be special in terms of its general complexity, any tenable theory for the basic Solar System must also be able to explain the existence of other planets. Estimates put the proportion of Sun-like stars with planets in the range 3–6% and a theory must accommodate this.

We begin by looking at the Solar System at various levels of detail and assembling a list of features that a theory might be expected to explain. Some of these features are so important, e.g. the existence of planets, that we shall call these *primary features* and it would be mandatory for a theory to explain these. Others may be more subtle, or features that might depend on evolutionary processes after the system formed. These we shall call *secondary features*; it would be a bonus if a theory could explain, at least, some of them but not fatal to the theory if it could not.

12.1. THE COARSE STRUCTURE OF THE SOLAR SYSTEM

Seen at the lowest resolution the Solar System consists of the Sun, planets divided into two groups and satellites around most of the planets. Some of the satellites are regular, suggesting a systematic relationship with the parent planet, but others are irregular and may have been captured after the system had formed. This suggests the primary features:

1. the existence of the Sun and of the planets in direct, nearly-coplanar orbits;
2. regular satellites for the major planets,

and the secondary features:

(i) the planets are divided into two groups;
(ii) the existence of non-regular satellites for the major planets and Mars;
(iii) the Earth–Moon relationship.

12.2. THE DISTRIBUTION OF ANGULAR MOMENTUM

When Galileo first saw Jupiter's satellite system with his telescope he was convinced that what he saw was a small-scale version of the Solar System. To a cursory inspection this may be so but there are

Table 12.1. *The ratio, R, of the intrinsic angular momentum in the secondary orbit to that of the primary spin at the equator.*

Primary	Secondary	Ratio R
Sun	Jupiter	7800
Sun	Neptune	18 700
Jupiter	Io	8
Jupiter	Callisto	17
Saturn	Titan	11
Uranus	Oberon	21

important differences in the two types of system in the way that angular momentum is distributed. The Sun contains 99.86% of the mass of the Solar System but only 0.5% of the angular momentum in its spin, the rest being in the planetary orbits. Thus solar material has very little intrinsic angular momentum while planets have very much more. In table 12.1 we show this as the ratio of the intrinsic angular momentum for a planet in orbit to that of equatorial solar material. In the same table we show the corresponding ratio for various satellite–planet pairs and the distinction between the two types of system is obvious.

Another feature involving angular momentum is the 7° tilt of the solar spin axis to the normal to the mean plane of the system. This value is too small to be just chance; the likelihood that two vectors chosen at random are aligned to within 7° is 0.004. On the other hand the 7° tilt is too large to be easily accommodated by some theories. Angular momentum considerations lead to three more primary features:

3. Why does the Sun (and perhaps similar stars) spin so slowly?
4. How does planetary material acquire its orbital angular momentum?
5. The 7° tilt of the solar spin axis.

12.3. OTHER FEATURES OF THE SOLAR SYSTEM

The primary features already given are so difficult for most theories that, by comparison, all other features can be put in the secondary category. These are:

(iv)   departures from planarity in the Solar System;
(v)   variable directions of planetary spin axes;
(vi)   commensurabilities of planetary orbital periods;
(vii) the existence and properties of asteroids;
(viii) the existence and properties of comets;
(ix)   the formation and survival of the Oort cloud;
(x)   the physical and chemical characteristics of meteorites;
(xi)   isotopic anomalies in meteorites;
(xii) Kuiper-belt and other small objects;
(xiii) Pluto and its satellite Charon.

While many of these features may relate to the evolution of the system, it may not be possible sharply to distinguish between origin and evolution. The initial state of the system may decide whether or not an evolutionary process is likely to operate.

12.4.   THE LAPLACE NEBULA THEORY

In 1796 Pierre Laplace (1749–1827) described the first well-formulated scientific theory of the origin of the Solar System, illustrated in figure 12.1. The starting point is a slowly rotating spherical nebula, a cloud of gas and dust at very low density, slowly collapsing under self-gravitational forces (figure 12.1a). As it collapsed so, to conserve angular momentum, it spun more rapidly and flattened along the spin axis (figure 12.1b). Eventually the nebula took on a lenticular form and material at the edges was in free orbit around the central mass (figure 12.1c). Thereafter, as the collapse proceeded, material was left behind in the equatorial plane. Laplace postulated that this happened in a spasmodic way so the material formed a set of annular rings (figure 12.1d). Clumps formed in each ring but they coalesced since they orbited at slightly different rates. The final outcome was one planet in each ring (figure 12.1e). A smaller version of the same process for each of the collapsing planets formed satellite systems. The central part of the nebula eventually formed the Sun.

12.4.1.   *Objections and difficulties*

The strongest criticism of the Laplace nebula theory related to the distribution of angular momentum. It could not be seen how the evolution of the nebula could lead to the partitioning of mass and angular momentum described in section 12.2. Most of the angular momentum should end up in the central body.

   Another criticism, made by Clerk Maxwell, was that differential rotation between the inner and outer parts of a ring would prevent material from condensing. To overcome this effect would have required rings with hundreds of times more mass than the planets they produced.

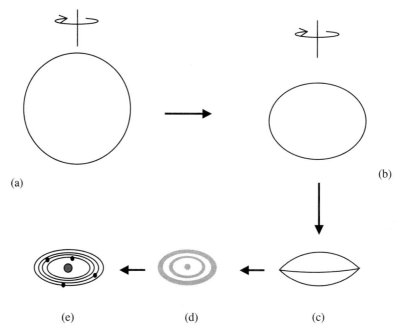

Figure 12.1. *An illustration of Laplace's nebula theory. (a) A slowly rotating and collapsing gas-and-dust sphere. (b) An oblate spheroid form as the spin rate increases. (c) The critical lenticular form. (d) Rings left behind in the equatorial plane. (e) One planet condenses in each ring.*

Although various attempts were made in the nineteenth century to rescue the theory these were all unsuccessful and by the beginning of the twentieth century the theory no longer had much support.

12.5.   THE JEANS TIDAL THEORY

The Laplace theory was a *monistic* theory in which the same body of material in a single process gave rise to both the Sun and the planets. In 1917 James Jeans (1877–1946) proposed a *dualistic* theory that separated the formation of the Sun from the formation of the planets. This theory involved a tidal interaction between the Sun and a very massive star. The mechanism is illustrated in figure 12.2. In figure 12.2a the massive star, passed close by, raised a tide on the Sun pulling out material in the form of a filament. This filament was gravitationally unstable and broke up into a series of blobs (figure 12.2b) that had greater than the Jeans critical mass (Topic D) and so condensed to form

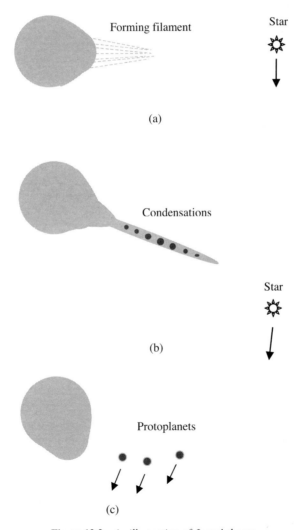

Figure 12.2. *An illustration of Jeans' theory.*

planets. These planets were attracted by the retreating massive star and were left in orbits around the Sun (figure 12.2c).

Jeans produced very convincing analyses to show that a tidally distorted star would lose material from a protuberance and also that the filament would be gravitationally unstable and break up into blobs (Topic AH). The theory enjoyed much support and many thought of it as the unquestionable way in which the Solar System had formed.

12.5.1. Objections and difficulties

Harold Jeffreys (1891–1989), an early supporter of the Jeans theory, raised the first difficulties in 1929. His first objection, that very massive stars are very rare, was no strong argument at the time since, as far as was known then, the Solar System could have been the only planetary system in the universe. Jeffreys' second objection was that since the Sun and Jupiter had about the same density and Jupiter's material came from the Sun then the *circulation* ($\nabla \times \mathbf{v}$, where $\mathbf{v}$ is the velocity of the material) and hence the rotational period of the two bodies should be the same. In fact the spin periods are different by a factor of 65 or so. The objection is not a strong one. If, for example, the material drawn from the Sun to produce Jupiter had 1% of the mean density of the Sun and then collapsed to produce the planet it would have spun only 20 times faster than the Sun. Then there could have been processes operating that slowed down the Sun's spin with time to the present value (Topic AJ).

The next objection was raised by Henry Norris Russell (1877–1957) in 1935. He argued that material from the Sun would have a perihelion no greater than the radius of the Sun and a semi-latus rectum (a measure of the intrinsic angular momentum) no greater than twice the radius of the Sun. It does not seem possible for the material to acquire enough angular momentum to explain Mercury, let alone the outer planets.

Another telling objection was made in 1939 by Lyman Spitzer (1914–1997). If material was taken from the Sun to make Jupiter it would have a density about that of the Sun and a temperature about 10^6 K. Inserting this in equation (D.4) gives a minimum mass for collapse some 100 times that of Jupiter, so that no planet would form.

More recently it has been noted that the Sun is deficient in the light elements lithium, beryllium and boron because they are consumed by nuclear reactions at solar temperatures. However, they are reasonably abundant on Earth so the Earth, and by implication the other planets, could never have never been at solar temperatures.

Two simple theories, one monistic and one dualistic, had both collapsed under critical review. However, the Solar System exists—so there *must* be at least one plausible theory to explain its existence. The search for such theories continued.

12.6. THE SOLAR NEBULA THEORY

In the 1960s it was realized that many characteristics of meteorites could be explained in terms of condensation from a silicate vapour. This stimulated a return to nebula ideas in the early 1970s through the Solar Nebula Theory (SNT) although the idea of condensation from a hot nebula was soon abandoned. It was thought that new theoretical approaches could solve the problems associated with the original Laplace model. Later the SNT was supported by observations of dusty disks around young stars—some of which have been directly imaged (figure 12.3) and others inferred by infrared excesses in the stellar spectra. Although low-temperature disks are inefficient radiators of energy per unit area, they are so large that they appreciably elevate the infrared end of the stellar spectra. Observations on young stars also show that the lifetimes of the disks are less than 10^7 years.

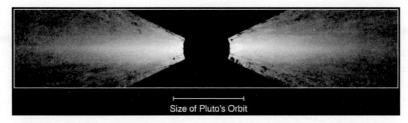

Size of Pluto's Orbit

Figure 12.3. *The dust disk around the star β-Pictoris. The dark regions are blocked out by the imaging equipment. The image is of a disk seen obliquely.*

12.6.1. The transfer of angular momentum

In 1974 Lynden-Bell and Pringle proposed a mechanism for transferring angular momentum from the core to the disk of a collapsing nebula. If the nebula was turbulent then heat would be generated and radiated away. Thus the nebula would evolve so that its energy became less while its angular momentum remained constant. This happens by inner material moving inwards while outer material moves outwards (Topic AI). This is equivalent to an outward flow of angular momentum.

The mechanism leads to material gradually spiralling in to join the central core; at the instant of joining the core it is in free orbit around the central mass. Since gravity is neutralized by the spin of the material, such a mechanism would produce a diffuse object with much more angular momentum than is observed in normal main-sequence stars; so some further process of removing angular momentum is required. Until this rapidly-spinning diffuse object can lose angular momentum it will not become what we can call a star, for the development of temperatures capable of setting off nuclear reactions requires the release of gravitational energy and hence collapse.

Observations of very young solar-type stars show a wide range of spin rates, some even a significant fraction of what is required for rotational breakup. Since these are stars on or near the main sequence they must be reasonably compact and once they have reached this state there are mechanisms that can operate to remove angular momentum and slow down their spins. These mechanisms are complex and depend on the interaction of mass flows of material and strong magnetic fields, which have actually been observed in T Tauri stars. The flows can be both inwards giving accretion and outwards carrying off angular momentum. One simple mechanism for losing angular momentum with little mass loss is described in Topic AJ but it may not be the most efficient of the mechanisms that have been proposed.

Work on the formation of stars from a nebula is an ongoing activity and while several mechanisms have been proposed there seems, as yet, to be no standard model for the process. It is quite possible that, depending on the conditions, there are many mechanisms that are possible, some still to be proposed.

12.6.2. The formation of planets

Four stages are proposed in the process of forming planets from disk material:

1. the dust component of the disk settles into the mean plane;
2. *planetesimals*, solid objects of dimension from hundreds of metres to a few kilometres, are produced from the dust disk;
3. the planetesimals collect together to form terrestrial planets and the cores of the major planets;
4. the cores of the major planets acquire gaseous envelopes from the gas component of the disk.

12.6.2.1. Settling of dust into the mean plane

Since the 1 μm dust particles, that originate in the ISM, are subjected to gas drag they would take of the order of 10^7 years to reach the mean plane of the disk—which is longer than an expected disk lifetime. It has been suggested by Weidenschilling (1980) that dust grains would stick together by a 'cold welding' process to form larger aggregates that would settle more quickly. However the CODAG (Cosmic Dust Aggregation Experiment) Shuttle experiment, carried out in 2000, has shown that dust actually tends to aggregate with individual dust particles in long thin chains rather than in compact clusters, and this may inhibit rapid settling.

12.6.2.2. Formation of planetesimals

In Topic AH the conditions for the instability of a filament are considered. The dust disk is a two-dimensional system and a condition of gravitational instability can occur to give a two-dimensional wave-like variation of density. If the clumps that form have a mean density of solid material that brings them outside the Roche limit (Topic AA) for disruption by the Sun then they may collapse to form planetesimals (Goldreich and Ward, 1973). This idea has been challenged by Weidenschilling (1995) who favours direct accumulation of dust by collision processes.

12.6.2.3. Planets and cores from planetesimals

The basic theory of forming planets from planetesimals was developed by Safronov (1972) and his model is described in Topic AK. It is necessary for planetesimal turbulence to die down so that they come together with a small hyperbolic excess (i.e. at little more than the mutual escape speed) so that with a small energy loss they can coalesce. The largest planetesimal in any region had the greatest collision cross section and so grew at the expense of its neighbours. Eventually, in each of a number of isolated regions there would be one dominant body that would be either a terrestrial planet or the core of a major planet.

Safronov's analysis gave times of formation for the planets. These were 1.1×10^6 years for the Earth, 2.5×10^8 years for a Jupiter core and 7.8×10^9 years for a Neptune core. The Jupiter core time is greater than the inferred disk lifetime and the Neptune core time is longer than the age of the Solar System. Various schemes for accelerated formation giving 'runaway growth' have been suggested (Stewart and Wetherill, 1988) but none of them solves the timescale problem for Uranus and Neptune.

12.6.2.4. Gaseous envelopes

If a solid core of $10M_\oplus$ formed for Jupiter with a mean density of 4×10^3 kg m^{-3} then the escape speed would be about 18 km s^{-1}. If the local temperature in the formation region is 100 K then the root-mean-square speed of hydrogen atoms would be only 1.6 km s^{-1}. It is clear that gas will accumulate on the core. The time to give a complete gaseous envelope to form a major planet is about 10^5 years, which is negligible compared with the times for other stages of planet formation.

12.6.3. General comments

The SNT is generally accepted as the most plausible theory for the origin of the Solar System. It has a number of problems, on which work continues, and some fairly complex processes are being considered, particularly for planet formation. For example, schemes are being explored for *planet migration* so that planets can form at distances from the Sun, or other stars, where formation timescales are acceptable. Migration inwards can then explain close orbits for extra-solar planets (table 2.1) and outward migration can give Uranus and Neptune in the Solar System (Topic AP).

A new observation, that of *free-floating* planets in the Orion Nebula, presents a further challenge in that either planets must be formed in the absence of a disk or be formed within a disk with some subsequent mechanism for escape.

12.7. THE CAPTURE THEORY

The SNT is a monistic theory with clear links to the Laplace nebula theory. Here we consider a dualistic theory, the capture theory (CT), with some links to Jeans tidal theory—although it apparently avoids the difficulties of the Jeans model.

12.7.1.   The basic scenario of the capture theory

Stars occur in clusters and Population I stars like the Sun, with high metallicity, are found in galactic, or open, clusters the formation of which is described in section 2.4. The stars in the clusters interact and occasionally one will have enough energy to escape and to become a field star. This process of evaporation will continue until a binary, or other small stable system of stars, remains. According to a theory given by S Chandrasekhar the complete process would take of the order of 10^9 years.

In recent years observations have shown that in the late development of a galactic cluster there occurs an *embedded stage* in which stellar densities are in the range 10^2–10^5 pc^{-3}. This compares with the local star density in the Sun's environment of 0.08 pc^{-3}. It is in the embedded stage when lower-mass stars and probably brown dwarfs are being produced in large numbers. In this environment interactions between condensed stars (those on or approaching the main sequence of more-or-less solar mass) and low-mass stars and brown dwarfs in a diffuse state can take place with reasonably high probability (Topic AQ). The type of interaction corresponding to the CT, first proposed by Woolfson (1964), is one where material drawn out of the diffuse protostar forms planets that are captured by the condensed star—hence the name of the theory.

12.7.2.   Modelling the basic capture theory

One of the most powerful tools for modelling astrophysical situations is smoothed particle hydrodynamics (SPH). The bodies are represented by a distribution of points. Each point possesses properties such as mass and internal energy which are spread out in a distribution around the point. Forces due to gravity and pressure gradients and the dissipative process of viscosity can be realistically simulated and the development of any astronomical system can be followed in time by suitable graphical output. Recently it has also been possible to simulate radiation transfer in a realistic way. Figure 12.4 shows an interaction between a diffuse lower-mass protostar and a condensed solar mass star with parameters given in the figure legend. The figure shows the initial state of the protostar as a diffuse spherical object and its state after 15 000 years where the whole protostar is drawn out into the form of a filament. As Jeans showed, this filament is gravitationally unstable and breaks up into a series of blobs. Starting from the left the first five blobs are captured by the star while remaining condensations escape from it to form free-floating planets.

None of the problems associated with the Jeans theory is associated with this model. The intrinsic angular momentum of the planetary orbit comes from the star–protostar orbit (Russell's objection) and the modelling shows that a planet actually forms (Spitzer's objection). In addition the planetary material is never at solar temperatures (light element problem).

The modelling has been done at a variety of scales, in particular with protoplanets of initial radius 100–2000 AU. From one to five captured planets have been produced in the various simulations that have been carried out.

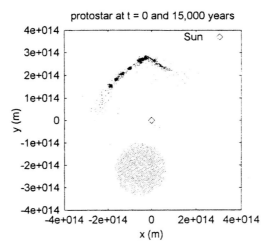

Figure 12.4. *An SPH simulation of the tidal disruption of a protostar with mass $0.35 M_\odot$, radius 800 AU and temperature 20 K. The initial orbit of the protostar around the condensed star had periastron distance 600 AU and eccentricity 0.95. The first five condensations, starting from the left are captured with masses (in Jupiter units), semi-major axes (in AU) and eccentricities (4.7, 1247, 0.835), (7.0, 1885, 0.772), (4.8, 1509, 0.765), (4.8, 1509, 0.726), (20.5, 2686, 0.902).*

A somewhat different kind of event, although related, has been simulated by Whitworth *et al.* (1998) in which, instead of a protostar, the donating body is a condensed star with a fairly massive surrounding disk. It is the disk that is disrupted and gives the planetary condensations.

12.7.3. *Planetary orbits and satellites*

The protoplanets produced as shown in figure 12.4 have orbits that are extensive and highly eccentric and are also surrounded by a disk of material a few AU in extent (figure 12.5). The presence of the disk around the protoplanet can lead to regular satellites forming by tidal interaction with the Sun at perihelion. This is similar, but on a smaller scale, to the planet formation process modelled by Whitworth *et al.*

From the calculations that give figure 12.4 it is found that a considerable amount of material from the protostar is captured in the form of a diffuse medium that surrounds the Sun. The behaviour of a planet moving in a resisting medium has been modelled by Dormand and Woolfson (1974, 1977) and is described in Topic AP. Figure 12.6 shows the variation of semi-major axis and eccentricity for a typical planet. The orbital evolution takes into account the decay of both the resisting medium and of the stellar wind, that affects the behaviour of the medium. Depending on the parameters of the medium the final outcome for the orbit of the protoplanet can be anything from a very close circular orbit to a more extended and sometimes eccentric orbit (table 2.1).

12.7.4. *General Comments*

Given a pre-existing Sun, for which there is a theoretical basis, the CT gives planets and regular satellites without any obvious problems. The low angular momentum of the Sun comes from its mode of formation (section 2.4). The high angular momentum of planetary orbits is derived from that of the Sun–protostar interaction. The material to form the protoplanets is in a comparatively

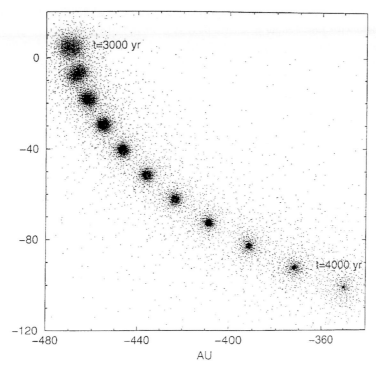

Figure 12.5. *The collapse of a protoplanet with mass 5 M$_J$. It ends up with an accompanying disk of radius several AU.*

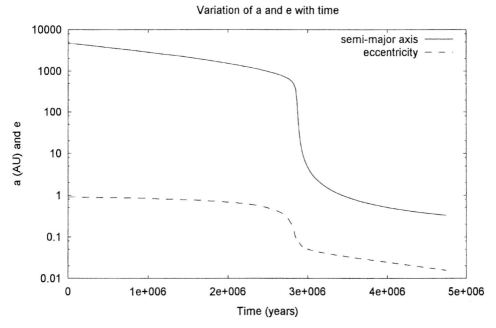

Figure 12.6. *The evolution of semi-major axis and eccentricity for a protoplanet in a slowly decaying resisting medium.*

high-density form and planets can be formed directly by gravitational instability. Again, the planets can round off or be left with fairly high eccentricities, as is observed in the Solar System and in planets around other stars. Calculations based on modest densities in the embedded state of a cluster give probabilities of planet formation for solar-type stars consistent with observations (topic AQ).

12.8.   IDEAS ON THE EVOLUTION OF THE SOLAR SYSTEM

Based on the CT scenario an integrated theory of the evolution of the Solar System has been developed, covering most of its important features. However, this evolutionary theory is not dependent on the validity of the CT; it applies to any theory that gives an initial system with protoplanets, accompanied by regular satellites, on highly elliptical and slightly inclined orbits rounding off in a resisting medium. So far the CT is the only theory giving such a starting point but one cannot exclude the possibility that other plausible theories will be advanced that do so.

12.8.1.   *Precession of elliptical orbits*

When computer simulations were made of the round-off of planetary orbits in a resisting medium it was found that in addition to changes in *a*, *e* and *i* there was precession of the orbits. In figure 12.7 a projected view of precession is shown although the three-dimensional form of the motion is quite complicated. It corresponds to a simultaneous rotation of the line of apses of the orbit, that is the line joining perihelion to aphelion, and the line of nodes (section 3.2). The reason for this

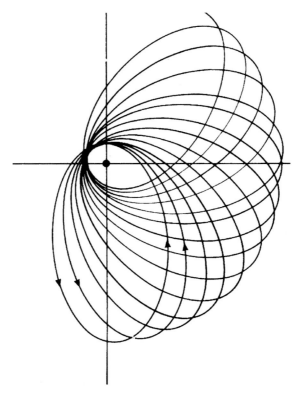

Figure 12.7. *The projected view of the precession of a protoplanet orbit.*

precessional motion is the gravitational force on the planet due to the medium. This force, coming from a flattened disk-like distribution of matter, is not centrally directed towards the Sun. This gives a torque trying to rotate the orbital axis, which is just the condition for gyroscopic motion, and the precession of the orbit is akin to the precession of the axis of a gyroscope spinning in the Earth's gravitational field.

12.8.2. Near interactions between protoplanets

Although orbits may intersect in projection, if they are inclined they are unlikely to intersect in space. However, if the precessions of the orbits are at different rates then from time to time orbits will intersect in space and then there is a possibility of strong interactions between protoplanets. It is also necessary for both the protoplanets to be close to the common region of the two orbits.

Due to tidal effects of the Sun imparting angular momentum to protoplanets while they are still extended objects, the initial spin vectors of the planets will be perpendicular to their orbits. If while they are still in an extended state a pair of protoplanets pass close to each other then their mutual tidal effects will impart extra angular momentum to each of them and hence change the directions of their spin axes. In one simulation (Woolfson, 2000) a proto-Uranus, with a radius 0.25 AU, interacted with a proto-Jupiter with a closest approach distance 1.15 AU. After the interaction the tilt of the spin axis of Uranus was 98.7° (see table 7.4). The tilts of the spin axes of other major planets can be explained in a similar way.

12.9.   A PLANETARY COLLISION

Tilting the spin axes of protoplanets involves fairly distant interactions but much closer interactions are also possible. One kind of interaction involves a close passage with an exchange of energy in which one of the protoplanets is completely expelled from the Solar System. Yet another kind of interaction is when the protoplanets actually collide. Dormand and Woolfson (1977), in considering the present state of the Solar System, postulated that initially the system consisted of six major protoplanets. These would have rounded off as follows: one close to the present orbit of Mars (protoplanet A), one in the asteroid belt (protoplanet B) and Jupiter, Saturn, Uranus and Neptune in their present orbits. On the basis of differentially-precessing orbits they derived the timescales for interactions between pairs of protoplanets that would lead either to one or other of the protoplanets being thrown out of the Solar System or to a direct collision. They found that the timescales for major interactions were very similar to round-off timescales and they concluded that, according to their postulate, the occurrence of one or more major interaction in the early Solar System was more likely than not.

In the following sections some possible outcomes of a planetary collision will be described. It should be stressed that this model is not one that is generally accepted and, where appropriate, the alternative accepted ideas will be mentioned.

12.9.1. The Earth and Venus

To explain the present Solar System, Dormand and Woolfson proposed that a direct collision took place between protoplanets A and B. They showed that it was possible for one of the protoplanets to be expelled from the system while the core of the remaining one sheared into two parts that would have rounded off in the orbits of Venus and the Earth. Another possibility, suggested by some later computational work (Woolfson, 2000), is that both planetary cores were retained and ended up as Venus and the Earth.

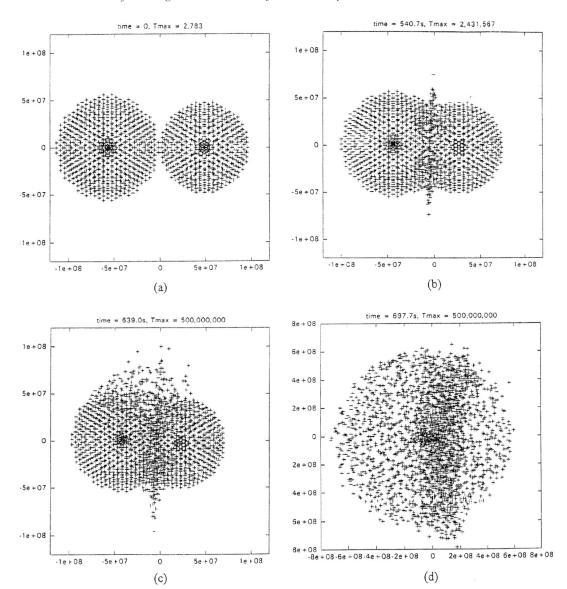

Figure 12.8. *Stages in the collision of two protoplanets. The cores remain intact but more volatile material completely disperses. Nuclear reactions are triggered when the temperature in the collision region reaches 2–3 million degrees.*

A smoothed particle–hydrodynamic simulation of a collision of two planets is shown in figure 12.8. Each planet is modelled in three layers—an iron core, a stony mantle and an extensive atmosphere, mainly of hydrogen and helium. The cores remain intact and they are separating at the end of the simulation although this is not apparent from the figure.

The normally accepted origin of the terrestrial planets is that they were formed by the accumulation of planetesimals in the inner reaches of the Solar System. Formation times as given by the basic theory (Topic AK) indicate that they would be formed on timescales of about one million years, that presents no problems with observational constraints.

12.9.2. Asteroids, comets and meteorites

Temperatures in the collision region would have been very high but other parts of the protoplanets, remote from the collision, would have remained cool and been thrown off by spallation effects. Spallation is caused by shock waves moving through the body; striking a brick at one end ejects flakes from the opposite side. The mantle and outer icy regions of the protoplanets would have been excellent sources of asteroid and comet material. Outer material, the more volatile part, would have been thrown out farthest and given potential comets. Much of this, at a distance of hundreds or thousands of AU, would have provided the inner reservoir of comets suggested by Bailey (section 9.3) that eventually gave the Oort cloud. Inner material would be thrown out less far and provided a large number of less volatile bodies, the residue of which now forms the asteroids. The fact that C-type asteroids, consisting of more volatile material, tend to be farther out than S-type asteroids (section 6.4) is consistent with this general pattern.

Between the extremely high temperatures in the collision region and the low temperatures in regions remote from the collision there would have been a complete range of temperatures. In particular, temperatures that vaporized silicates would have been available to explain the condensation sequences that gave rise to the revival of nebula ideas for the formation of the Solar System. Again, the dimensions of the vaporized regions would have been small by astronomical standards and they would have cooled rapidly—which explains the fast cooling of chondrules noted in section 10.2.1.

In the Solar Nebula Theory comets and asteroids are considered as the building blocks from which planets are made. Because asteroids show signs of having been heated, and some of them have shock features, it is taken that larger bodies, called *parent bodies*, were first formed. Collisions of these gave fragments that gave rise to the smaller asteroids.

12.10. THE ORIGIN OF THE MOON

The origin of the Moon is obviously a matter of some importance in planetary science and a number of theories have been put forward. Five of these will be described, the last of which is related to the idea of a planetary collision.

12.10.1. Darwin's fission hypothesis

In 1878 George Darwin suggested that when the Earth was formed it was spinning so quickly that it became rotationally unstable and broke up to give the Earth–Moon system. Tidal interaction between the Earth and the Moon then caused the Moon gradually to recede from the Earth. As it did so it gained orbital angular momentum and, to compensate for this, the Earth spun more slowly. Ignoring the tidal influence of the Sun we can, for present purposes, take the angular momentum of the Earth–Moon system as remaining approximately constant. It is possible, from this conservation principle alone, to deduce the closest distance of the Moon could ever have been to the Earth. At the moment of fission the Moon would have been locked with one face towards the Earth so that Earth plus Moon would be spinning as a rigid system (figure 12.9). It turns out that this period is about 5.5 hours (Topic Z).

A difficulty with Darwin's idea is that even if the Moon was absorbed into the Earth the spin period of the augmented Earth, 4 hours or so, would still be very far from what was

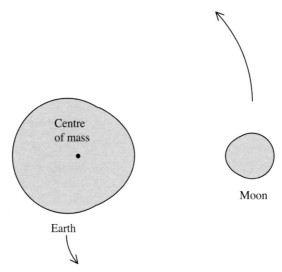

Figure 12.9. *The early Earth–Moon system according to Darwin. The spin of both bodies is locked to the orbital rate as both bodies orbit the centre of mass.*

required to give rotational instability. Kelvin suggested that this was due to a resonance with the tidal influence of the Sun. Any particular point on the Earth's surface would have experienced a periodic tidal force with a period of 2 hours (two tidal peaks per 'day'). If the Earth had been a fluid body, slightly larger than at present, then a resonance oscillation with the solar tide would have been possible.

The idea has an appealing simplicity and the Pacific Ocean was regarded as a probable region from which the Moon's material could have been pulled out of the Earth. However, the geophysicist Harold Jeffreys showed in 1930 that for an oscillating fluid Earth the internal dissipation of energy would have been very high so that large amplitude oscillations leading to disruption could not have occurred.

12.10.2. *Co-accretion of the Earth and the Moon*

A theory of Moon formation was advanced by Evgenia Ruskol in 1960 based on the idea that planets and satellites formed from a swarm of small bodies. Ruskol imagined that the Earth had largely accumulated but that a large number of bodies, with individual diameters between 10 and 100 km and with total mass $0.01–0.1 M_{\oplus}$, remained in its vicinity. Collisions between these bodies could have led to some of them being deflected towards the Earth while others could have aggregated to form a large body in orbit—the Moon. Ruskol suggested that the Moon would have formed in an orbit with radius between 5 and 10 Earth radii.

A problem with this theory is that it is not immediately obvious why the densities of the Moon and Earth should be so different. The theory is no longer supported.

12.10.3. *Capture of the Moon*

It has been suggested that the Moon was captured by the Earth from an independent heliocentric orbit. There are severe dynamical problems with this hypothesis. If two isolated bodies approach each other then the total energy of the two-body system is always positive, and a bound system cannot form, unless energy is somehow removed. There are two possible ways of removing energy. The first is by the presence of a third body and for this case the Sun is an obvious candidate. The second is by the interaction of the two bodies themselves which could be either tidal in form or, in an extreme case, a direct collision.

If capture of the Moon is to take place then it would clearly be an advantage if the original energy of the two-body system was only slightly positive. This requires that the Earth and the Moon were on very similar orbits so that they approached one another as slowly as possible. How the Moon would achieve such an orbit is difficult to imagine. The tidal interaction caused by a close passage could then remove some energy although it would be very small in quantity unless the passage was very close. If the passage *was* close then the Moon might have been within the Roche limit and torn apart by strong tidal effects (Topic AA). It turns out that for an Earth–Moon interaction outside the Roche limit the energy dissipation would be very small as would be the chances of a capture event. Alternatively if the Moon approached the Earth closely, or even collided with it, then it would have been disrupted. Debris in orbit around the Earth might then have recombined, as envisioned by Ruskol.

Perturbation by the Sun is another means of obtaining Moon capture but the resulting orbit is, perforce, of large extent and very eccentric. Just as the Sun can extract energy from the Earth–Moon combination to obtain a capture so it can also, with equal probability, add energy to disrupt the system again.

Capture of the Moon from a heliocentric orbit cannot be completely ruled out but it is so unlikely that it is not a possibility that is seriously considered.

12.10.4. A single impact theory

In the mid 1970s an idea was put forward that the formation of the Earth–Moon system involved a sideswipe collision on the Earth by a large body with a mass about 10% of that of the Earth itself. The scenario has been convincingly modelled by the computing technique of SPH. The sequence of events for one particular impact, as calculated in 1987 by Benz, Slattery and Cameron, is shown in figure 12.10. In the first frame the tidally distorted impactor is seen striking the Earth with a glancing blow. The impactor material is seen to spread out and then much of it, including virtually all the iron component, is captured by the Earth. However, some of the mantle material from the impactor coalesces and goes into orbit around the Earth.

The parameters of the collision have to be within fairly tight limits for the result seen in figure 12.10 to be obtained. If the collision is too head-on or at too low a speed then all the impactor is accreted by the Earth. On the other hand if the impact is too fast then the impactor is completely vaporized and lost. A side effect of the collision is to give spin angular momentum to the Earth, a desirable characteristic since some theories of the origin of the Earth would have it spinning very slowly when it formed. In addition the model gives an initially molten Moon with little iron and a lack of volatile materials—all consistent with expectation.

The impact model is not completely without problems. Modelling gives 'Moons' that are too massive and also that have insufficient iron to explain the Moon's expected internal structure. Another, although less serious, factor is that the Moon fits comfortably in its physical characteristics with other satellites in the Solar System. From tables 7.1 and 7.2 it will be seen that the Moon is intermediate in its mass and density between Io and Europa. The large satellites of Jupiter could not have been formed by impacts so it might be considered fortuitous that quite different processes gave the Moon with properties between those of the two rocky Galilean satellites.

12.10.5. Capture in a collision scenario

The idea that the Moon was captured by the Earth from a heliocentric orbit was considered in section 12.10.3 and shown to be unlikely. However, the colliding protoplanets A and B, because they were

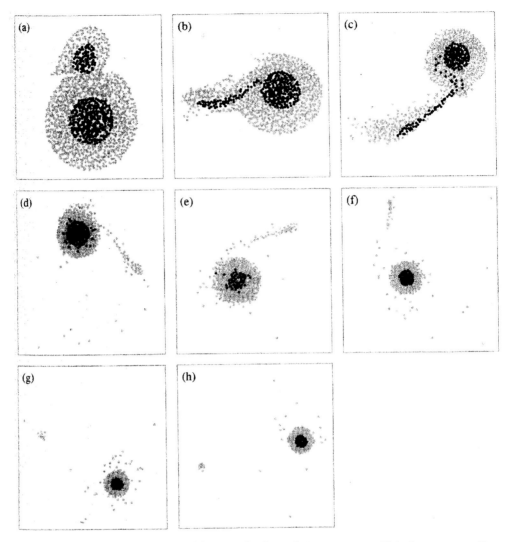

Figure 12.10. *Stages in the formation of the Moon by the single-impact process. Light dots represent silicate (granite) and dark dots are iron. The scale of the individual stages is progressively reduced to show the formation of the Moon.*

closest to the Sun and subjected to large solar tidal forces, would have had numerous large satellites. Dormand and Woolfson (1977) showed by numerical simulations that one of these could have been captured by the newly formed Earth into a very stable orbit with low eccentricity. The reason why this capture is possible is because it is taking place in a many-body environment and large close bodies take energy away from the Earth–Moon system.

This scenario also explains the hemispherical asymmetry of the Moon. The side facing the collision would have been abraded by large projectiles so thinning the crust on that side (section 6.5.3). For dynamical reasons associated with the distribution of matter in the Moon, when the Moon's spin became locked to its orbital period the same hemisphere that faced its original parent planet now faces the Earth.

12.11. OTHER BODIES IN THE SOLAR SYSTEM

12.11.1. *Mars and Mercury*

The densities of various solid bodies in the Solar System are illustrated in figure 12.11 and it seems that, in terms of density, Mars is more like a rocky satellite than a terrestrial planet. In particular its uncompressed density would be somewhere in the range 3700–$3800 \, \mathrm{kg \, m^{-3}}$. Dormand and Woolfson (1977) suggested that Mars was a satellite of one of the colliding planets that was released into a heliocentric orbit. Mars, just like the Moon, has hemispherical asymmetry although, unlike the Moon, its spin axis is not contained in the plane of asymmetry. Connell and Woolfson (1983) showed that the distribution of material in the crust of Mars would lead to polar wander and they explained the present arrangement of spin axis and surface features. An early bombardment of Mars, giving rise to the asymmetry, was also shown to explain the presence and evolution of water on the planet.

Mercury too is readily explained as an ex-satellite—in this case one that was closer to the planet and so heavily abraded that much of the mantle was removed leaving a dense body. Abrasion of a satellite similar to Mars could give a body with the characteristics of Mercury (Woolfson, 2000). With a large proportion of the mantle removed there would have been a massive redistribution of material to create a body with minimum potential energy (section P.7.1). Even so there is still a vestigial residue of hemispherical asymmetry similar to that found in the Moon and Mars.

Mars and Mercury are usually considered to have been formed by planetesimal accumulation. Its vicinity to the Sun has been cited as a reason why Mercury has such a high complement of iron, which has a higher vaporization temperature than silicates. Alternatively, as mentioned previously, a collision has been postulated to strip away a large proportion of the mantle of an originally less dense but more massive Mercury.

12.11.2. *Neptune, Triton, Pluto and Charon*

The relationship of Pluto's orbit to that of Neptune and the anomalous nature of Triton, Neptune's satellite, have been described in sections 5.5 and 7.7. Woolfson (1999) described a computational simulation in terms of the planetary collision scenario that explains the relationship between the four bodies Neptune, Triton, Pluto and Charon. This is illustrated in figure 12.12. The model begins with Pluto as a regular satellite of Neptune and Triton as a moderate-size satellite of a colliding planet released into a very eccentric orbit. Triton collides with Pluto sending it into a heliocentric orbit similar to what it has at present. The collision is an oblique one shearing off part of Pluto to

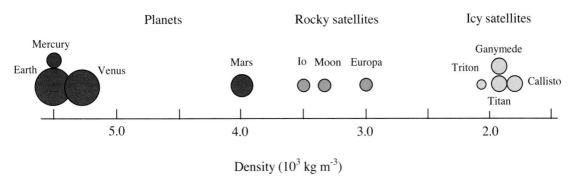

Figure 12.11. *The densities of some small solar-system bodies.*

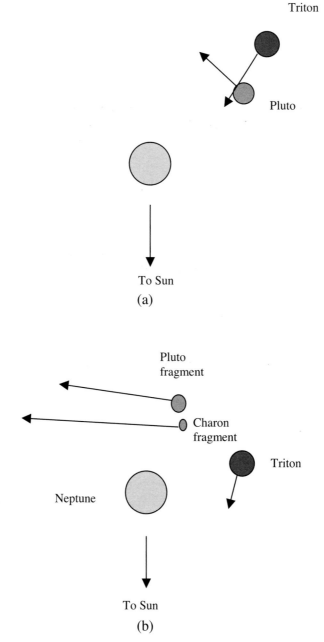

Figure 12.12. *(a) Pluto is in a near-circular orbit about Neptune. Triton is coming from aphelion on a heliocentric orbit with (a, e) = (29.09 AU, 0.912). (b) After an oblique collision the main fragment of the original Pluto is on a heliocentric orbit with (a, e) = (39.49 AU, 0.253). Triton is in a retrograde orbit around Neptune with (a, e) = (436 500 km, 0.8805) and the Charon fragment is in a retrograde orbit around Pluto.*

give Charon. Triton is captured by Neptune into a retrograde orbit that would evolve to what is seen today.

12.12. ISOTOPIC ANOMALIES IN METEORITES

An isotopic anomaly not mentioned in section 10.6 is that of deuterium. The ratio D/H in Jupiter is 2×10^{-5}, which is generally accepted as a cosmic value. On the Earth it is 1.6×10^{-4}, eight times the Jupiter value, and in some meteorites it is up to eight or so times the Earth value. The ratio in Venus is 1.6% or 100 times the Earth value. The process by which this occurred is well understood. The dissociation of water vapour in the early atmosphere of Venus, due to solar radiation, gave

$$H_2O \Rightarrow OH + H \qquad \text{and} \qquad HDO \Rightarrow OH + D$$

The combination of temperature and gravitational field enabled H to escape and D to be retained. Venus is a very dry planet and in losing most of its water the D/H ratio was greatly enhanced. Michael (1990) showed that under a wide range of conditions it would be possible for a lower-mass major planet to reach a similar D/H ratio to that in Venus. Holden and Woolfson (1995) considered the implication of one, or even both, the colliding planets having such a high D/H ratio.

It can be shown by approximate calculations, but also by numerical modelling, that the temperature in the planetary collision region would be a few million degrees, typically 3×10^6 K. At such a temperature nuclear reactions involving deuterium can take place and, with a high D/H ratio, this will raise the temperature to set off a chain of nuclear reactions. The original composition

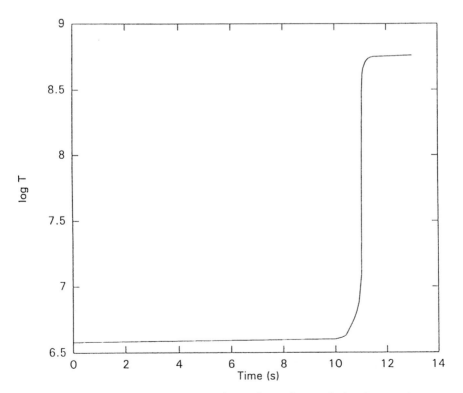

Figure 12.13. *Temperature variation with time during the period of nuclear reactions.*

in the collision region corresponded to a mixture of silicate minerals containing sodium, magnesium, aluminium and iron, more iron as a free metal, oxide and sulphide, water ammonia and methane as frozen components of surface material and atmospheric constituents such as hydrogen, helium, methane and neon. There were 283 nuclear reactions, plus corresponding reverse reactions, in the simulation. After about 11 s an explosion took place with the temperature rising to 5×10^8 K as seen in figure 12.13. At this stage the colliding planets would be blown apart, the reaction region would cool and the reactions cease. To illustrate the kind of results that were obtained the concentration of the three stable oxygen isotopes and of ^{17}F and ^{18}F, that quickly decay to ^{17}O and ^{18}O are shown in figure 12.14. The explosion takes place where the concentrations change most rapidly. Taking note of the fact that log(concentration) is being shown it is clear that, even allowing for radioactive fluorine decay, at the time of the explosion the composition is virtually pure ^{16}O. Mixtures of this material with unprocessed material can explain the oxygen anomaly described in section 10.6.1.

The products of the simulation are found to explain all the anomalies described in section 10.6, including the production of ^{26}Al and ^{22}Na. Because of the small-scale nature of the event in which the ^{22}Na is produced, condensation and its inclusion in cold material takes place on a timescale short compared to its 2.6 year half-life. There is no difficulty in explaining the production of neon-E with this model.

The more normal explanations for these anomalies, involving either the addition of grains from outside the Solar System or particle irradiation, have been given for each of the anomalies in the appropriate sections of chapter 10.

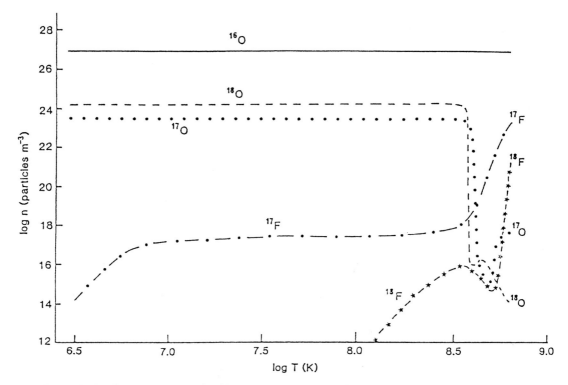

Figure 12.14. *The concentrations of stable oxygen isotopes and radioactive fluorine isotopes as a function of temperature during the period of nuclear reactions (Holden and Woolfson, 1995).*

12.13.   GENERAL COMMENTS ON A PLANETARY COLLISION

The planetary-collision hypothesis explains a large number of disparate features of the Solar System in terms of a single event, and this is its great attraction. A good feature of any theory is to have the minimum number of *ad hoc* assumptions and here there is only one—and that is that a planetary collision occurred. Even this assumption is based on the precession of planetary orbits which derives from a theory explaining, without obvious difficulty, the existence of planets and regular satellites. The coherence of the CT plus planetary collision is a strong feature but, like any other theory, it may be shown to be wrong in due course.

Problem 12

12.1 Mars, with mass 6.42×10^{23} kg and radius 3397 km, consists of an iron core of density 7.8×10^3 kg m^{-3} surrounded by a mantle of density 3.3×10^3 kg m^{-3}. What is the radius of the core?

Part of the mantle is abraded off until the total radius equals that of Mercury, 2440 km. What is the mass of the remaining body? How does this compare with the mass of Mercury?

TOPIC A

BASIC MINERALOGY

A.1. TYPES OF ROCK

Mineralogy has its origins in the study of rocks on Earth but has also now extended to rocky materials that come from outside the Earth. Rocks are distinguished by the way that they form and by the minerals they contain. It is important to recognize that common forms of rock contain mixtures of minerals and that rocks of the same general type may differ markedly in the proportions of the different minerals they contain. For example, the well-known rock-type granite consists mainly of the minerals quartz, feldspar and mica but the proportions of these and of the other components can vary. The minerals themselves are characterized by their crystal structure—the arrangement of atoms and the pattern of chemical bonds between them—but a given mineral can have a wide range of chemical composition. In some minerals iron and magnesium can readily substitute for each other without changing the crystal structure and the only effect may be a small change in size and/or shape of the unit cell, the unit of structure which repeats periodically in three dimensions to give the whole crystal. Such forms, of similar structure but of different chemical composition, are called *isomorphous* and for many minerals there is an isomorphous series where the relative amounts of two or more components can vary—for example from 100% iron to 100% magnesium with every intermediate composition also possible.

Most, but not all, rocks that come from outside the Earth have terrestrial equivalents so by understanding how rocks form on Earth we may obtain insights into processes that occurred on other bodies.

There are three types of process by which rocks can form. In the first of these *igneous rocks* are formed by the crystallization of molten rock, or *magma*, as it cools. If this happens on the surface then the rocks are *volcanic* and, since cooling will be relatively fast, the small crystallites will not have time to grow substantially and volcanic rocks are typically fine-grained. However, the rocks might cool at a moderate depth below the surface, giving *plutonic* rocks. In this case the cooling will be slower and crystals will grow to a larger size giving a coarse-grained material, an example of which is granite, referred to previously. We can see that texture is an important guide to cooling rate and for this reason is always of interest to mineralogists.

The second kind of process, which occurs at low temperature, gives *sedimentary rocks*. These are produced by deposits, or sediments, of mineral grains and subsequent compaction. Thus the compaction of sandy deposits gives rise to *sandstone* and fine mud sediments give rise to *shale*. If the mud contained organic material then the compaction under high pressure could transform this into hydrocarbons. Shale oil from Scotland was an important source of fuel in the early days of the motor industry in the United Kingdom and vast untapped deposits of shale oil exist in Canada. Another kind of sedimentary rock is that produced by crystallization of material from a solution; rock salt is made by the crystallization from the sea mainly of common salt, NaCl. Sometimes the material giving rise to the sedimentary rock may be of igneous origin, e.g. material deposited from an erupting volcano and such rocks are termed *pyroclastic*.

If an igneous or sedimentary rock is subjected to high temperature or pressure or both then its structure will be changed (metamorphosed). This gives what are called *metamorphic rocks*, examples being marble and slate.

Each of the three forms of rock can be converted into both other forms. Thus grains of igneous or metamorphic rocks can give sediments which eventually transform by compaction into a sedimentary rock. As is described in section 4.3, the Earth is a very active body and convective processes take surface material into the interior and material from the interior to the surface. Sedimentary or igneous rocks subjected to high temperature or pressure will be transformed to metamorphic forms and metamorphic rocks brought to, or near, the surface at high temperature will cool to an igneous form. It can also happen that metamorphic rocks reach the surface without transformation, or are formed close to the surface, which is why marble and other metamorphic rocks are available for quarrying.

Similar conditions of rock formation in different parts of the world give similar outcomes in terms of the types of rock formed. Consequently, when distinctive rocks are found from an extra-terrestrial source that are similar to those found on Earth then we may be able to make confident statements about the conditions under which they originated.

A.2. TYPES OF MINERALS

The total number of minerals is very large, perhaps tens of thousands in all but, fortunately, there are comparatively few which make up the bulk of the Earth, the Moon and meteorites, as far as we can explore these bodies, and they also fall into a few systematic groups. The following account is therefore certainly not comprehensive but it will cover all the major minerals. We shall deal with them under the headings: (i) silicates, (ii) carbonates, (iii) oxides and (iv) others.

Minerals of the same group, which are similar in the way that atoms are bound together, may nevertheless show different symmetries in their crystal form. Crystals are periodic structures in which a basic unit, the *unit cell*, is repeated in three dimensions. The symmetry of the crystal as a whole is a macroscopic manifestation of the underlying symmetry within the microscopic unit cell and there are seven kinds of basic symmetry which may exist for the whole crystal. For our present purpose these seven *crystal systems* can be described in terms of the characteristics of the parallelepiped (figure A.1) which defines the unit cell:

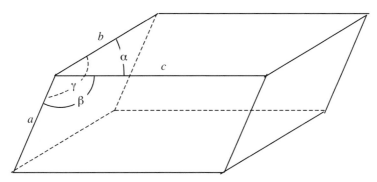

Figure A.1. *The parallelepiped defining a unit cell.*

1.   *Triclinic*, in which there are no special relationships between either the cell edges, a, b, c or the angles α, β, γ.
2.   *Monoclinic*, in which b is orthogonal to the a–c plane so that $\alpha = \gamma = \pi/2$.
3.   *Orthorhombic*, in which $\alpha = \beta = \gamma = \pi/2$.
4.   *Tetragonal*, in which $a = b \neq c$ and $\alpha = \beta = \gamma = \pi/2$.
5.   *Trigonal*, in which $a = b = c$ and $\alpha = \beta = \gamma \neq \pi/2$.
6.   *Hexagonal*, in which $a = b \neq c$ and $\alpha = \beta = \pi/2$; $\gamma = 2\pi/3$.
7.   *Cubic*, in which $a = b = c$ and $\alpha = \beta = \gamma = \pi/2$.

A.2.1.   Silicates

Silicates form the most extensive and important types of mineral. They show a wide variety of structural forms that are dependent on a basic group SiO_4 which is common to most of them. This group has the form of a tetrahedron, as shown in figure A.2, which is an extremely stable entity. In the mineral *olivine*, in the orthorhombic crystal system, the SiO_4 groups exist in independent units so that olivine is, in chemical terms, $(Mg,Fe)_2SiO_4$. Since magnesium and iron freely substitute for each other in this structure a general olivine mineral may be described as $Mg_xFe_{2-x}SiO_4$ in which x may be anywhere in the range 0 to 2. The end members of this isomorphous series are Mg_2SiO_4, *fosterite*, and Fe_2SiO_4, *fayalite*. Olivine is believed to be the major component of the Earth's mantle that stretches from below the crust to the Earth's core (section 4.3.6.2).

In other types of silicate mineral the SiO_4 units are linked together by sharing oxygen atoms. The most extreme form of this sharing is shown by *quartz*, a common mineral consisting only of silicon and oxygen, which shows hexagonal symmetry. For the quartz structure there is a space-filling framework in which each oxygen atom is part of two SiO_4 tetrahedra. One can think of each silicon atom as having only one half of the four bound oxygen atoms associated with it exclusively so the overall chemical description of quartz is SiO_2. A pure colourless form of quartz is *rock crystal* but with various trace impurities it can be coloured as in the precious stone *amethyst*.

Another type of mineral where each oxygen is shared by two tetrahedra is the *feldspars*, which are aluminosilicates of potassium, sodium and calcium. However, they do not show the ratio $Si:O = 1:2$ as some of the silicon atoms are replaced by aluminium. Thus in the alkali feldspars one quarter of the silicons are replaced by aluminium giving the minerals *albite*, $NaAlSi_3O_8$, and *orthoclase*, $KAlSi_3O_8$. On the other hand in the calcium feldspar *anorthite* one half of the silicons are replaced by

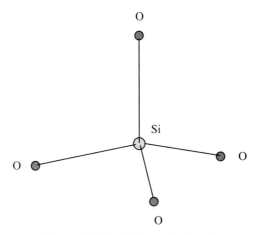

Figure A.2. *The SiO_4 structural unit.*

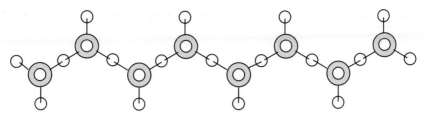

Figure A.3. *The linked chains of SiO$_4$ groups in pyroxenes.*

aluminium to give $CaAl_2Si_2O_8$. Albite and anorthite are the end members of a more complex mineral form, the *plagioclase* feldspars (triclinic system). Depending on the relative amounts of sodium and calcium the proportion of silicate atoms in the tetrahedra that are substituted by aluminium varies between one quarter and one half.

The three-dimensional framework arrangement of linked strongly bonded tetrahedral in quartz and the feldspars units creates very hard crystals which are extremely resistant to being fractured in any way, including by cleavage, which is a way of breaking crystals with planes containing weak or few chemical bonds.

The final type of silicate mineral we shall consider in detail is the *pyroxene* family, which can generally be described as $(Mg,Fe,Ca)SiO_3$. In this structure the tetraheda occur in chains so that each SiO_4 group has two of its oxygen atoms shared by other groups which gives the ratio $Si:O = 1:3$ (figure A.3). If less than about 5% of the combined number of metallic atoms is Ca then the mineral is usually an orthorhombic form called *orthopyroxene*. An important orthopyroxene, which has particular significance for meteorite studies (section 10.2.1), is *enstatite*, which contains more than 90% magnesium. With more than 5% Ca the mineral takes on a monoclinic form, *clinopyroxene*, which includes *pigeonite* (5–15% Ca) and *diopside* (45–50% Ca and less than 10% Fe). The pyroxenes are quite hard minerals but since the silicate units occur in disconnected chains they can be cleaved parallel to the chains.

The adaptable SiO_4 tetrahedron is capable of linking together in many more ways than we have discussed above to give $Si:O$ ratios such as $2:5$ and $4:11$, but we have dealt with all the common and more interesting silicate minerals from the point of view of describing the main features of bodies in the Solar System.

A.2.2. Carbonates

The main rocks in this category are *calcite*, $CaCO_3$, and *dolomite*, $CaMg(CO_3)_2$. They occur only in small amounts in igneous and metamorphic rocks but are quite common in sedimentary rocks. Carbon dioxide is a constituent of the atmosphere at the 0.03% level but it dissolves in sea-water to give carbonic acid, H_2CO_3, which reacts with soluble calcium-containing materials to give the precipitation of calcite, a hexagonal form of $CaCO_3$, or sometimes *aragonite*, an orthogonal form of $CaCO_3$. At the same time as this purely chemical formation process is going on marine creatures are exploiting the presence of CO_2, to manufacture calcium and some other carbonates to make protective shells. When these creatures die their shells add a biological source of $CaCO_3$ to the sediments and fossil shells are frequently found in limestone deposits, rocks containing more than about 50% $CaCO_3$.

A.2.3. Oxides

By far the most important oxide is the black oxide of iron, *magnetite*, Fe_3O_4, which forms cubic crystals. It is a common minor component of igneous and metamorphic rocks, of lunar maria

Table A.1. *Some commonly occurring minerals other than silicates, carbonates or oxides.*

Sulphides	Halides	Sulphates	Elements
Galena, PbS	Halite, NaCl	Gypsum, $CaSO_4 \cdot 2H_2O$	Diamond C
Pyrite, FeS_2	Fluorite, CaF_2	Anhydrite, $CaSO_4$	Graphite C
Troilite, FeS		Barites, $BaSO_4$	Gold Au
Chalcopyrite, $CuFeS_2$			Silver Ag
			Sulphur S

material and it also occurs in some meteorites. It is a ferromagnetic material which means that it can be magnetized in a magnetic field and will remain magnetized after the field is removed. The remnant magnetism of rocks containing magnetite gives information about extinct magnetic fields, both in strength and direction.

There are other oxides which occur such as the common oxide of iron, *haematite*, Fe_2O_3, *rutile*, TiO_2, and *ilmenite*, $FeTiO_3$, but they are less important than magnetite in the study of the Solar System.

A.2.4.   Other minerals

Other minerals that occur commonly enough to be significant are listed in table A.1 under various chemical headings. Some of these minerals are quite important for understanding the Solar System. The mineral troilite occurs as nodules in many meteorites and it may be an important component of planetary cores, including that of the Earth. Again, diamonds can only be produced from other forms of carbon, e.g. graphite, by very high pressure that might be due to being at some depth within a planet or due to a collision.

A.3.   ROCK COMPOSITION AND FORMATION

An individual rock normally contains a variety of minerals which have crystallized from the same original melt. For this reason the whole-rock content of rocks will sometimes be categorized by their chemical composition and this is often done by breaking it down into components, mainly oxide components. How this is done is best seen by looking at some of the common silicates. Thus

$$MgSiO_4 = MgO_2 + SiO_2$$

$$2(KAlSi_3O_8) = K_2O + Al_2O_3 + 6SiO_2$$

$$CaMgSi_2O_6 = CaO + MgO + 2SiO_2.$$

It is important to realize that when the composition of rocks is described in this way the components do not usually exist as the oxides but are assembled to form normal minerals.

There is a vast number of minerals and rock types but the following treatment will deal with only the major types of rock, the main processes by which they form and a sufficient number of the most common minerals to illustrate the principles governing rock composition and formation.

A.3.1.   Igneous rocks

Nearly all igneous rocks are silicates plus some oxides such as magnetite and ilmenite. The mixture of minerals present in an igneous rock depends very much on the composition of the magma from which it

Table A.2. *Percentage composition by weight of some common minerals in igneous rocks.*

Type of mineral	Mineral	SiO_2	$MgO + FeO$	Al_2O_3	$Na_2O + K_2O$	CaO
Ferromagnesian	Olivine	40	60	0	0	0
	Augite	50	23	3	0	20
	Hornblende	40	30	10	~0	12
Plagioclase	Albite 40%	54	0	29	5	12
	Albite	68	0	20	12	0
Feldspars	Orthoclase	65	0	18	17	0
	Leucite	55	0	23	22	0
Silica	Quartz	100	0	0	0	0

forms. To assist our discussion of this dependency we show in table A.2 the compositions of some common minerals which occur in igneous rocks and in table A.3 the compositions of some common rock types which together account for about 90% of all igneous rocks.

Granite is a mineral that is very commonly formed as an inclusion by flowing in a liquid state into the interstices of a previously formed rock. The consequent slow cooling due to the insulating effect of the surrounding rock is responsible for the formation of larger crystals and hence its coarse structure. We notice that compositionally it contains about 70% SiO_2 which means that there is a tendency to produce silicon-rich minerals. However, there is too much SiO_2 to produce a complete silicate mixture and some 10–30% of the SiO_2 will be present as quartz. We see from table A.2 that orthoclase and albite are SiO_2-rich minerals and these tend to be very common in granite. However, since these minerals have a high content of sodium and potassium and are correspondingly poor in calcium, magnesium and iron, then this too is reflected in the overall composition of granite.

Another kind of igneous rock, not given in table A.3, is *ultramafic rock*, which is usually dark coloured and is rich in iron and magnesium. These rocks are composed almost exclusively of the ferromagnesian minerals with olivine as a principal component. The relationship between type of rock and mineral composition is shown in figure A.4 which shows that the three types of rock given in table A.3 plus ultramafic rocks actually constitute a continuum of composition. The relative amounts of the components shown in the figure will link with the overall composition in table A.3,

Table A.3. *Mean compositions of the three most common types of igneous rock. The upper name applies to the coarse-grained plutonic material and the lower to the corresponding fine grain material produced by rapid surface cooling.*

Component	Type of rock		
	Granite Rhyolite	Diorite Andesite	Gabbro Basalt
SiO_2	70.8	57.6	49.0
Fe_2O_3	1.6	3.2	3.2
FeO	1.8	4.5	6.0
MgO	0.9	4.2	7.6
Al_2O_3	14.6	16.9	18.2
CaO	2.0	6.8	11.2
Na_2O	3.5	3.4	2.6
K_2O	4.2	2.2	0.9
TiO_2	0.4	0.9	1.0

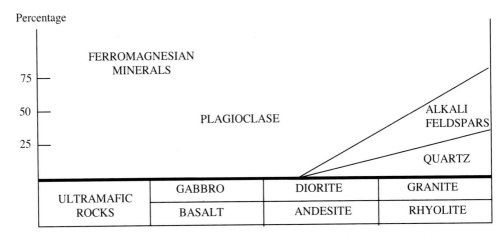

Figure A.4. *The continuum of mineral compositions from ultramafic rocks to granite/rhyolite.*

which reflects the composition of the melt from which the rock formed, and also with the composition of the individual mineral components of the rock as given in table A.2.

Another aspect of igneous rock formation relates to the rate of cooling of the melt and the environment in which it takes place. If the cooling is close to or at the Earth's surface and microcrystallites are forming rapidly throughout the melt and on a short timescale, then there is local migration of atoms and groups of atoms to organize themselves into mineral assemblages, which are low-energy, and hence favourable arrangements. This will tend to produce a more-or-less homogeneous rock. If the cooling is *extremely* rapid then the migrations of atoms and groups of atoms may not have time to take place, in which case an amorphous glassy material may be produced.

At the other extreme if cooling is slow so that larger crystals of higher-melting-point minerals form early within the fluid then these will usually be denser than the fluid and sink down to the bottom of the chamber in which the solidification is taking place. This tends to form layers of tightly packed crystals, or *cumulates*, and the final rock will show a layered structure, similar in some ways to that shown by sedimentary rocks, although the two types of rock are readily distinguishable. An example of this kind of rock formation is the layered gabbro in figure A.5. In practice, some igneous rocks cooling at intermediate temperatures may show mixed features with moderate-size crystals, giving a neither particularly coarse nor fine grained rock with some slight cumulate characteristics.

The commonest form of *extrusive* rock, i.e. one cooling at the surface, is basalt and since it is the product of volcanism it is the only one we can observe in the natural liquid state. In fact most basalts come from underground regions which are at temperatures at which solid and liquid may co-exist so that in the magma which comes to the surface quite large pre-existing crystals (*phenocrysts*) may be present which will end up embedded in a fine-grained or even glassy basalt.

This rather simplified picture of igneous rocks has brought out their major features but it must be emphasized that there are many kinds of production process, some not very common, which give igneous rocks not fitting into any of the patterns that have been described here.

A.3.2. Sedimentary rocks

The raw material for the formation of sediments comes from igneous and metamorphic rocks, most of which have been produced at high temperatures and end up on or close to the surface at low

Figure A.5. *A layered gabbro, the scale being given by the geologist's hammer.*

temperatures. If they cooled from a hot but solid state on a very long time scale then the atoms in the minerals are mobile and the minerals become reorganized to conform to an equilibrium mixture at the final temperature. On the other hand if cooling is rapid then the mineral assemblage is quenched and the final mineral content is in a non-equilibrated and metastable state. In such a state these mineral assemblages are easily affected by various processes which affect them chemically, such as chemical weathering in the atmosphere or from being in water.

The action of water is mainly by the process of *hydrolysis* which acts upon silicates to form silicic acid, H_4SiO_4, and a new mineral. If the initial mineral is an aluminosilicate then the end product is usually a *clay mineral*, which are hydrated aluminium silicates containing the groups Si_2O_5 where the silicate structure is in the form of extended sheets. As the most important constituents of igneous rocks are feldspars so the commonest components of sedimentary rocks are clay minerals. As an example of hydrolysis the action of water on orthoclase gives *kaolinite* by

$$4KAlSi_3O_8 + 22H_2O \rightarrow 4K^+ + 4OH^- + Al_4Si_4O_{10}(OH)_8 + 8H_4SiO_4.$$

By considering kaolinite in the form $2Al_2O_3 + 4SiO_2 + 4H_2O$ we can see the significance of it being considered as a hydrated mineral.

Sediments are formed from fragments of igneous and metamorphic rock which may come about in several different ways. There are the processes of erosion both by the weather, for example by frost, and by mechanical processes, such as the pounding of waves on a coastline. A rock acted on chemically by water, as previously described, will also be weakened mechanically and so be more prone to provide the fragments for a sediment. The fragments producing the sediment, referred to as *clasts*, come

Table A.4. *The chemistry of some sedimentary rocks (not all constituents are included).*

	Greywacke	Orthoquartzite	Silty clay	Clay from gneiss	Allochemical limestone
SiO_2	74.14	99.14	58.10	55.02	7.61
Al_2O_3	10.17	0.40	16.40	22.17	1.55
$Fe_2O_3 + FeO$	4.71	0.12	6.47	8.00	1.90
MgO	1.43	0	2.44	1.45	2.70
CaO	1.49	0.29	3.11	0.15	45.44
Na_2O	3.56	0.01	1.30	0.17	0.15
K_2O	1.36	0.15	3.24	2.32	0.25
H_2O	2.66	0.17	5.00	9.86	0.68
CO_2	0.14	0	2.63	0	39.27

together transported by wind and water and in size may consist of anything from small framents to tiny grains and in composition anything from the original minerals, known as the *resistate fraction*, to transformed materials such as clays or carbonates. The chemistry of some typical sedimentary rocks is given in table A.4.

Greywacke is a sedimentary rock rich in quartz grains but also containing significant amounts of resistate feldspar and other rocky materials embedded in a clay matrix. The clay from gneiss (section A.3.3.3) in the penultimate column of table A.4 indicates that the material is a weathered metamorphic rock and the allochemical limestone indicates that the material was formed within the area of deposition and will have an important component due to shell deposition.

Some of the compositional features are clearly related to the description of the sediment. The orthoquartzite is almost pure quartz, or SiO_2, while the silty clay and the clay from gneiss have high aluminium content because that is an essential component of a clay mineral. The limestone is largely $CaCO_3$ with some silicate fraction including clay minerals.

Minerals containing sodium and potassium are very prone to break down by hydrolysis and the alkali metals readily transfer into solution and are lost. In igneous rocks sodium is more common than potassium, but in sedimentary rocks this tends to reverse as will be seen in all but one of the sediments in table A.4. This is because sodium passes into solution much more readily than does potassium, while potassium is more readily incorporated into clay minerals than is sodium. By contrast aluminium is very insoluble, does not pass into solution and is the major non-silicate component of the clay minerals. Iron in the form of haematite or a hydrated oxide, *goethite*, is also extremely insoluble and when silicates are broken down by hydrolysis the iron is usually precipitated in these forms, which is how they predominantly occur in sediments. Calcium and magnesium readily pass into solution and so they are not common in most sediments, although the sea is saturated in $CaCO_3$ which is precipiated as limestone with magnesium also precipitating as *dolomite*, $CaMg(CO_3)_2$.

Sedimentary deposits have been extensively studied, not least because they contain many economically important materials. These materials can be either the sediments themselves, e.g. iron ores, or regions able to contain other materials, such as salt domes within which oil is trapped.

A.3.3.   Metamorphic rocks

Metamorphic rocks are those which have been formed under conditions of high temperature, high pressure or both. Where heating is a factor it is usual for it to be in an environment where the temperature is sustained for a long time and the rocks will be in an equilibrated state appropriate to the high temperature and may retain the high-temperature mineralogy even when they cool. We now consider the various types of process leading to the formation of metamorphic rocks.

A.3.3.1. Thermal metamorphism

If a hot magma intrudes into existing rocks then the magma is thermally insulated and so cools slowly while at the same time the previously existing rock is heated for a considerable period. Sedimentary rocks tend to be rather porous and so the intrusion of a molten granite into a region of sedimetary rock could produce this condition. The existing rock will become metamorphosed to take up the mineral composition more in equilibrium with the new prevailing temperature. Since the surrounding rocks will vary in temperature, depending on their distance from the molten material, the degree of metamorphic change will be variable and depend on position. If the surrounding rock is a clay then the metamorphism will form coarse grained high temperature minerals such as *andalusite*, Al_2SiO_5, and *cordierite*, $(Mg,Fe)_2Al_3Si_5AlO_{18}$. At very high temperatures another form of Al_2SiO_5, *sillimanite*, may occur. Another metamorphic transformation brought about by high temperature is that of limestone to marble.

A.3.3.2. Pressure metamorphism

Minerals are affected far more easily by temperature than by pressure but there are some high-density minerals which can only be formed at very high pressure. If such rocks are then transported to the surface they maintain their high-pressure forms. One way in which such material comes to the surface is through deep volcanic vents in the Earth's surface known as *kimberlite pipes*, since the prototype of such pipes was found in South Africa at Kimberley, a great diamond centre. The minerals are transported towards the surface in the form of nodules and are commonly rich in olivine with smaller amounts of pyroxene and magnesium-rich *garnet*, $(Mg,Fe)_3Al_2Si_3O_{12}$. Diamonds occur within the minerals brought up in kimberlite pipes and with the diamonds there occasionally occur small inclusions of *coesite*, a high-pressure polymorph of quartz, but there is an absence of *stishovite*, another polymorph of quartz which forms at an even higher pressure. This indicates that some of the material in the kimberlite pipes originates as deep as 150 km within the Earth, where the pressure is of order 50 000 kb (1 kilobar $\approx$ 990 atmospheres).

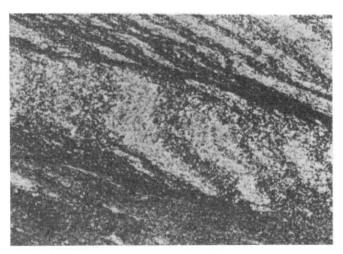

Figure A.6. *A cross section of a schist showing almost parallel elongated mica grains (dark grey). The portion shown is about 1 cm from top to bottom.*

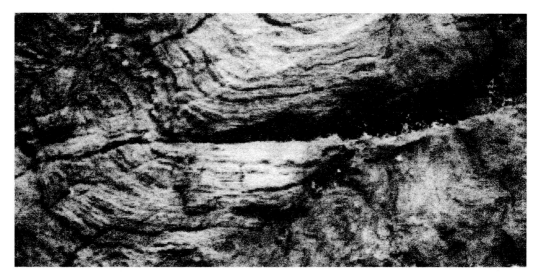

Figure A.7. *A portion of a gneiss distorted by deformation forces. This part of a rock face is about 2 m from top to bottom.*

Although kimberlite pipes are important in terms of understanding metamorphism they are a minority source of metamorphic material at or near the surface of the Earth. The large regions of metamorphic rocks which are brought up from below due to various convulsions of the Earth and exposure by weathering come from less deep regions where the pressure is less than 10 000 kb.

A.3.3.3.   Regional metamorphism

Metamorphic rocks are often found covering large regions, thousands of square kilometres in extent, which are due to the compressions and shear forces generated by motions of the Earth's crust, particularly in events such as mountain building. The motions involved in forming the rocks also lead to them showing distinctive structural features. For example the metamorphosed material may form as grains elongated in one direction, or as thin plates, with the grains or plates taking up parallel orientations. A very common structure of this kind is *schistosity* and the appearance of a schist is shown in figure A.6.

Another characteristic seen in rocks produced by regional metamorphism is *foliation*, where regions of the same mineral form patterns in the rock which may be streaks, layers or extended oval shapes (figure A.7). Where the foliation is on a larger scale, with individual pattern features up to tens of centimetres in size, then the rocks are called *gneisses*. To produce such large scale patterning, in which relatively large regions of neighbouring material were metamorphosed in the same way, requires high temperatures and/or pressures. This again shows that from the study of rocks we can learn a great deal about their history and what we learn from terrestrial rocks can be applied to any rocks obtained from outside the Earth.

Problems A

A.1   A granite is known to contain four component minerals—quartz, orthoclase, albite and olivine—the compositions of which are given in table A.2. A chemical analysis of the rock shows the following composition of chemical units:

SiO_2	65.85%
Al_2O_3	10.65%
$MgO + FeO$	15.00%
$Na_2O + K_2O$	6.70%
CaO	1.80%

Find the proportion of each kind of mineral in the granite. In the table use the data for albite 40%.

A.2   In a class of minerals called phyllosilicates the SiO_4 entities are formed by linkages of oxygen atoms as shown below. The chain may be taken as having infinite extension. The small clear circles indicate oxygen and there are a number of them directly above the shaded silicon atoms. If this arrangement gives the structural composition Si_xO_y then find x and y.

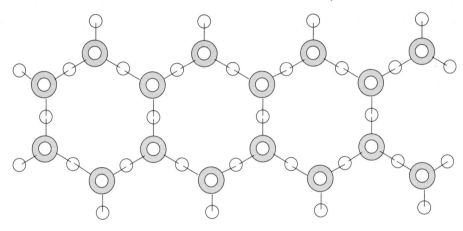

TOPIC B

GEOCHRONOLOGY—RADIOACTIVE DATING

Atomic phenomena are used to provide the most accurate dating of rocks. This involves the transformation of atoms either spontaneously (*natural radioactivity*) or artificially (*induced radioactivity*) in nuclear reactors. Henri Becquerel discovered natural radioactivity in 1896 in connection with the spontaneous fission of uranium salts but we know now that it is associated with some isotopes of all the chemical elements (section B.1.1).

B.1. COMMENTS ON ATOMIC STRUCTURE

The atom consists of a small central nucleus carrying a positive electric charge surrounded by negatively charged *electrons*. The positive charge is carried by the *proton* (p) which is about 1836 times as massive as the electron. The magnitude of the unit of charge (the charge on the electron) is the same for the proton as for the electron and is 1.602×10^{-19} coulomb (C). The nucleus also contains an electrically neutral particle called the *neutron* (n) with a mass slightly greater than that of the proton. The nucleus, therefore, carries almost all the mass of the atom.

B.1.1. Nuclear structure

The atom is electrically neutral so the number of protons in the nucleus must equal the number of electrons outside. This number is the *atomic number*, usually denoted by Z. It is characteristic of a particular chemical element and the *periodic table* displays a continuous range of the elements in terms of their Z numbers. The naturally occurring elements have Z numbers in the range $1 \leq Z \leq 92$. Elements with $Z > 92$ have been produced artificially in nuclear reactors. Neutrons are also present in the nucleus but their presence does not affect the electrical conditions of the atom. For the heavier atoms the number of neutrons is about one and a half times the number of protons but the precise number of neutrons for a particular element can vary. The number of protons and neutrons combined in a nucleus is called the *atomic mass*, denoted by A. It can be expressed in kg by remembering that each nucleon (proton or neutron) has a mass of 1.67×10^{-27} kg.[†] The number of neutrons in the nucleus is given by $(A - Z)$. The structure of an atom of element X can be represented by the symbol $_Z^A X$, which gives information simultaneously about Z and A. Since the symbol representing the element also indicates the value of Z this can be excluded to just give $^A X$.

[†] Actually the mass of the nucleus is less than the sum of the masses of the consituent nucleons since some mass is converted into binding energy. The atomic mass unit (amu) is one-twelfth of the mass of a neutral ^{12}C atom and is 1.66×10^{-27} kg.

Atoms with a given Z but different A are called *isotopes* of the particular chemical element. Because the Z number is the same, all the isotopes of a particular atom have exactly the same chemical properties and cannot be distinguished by chemical means alone but they have different masses. They can, therefore, be separated by the mechanical means of the mass spectrometer. All elements have isotopes though the number varies from element to element. Not all the isotopes occur naturally, some having to be made artificially, usually using nuclear reactors. As examples, hydrogen has two naturally occurring isotopes, H (or $^1_1$H) with a single proton and electron ($Z = 1$, $A = 1$) and $^2_1$H (deuterium or *heavy hydrogen*) with a proton and a neutron so that $Z = 1$ but $A = 2$. These isotopes occur naturally in terrestrial material in the proportions 6250:1. There is a third isotope, $^3_1$H, *tritium*, which is not naturally occurring. It decays into deuterium with the release of a neutron and so is *radioactive*. At the other extreme, uranium has $Z = 92$ and the most abundant isotope has $A = 238$, giving 146 neutrons in the nucleus. The naturally occurring isotopes are $^{238}_{92}$U (99.275%), $^{235}_{92}$U (0.720%) and $^{234}_{92}$U (0.005%). Another isotope, $^{236}_{92}$U, does not occur naturally but can be made artificially as a by-product of other nuclear reactions as can $^{233}_{92}$U, an isotope that is of use in nuclear reactor technology.

Radioactive nuclei are not stable over time and *decay* into other elements. The initial nucleus is called the *parent* and the resulting nucleus after the decay is called the *daughter*. There may be one stage of decay or several in sequence, the latter case being referred to as a *decay series*. Uranium is a good example where each of the isotopes passes through a series of decays with the *end point* being lead; the particular isotope of lead depends on the particular decay chain. The details of the chains terminating in lead are as follows:

Original element		End point	No. of emitted particles	
			α ($^4_2$He)	β^-
$^{238}_{92}$U	$\Rightarrow$	$^{206}_{82}$Pb	8	6
$^{235}_{92}$U	$\Rightarrow$	$^{207}_{82}$Pb	7	4
$^{232}_{92}$U	$\Rightarrow$	$^{208}_{82}$Pb	6	6

Lead is not the inevitable end of a radioactive series. For example, the neptunium series with 14 emissions has the end point $^{209}_{82}$Bi, an isotope of bismuth.

B.1.2.   The emissions

The transition from a parent to a daughter nucleus comes about by the emission of particles and of radiation. Three distinct emanations were recognized in the initial discovery. Their composition was not then known so they were called α, β and γ rays. It soon became clear that these are respectively (i) a helium nucleus ($Z = 2$, $A = 4$) which is a strong combination of nucleons, (ii) fast negatively charged electrons (β^- particles) or positively charged positrons (β^+ particles) and (iii) the very high energy electromagnetic γ radiation. The loss of an α particle means that the nucleus *loses* both two units of positive charge and four units of mass. The loss of a β^- particle means the nucleus *gains* one unit of positive charge, the mass remaining the same. The process by which this happens is that in the nucleus a neutron decays into a proton and a negative electron. Electrons, both of positive and negative charge, that come from radioactive decay have very high energies and are conventionally referred to as β particles with the appropriate sign. The process that goes on in the nucleus can be described as

$$\text{n} \Rightarrow \text{p} + \beta^- + \bar{\nu}, \qquad (\text{B.1})$$

where $\bar{\nu}$ is a massless particle, that nevertheless possesses energy and momentum, called an antineutrino. In fact, a naked neutron is a radioactive particle in its own right so that the transformation equation (B.1) can take place outside the nucleus. An example of this kind of radioactive transformation, important for dating on an archaeological timescale, is

$$^{14}_{6}\text{C} \Rightarrow {}^{14}_{7}\text{N} + \beta^-.$$

The emission of a β^+ particle is due to

$$\text{p} \Rightarrow \text{n} + \beta^+ + \nu, \qquad (\text{B.2})$$

where now ν is a neutrino. An example of such a radioactive transformation is

$$^{22}_{11}\text{Na} \Rightarrow {}^{22}_{10}\text{Ne} + \beta^+.$$

The decay mode (B.2) is *decay by positron emission*. Such a decay for the proton is not found outside the nucleus and the proton is not, therefore, regarded as a radioactive particle.

Another way a proton in the nucleus can transform into a neutron is by

$$\text{p} + \beta^- \Rightarrow \text{n} + \nu$$

and this process, *decay by electron capture*, can occur within the confines of an atom by absorption of a valence electron inside the nucleus. It can also occur when atoms are bombarded by high-energy electrons such as in the reaction

$$^{7}_{4}\text{Be} + \beta^- \Rightarrow {}^{7}_{3}\text{Li}.$$

B.2.   THE LAWS GOVERNING RADIOACTIVE DECAY

B.2.1.   *The physical principles*

Radioactivity is a random process and the number of radioactive atoms, δN, that will decay during a given time interval, δt, is proportional to the number of atoms, N, present in the sample. Introducing a decay constant, λ, characteristic of the particular radioactive atoms involved, the law of radioactivity can be written in the form

$$\delta N = -\lambda N \, \delta t. \qquad (\text{B.3})$$

The negative sign here indicates that N decreases with the time. The ratio $\delta N/N$, referred to a unit time interval, is often called the *activity*. The number of atoms that remain after the time interval δt is obtained by integrating equation (B.3) in the differential form

$$\frac{\mathrm{d}N}{\mathrm{d}t} = -\lambda N.$$

Integration over the time interval $t = 0$ to τ gives the result

$$N_\tau = N_0 \exp(-\lambda\tau). \qquad (\text{B.4})$$

Where N_0 and N_τ are the numbers of radioactive atoms at the the beginning and end of the time interval. The N_0 original atoms are called the *parent* atoms. The atoms formed by the decay are called *daughter* atoms; since the total number of atoms remains constant the number, D_τ, of daughter atoms after time τ is $D_\tau = N_0 - N_\tau$.

A useful indicator of the timescale of the decay is the half-life, $T_{1/2}$, of the isotope, the time for one half of the initial atoms to decay. With $\tau = T_{1/2}$, and $N_\tau = N_0/2$, equation (B.4) gives immediately

$$T_{1/2} = \frac{\ln 2}{\lambda} = \frac{0.693}{\lambda}$$

where ln refers to the natural logarithm. A short half-life is associated with a large decay constant and vice versa (see Problem B.1).

B.2.2. A simple age measurement

The previous formulae are used with the initial values and those at the later time:

$$t = 0 \qquad N = N_0$$
$$t = \tau \qquad N_\tau = N_0 - D_\tau.$$

Then

$$N_0 - D_\tau = N_0 \exp(-\lambda\tau)$$

which gives

$$D_\tau = N_0[1 - \exp(-\lambda\tau)] = N_\tau[\exp(\lambda\tau - 1)] \qquad (\text{B.5a})$$

or

$$\tau = \frac{1}{\lambda} \ln\left[1 + \frac{D_\tau}{N_\tau}\right]. \qquad (\text{B.5b})$$

The ratio D_τ/N_τ can be measured using a mass spectrograph and, with λ known, the age τ of the specimen follows immediately. The method depends upon knowing that the only source of daughter atoms is the radioactivity being considered.

B.2.3. Decay in a radioactive chain

Equation (B.5b) applies to the simplest case when there is a single daughter but we saw in section B.1.1 that for the heavier radioactive elements the daughters themselves have daughters. Then it is necessary to repeat the decay chain an appropriate number of times until the stable end product is achieved. Suppose the initial element, with number of atoms N_1, decays through n stages to form the stable end product of number D. For the ith stage of the chain the number of atoms is N_i and the decay constant is λ_i. There is, then, a set of coupled differential equations with each one, other than the first and last, involving a loss of atoms due to decay and a gain of atoms due to the previous stage. These equations are:

$$\frac{dN_1}{dt} = -\lambda_1 N_1, \qquad \frac{dN_2}{dt} = \lambda_1 N_1 - \lambda_2 N_2, \qquad \frac{dN_m}{dt} = \lambda_{m-1} N_{m-1} - \lambda_m N_m, \qquad \frac{dD}{dt} = \lambda_n N_n. \qquad (\text{B.6})$$

This set of linked equations occurs in other physical and chemical situations and is often called the *Bateman relations*. It is complicated to solve for an extended chain and is best done using numerical methods with a computer.

There is one situation, however, where the solution of the set (B.6) is simple. This is when the half-life of one of the decay steps is substantially larger than any of the others in the chain. This is the case, for example, for the decay of $^{238}_{92}$U (section B.1.1) with a half-life of 4.51×10^9 years, or about the present age of the Solar System: equation (B.5b) can be applied as though the final product comes from the original element in a single stage with the dominant half-life.

B.2.4.   Bifurcated decay

The example of $^{40}_{19}$K is complicated in that it can decay either by β^- emission to a calcium isotope, $^{40}_{20}$Ca (89% probability), or by electron capture to the argon isotope, $^{40}_{18}$Ar (11% probability). This *bifurcated decay* introduces new features into the analysis. Using an obvious notation

$$\frac{\mathrm{d}N}{\mathrm{d}t} = -(\lambda_A + \lambda_C)N$$

so that

$$N_\tau = N_0 \exp[-(\lambda_A + \lambda_C)\tau]$$

in place of equation (B.4). The rate of increase of the daughter argon atoms and of the daughter calcium atoms is respectively

$$\frac{\mathrm{d}D_A}{\mathrm{d}t} = \lambda_A N_t, \qquad \frac{\mathrm{d}D_C}{\mathrm{d}t} = \lambda_C N_t.$$

Then

$$\frac{\mathrm{d}D_A}{\mathrm{d}t} = \lambda_A N_0 \exp[-(\lambda_A + \lambda_C)t]$$

so that, if $D_A = 0$ at $t = 0$

$$(D_A)_\tau = \frac{\lambda_A N_\tau}{[\lambda_A + \lambda_C]} [\exp(\lambda_A + \lambda_C)\tau - 1]. \tag{B.7}$$

Assuming that there were no daughter atoms at zero time, the age of the mineral is readily shown to be

$$\tau = \frac{1}{(\lambda_A + \lambda_C)} \ln \left[1 + \frac{(\lambda_A + \lambda_C)}{\lambda_A} \frac{(D_A)_\tau}{N_\tau} \right]. \tag{B.8}$$

A companion expression (with A and C interchanged) results from the use of calcium instead of argon. The use of potassium decay to calcium is rarely used in practice because calcium occurs so widely in rocks that there is always considerable ambiguity as to how much calcium has been produced during the decay period and how much was there initially. On the other hand if the argon is to be measured then there is the problem of the extent to which the containing rock is impermeable to argon. If some of the argon escapes then $(D_A)_\tau$ in equation (B.8) would be too small and hence the determined age would also be too small. We now consider this problem of the retention of gases in rocks.

B.2.5.   Age determination: the closure temperature

The first age determinations using radioactivity were made by Rutherford and Boltwood in 1907 using the parent uranium and the daughter helium (α particles) which had collected in a rock cavity. The rock was permeable to helium so the age determinations were not reliable, being too low. This was later corrected using lead (the end product of the radioactive chain (section B.2.6.4)) as the daughter, the results being published in 1907. It was immediately clear that the geological age scale of the Earth accepted at that time was a gross underestimate. The method has since been developed.

The most basic uncertainty is whether the measured number of daughter atoms has arisen entirely from the parent decay or whether there might be an earlier source. Again, gases can percolate slowly through rocks even at lower temperatures but the speed of percolation rises with

Table B.1. *Some closure temperatures for three dating method.*

Parent/daughter	Mineral	Closure temperature (°C)
K–A	Hornblende	540
	Biotite	280 ± 40
	Muscovite	≈ 350
U–Pb	Zircon	750
	Monazite	650
	Apatite	≈ 350
Rb–Sr	Biotite	320
	Muscovite	>500
	Feldspar	≈ 350

the temperature. There is a temperature above which trapped gases cannot be retained by the mineral. This is the *closure temperature*. Some values with various minerals are collected in table B.1. Age determinations using trapped gases evolved by the rock then refer back to the time the rock had cooled to this temperature. It is seen that these temperatures vary over a wide range: it is some 700°C in the mineral zircon for uranium–lead measurements (section B.2.6.4) down to 280°C for biotite using the potassium–argon method (section B.2.4.). The most accurate radioactive decay scheme is where the half-life is comparable with the age being measured. Then, the number of parent and daughter atoms is similar so the ratio N/D is of order unity. The decay characteristics for a range of geologically interesting parent–daughter combinations are collected in table B.2. It is seen that a wide range of half-lives occur covering the range from a few times 10^{11} years down to a little less than 300 years. It is important to choose a scheme with the greatest concentrations of parent and daughter atoms for the most accurate measurements. Estimates of the quantities of the most common radioactive elements in four classes of rocks are collected in table B.3.

It is always desirable, for the greatest accuracy, to use several independent methods for the age determination of a given sample if at all possible. The accuracy can be tested by the spread of the values obtained, the aim being to achieve the minimum spread. The method should be carefully chosen to provide the best precision, as judged by repeatability.

Table B.2. *The details of the parent/daughter decays of geologically interesting radioactive minerals.*

Parent	Daughter	Decay constant (10^{-11} years^{-1})	$T_{1/2}$ (10^6 years)	Decay products
^{238}U	^{206}Pb	15.5	4468	α, β^-
^{235}U	^{207}Pb	98.5	704	α, β^-
^{232}Th	^{208}Pb	4.95	14010	α, β^-
^{87}Rb	^{87}Sr	1.42	48800	β^-
^{147}Sm	^{143}Nd	0.654	106000	α
^{40}K	^{40}Ca	49.6	1250	β^-
	^{40}Ar	5.81	(combined)	electron capture
^{39}Ar	^{39}K	2.57×10^8	2.69×10^{-4}	β^-
^{176}Lu	^{176}Hf	1.86	37300	β^-
^{187}Re	^{187}Os	1.65	42000	β^-
^{14}C	^{14}N	1.21×10^7	5.73×10^{-3}	β^-

Table B.3. *Estimates of the concentrations of the common radioactive elements in four classes of rock.*

Element	Content of element in rock type			
	Granitoid (ppm)	Basalt (ppm)	Ultramafic (ppm)	Shale (ppm)
Uranium	4	0.3	0.02	4
Thorium	15	1	0.1	12
Lead	20	4	0.1	20
Rubidium	200	30	0.5	140
Strontium	300	500	50	300
Samarium	8	10	0.5	10
Neodymium	44	40	2	50
Potassium (%)	3	3	3	3

B.2.6.   The isochron diagram

An age determination using a single measurement is unlikely to be very accurate and it is preferable to use a method requiring several measurements and where the age can be derived from some least-squares procedure. This is achieved by the method of the isochron. It is not possible to derive a precise measurement of the age from a single measurement when an unknown amount of the daughter product was present initially but this problem can be solved by the isochron approach. We now illustrate the principle for various parent/daughter combinations.

B.2.6.1.   Rubidium $\rightarrow$ strontium

The half-life of rubidium is 4.88×10^{10} years so this decay is best used for old rocks. Rubidium has two isotopes of masses 85 and 87 while strontium has four isotopes of masses 84, 86, 87 and 88. The decay process of interest is that of $^{87}_{37}\mathrm{Rb}$ to $^{87}_{38}\mathrm{Sr}$. In an obvious and conventional notation with $N_\tau = [^{87}\mathrm{Rb}]_{\mathrm{now}}$ and $D_\tau = [^{87}\mathrm{Sr}]_{\mathrm{now}}$ in equation (B.5a)

$$[^{87}\mathrm{Sr}]_{\mathrm{now}} = [^{87}\mathrm{Rb}]_{\mathrm{now}}[\exp(\lambda\tau) - 1]. \tag{B.9}$$

With the decay constant for rubidium known, it is necessary only to measure the present-day ratio of the involved isotopes of strontium to rubidium to determine the age of the sample. Unfortunately, strontium occurs fairly widely in rocks and it is quite likely that there was some initial strontium present, although it is not known how much. This can be found. Suppose this unknown quantity was $[^{87}\mathrm{Sr}]_0$. Then

$$[^{87}\mathrm{Sr}]_{\mathrm{now}} = [^{87}\mathrm{Sr}]_0 + [^{87}\mathrm{Rb}]_{\mathrm{now}}[\exp(\lambda\tau) - 1].$$

At the present time the ratios of the strontium isotopes are: $84 : 86 : 87 : 88 = 0.6\% : 10\% : 7\% : 83\%$. Since $^{86}\mathrm{Sr}$ is not a product of radioactive decay the amount at the present time should the same as was there initially, so that $[^{86}\mathrm{Sr}]_{\mathrm{now}} = [^{86}\mathrm{Sr}]_0$. The measured values of the isotopes can be normalized to this initial value to give

$$\frac{[^{87}\mathrm{Sr}]_{\mathrm{now}}}{[^{86}\mathrm{Sr}]_{\mathrm{now}}} = \frac{[^{87}\mathrm{Sr}]_0}{[^{86}\mathrm{Sr}]_{\mathrm{now}}} + \frac{[^{87}\mathrm{Rb}]_{\mathrm{now}}}{[^{86}\mathrm{Sr}]_{\mathrm{now}}}[\exp(\lambda\tau) - 1] = \frac{[^{87}\mathrm{Sr}]_0}{[^{86}\mathrm{Sr}]_0} + \frac{[^{87}\mathrm{Rb}]_{\mathrm{now}}}{[^{86}\mathrm{Sr}]_{\mathrm{now}}}[\exp(\lambda\tau) - 1]. \tag{B.10}$$

The expression in the final bracket is the same for all minerals in a rock. However, the strontium to rubidium ratio will differ from one mineral to another although these crystallized originally from the same molten material. If, for different mineral samples, the left-hand side of equation (B.10),

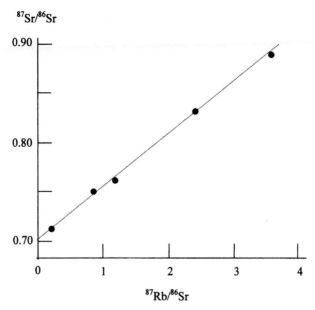

Figure B.1. *A whole-rock isochron for the Amitsoq gneiss of the Godthåb district of Western Greenland. The age estimated from the slope of the graph is 3660 ± 99 Ma. The initial ratio estimated from the intercept is 0.7009 ± 0.0011.*

$[^{87}Sr]/[^{86}Sr]$, is plotted against $[^{87}Rb]/[^{86}Sr]$ then the result is a straight line where the slope of the line can be measured to give $[\exp(\lambda\tau) - 1]$. This is called the *rubidium–strontium whole-rock isochron*. With λ known, the age, τ, follows at once. There will, of course, be a spread of points on the graph and the appropriate straight line is determined by a least-squares fit.

The line does not pass through the origin and the intercept gives the ratio $[^{87}Sr]_0/[^{86}Sr]_0$ referring to the state of the rock when it formed: this is called the *initial ratio*.

An example is shown in figure B.1 which is a whole-rock isochron for the Amîtsoq gneiss (section A.3.3.3) of the Godthåb district of western Greenland, among the oldest rocks yet found on Earth. The age determination is 3660 My. This is the time the gneiss was formed from the metamorphosis of granite and the initial rock must have been older.

The decay constant for rubidium–strontium (table B.2) is of order 10^{-11} yr^{-1} so the maximum value of $\lambda\tau$ is of order 10^{-2}. This means that to good approximation $[\exp(\lambda\tau) - 1] \approx \lambda\tau$. Then, to this approximation equation (B.10) becomes alternatively

$$\frac{[^{87}Sr]_{now}}{[^{86}Sr]_{now}} = \frac{[^{87}Sr]_0}{[^{86}SR]_0} + \frac{[^{87}Rb]_{now}}{[^{86}Sr]_{now}}\lambda\tau.$$

There is an assumption that the molten material of the Earth's mantle and similar molten material that gave rise to chondritic meteorites and eucrites (section 10.2.2), began their existence at the same time and with the same concentrations of strontium and rubidium. While the material was molten ^{87}Rb was decaying and increasing the amount of ^{87}Sr present. Since the half-life is so long the rate of ^{87}Sr production has been roughly constant so the initial concentration of ^{87}Sr and hence the ratio $[^{87}Sr]/[^{86}Sr]$ would have increased linearly with time. The intercept of a whole-rock isochron indicates the initial value of $[^{87}Sr]/[^{86}Sr]$ when the material solidified and so, as long as the rock is a representative sample of the melt, the intercept will be another indicator of age. The very oldest age measurements, for achondrites, give an age of about 4500 million years and a value of $[^{87}\mathrm{Sr}]/[^{86}\mathrm{Sr}]$ equal to 0.699, known as BABI (Basic Achondrite Best Initial). This is taken to be the value of the ratio when the Solar System was formed. A plot showing the increase in the strontium ratio with time in the mineral samples spanning a time interval is called the *mantle growth curve* (mgc) for strontium. The oldest known terrestrial rocks are the Archaean and these have mgc = 0.700. For modern oceanic basalts mgc = 0.704.

However, not all samples give intercepts falling on the mantle growth curve. Consider a reservoir of molten material that has become isolated from the effectively infinite mantle source. If granitic rocks solidify from the melt then, since they contain a high ratio of rubidium to strontium with a large content of both elements then the residual melt is effectively depleted in rubidium. Thus the ratio $[^{87}\mathrm{Sr}]/[^{86}\mathrm{Sr}]$ will be lower in the residual melt than in the main mantle material and any subsequent solid products from the rubidium-depleted source will show a lower mgc value. A comparison of the mgc value with age from the whole-rock isochron can indicate something about the history of the rock.

The isochron method is reliable in general, but the application has certain restrictions. Rubidium and strontium are often mobile within the rock as the result of chemical processes and rubidium does not occur in all rocks—it has a very low abundance in ultramafic rocks. Again, the half-life of rubidium is very long so the method is not suitable for the study of young rocks.

B.2.6.2.   *Samarium → neodymium*

These are rare earth elements with very similar chemical properties and can be used to construct an isochron as described in section B.2.6.1. Neodymium has seven naturally occurring isotopes, six of which are stable. These are: $^{142}_{60}\mathrm{Nd}$ (27.16%), $^{143}_{60}\mathrm{Nd}$ (12.18%), $^{145}_{60}\mathrm{Nd}$ (8.29%), $^{146}_{60}\mathrm{Nd}$ (17.19%), $^{148}_{60}\mathrm{Nd}$ (5.75%) and $^{156}_{60}\mathrm{Nd}$ (5.63%). The radioactive isotope is $^{144}_{60}\mathrm{Nd}$ (23.80%) which decays by α decay with a half-life of 2.4×10^{15} years and so is effectively also stable. This isotope is taken as the normalizer. There are nine isotopes of samarium of which $^{147}_{62}\mathrm{Sm}$ decays to $^{143}_{60}\mathrm{Nd}$ by α decay. The samarium–neodymium reaction is normalized by comparison with the present proportion of $^{144}\mathrm{Nd}$. The decay expression equivalent to equation (B.10) is

$$\frac{[^{143}\mathrm{Nd}]_{\mathrm{now}}}{[^{144}\mathrm{Nd}]_{\mathrm{now}}} = \frac{[^{143}\mathrm{Nd}]_0}{[^{144}\mathrm{Nd}]_0} + \frac{[^{147}\mathrm{Sm}]_{\mathrm{now}}}{[^{144}\mathrm{Nd}]_{\mathrm{now}}}[\exp(\lambda\tau) - 1]. \tag{B.11}$$

The isochron gives τ and the initial $[^{143}\mathrm{Nd}]/[^{144}\mathrm{Nd}]$ ratio. In general $[^{147}\mathrm{Sm}/^{144}\mathrm{Nd}]_{\mathrm{now}} \approx 0.1\text{--}0.2$. The concentrations of Sm and Nd in rocks are not very high (typically 10 ppm) so high precision apparatus is required for measurements. But these elements are essentially unaffected by weathering or metamorphism. The decay constant for samarium is very small so the combination is particularly good for dating old rocks. Actually the high precision of the measurements that is possible and the stability of the elements make it also useful for dating younger rocks.

B.2.6.3.   *Rhenium → osmium : lutetium → hafnium*

Rhenium decays to osmium ($^{187}_{75}\mathrm{Re} \rightarrow ^{187}_{76}\mathrm{Os}$) through beta decay, as does lutetium to hafnium ($^{176}_{71}\mathrm{Lu} \rightarrow ^{176}_{72}\mathrm{Hf}$). The processes can be used to construct isochrons as for Sm and Nd. These have proved useful in the study of the origin of magmas and for studies of mantle evolution.

B.2.6.4.   *Uranium → lead*

This uses the decay chains $^{238}_{92}\mathrm{U} \rightarrow ^{206}_{82}\mathrm{Pb}$ and $^{235}_{92}\mathrm{U} \rightarrow ^{207}_{82}\mathrm{Pb}$. Lead has six naturally occurring isotopes, four of which are stable. Isochrons can be constructed in the usual way and $^{204}\mathrm{Pb}$ is taken as the normalizer. Alternatively, the formulae of section B.2.2 can be applied to give the ratio

$$\frac{[^{207}\text{Pb}]_{\text{now}}}{[^{206}\text{Pb}]_{\text{now}}} = \frac{[^{235}\text{U}]_{\text{now}}}{[^{238}\text{U}]_{\text{now}}} \frac{[\exp(\lambda_{235}\tau) - 1]}{[\exp(\lambda_{238}\tau) - 1]}. \qquad (B.12)$$

It is known that $[^{235}\text{U}]_{\text{now}}/[^{238}\text{U}]_{\text{now}} = 1/137.88 = 7.2526 \times 10^{-3}$ and the Pb ratio can be measured. This leads to a value of τ from one measurement.

B.2.6.5.   *Thorium → lead*

The thorium decay is $^{232}_{90}\text{Th} \to ^{208}_{82}\text{Pb}$. The isochron can be constructed as before. Using equation (B.10) and choosing non-radiogenic ^{204}Pb as the normalizer:

$$\frac{[^{208}\text{Pb}]_{\text{now}}}{[^{204}\text{Pb}]_{\text{now}}} = \frac{[^{208}\text{Pb}]_0}{[^{204}\text{Pb}]_0} + \frac{[^{232}\text{Th}]_{\text{now}}}{[^{204}\text{Pb}]_{\text{now}}} [\exp(\lambda\tau) - 1].$$

Thorium and lead are less easily lost from minerals than is uranium making this an attractive dating procedure. The very long half-life for thorium means, however, that the method is not very useful for young rocks.

B.2.6.6.   *Potassium → argon*

It is seen from table B.2 and section B.2.4 that there is ambiguity in the decay of potassium 40. The decay relation, using ^{36}Ar as the normalizer, is:

$$\frac{[^{40}\text{Ar}]_{\text{now}}}{[^{36}\text{Ar}]_{\text{now}}} = \frac{[^{40}\text{Ar}]_0}{[^{36}\text{Ar}]_0} + \frac{[^{40}\text{K}]_{\text{now}}}{[^{36}\text{Ar}]_{\text{now}}} \frac{\lambda_A}{(\lambda_A + \lambda_C)} [\exp(\lambda_A + \lambda_C)\tau - 1]. \qquad (B.13)$$

It is usually sufficient to assume the initial argon ratio $[^{40}\text{Ar}/^{36}\text{Ar}]_0$ has its present-day atmospheric value of 295.5. It is then possible to date a sample from a single measurement. The short half-life of potassium means this is a suitable method for dating young, and very young, rocks. Argon, a gas, is often lost from a mineral and the spread of closure temperatures is wide for a range of minerals. On the positive side, this makes the method useful for exploring the thermal history of the rocks.

B.2.7.   *The concordant diagram*

From equation (B.5b) it is seen that if both the daughter and parent material are retained in a rock then its age can be found from the ratio of daughter to parent atoms. Thus ^{206}Pb is the daughter product of ^{238}U and the ratio $^{206}\text{Pb}/^{238}\text{U}$ can be used to date a rock specimen. However, rocks under investigation have very often suffered one or more metamorphic events, for example by being subjected to high temperature for a short period. A volatile element, such as lead, would be lost during the heating episode and the ratio $^{206}\text{Pb}/^{238}\text{U}$ would then no longer be a useful indicator of age.

Uranium contains two radioactive isotopes, the second one being ^{235}U that decays to ^{207}Pb. If a rock remains a closed system then the plot of the ratio $^{206}\text{Pb}/^{238}\text{U}$ against $^{207}\text{Pb}/^{235}\text{U}$ for different ages gives a theoretical curve known as a *concordia* or *concordant diagram* (figure B.2). If the ratios for a particular rock fall on this curve then it can be deduced that the rock *has* remained a closed system since it first formed and its age can be deduced from one or other of the two ratios. If the rock has suffered from some metamorphic event which has led to a loss of lead then the lead : uranium ratios would both be lessened and a point representing the ratios would fall below the concordant line.

As was pointed out in section A.1, rocks contain many different mineral grains, and these are located differently within the rock so the fraction of lead loss will vary for different grains and different locations. What can be assumed is that, for any particular mineral grain, the fractional loss

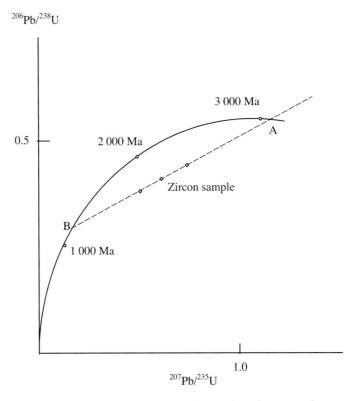

Figure B.2. *The concordant diagram for uranium showing discordant points for a zircon sample.*

of ^{206}Pb and ^{207}Pb will be the same; the difference of mass of the two isotopes is too small to give them substantially different mobility. We shall now see how to exploit this characteristic to find both an age for the rock and also to estimate the time of the metamorphic event.

The assumed starting condition is that the rock has formed a closed system with no initial ^{206}Pb or ^{207}Pb. After a time t_m an event takes place leading to a loss of lead. Just before the event the amount of lead produced in a grain is

$$^d\mathrm{Pb} = {}^p\mathrm{U}_0(1 - \mathrm{e}^{-\lambda_p t_m}), \tag{B.14}$$

where d is 206 or 207 and p is, correspondingly, 238 or 235. If a fraction f of this is lost then after a total time τ, the age of the rock, the amount of lead in the grain will be

$$^d\mathrm{Pb}' = {}^p\mathrm{U}_0(1 - \mathrm{e}^{-\lambda_p \tau}) - f{}^p\mathrm{U}_0(1 - \mathrm{e}^{-\lambda_p t_m}). \tag{B.15}$$

Using ${}^p\mathrm{U}_0 = {}^p\mathrm{U}\,\mathrm{e}^{\lambda_p \tau}$ the measured ratios of lead to uranium is found as

$$\frac{^d\mathrm{Pb}'}{^p\mathrm{U}} = \mathrm{e}^{\lambda_p \tau} - 1 - f\,\mathrm{e}^{\lambda_p \tau}(1 - \mathrm{e}^{-\lambda_p t_m}). \tag{B.16}$$

Without any lead loss the ratios would give the point A on the concordant line (figure B.2) corresponding to the age of the rock. From equation (B.16)

$$f = \frac{\mathrm{e}^{\lambda_p t} - 1 - ({}^d\mathrm{Pb}'/{}^p\mathrm{U})}{\mathrm{e}^{\lambda_p t}(1 - \mathrm{e}^{-\lambda_p t_m})}.$$

Equating the two values of f for $(d, p) = (207, 235)$ and $(206, 238)$ gives a relationship between the two lead : uranium ratios as a straight line of the form

$$\frac{^{206}\text{Pb}'}{^{238}\text{U}} = s\frac{^{207}\text{Pb}'}{^{235}\text{U}} + c$$

where

$$s = \frac{e^{\lambda_{238}\tau}(1 - e^{-\lambda_{238}t_m})}{e^{\lambda_{235}\tau}(1 - e^{-\lambda_{235}t_m})}.$$
(B.17)

Three such *discordant points* for a zircon specimen are shown in figure B.2. The best straight line through these points intersects the concordant line at A (corresponding to $f = 0$) and so gives an age for the total rock. The slope of the straight line is s from which the time of the metamorphic event may be found since t_m is the only unknown quantity. Some sources say that t_m is given by the intercept at B in figure B.2 but this is not so.

B.3. USING NUCLEAR REACTORS

There are two methods of dating that involve the use of neutrons from reactors.

B.3.1. *Argon–argon dating*

When the mineral contains both potassium and argon then the method described in section B.2.4 can be used. There is an alternative method that offers some advantages. The most abundant isotope of potassium is ^{39}K (93.3%) and the present method involves its conversion to ^{39}Ar, a radioactive isotope that does not occur naturally. There are three naturally occurring isotopes of argon, ^{36}Ar (0.34%), ^{38}Ar (0.06%) and ^{40}Ar (99.6%). The isotope ^{39}Ar is produced artificially in a reactor as the result of a collision between a fast neutron (energy 2 Mev) and the naturally occurring potassium isotope ^{39}K according to the scheme

$$^{39}\text{K} + \text{n} \rightarrow {}^{39}\text{Ar} + \text{p}.$$

Although ^{39}Ar is unstable and decays through β^- decay to revert to ^{39}K, because of its half-life of 269 years, the isotope can be regarded as entirely stable and not changing its concentration over the normal laboratory experimentation time intervals. From equation (B.7) with $D_A \equiv [^{40}\text{Ar}]_{\text{now}}$ and $N_t \equiv [^{40}\text{K}]_{\text{now}}$

$$[^{40}\text{Ar}]_{\text{now}} = [^{40}\text{K}]_{\text{now}}\frac{\lambda_A}{(\lambda_A + \lambda_C)}\{\exp[(\lambda_A + \lambda_C)\tau] - 1\}.$$

The number of ^{39}Ar atoms produced in the reactor during a given time interval will be directly proportional to the number of ^{39}K atoms there initially, that is $[^{39}\text{Ar}]_{\text{now}} = a[^{39}\text{K}]_{\text{now}}$, where a is a constant dependent on the neutron energy spectrum, on the interaction cross-section between potassium atoms and neutrons and on the length of time of the irradiation. Normalizing using $[^{39}\text{Ar}]_{\text{now}}$ gives

$$\frac{[^{40}\text{Ar}]_{\text{now}}}{[^{39}\text{Ar}]_{\text{now}}} = C[\exp(\lambda_A + \lambda_C)t - 1]$$

where C is a constant given by

$$C = \frac{[^{40}\text{K}]_{\text{now}}}{a[^{39}\text{K}]_{\text{now}}}\frac{\lambda_A}{\lambda_A + \lambda_C}.$$

Although the constant a is unknown, it can be found by repeating the dating operation using the same irradiation conditions but with a specimen of known age for which the ratio of potassium isotopes is known.

The measurement of the ratio $[^{40}Ar]_{now}/[^{39}Ar]_{now}$ can be made with a mass spectrometer within the same experiment and the ratio can be determined without the need for absolute measurements of the individual concentrations. In addition by heating the irradiated rock sample in a vacuum through a number of temperature steps the ratio $[^{40}Ar]/[^{39}Ar]$ can be measured at each temperature step. If the rock sample has remained closed throughout its history the age determined at each step would be the same but if the system has been open then the age values will be different. This age spectrum can be interpreted to give information about the thermal history of the specimen. There are technical problems with this method associated with argon being a gas and extreme care is required if reliable data are to be obtained.

B.3.2. Fission-track dating

This method uses the fact that fast, heavy charged particles (heavier than an α particle) passing through a mineral structure cause damage that can be detected and measured using appropriate techniques. Radioactive ^{238}U occurs widely in rocks and the atoms undergo spontaneous fission, a distinct process from radioactive decay. The atom breaks into two comparable parts (the total energy release being about 200 Mev) and also releases 2 or 3 high-energy (2 MeV) neutrons. The decay constant for the process is $\lambda_d \approx 8.5 \times 10^{-17}\,yr^{-1}$, less than the decay constant for radioactive decay by a factor about 10^{-7}. When a fission occurs the heavy particles move through the mineral causing damage by multiple collisions until it is brought to rest. The surface of the specimen is polished and etched, and then the damaged regions in the solid can be seen because they are weaker, and more strongly etched, than the undamaged surroundings (figure B.3). Assuming that the uranium is distributed uniformly throughout the sample and knowing the rate of radioactive decay, a measurement of the number of decays within a region allows the time the region has remained undisturbed (that is its age) to be estimated.

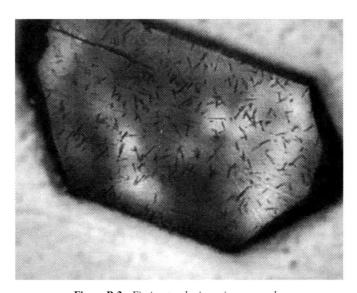

Figure B.3. *Fission tracks in a zircon sample.*

The counting of the uranium fissions need not be made in absolute terms. Another isotope, ^{235}U suffers (induced) fission by collision with slow neutrons (energy ≈ 0.025 eV) at a rate determined by the collision cross-section and the flux of the neutrons. The natural fission decay constant for this isotope is much smaller than for ^{238}U. Inserting the sample in a nuclear reactor with slow neutrons for a specified length of time provides a series of tracks that can also be viewed by the etching process. Comparison between the number of tracks due to natural fission and to induced fission allows the age of the sample to be deduced.

Care must be taken in applying this method in practice. It is presumed that the sample has not been permeated by neutrons in the past as would have happened in certain areas in the Precambrian era, such as the set of natural nuclear reactors found in the Gabon. The ^{238}U in swampy conditions was enriched by about 3% with ^{235}U and so behaved very much like a water-moderated nuclear reactor of today. The enrichment was not, of course, sufficient to lead to a thermonuclear explosion—the enrichment for this has to be close to 100%, and this cannot occur naturally on Earth. Again, it is assumed that no other isotopes make a contribution to the tracks. Studies of the oldest material, such as Archaean rocks and many meteorites, need to take account of the existence of plutonium in the early Solar System which then was probably some 10% of the uranium abundance. Arguments involving events in the distant past must always take account of the possible effects of the shorter-lived radioactive isotopes that are not significant today but could have been at some earlier time.

The method is especially valuable because the fission tracks fade by an annealing process at higher temperatures and the distribution of the number and the quality of tracks in different mineral samples can allow their temperature histories to be inferred.

Problems B

B.1   (a) A radioactive material has a decay constant of 5×10^{-11} years^{-1}. Calculate its half-life.

     (b) The isotope ^{42}K has a half-life of 12.4 hours. Calculate its decay constant.

B.2   Four mineral grains in a rock contain the following isotopes, expressed in ppm.

^{86}Sr	^{87}Sr	^{87}Rb
29.6	21.1	5.93
40.2	29.4	21.7
19.7	14.7	15.5
33.4	25.5	36.5

By plotting a Rb–Sr whole-rock isochron find the age of the rock.

TOPIC C

THE VIRIAL THEOREM

The Virial Theorem applies to any system of particles with pair interactions for which the distribution of particles, in a statistical sense, does not vary with time. The theorem states that

$$2K + \Omega = 0 \tag{C.1}$$

where K is the total translational kinetic energy and Ω is the potential energy. Here we show the validity of the theorem for a system of gravitationally interacting bodies.

We take a system of N bodies for which the ith has mass m_i, coordinates (x_i, y_i, z_i) and velocity components (u_i, v_i, w_i). We define the *geometrical moment of inertia* as

$$I = \sum_{i=1}^{N} m_i(x_i^2 + y_i^2 + z_i^2). \tag{C.2}$$

Differentiating I twice with respect to time and dividing by two

$$\tfrac{1}{2}\ddot{I} = \sum_{i=1}^{N} m_i(\dot{x}_1^2 + \dot{y}_i^2 + \dot{z}_i^2) + \sum_{i=1}^{N} m_i(x_i\ddot{x}_i + y_i\ddot{y}_i + z_i\ddot{z}_i). \tag{C.3}$$

The first term is $2K$; the second can be transformed by noting that $m_i\ddot{x}_i$ is the x component of the total force on the body i due to all the other particles or

$$m_i x_i \ddot{x}_i = \sum_{\substack{i=1 \\ i \neq j}}^{N} G m_i m_j \frac{x_i(x_j - x_i)}{r_{ij}^3}, \tag{C.4}$$

where r_{ij} is the distance between particle i and particle j.

Combining the force on i due to j with the force on j due to i the second term on the right-hand side of equation (C.3) becomes

$$\sum_{i=1}^{N} m_i(x_i\ddot{x}_i + y_i\ddot{y}_i + z_i\ddot{z}_i) = -\sum_{\text{pairs}} G m_i m_j \frac{(x_i - x_j)^2 + (y_i - y_j)^2 + (z_i - z_j)^2}{r_{ij}^3}$$

$$= -\sum_{\text{pairs}} \frac{G m_i m_j}{r_{ij}} = \Omega. \tag{C.5}$$

Equation (C.3) now appears as

$$\tfrac{1}{2}\ddot{I} = 2K + \Omega. \tag{C.6}$$

If the system stays within the same volume with the same general distribution of matter, at least in a time-averaged sense, then $\langle \ddot{I} \rangle = 0$ and the Virial Theorem is verified. The Virial Theorem has a wide range of applicability and can be applied to the motions of stars within a cluster of stars or to an individual star where the translational kinetic energy is the thermal motion of the material.

Problems C

C.1 A galactic cluster of 300 stars, with average mass $0.5M_{\odot}$, occupies a spherical region of radius 1.6 pc with approximately uniform density. Estimate the root-mean-square speed of the stars relative to the centre of mass. (The self-gravitational potential energy of a uniform sphere of mass M and radius R is $3GM^2/5R$.)

C.2 Estimate the mean temperature in a star, assumed of uniform density, with twice the mass of the Sun and 1.2 times its radius. Would your answer be larger or smaller with a more realistic density distribution? (You should assume that the mean mass of the particles comprising the star is 10^{-27} kg.)

TOPIC D

THE JEANS CRITICAL MASS

Many astronomical bodies begin their existence as large gaseous objects of an approximately spherical shape. They may end up as gaseous bodies, for example as stars or major planets. For a major planet the solid grains within them may collect to form a solid body at the centre, the core of the planet, variously estimated to have a mass in the range $5–15M_\oplus$ in the case of Jupiter. Alternatively, after the solid core has formed, the outer gas may be lost over a long period to leave behind a solid body.

If a gaseous sphere is formed then there are two influences acting on it in opposite directions. The first of these is the force of gravity that tends to hold the sphere together and to prevent it from dispersing. The second is the kinetic energy associated with the motion of the material of the sphere that is tending to cause it to fly apart. If these two influences are in balance then the sphere is in a critical state. For a static sphere of gaseous material at a particular density and temperature if the mass is below a certain limit then the sphere will disperse while for a mass above the limit the sphere will collapse. This limiting mass is known as the *Jeans critical mass*.

D.1. AN APPLICATION OF THE VIRIAL THEOREM

One approach to finding the Jeans critical mass is to assume that it corresponds to the situation when the Virial Theorem (Topic C) is just satisfied. If the left-hand side of equation (C.6) is positive then the geometrical moment of inertia, and hence the mean value of r^2, is increasing. This suggests that the material is moving outwards, or dispersing. Conversely if the left-hand side is negative then the sphere will be in a state of collapse. Making the left-hand side equal to zero, the usual form of the Virial Theorem, gives the critical state.

For a static sphere where there is no mass motion of the material, then the only source of kinetic energy is the thermal motion of the molecules. The part of the kinetic energy of the molecules that is relevant here is the translational component, $\frac{1}{2}kT$, for each of the three degrees of freedom. Molecules, with more than a single atom in each entity also have kinetic energy associated with tumbling modes of motion but this energy is not relevant in the present context. If the total mass of the gas sphere has the Jeans critical mass M_c then its total translational kinetic energy is

$$K = \frac{3kTM_c}{2\mu},\qquad (D.1)$$

where T is the mean temperature of the gas sphere and μ is the mean mass of the particles forming the gas.

The gravitational potential energy of the sphere will depend on the distribution of matter within it, which we assume to be spherically symmetric. This has the general form

$$\Omega = -\alpha \frac{GM_c^2}{R} \tag{D.2}$$

where α is a numerical constant, equal to 0.6 for a uniform sphere. Using the Virial Theorem for a uniform sphere the Jeans critical mass is found as

$$M_J = \frac{5k}{\mu G} RT. \tag{D.3}$$

Substituting $R = (3M_J/4\pi\rho)^{1/3}$ and rearranging gives

$$M_J = \left(\frac{375k^3}{4\pi\mu^3 G^3}\right)^{1/2} \left(\frac{T^3}{\rho}\right)^{1/2} = Z\left(\frac{T^3}{\rho}\right)^{1/2}. \tag{D.4}$$

For a mix of gases consisting of atomic hydrogen, molecular hydrogen and helium a value of μ around 4×10^{-27} kg gives $Z = 2.0 \times 10^{21}$ in SI units. The corresponding Jeans critical masses for various combinations of temperature and density are shown in figure D.1. It will be seen that to produce a condensing mass equivalent to that of a globular cluster, $\sim 10^{36}$ kg, from ISM material with temperature 10^4 K and density 10^{-21} kg m^{-3} requires either a more than tenfold reduction of temperature or more than a thousand-fold increase of density. At the other extreme, to produce a condensation with the mass of Jupiter, densities of order 10^{-7} kg m^{-3} with temperatures below 30 K

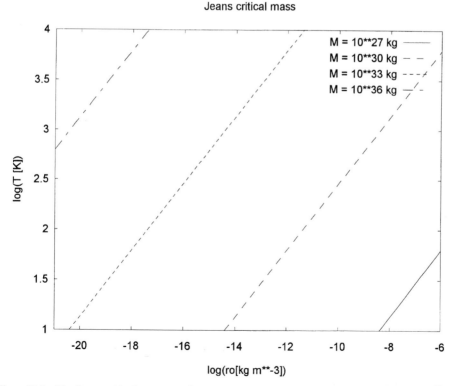

Figure D.1. *The Jeans critical mass as a function of temperature and density for $\mu = 4.0 \times 10^{-27}$ kg.*

are needed. A combination such as $T = 50\,\text{K}$ and $\rho = 10^{-12}$ can produce a solar-mass condensation while $T = 100\,\text{K}$ and $\rho = 10^{-17}\,\text{kg}\,\text{m}^{-3}$ can produce a condensing mass equal to that of a galactic cluster.

D.2. FROM CONDENSATIONS TO CONDENSED BODIES

It is often assumed in theoretical work that an isolated static uniform sphere of gas equal to or greater than a Jeans critical mass will completely collapse, while if it has less than that mass it will completely disperse. This is a misunderstanding of what would actually occur. The derivation we have used for the Jeans critical mass involves the Virial Theorem in the form $2K + \Omega = 0$, but this is a special case of the more general form given in equation (C.6) as

$$2K + \Omega = \tfrac{1}{2}\ddot{I}. \tag{D.5}$$

The quantity I is the geometric moment of inertia given for a distribution of masses by

$$I = \sum_{j=1}^{N} m_j r_j^2 \tag{D.6}$$

where m_j is the mass of the jth particle and r_j its distance from the centre of mass of the system. Clearly if the right-hand side of equation (D.5) is zero then the mass-averaged value of r^2 does not change with time but that does not mean that the distribution of masses is static. What happens in practice is that the central material starts moving inwards since, with a uniform density and temperature, there are no differential pressure forces acting—only gravity—while outside material moves outwards. If this did not happen then the Virial Theorem would indicate that a Jeans critical mass could not collapse at all. The final outcome will depend on the way that energy is dissipated as heat radiation but it could be a condensed body with less than a Jeans critical mass for the original state of the material. The same pattern occurs with a body somewhat less than, or somewhat more than, a Jeans critical mass. This form of behaviour has been described by Ruskol (1955) and by Woolfson (1964). In figure D.2

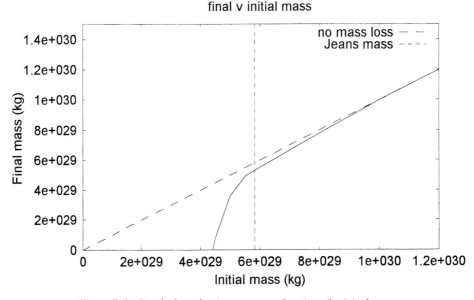

Figure D.2. *Residual condensing mass as a function of original mass.*

we show the results from SPH (section 12.7.2) simulations of the collapse of gas spheres of radius 1000 AU, initial temperature 15 K and with $\mu = 4 \times 10^{-27}$ kg. The simulation includes radiation transfer both within the body and to the outside. The Jeans critical mass is 5.82×10^{29} kg ($\sim 0.3 M_\odot$) and the figure shows the relationship of the final collapsed mass to the initial mass. It will be seen that there are some slight losses of material for initial masses at and above the Jeans critical mass and below 4.4×10^{29} kg ($0.76 M_c$) no final condensation is formed. Although the SPH simulation is not a perfect representation of physical reality the general conclusion of the calculation is clear and valid. However, in practice the assumption that the Jeans critical mass denotes a boundary between two different behaviour patterns is reasonably valid. The range of masses below M_c that give a final condensation is not too great and in much theoretical work, where simulations are being made with rather uncertain parameters, the assumption is acceptable.

Problem D

D.1   Material in a cool cloud has density 2.3×10^{-15} kg m^{-3} and temperature 12.5 K. If the material is pure atomic hydrogen then what is the Jeans critical mass? Assuming that this mass is contained within a sphere then calculate separately the potential and kinetic energies of the material and show that they satisfy the Virial Theorem.

TOPIC E

FREE-FALL COLLAPSE

The model we are taking here to illustrate free-fall collapse is that of a uniform sphere of mass M, initial radius r_0 and initial uniform density ρ, collapsing from rest under the force of gravity alone. If the material of such a sphere were a gas then clearly there would be pressure forces acting as well, so the model must be thought of as small solid bodies uniformly distributed throughout the sphere.

In figure E.1 we show the sphere at the beginning of the collapse. A particle at a distance x from the centre experiences a force due to the shaded volume that gives it an acceleration

$$\frac{\mathrm{d}^2 x}{\mathrm{d}t^2} = -\frac{G \times \frac{4}{3}\pi x^3 \rho}{x^2} = -\frac{4}{3}\pi\rho G x. \tag{E.1}$$

This equation is linear in x that implies that the acceleration and also the velocity at different points at any time will be proportional to the distance of the point from the centre. Hence during the collapse the increasing density remains uniform throughout the sphere.

Since there are no dissipative forces acting, the total energy of any particle must remain uniform throughout the collapse. Equating the potential energy at the beginning of the collapse to the total

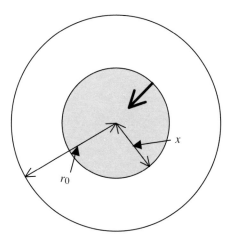

Figure E.1. *The shaded volume of the sphere providing the effective gravitational force at distance x from the centre of the sphere.*

energy at some subsequent time for a boundary particle we have

$$-\frac{GM}{r_0} = -\frac{GM}{r} + \frac{1}{2}\left(\frac{dr}{dt}\right)^2,$$ (E.2)

where r is the radius of the sphere at time t. Rearranging we find

$$\frac{dr}{dt} = -\left[2GM\left(\frac{1}{r} - \frac{1}{r_0}\right)\right]^{1/2},$$ (E.3)

where the negative sign is taken because the sphere is collapsing.
  Substituting $r = r_0 \sin^2 \theta$ gives

$$\frac{d\theta}{dt} = -\left(\frac{GM}{2r_0^3}\right)^{1/2} \frac{1}{\sin^2 \theta}.$$ (E.4)

If in time t the radius of the sphere is r_t then, from equation (E.4)

$$t = -\left(\frac{2r_0^3}{GM}\right)^{1/2} \int_{\pi/2}^{\theta_t} \sin^2 \theta \, d\theta,$$ (E.5)

where $\theta_t = \sin^{-1}[(r_t/r_0)^{1/2}]$. This gives

$$t = \left(\frac{r_0^3}{2GM}\right)^{1/2} \left\{\frac{\pi}{2} - \sin^{-1}\left[\left(\frac{r_t}{r_0}\right)^{1/2}\right] + \left(\frac{r_t}{r_0}\right)^{1/2}\left(1 - \frac{r_t}{r_0}\right)^{1/2}\right\}.$$ (E.6)

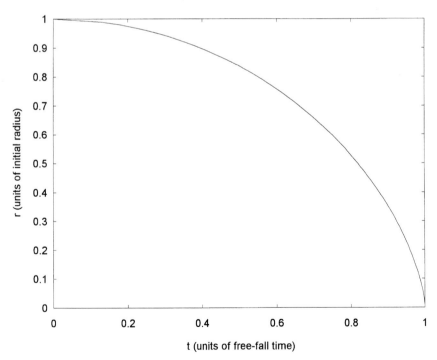

Figure E.2. *The relationship between radius and time in free fall.*

The time for complete collapse, the free-fall time, t_{ff}, is found by making $r_t = 0$ in (E.6) giving

$$t_{ff} = \frac{\pi}{2}\left(\frac{r_0^3}{2GM}\right)^{1/2} = \left(\frac{3\pi}{32\rho G}\right)^{1/2}. \qquad (E.7)$$

From equations (E.6) and (E.7)

$$\frac{t}{t_{ff}} = 1 - \frac{2}{\pi}\left\{\sin^{-1}\left[\left(\frac{r_t}{r_0}\right)^{1/2}\right] - \left(\frac{r_t}{r_0}\right)^{1/2}\left(1 - \frac{r_t}{r_0}\right)^{1/2}\right\}. \qquad (E.8)$$

Figure E.2 shows the variation of r_t/r_0 with time expressed in units of the free-fall time. Even for a collapse of solid particles, so that pressure forces do not occur, the final radius cannot be zero since that would imply infinite density. However, as is seen from the figure, the collapse is slow at first, the radius decreasing by only 10% or so after 40% of the free-fall time. Thereafter the collapse speeds up and then is very rapid at the end. The difference of time between collapsing to zero radius and to a small finite radius is negligible.

For an initially uniform gas sphere the pressure forces will be small at first because there will be no pressure gradient except at the boundary. Thus the first part of the collapse of a gas sphere will approximate to free fall. If the heat energy generated by the compression of the gas is efficiently radiated away then the free-fall time may not be a serious underestimate of the time to collapse to high density.

It is interesting to calculate free-fall times for some astronomical objects based on the densities suggested in Topic D. For a galactic cluster, with an initial density of $10^{-17}\,\mathrm{kg\,m^{-3}}$, the time from equation (E.7) is 6.6×10^5 years. Taking $10^{-12}\,\mathrm{kg\,m^{-3}}$ for a star the time is 2.1×10^3 years while for a planet with initial density $10^{-7}\,\mathrm{kg\,m^{-3}}$ the time is 7 years. Because of the neglected pressure forces these times are under-estimates but they do give a good ballpark figure.

Problem E

E.1 A uniform sphere of material undergoes free-fall collapse. After what fraction of the free-fall time, t_{ff}, is the radius (i) 0.9, (ii) 0.6 and (iii) 0.3 of the initial radius? Estimate r/r_0 when $t/t_{ff} = 0.5$.

TOPIC F

THE EVOLUTION OF PROTOSTARS

Before embarking on a description of how a protostar evolves towards becoming a normal star we must first describe some tools for defining the state of a star. Apart from its composition, which we here always take as Population I material, the quantities that characterize a star from an external point of view are its mass, radius, temperature and luminosity (i.e. its power output). As the star evolves so these quantities vary and we now describe a useful device for recording the variations of the three quantities radius, temperature and luminosity, which are interrelated.

F.1. THE HERTZSPRUNG–RUSSELL DIAGRAM

A very convenient way of representing the state of a star is by the *Hertzsprung–Russell (H-R) diagram*, proposed by Ejnar Hertzsprung and, independently, by Henry Norris Russell. It is a two-dimensional plot of luminosity against temperature. From the normal radiation laws, assuming a black body, the stellar luminosity

$$L_* = 4\pi R^2 \sigma T^4, \tag{F.1}$$

where σ is Stefan's constant, $5.67 \times 10^{-8}\,\mathrm{W\,m^{-2}\,K^{-4}}$, from which we see that luminosity and temperature also indicate the radius of the star.

Often in place of luminosity and temperature the quantities plotted are the related quantities *absolute magnitude* and *spectral class*. The *magnitude* of a star is related to the *perceived* brightness and, to allow for changes of brightness with distance, *absolute magnitude* is defined as the magnitude that would be found if the star were at a standard distance of 10 pc. Since the eye has a logarithmic response to light intensity, the absolute magnitude is linearly related to log(luminosity). Magnitudes were once estimated by eye but are now determined by photometric measurements.

Spectral classification is based on the characteristics of the stellar spectrum, which are heavily dependent on temperature. Thus at very low temperature virtually none of the particles in the star have sufficient energy to excite a hydrogen atom by collision. The ground state energy of the hydrogen atom is $-13.6\,\mathrm{eV}$ or $-2.2 \times 10^{-18}\,\mathrm{J}$ and it requires a minimum of $10.2\,\mathrm{eV}$ ($1.6 \times 10^{-18}\,\mathrm{J}$) of energy to excite it. The mean energy of a particle at 3000 K, is $1.5kT$ or $6.2 \times 10^{-20}\,\mathrm{J}$. Thus at such a low temperature hydrogen spectral features are very weak. With increasing temperature more particles in the Maxwell tail of the energy distribution have enough energy to excite hydrogen and so hydrogen spectral lines strengthen. At very high temperatures, say 30 000 K, there are many particles available with energies that are able completely to ionize hydrogen by collision and so the hydrogen spectrum again becomes weak. This rise and then fall of strength of hydrogen lines with increasing temperature happens similarly, but with different patterns of rise and fall, for many atoms and ions. As the temperature changes so does the complete spectral pattern and the spectra for a large number of stars can be ordered such that there are gradual changes from one to the next. This enabled astronomers to classify stars. The spectral classification, from hot to cool, is O, B, A, F, G, K, M with temperatures running from above 50 000 K for the hottest O stars to below 3000 K for the coolest M stars. Within each letter class there are 10 divisions, e.g. A0, A1, . . ., A9 so, for example, the Sun is a G2 star. Another property of stars related to the spectral classification is mass since the power generated per unit mass, and so the temperature of a star, increases with increasing mass.

An H-R diagram for stars close to the Sun is shown in figure F.1. The well populated band of stars marked with filled-in circles is the main sequence (Topic G). These are stars deriving their energy from the conversion of hydrogen to helium within the stellar core. In the lower left-hand part of the diagram there are marked stars of low luminosity but high temperature; from equation (F.1) we see that this indicates a star with a very small radius. These are white dwarf stars the physics of which is described in Topic J. In the top right-hand part of the diagram there are stars of high luminosity but with temperatures ranging down to low values. Again from equation (F.1) we

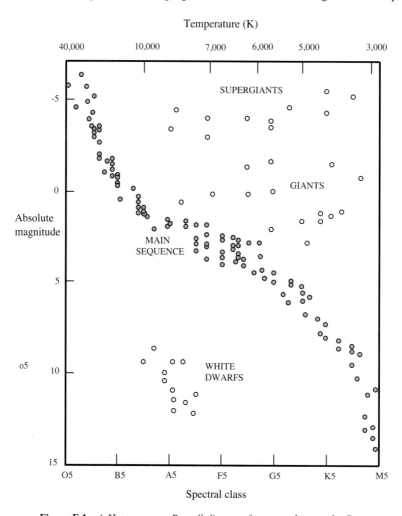

Figure F.1. *A Hertzsprung–Russell diagram for stars close to the Sun.*

can determine that these stars have a large radius. These giant and supergiant stars will be mentioned in Topic I as part of a description of the evolution of stars away from the main sequence.

F.2.   THE EVOLUTION OF A PROTOSTAR

It is generally accepted in most theories of star formation that stars begin their existence as protostars, low density objects that exceed the Jeans critical mass and so are able to collapse. A typical temperature in a dense cool cloud within which stars are forming is 30 K and for that temperature figure D.1 gives a density 10^{-13} kg m^{-3} as the Jeans critical mass for a solar-mass star. Such a protostar would have a radius about 1.1×10^3 AU.

For such a protostar the luminosity from section F.1 is 1.6×10^{28} W—or about 40 times the luminosity of the Sun. Although the protostar has a low temperature it also has a very large surface area. The radiation is in the infrared part of the electromagnetic spectrum. The wavelength corresponding to the peak of the emission is given by Wien's displacement law

$$\lambda_{max} T = 2.898 \times 10^{-3} \text{ m K} \tag{F.2}$$

and for $T = 30$ K this gives $\lambda_{max} = 97$ μm.

Hayashi (1961) described the evolution of a protostar of one solar mass on to the main sequence. The evolutionary path is shown on an H-R diagram in figure F.2. For an H-R diagram giving $\log(L/L_\odot)$ against $\log(T)$, lines of constant radii are straight. The start of the evolution, at point A, corresponds to $3 \times 10^4 R_\odot$, where $R_\odot$ is the radius of the Sun, or about 140 AU. Although, from the analysis given in section F.1 the starting radius could be as large as 1000 AU, the essential character of the evolutionary process would still be much as described by Hayashi.

The first stage of the evolution, AB, represents a free-fall that starts off very slowly, as indicated in figure E.2, and then speeds up as the collapse progresses. For most of this stage the protostar is very diffuse and transparent to radiation so that the heat energy generated in the collapse is radiated away, so keeping the temperature fairly constant. At the end of this stage, at B, the collapse has become rapid, the protostar has become denser and opaque, and heat energy is being retained with a consequent increase in temperature. The time for the path AB will be of the same order as the free-fall time. Assuming a larger starting radius and density 10^{-13} kg m^{-3}, from equation (E.7) this is about 6600 years.

The stage BC is still rapid but is being slowed down by the pressure forces that build up as the temperature increases. In the region of C there is a bounce as the protostar overshoots through an equilibrium configuration and then returns to it. Modelling shows that the whole stage BC takes about 20 years. During the bounce, which takes about 100 days, there is a rise and then fall in luminosity. In 1936 Herbig observed an increase in luminosity by a factor of 200 of the young evolving star FU-Orionis that lasted somewhat less than a year and this may be related to the bounce event.

The final stage of the collapse towards D, on the main sequence, is known as Kelvin–Helmholtz contraction. During this stage the star is in a state of quasi-equilibrium in which it slowly collapses as it radiates energy. Since it is virtually in equilibrium it satisfies the Virial Theorem

$$2K + \Omega = 0 \tag{F.3}$$

where K, the translational kinetic energy, is an expression of temperature. The total energy of the star is

$$E = K + \Omega = -K \tag{F.4}$$

and this is reducing as energy is radiated. Since E is reducing then K must be increasing, because of the negative sign in equation (F.4). This means that as a consequence of radiating energy the star is heating

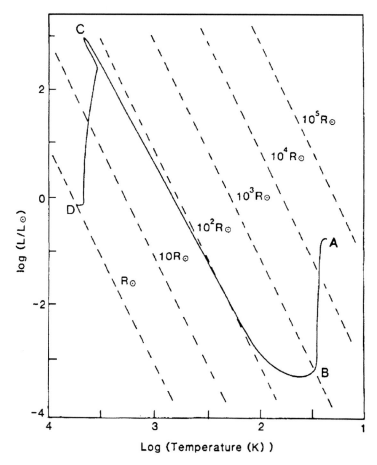

Figure F.2. *The Hayashi track for a solar-mass star.*

up! The increase in thermal energy plus the radiated energy comes from the release of potential energy. From equation (F.3), if K increases then Ω must decrease, which means that the star must collapse.

The Kelvin–Helmholtz contraction stage takes about 5×10^7 years for a star like the Sun. The dependence of the time on mass is shown in table F.1

Table F.1. *Kelvin–Helmholtz contraction times for different stellar masses.*

Mass ($M_\odot$)	Time (10^6 years)
15	0.062
9	0.15
5	0.58
3	2.5
2.25	5.9
1.5	18
1.25	29
1	50
0.5	150

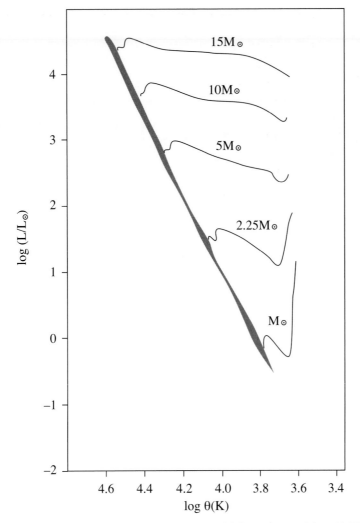

Figure F.3. *Kelvin–Helmholtz tracks for stars of different masses (Iben, 1965).*

The form of the Kelvin–Helmholtz path has been calculated by many workers whose results differ in detail but are generally similar. Paths calculated by Iben (1965) are shown in figure F.3.

Problems F

F.1 A protostar of solar mass begins with a radius of 1000 AU. What is the ratio of the Kelvin–Helmholtz contraction time (table F.1) to the free-fall time (equation E.7) for the protostar?

F.2 From figure F.2 it appears that in an average position on the Kelvin–Helmholtz path a solar-mass star has a luminosity three times that of the Sun. What is the rate of increase of its average temperature (in K year^{-1}) at this position? Take the average particle mass in the star as 10^{-27} kg.

TOPIC G

THE EQUILIBRIUM OF STARS ON THE MAIN SEQUENCE

A star on the main sequence is converting hydrogen to helium in its high-temperature core. Over the main-sequence period, lasting about 10^{10} years for a G2 star like the Sun, it will slowly evolve but at such a slow rate that the assumption that it is in a state of equilibrium is a valid one. The lifetime on the main sequence is heavily mass dependent. The observed relationship of luminosity to mass is approximately

$$L_* = CM_*^{7/2} \tag{G.1}$$

where C is a constant. If the proportion of hydrogen available for conversion to helium is a fraction β of the total mass of the star and the duration of the main sequence is t_{ms} then, since luminosity is proportional to the *rate* of conversion of hydrogen, we may write

$$L_* = \xi \frac{\beta M_*}{t_{\mathrm{ms}}} = CM_*^{7/2}$$

where ξ is another constant. Scaling to the given main-sequence duration for the Sun gives

$$t_{\mathrm{ms}} = \left(\frac{M_\odot}{M_*}\right)^{5/2} \times 10^{10} \text{ years.} \tag{G.2}$$

This relationship is only approximate since the power in equation (G.1) and the constants of proportionality vary somewhat with mass. Nevertheless the conclusions that massive stars spend little time on the main sequence, while stars of very low mass spend times that are much longer than any estimated present age of the universe, are valid. From equation (G.2) the duration on the main sequence of a star of mass $10M_\odot$ is 3.2×10^7 years while for one of mass $0.1M_\odot$ it is 3.2×10^{12} years.

G.1. CONDITIONS FOR MODELLING A MAIN-SEQUENCE STAR

It is assumed in modelling a star like the Sun that it is spherically symmetric. Since the centripetal acceleration at the solar equator, $5 \times 10^{-3} \mathrm{m\,s^{-2}}$, is much smaller than the acceleration due to the Sun's mass, $274 \mathrm{m\,s^{-2}}$, the assumption is clearly valid. Thus to describe the equilibrium state of a star it is only necessary to express the conditions in it as a function of the distance from its centre. The quantities that we shall be considering to define the state of the material at distance r from the centre are the pressure, $P(r)$ and temperature, $T(r)$. For any particular composition of star these

230

two quantities will also give the density since, under the conditions in a star, the material will behave like a perfect gas. We shall be assuming that the normal gas pressure is dominant over radiation pressure so that we can ignore the latter. The two pressures are given by

$$P_{\text{gas}} = \frac{\rho k T}{\mu} \quad \text{and} \quad P_{\text{rad}} = \frac{4\sigma T^4}{3c}, \tag{G.3}$$

where ρ is the density of the material, μ is the mean particle mass, σ is Stefan's constant and c the speed of light. For conditions deep in the Sun the density will be of order 10^4 kg m^{-3}, the mean particle mass $\sim 10^{-27}$ kg and the temperature $\sim 10^7$ K. Under these conditions the values of P_{gas} and P_{rad} are 1.4×10^{15} and 2.5×10^{12} N m^{-2} respectively showing that the radiation pressure can be ignored. However, for more massive stars, where higher temperatures prevail, radiation pressure can be important.

Apart from pressure and temperature the variation of two other quantities are considered. These are defined in relation to a spherical surface of radius r and are the *included mass* $M_{\text{I}}(r)$, the total mass contained within the spherical surface, and $L(r)$, the luminosity or the rate of flow of radiation energy across the surface.

Two other quantities, related to the physics of the stellar material, are involved in setting up the equations for equilibrium. The first of these is the *intrinsic energy generation function*, ε. At higher temperatures nuclear reactions will take place (Topic H) that generate energy at a rate that depends on the composition of the material and its density and temperature. It is expressed in terms of power output per unit mass, or W kg^{-1}, in standard units. The second quantity is *opacity* that describes the resistance of the material to the flow of radiation through it. If a beam of radiation of intensity I travels a distance dx through an absorbing medium then the intensity is changed by an amount

$$\mathrm{d}I = -\zeta I \,\mathrm{d}x$$

where ζ is the *linear absorption coefficient*, with dimensions m^{-1}. Opacity, a quantity preferred by astronomers, is defined as $\kappa = \zeta/\rho$ so that

$$\mathrm{d}I = -\kappa \rho I \,\mathrm{d}x. \tag{G.4}$$

Opacity has the dimensions m^2 kg^{-1} and expresses the resistance to the passage of radiation as an effective opaque, i.e. impenetrable, area per unit mass of material in the path. The action of opacity is to subtract energy from the radiation and to convert it to some other form of energy. This can be by heating material, by exciting material in some way or by scattering the oncoming radiation.

We are now in a position to generate the basic equations for the equilibrium of a star.

G.2.   THE PRESSURE GRADIENT

Figure G.1 shows a slice of stellar material with surfaces at distance r and $r + \mathrm{d}r$ from the centre and with unit area. Since the slice is in equilibrium the total force on it must be zero. The pressure forces, P and $P + \mathrm{d}P$, are shown in the figure. The other force acting is that of gravity due to the interior mass, M_{I}, acting on the slice. The equation of equilibrium is

$$P - (P + \mathrm{d}P) - \frac{GM_{\text{I}}\rho \,\mathrm{d}r}{r^2} = 0 \quad \text{or} \quad \frac{\mathrm{d}P}{\mathrm{d}r} = -\frac{GM_{\text{I}}\rho}{r^2} = -\frac{GM_{\text{I}}P\mu}{kTr^2}. \tag{G.5}$$

The boundary condition for pressure is that it must be zero (actually very small) at the boundary of the star where ρ becomes effectively zero (actually very small).

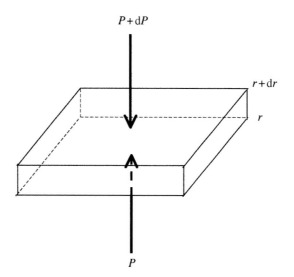

Figure G.1. *The pressure forces on a thin slab of thickness* dr.

G.3.   THE INCLUDED-MASS GRADIENT

Between the spherical surfaces with radii r and $r + \mathrm{d}r$ the contribution to the included mass is

$$\mathrm{d}M_\mathrm{I} = 4\pi r^2 \rho \,\mathrm{d}r = \frac{4\pi r^2 P \mu}{kT}\,\mathrm{d}r \qquad \text{or} \qquad \frac{\mathrm{d}M_\mathrm{I}}{\mathrm{d}r} = \frac{4\pi r^2 P \mu}{kT}. \qquad (\text{G.6})$$

The boundary conditions for M_I are that $M_\mathrm{I} = 0$ for $r = 0$. Another condition is that $M_\mathrm{I} = M_*$ for $r = R$, the radius of the star, but for some methods of solution of the equations, M_* is a determined quantity from the calculations.

G.4.   THE LUMINOSITY GRADIENT

In a state of equilibrium the energy content of the material between spherical surfaces of radius r and $r + \mathrm{d}r$ remains constant. Hence, by energy conservation, the radiation flux through the outer surface exceeds that through the inner surface by the power generated within the shell of thickness $\mathrm{d}r$. This gives

$$L + \mathrm{d}L = L + 4\pi r^2 \rho \varepsilon \,\mathrm{d}r.$$

or

$$\frac{\mathrm{d}L}{\mathrm{d}r} = 4\pi r^2 \rho \varepsilon = \frac{4\pi r^2 P \mu \varepsilon}{kT}. \qquad (\text{G.6})$$

The boundary conditions for luminosity are $L = 0$ for $r = 0$. At the surface $L = L_*$, the luminosity of the star, but as for total mass this is normally a derived quantity from the calculations.

G.5.   THE TEMPERATURE GRADIENT

Consider a thin slab of material illustrated in figure G.2. Radiation is passing through it and, indeed, some of that radiation may be generated within it by nuclear reactions. If the *mean* intensity of the

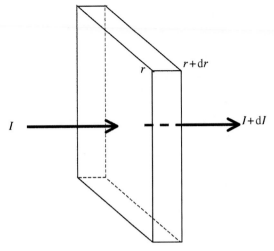

Figure G.2. *The change of intensity of a beam passing through a thin slab of thickness* dr.

radiation within it is I then the energy absorbed per unit area in passing from the inner to outer surfaces of the slab is $I\kappa\rho\,dr$. This will be true for a thin slab even if I is being augmented by energy generation during its passage. The intensity of the radiation passing through the slab may be expressed in terms of the local luminosity by

$$I = \frac{L}{4\pi r^2}.$$
(G.7)

Because of the relationship between energy and momentum for radiation the change in momentum of the radiation per unit time between the inner and outer surface is

$$dp = -\frac{I\kappa\rho}{c}\,dr = -\frac{L\kappa\rho}{4\pi r^2 c}\,dr.$$
(G.8)

This rate of change of momentum in the section of unit area represents a difference in force per unit area, or radiation pressure between the inner and outer surfaces of the slab, or

$$\frac{dP_{\mathrm{rad}}}{dr} = -\frac{L\kappa\rho}{4\pi r^2 c}.$$
(G.9)

Differentiating the expression for P_{rad} in equation (G.3) with respect to r gives

$$\frac{dP_{\mathrm{rad}}}{dr} = \frac{16\sigma T^3}{3c}\frac{dT}{dr}$$

and substituting this in equation (G.9) we find

$$\frac{dT}{dr} = -\frac{3L\kappa\rho}{64\pi\sigma T^3 r^2}.$$
(G.10)

The boundary condition for temperature is that $T = 0$ at $r = R$. The conventional temperature of a star, as determined by observation, is the temperature of the photosphere, the visible boundary of the star, which corresponds to a density and temperature that are very small compared with the corresponding quantities within the star. For this reason it is usual to take them as zero. The estimates of total mass, radius and luminosity, as found by computation, are insensitive to the actual boundary values taken for ρ and T, as long as they are small compared with interior values.

G.6. MAKING MODELS OF STARS

We have now defined the four basic equations to describe the equilibrium of a star. In developing them we have assumed that there is no convection in the star. In most stars there are convective regions so, inevitably, any stellar models derived from the basic equations will be flawed to some extent. However, the distributions of density and temperature deduced from the equations indicate the general conditions within stars reasonably well.

To carry out the calculations it is necessary to know the composition of the stellar material and to have available the values of ε and κ as functions of density and temperature. One method of solution is to fix a radius and to guess the values of P, M_1, L and T at some intermediate value of r, say at $r = \frac{1}{2}R$. The basic equations are then integrated both inwards towards $r = 0$ and outwards towards $r = R$. In general the values of L and M_1 will not be zero at $r = 0$ and the values of P and T will not be zero at $r = R$. By changing the initially-guessed values by a small amount, one at a time, the rates of change of the boundary values can be found for rates of change of the dependent variables at the intermediate point. In this way better values can be found for the starting point dependent variables and the process is repeated until convergence to an acceptable solution is reached. There are other, and probably better, ways of solving the equations but the simple method described here shows that it is possible in principle.

Problem G

G.1   At a point within a star the conditions are as follows: density, $\rho = 10^4 \, \mathrm{kg \, m^{-3}}$; temperature, $T = 6 \times 10^6 \, \mathrm{K}$; opacity, $\kappa = 5 \, \mathrm{m^2 \, kg^{-1}}$; distance from star centre, $r = 2.2 \times 10^8 \, \mathrm{m}$; included mass, $M_1 = 1.3 \times 10^{30} \, \mathrm{kg}$; energy production rate, $\varepsilon = 1.5 \times 10^{-4} \, \mathrm{W \, kg^{-1}}$; luminosity, $L = 2 \times 10^{26} \, \mathrm{W}$. Assuming that the local material, including the effect of ionization, has mean particle mass $10^{-27} \, \mathrm{kg}$ then estimate the pressure at the point. What would be the approximate pressure, include mass, temperature and luminosity at a point 10 000 km closer to the centre?

TOPIC H

ENERGY PRODUCTION IN STARS

In section F.2 it was shown that a star in the Kelvin–Helmholtz stage of its pre-main-sequence evolution was slowly contracting and becoming hotter. If the body had a planetary, rather than a stellar, mass then the internal pressure would halt the collapse well before internal temperatures had reached the stage where any nuclear reactions could occur. However, at a mass of about $0.013M_{\odot}$, internal temperatures of a few million degrees occur and this is sufficient to set going reactions just involving deuterium, $D \equiv {}_1^2H$, that produce tritium, $T \equiv {}_1^3H$. This reaction is $D(D, p)T$ which is equivalent to

$$D + D \rightarrow T + p,$$

where p is a proton. Once tritium is produced then reactions $T(D, n)^4He$ and $T(T, 2n)^4He$ can occur, where n is a neutron. However, the normal concentration of deuterium in stellar material, with $D/H \approx 2 \times 10^{-5}$, gives insufficient energy when all possible deuterium and tritium reactions have taken place to raise the temperature sufficiently for other reactions to occur. This state of just having deuterium-based reactions is what defines a brown dwarf.

With a mass above $0.075M_{\odot}$ the temperature rises to the point where reactions involving hydrogen, ${}_1^1H$, can take place and then a feedback mechanism occurs where hydrogen reactions generate energy that raise the temperature, which then increases the rate of hydrogen reactions and so on. When this happens the star is on the main sequence and, because there is so much hydrogen present and it is transformed so slowly, the main-sequence stage is very long lasting. We shall now consider various aspects of the processes in the main-sequence stage in rather more detail.

H.1. PROTON–PROTON (P-P) REACTIONS—A CLASSICAL VIEW

The temperature in the core of the Sun is about 1.5×10^7 K and in this environment reactions take place involving two protons, or hydrogen nuclei. The main reaction that occurs is $H(p, e^+ + \nu)D$ where e^+ is a positive β particle and ν is a neutrino. It is conventional to write H outside the bracket and p inside, although they represent the same type of particle. We shall now consider the conditions under which a nuclear reaction can take place between two protons, although the same general principles will apply for nuclear reactions involving other nuclei.

The radius of a proton is about 10^{-15} m. Outside this distance it acts like a simple charged particle giving a Coulomb potential at exterior points but if another proton, or atomic nucleus, manages to approach closer than this then a strong force of attraction operates. Figure H.1 represents in a schematic form the potential energy corresponding to one proton being at a distance r from

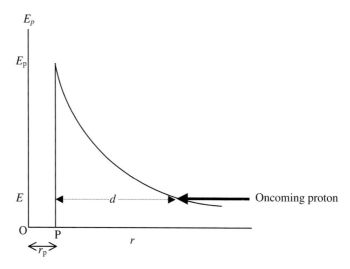

Figure H.1. *Representation of a proton of energy E approaching a Coulomb barrier of maximum height E_p.*

another. It will be seen that there is a potential barrier, of height E_p, that must be overcome by an approaching particle. Once this barrier is crossed then the particles may interact to give a nuclear reaction. The maximum height of the barrier is given by

$$E_p = \frac{e^2}{4\pi\varepsilon_0 r_p} \tag{H.1}$$

where e is the charge of the proton (1.602×10^{-19} C) and ε_0 is the permittivity of free space (8.854×10^{-12} F m^{-1}). With $r_p = 10^{-15}$ m this potential barrier is 2.31×10^{-13} J or 1.44×10^6 eV. If the kinetic energy of the approaching particle is less than this then, according to classical ideas, it will not penetrate the barrier but simply undergo an elastic repulsion. We may translate this energy into a temperature by

$$E_p = kT$$

and the corresponding temperature is 1.7×10^{10} K. This is three orders of magnitude greater than the actual temperature inside the Sun. Even allowing for some hydrogen atoms having energies greater than the average by being in the 'Maxwell tail' (section O.2.2.1) it is evident that classical theory tells us that proton–proton reactions could not take place within the Sun—but they do.

H.2.   A QUANTUM-MECHANICAL DESCRIPTION

The answer to this apparent difficulty comes from quantum mechanics in the mechanism of tunnelling. In figure H.2 a particle of mass m and with energy E approaches a rectangular potential barrier of thickness d and height E_m ($>E$). According to quantum mechanics the particle has a finite probability of penetrating the barrier

$$P_{pen} = \exp\left(-2\sqrt{\frac{2m(E_m - E)}{\hbar^2}}\,d\right). \tag{H.2}$$

A particle that penetrates the barrier comes through it with the same energy, E, with which it entered the barrier. The remaining particles are reflected backwards with energy E, as in the classical case.

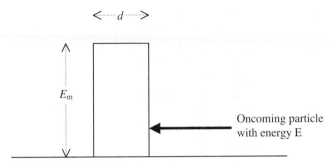

Figure H.2. *A particle of energy E approaching a rectangular potential barrier of height E_m and thickness d.*

To apply this process to a pair of protons, as an approximation we substitute $E_\mathrm{m} = \frac{1}{2}E_\mathrm{p}$ in equation (H.2), use for m the *reduced mass* of the two-particle system

$$m_r = \frac{m_1 m_2}{m_1 + m_2} \quad (= \tfrac{1}{2}m_\mathrm{p}, \text{ where } m_\mathrm{p} \text{ is the proton mass, in this case})$$

and make d equal the distance shown in figure H.1. This will give a crude estimate of the probability of barrier penetration. From figure H.1

$$d = \frac{e^2}{4\pi\varepsilon_0 E} - r_\mathrm{p} \tag{H.3}$$

where r_p is the radius of the proton, taken as 10^{-15} m.

For a temperature 1.5×10^7 K, such as exists in the Sun's core, the mean translational energy of a proton is $3kT/2$ or 3.1×10^{-16} J. We may assume that there are many protons with three times the mean energy, or more, so for our illustration we take $E = 10^{-15}$ J. From equation (H.3) we find $d = 2.29 \times 10^{-13}$ m. The value of E_m, found from equation (H.1) with $r = r_\mathrm{p}$ is 2.31×10^{-13} J. Putting all these values into equation (H.2) we find $P \approx 10^{-37}$. In view of the crudeness of the model the value found for P is only indicative of the fact that the fusion of hydrogen takes place very slowly in the interior of the Sun and other stars. The basic reason is that it is a rare occurrence for protons to approach sufficiently closely for a reaction to occur. To get a more precise quantitative assessment of reaction rates we now develop a more detailed theory of how p-p reactions occur.

H.2.1. The distribution of proton relative energies

In determining the tunnelling estimate in the previous section we took the proton with three times the mean energy. Since $E \ll E_\mathrm{m}$ the value of $E_\mathrm{m} - E$ is little affected by large relative changes of E. On the other hand since $r_\mathrm{p} \ll d$, the value of d is inversely proportional to E. From this it can be seen that penetration increases rapidly with particle energy. However, the numbers of high-energy particles falls rapidly with increasing energy so, clearly, the energy distribution is important.

In the above approximate treatment, reference was made to the energy of a single proton. What is actually important in these two-particle interactions is the relative energy of pairs of particles, i.e. the kinetic energy of one particle as seen by the other, considered at rest. This distribution is the well-known Maxwell–Boltzmann distribution that gives the proportion of particles with relative energy between E and $E + \mathrm{d}E$ as

$$P(E)\,\mathrm{d}E = 2\pi \left(\frac{1}{\pi kT}\right)^{3/2} E^{1/2} \exp\left(-\frac{E}{kT}\right) \mathrm{d}E. \tag{H.4}$$

H.2.2.   The rate of making close approaches

In calculating the rate at which classical particles collide, for example where there are a large number of billiard balls moving randomly on a table, the factors governing the collision rate for a particular ball would be its speed and the radius of the balls. When we consider protons we must take into account the fact that their behaviour is governed by quantum mechanics and that in some circumstances they have wave-like properties. Two protons will be deemed to have 'come together' when, considered as particles, their distance apart is a distance related to their wave-like properties. We take this distance equal to the De Broglie wavelength for a particle of mass m_r moving at the relative speed of the two particles, v_r, i.e.

$$\lambda = \frac{h}{m_r v_r} = \frac{h}{\sqrt{2Em_r}}. \tag{H.5}$$

To define the domain of the model we now imagine that we have N protons within a volume V. We shall consider two particular protons and calculate the probability per unit time that they will give rise to a reaction. From this, by considering all possible pairs, we shall calculate the number of reactions per unit time per unit mass of hydrogen.

Relative to one of the pair of protons the other sweeps out a volume $\pi\lambda^2 v_r$ per unit time such that if the other falls within this volume then a close approach will have taken place. The probability of this is

$$P_{ap} = \frac{\pi\lambda^2 v_r}{V} = \pi\frac{h^2}{V}\sqrt{\frac{1}{2Em_r^3}}, \tag{H.6}$$

where we have used $v_r = \sqrt{2E/m_r}$.

The close approach we have defined here is to within a distance λ that is much larger than r_p. Now we must consider the probability that the approach leads to a tunnelling event.

H.2.3.   The tunnelling probability

Consider that the potential barrier shown in figure H.1 is divided up into s narrow strips, each of width Δx. The probability that the particle penetrates the ith strip is, from equation (H.2),

$$P_i = \exp\left(-2\sqrt{\frac{2m_r(E_i - E)}{\hbar^2}}\Delta x\right).$$

The probability that the particle penetrates the barrier is the product of the probabilities for the s strips

$$P_{pen} = \prod_{i=1}^{s} P_i = \prod_{i=1}^{s} \exp\left(-2\sqrt{\frac{2m_r(E_i - E)}{\hbar^2}}\Delta x\right). \tag{H.7}$$

Taking the natural logarithm of each side of equation (H.7)

$$\ln(P_{pen}) = -2\sqrt{\frac{2m_r}{\hbar^2}}\sum_{i=1}^{s}\sqrt{(E_i - E)}\Delta x. \tag{H.8}$$

Taken to the limit, with an infinite number of strips

$$\ln(P_{pen}) = -2\sqrt{\frac{2m_r}{\hbar^2}}\int_{r_p}^{r_E}\sqrt{(E_x - E)}\,dx. \tag{H.9}$$

Changing the variable to

$$E_x = \frac{e^2}{4\pi\varepsilon_0 x}$$

we find

$$\ln(P_{\text{pen}}) = \frac{e^2}{2\pi\varepsilon_0}\sqrt{\frac{2m_{\text{r}}}{\hbar^2}}\int_{E_{\text{p}}}^{E}\frac{\sqrt{(E_x - E)}}{E_x^2}\,\mathrm{d}E_x.$$

Substituting $E_x = E\sec^2\theta$ in the integral with the condition $E_{\text{p}} \gg E$ gives the value of the integral as $-\pi/(2E^{1/2})$ so that

$$\ln(P_{\text{pen}}) = -\sqrt{\frac{e^4 m_{\text{r}}}{8\varepsilon_0^2\hbar^2}}\frac{1}{E^{1/2}}. \tag{H.10}$$

We now define the *Gamow energy*

$$E_{\text{G}} = \frac{e^4 m_{\text{r}}}{8\varepsilon_0^2\hbar^2}$$

which is about 493 keV. In terms of the Gamow energy, equation (H.10) becomes

$$\ln(P_{\text{pen}}) = -\frac{E_{\text{G}}^{1/2}}{E^{1/2}} \quad\text{or}\quad P_{\text{pen}} = \exp\left(-\frac{E_{\text{G}}^{1/2}}{E^{1/2}}\right). \tag{H.11}$$

H.2.4. The cross-section factor

The cross section for an interaction depends on three separate quantities. The first of these is the size of the target for approach, given by $\pi\lambda^2$, the second is the probability that barrier penetration will take place (equation H.11) and the third is the probability that after barrier penetration a reaction will actually occur. This last quantity, the *cross-section factor*, is a slowly-varying function of energy, $S(E)$, that can be determined by laboratory experiments. In practice, to make reasonable measurements the experiments are done at energies much higher than those prevailing in stars and the results extrapolated into the stellar region.

H.2.5. The energy generation function

We are now in a position to write down an expression for the probability per unit time that a particular pair of protons will give a reaction. This is

$$P_{\text{pair}} = \int_0^\infty P(E)P_{\text{ap}}P_{\text{pen}}S(E)\,\mathrm{d}E = \left(\frac{2\pi}{k^3T^3m_{\text{r}}^3}\right)^{1/2}\frac{h^2}{V}\int_0^\infty S(E)\exp\left(-\frac{E}{kT} - \frac{E_{\text{G}}^{1/2}}{E^{1/2}}\right)\mathrm{d}E. \tag{H.12}$$

The total number of interactions for the N protons is $\frac{1}{2}N^2 P_{\text{pair}}$, since the number of pairs of protons is $\frac{1}{2}N(N-1)$ which is approximately $\frac{1}{2}N^2$. The total mass of protons (hydrogen) is Nm_{p} and the density $\rho = Nm_{\text{p}}/V$. Taking $m_{\text{r}} = \frac{1}{2}m_{\text{p}}$ gives the number of reactions per unit time per unit mass as

$$P_{\text{reac}} = \rho h^2\sqrt{\frac{4\pi}{k^3T^3m_{\text{p}}^7}}\int_0^\infty S(E)\exp\left(-\frac{E}{kT} - \frac{E_{\text{G}}^{1/2}}{E^{1/2}}\right)\mathrm{d}E. \tag{H.13}$$

Without detailed knowledge of $S(E)$ the form of the integrand is not defined but over the range of energies of interest $S(E)$ does not vary greatly and so can be taken outside the integration. The exponential term then remains as the integrand and its form is shown in figure H.3. The combination of terms in the integrand one of which increases with E and the other of which decreases with E leads to the *Gamow peak*. This occurs when the argument of the exponential is a

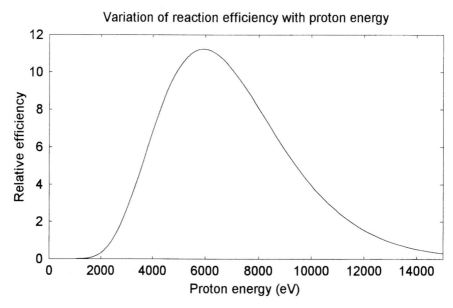

Figure H.3. *The p-p reaction efficiency as a function of the relative energy of the protons, showing the Gamow peak.*

maximum, when

$$E = (\tfrac{1}{2} E_G^{1/2} kT)^{2/3}. \tag{H.14}$$

For a temperature 1.5×10^7 K, corresponding to the interior of the Sun, E_{max} is about 9.5×10^{-16} J which is about three times the mean energy of the protons. These are the most effective protons. At lower energies there are more protons but they are less effective in producing reactions. At higher energies the protons are more effective in producing reactions but there are fewer of them.

The derivation of the energy generation rate for p-p reactions can be generalized to reactions involving other nuclei when the appropriate nuclear charges and masses are incorporated.

H.3. NUCLEAR REACTION CHAINS IN THE SUN

The p-p reaction described in section H.2, $H(p, e^+ + \nu)D$, is just the first step in a chain of reactions. The energy released in a nuclear reaction is known as its Q *value*, and for this particular reaction it is 1.442 MeV. However, not all of this energy goes into providing heat locally because the neutrino carries off 0.263 MeV. Neutrinos have a very small cross section for interacting with normal matter and virtually all those produced within the Sun leave that body without interacting in any way. A summary of what happens in the Sun is that four protons are converted into a single ^{4}He nucleus (an α particle). Since the helium nucleus has less mass than the four protons the mass loss appears as energy according to Einstein's well-known equation $E = mc^2$. The reactions that are required to give this outcome are as follows:

$$
\begin{aligned}
&2 \times (p + p \rightarrow e^+ + \nu + D) &&Q = 1.442 \text{ MeV with } 0.263 \text{ MeV taken by neutrino}\\
&2 \times (D + p \rightarrow \gamma + {}^3\text{He}) &&Q = 5.493 \text{ MeV}\\
&{}^3\text{He} + {}^3\text{He} \rightarrow 2p + {}^4\text{He} &&Q = 12.859 \text{ MeV}
\end{aligned}
$$

Examining the input and output of these five $(2 + 2 + 1)$ reactions shows that four protons are indeed converted into helium with a production of energy

$$2 \times 1.442\,\text{MeV} + 2 \times 5.493\,\text{MeV} + 12.859\,\text{MeV} = 26.73\,\text{MeV}$$

of which $2 \times 0.263\,\text{MeV} = 0.53\,\text{MeV}$ is carried off by neutrinos.

At the core temperature of the Sun the half-life of a proton is about 8×10^9 years, which explains why the Sun spends so long on the main sequence. By contrast the half lives for D and ^{3}He are a fraction of a second and 2.4×10^5 years respectively.

There is another process for conversion of protons into α particles—the CNO cycle. This is described by the following reactions, with the Q value in parentheses.

$$^{12}\text{C} + \text{p} \rightarrow {}^{13}\text{N} + \gamma \qquad (1.944\,\text{MeV})$$

$$^{13}\text{N} + \text{e}^+ \rightarrow {}^{13}\text{C} + \gamma \qquad (2.221\,\text{MeV})$$

$$^{13}\text{C} + \text{p} \rightarrow {}^{14}\text{N} + \gamma \qquad (7.550\,\text{MeV})$$

$$^{14}\text{N} + \text{p} \rightarrow {}^{15}\text{O} + \gamma \qquad (7.293\,\text{MeV})$$

$$^{15}\text{O} + \text{e}^+ \rightarrow {}^{15}\text{N} + \gamma \qquad (2.761\,\text{MeV})$$

$$^{15}\text{N} + \text{p} \rightarrow {}^4\text{He} + {}^{12}\text{C} \qquad (4.965\,\text{MeV})$$

Again the net result is that four protons are converted into a helium nucleus with the production of 26.73 MeV of energy but, for this system, none is taken off by neutrinos. It will be seen that ^{12}C acts as a catalyst for this process since it is restored at the end of the chain.

When incorporating energy generation rates into numerical modelling of stellar structures it is convenient to have some simple analytical expression. The one normally employed for an interaction between particles A and B is of the form

$$\varepsilon = c\rho X_A X_B T^\eta. \qquad (\text{H.15})$$

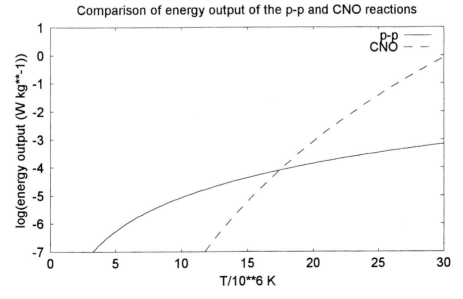

Figure H.4. *Comparison of the p-p and CNO cycles.*

In this expression ε is the energy generation per unit mass of material per unit time, c is a constant, ρ the overall density of the material, X_A is the fractional content of A in the material by mass, T the temperature and η another constant. For the p-p reaction $\eta = 4$ but for the CNO reaction $\eta = 17$. The variation of ε with temperature for the two types of reaction is shown in figure H.4 for solar material. At the temperature of the solar core, $\sim 1.5 \times 10^7$ K, the p-p reaction dominates but at $\sim 1.7 \times 10^7$ K the CNO reaction reaches equality. For stars considerably more massive than the Sun, where higher temperature prevail, the CNO reactions would certainly dominate.

Problem H

H.1 The energy generation rate for the p-p reaction is $\varepsilon_{pp} = 9.5 \times 10^{-37} X_p^2 \rho T^4$ and that for the CNO cycle is $\varepsilon_{CNO} = 2.3 \times 10^{-8} X_p X_C \rho (T/10^7)^{17}$. Given the fractional content of hydrogen, X_p, is 0.7 and that of carbon, X_C, is 0.0003 then find the temperature at which the CNO generation rate is (i) one quarter of, (ii) equal to and (iii) four times that of the p-p cycle. By what percentage is the total output of energy increased if the temperature is raised from 1.50×10^7 K to 1.51×10^7 K?

EVOLUTION OF STARS AWAY FROM THE MAIN SEQUENCE

Hydrogen conversion to helium, which characterizes the main sequence, is a slow process for a star like the Sun but eventually the supply of hydrogen in the stellar core becomes exhausted. When this occurs the star moves off the main sequence. The progress of the star thereafter depends on its mass since new sources of nuclear energy come into play and these depend on the temperatures attained within the star. The more massive the star the higher the temperatures that become available and the greater the number of nuclear reactions that become possible.

Here we shall be concerned mainly with the post-main-sequence evolution of stars of about solar mass and the detailed description of evolution will be limited to such stars. However, in a final section the features of the evolution of more massive stars will be briefly described.

I.1. AN OVERVIEW OF THE EVOLUTIONARY PATH

The eventual outcome of the evolution of a solar-type star is that it becomes a white dwarf (Topic J). The path from the main sequence to this conclusion is conveniently illustrated on a Hertzsprung–Russell diagram (Topic F) and is shown schematically in figure I.1. It will be seen that the path is not a simple one but meanders to and fro across the diagram. During this journey the temperature and pressure within the star vary greatly, affecting both the nuclear reactions that can take place (Topic H) and also the properties of the stellar material. In particular the material sometimes becomes degenerate, a state described in Topic J and associated with high density and/or low temperature.

An important characteristic of degenerate material is that pressure is virtually independent of temperature. There are three sources of pressure within a star—normal gas pressure, radiation pressure and electron degeneracy pressure. Their dependence on the properties of the material are given by

$$P_{\text{gas}} = \frac{\rho k T}{\mu} \tag{I.1a}$$

$$P_{\text{rad}} = \frac{4\sigma T^4}{3c} \tag{I.1b}$$

$$P_{\text{ed}} = \frac{3h^2}{2m_{\text{e}}} \left(\frac{\rho}{\mu}\right)^{5/3} \tag{I.1c}$$

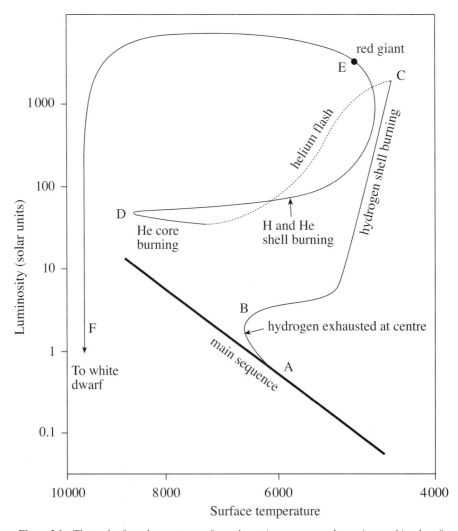

Figure I.1. *The path of a solar-mass star from the main sequence to becoming a white dwarf.*

where μ is the particle mass for the stellar material, σ is Stefan's constant, c the speed of light and m_e the electron mass, with other symbols having their usual meaning.

The most obvious pressure source is normal gas pressure due to the motion of the constituent particles moving according to classical rules and it is this that has been included in the description of the equilibrium of a star in Topic G. Radiation pressure is due to the momentum associated with photons. For the Sun, radiation pressure accounts for less than 1% of the pressure in the core and so is usually ignored in theoretical treatments. For massive stars this may not be so. Because of their respective temperature dependence, with an internal temperature five times that in the solar core, radiation pressure becomes comparable with gas pressure.

Electron degeneracy pressure has no dependence on temperature but varies as $\rho^{5/3}$. It is clear from equations (I.1) that, at any density, radiation pressure will become dominant at a high enough temperature and that, at any temperature, electron degeneracy pressure will become dominant at a

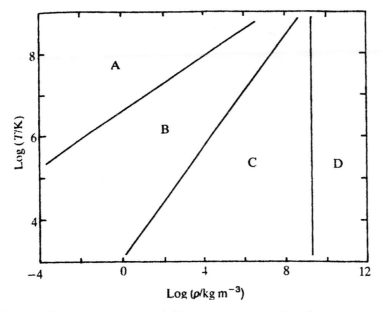

Figure I.2. *Regions of pressure dominance. A, kinetic gas pressure; B, radiation pressure; C, electron degeneracy (non-relativistic); D, electron degeneracy (relativistic).*

high enough density. All sources of pressure are always present but, usually, one or other of them is dominant. The regions of dominance are shown in figure I.2.

I.2.   HYDROGEN-SHELL BURNING

The temperature of a main-sequence star is highest in the core region and, since nuclear-reaction rates have such a high dependence on temperature (equation H.15), hydrogen is being consumed more rapidly in the core than in the surrounding regions. The first effect of this dilution of core hydrogen is, paradoxically, to increase the rate of burning and hence the luminosity of the star. As the hydrogen is burnt equilibrium is maintained by slow collapse of the core, thus increasing both density and temperature and hence the rate of energy generation. The extra internal pressure that results leads to a small expansion of the outer regions of the star so that, although its black-body temperature falls, its luminosity increases. This stage of the evolution of the star is AB in figure I.1.

Hydrogen eventually becomes exhausted in the core while surrounding regions are still hydrogen-rich. Without a nuclear energy source available to add to the resisting pressure, the core slowly collapses so releasing gravitational energy. Hydrogen surrounding the core continues to burn, giving the hydrogen-shell burning stage. This is illustrated in figure I.3. The region of shell burning gradually moves outwards as the inner hydrogen is progressively exhausted.

As the source of energy is removed so the centre of the star collapses under gravity and becomes both hotter and denser. The compression of the centre is augmented by the pressure exerted by the shell burning, which pushes both inwards and outwards. The outwards pressure increases the size of the star, the expansion of which reduces its external temperature. With greater internal production of energy the luminosity of the star increases and it becomes a large object with a lower effective black-body temperature—a red giant. This part of the evolutionary process is shown as BC in figure I.1.

Figure I.3. *A representation of hydrogen-shell burning. The boundary is not actually sharp but reactions fall off inwards and outwards from some peak.*

I.3.   HELIUM IGNITION AND HELIUM CORE BURNING

Throughout stage BC in figure I.1 the collapse of the core had been moderated by gas pressure. Collapse induced a higher temperature, a higher temperature increased the pressure and a higher pressure resisted collapse. The gas pressure acted as a safety-valve. Assuming an adiabatic gas law

$$PV^\gamma = \text{constant}$$

with the ratio of specific heats, $\gamma = 5/3$, the pressure varies as $\rho^{5/3}$. At some stage the material may become degenerate and, if so, the collapse is resisted by the degeneracy pressure also varying as $\rho^{5/3}$. Nevertheless, in the absence of energy generation, the core continues to collapse and the temperature continues to rise.

   The next significant event in the evolution is the onset of helium burning at a temperature somewhere above 10^8 K. This takes place through the *triple-α* reaction. The first step in this process is

$$^4\text{He} + {}^4\text{He} \Leftrightarrow {}^8\text{Be}.$$

The symbol $\Leftrightarrow$ indicates that the reaction is reversible since ^{8}Be is an unstable nucleus and spontaneously decays back into two helium nuclei. However, in a very dense environment, before that decay takes place there is a reasonable probability of the reaction

$$^4\text{He} + {}^8\text{Be} \Leftrightarrow {}^{12}\text{C}^*.$$

Again, $^{12}C^*$ is an excited form of ^{12}C that would spontaneously decay back to ^4He plus ^8Be but a small amount of the excited carbon nuclei de-excite by

$$^{12}C^* \rightarrow {}^{12}C + (\text{gamma radiation or } e^+ + e^-).$$

The net effect of this chain of reactions is that three helium nuclei (α particles) are converted into a carbon-12 nucleus.

With this new source of nuclear energy available there is a rapid rise of temperature but, since the material is now degenerate, there is no significant increase of pressure and hence 'safety-valve' behaviour. The temperature dependence of the triple-α reaction goes as T^{40} so a runaway effect occurs in which a small temperature rise gives a large increase in the nuclear-reaction rate giving a greater temperature and so on. The helium burning begins with what is known as the *helium flash*, an explosive episode lasting of the order of 100 s. During this time the temperature rises to the level where the degeneracy is removed and pressure once again depends on temperature. Thereafter the star has a helium-burning core that mimics to some extent the original state when there was a hydrogen-burning core. Because of this the star tends to evolve towards a configuration resembling that of the original main sequence. The complete stage of helium ignition and helium core burning is represented as CD in figure I.1. The dashed part of the path is one of comparatively short duration and represents the period of readjustment after the helium flash. The horizontal-branch region represents the longer-lasting period of helium core burning with the star in a quasi-stable state. Since it is a long-lasting stage numbers of stars are found in this part of the H-R diagram.

I.4. HYDROGEN AND HELIUM SHELL BURNING

Just as hydrogen burning led to a deficiency of hydrogen in the core so a helium deficiency eventually occurs. This leads to a period of helium shell burning, while at the same time residual hydrogen shell burning will also be taking place. This gives a new move towards the red-giant part of the H-R diagram (region E) with the formation of a dense degenerate core and outward pressure due to the large rate of generation of energy. During the final part of the shell-burning stage material is being pushed out so violently that it escapes from the star and forms a planetary nebula (figure I.4). This is a shell of

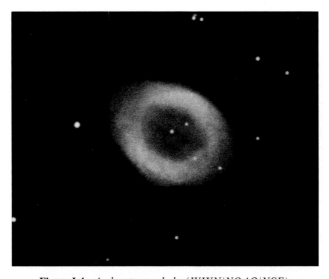

Figure I.4. *A planetary nebula (WIYN/NOAO/NSF).*

material leaving the star and illuminated by it; its appearance as a ring is because, to a viewer, the optical depth is greater at the edge than at the centre—a natural outcome of a shell-like structure.

A considerable part of the outer region of the star is lost in this way and, as it is lost, higher temperature regions in the interior become exposed and the effective radius becomes less. The net luminosity is high but relatively constant at this stage. Eventually only the dense degenerate core remains; this is a white dwarf (Topic J). This final path in the evolutionary process corresponds to EF in figure I.1.

I.5. THE EVOLUTION OF HIGHER MASS STARS

For higher-mass stars, with much greater reservoirs of gravitational potential energy, the temperature at which helium burning is initiated occurs before the core becomes dense enough to be degenerate so the gas-pressure safety valve operates and there is no helium flash. It is likely that after the motion from point E in figure H.1 the temperature will rise to a level where reactions can take place involving elements heavier than helium. In that case the scenario of moving to and fro from the red giant to near the main sequence region of the H-R diagram may be repeated a few time, although not necessarily with an amplitude as large as that for the first two swings.

A selection of the reactions that can take place is

$$^4\text{He} + {}^{12}\text{C} \rightarrow {}^{16}\text{O} + \gamma$$

$$^{12}\text{C} + {}^{12}\text{C} \rightarrow {}^{20}\text{Ne} + {}^4\text{He}$$

$$^{12}\text{C} + {}^{12}\text{C} \rightarrow {}^{23}\text{Na} + \text{p}$$

$$^{12}\text{C} + {}^{12}\text{C} \rightarrow {}^{23}\text{Mg} + \text{n}$$

$$^{16}\text{O} + {}^{16}\text{O} \rightarrow {}^{28}\text{Si} + {}^4\text{He}$$

Reactions that build up nuclei of larger and larger atomic number are exothermic, that is they produce heat energy, up to the production of ^{56}Fe beyond which reactions are endothermic, that is they require an input of energy if they are to occur. Exothermic reactions provide a positive feedback mechanism where reactions taking place produce heat and higher temperature and the higher temperature allows even more reactions to take place. In the normal development of a star considerably more massive than the Sun, say $10–20M_\odot$, the end of the sequence of reactions is the formation of an iron core. At this stage there may be several shells of burning with different major constituents in each shell, as illustrated in figure I.5. The central iron core would have a temperature more than 4×10^9 K with a density several times 10^{10} kg m^{-3}.

The pressure within the iron core would be extremely high. We can get a measure of this by just considering the pressure at the centre of the iron core assuming that it was isolated. The pressure at the centre of a body of mass M and radius R is of order

$$P = \frac{GM^2}{R^4}$$

and for a core of mass $M_\odot$ and density 5×10^{10} kg m^{-3} it would be greater than 10^{25} N m^{-2}. Taking into account the overlying shells in an actual star this would be a gross underestimate. The electron degeneracy pressure, as estimated from equation (I.1c) is just over 10^{23} N m^{-2} and even allowing for more than one free electron per iron atom it is clear that it cannot support the pressure of the iron core. This lack of balance of pressures is resolved by the combination of protons and electrons in the core to form neutrons. The atomic structure collapses and the core rapidly converts into a

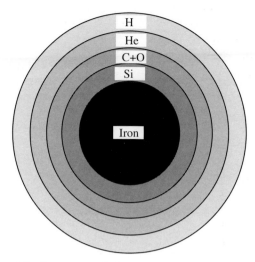

Figure I.5. *Composition shells in a highly evolved massive star.*

sphere of neutrons of density $\sim 10^{18}\,\mathrm{kg\,m^{-3}}$ supported by neutron degeneracy pressure (Topic J). If neutron degeneracy pressure is insufficient, that will happen for a large core mass, then a black hole will form (Topic J).

During the period before the core collapses there will be a flow of material towards the core. When the core collapses to form a small incompressible sphere the incoming material bounces outwards, reacts violently with material still flowing inwards and the enormous energy produced throws off most of the mass of the star exterior to the neutron core. The event is a Type II supernova and the residual core is a neutron star. The energy available in this event is sufficient to allow endothermic reactions to occur, thus producing elements heavier than iron, and all such material in the universe has been produced in this way in supernovae. Material thrown out by a supernova is visible long thereafter; figure I.6 shows the Crab Nebula, the material thrown out of a

Figure I.6. *The crab nebula—a supernova remnant (Jay Gallagher, U. of Wisconsin; NA Sharp (NOAO)/ WIYN/NDAO/NSF).*

supernova observed in 1054 AD. This is associated with the Crab pulsar, a neutron star emitting a regular series of radio pulses.

Type II supernovae only occur in regions of active star formation and they are due to the evolution of massive stars that go through all their evolutionary stages in a period of a few times 10^7 years. Thus a Type II supernova occurs while nearby lower-mass stars are still forming and evolving. There are other supernovae, known as Type I, that may occur in any part of any galaxy and are obviously related to the behaviour of lower-mass stars. One possibility is that they represent the merging of two white dwarf stars, perhaps members of a decaying binary system. Another is that material is transferred from a normal star to a white dwarf in a binary system until the mass of the white dwarf exceeds the Chandrasekhar limit (topic J). It will then rapidly collapse into a neutron star and a supernova event will then occur in much the same way as it does for a Type II supernova.

I.6. FINAL COMMENTS

The standard models of stellar evolution are fairly convincing in their general outline but they must be described with a 'health warning'. Many aspects of these evolutionary models are pure conjecture and are not supported by any observational evidence or analysis, either theoretical or numerical.

Problem I

I.1   Calculate the kinetic gas pressure, the radiation pressure and the electron degeneracy pressure for atomic hydrogen under the following conditions:

	ρ (kg m^{-3})	T (K)
(i)	10^3	10^4
(ii)	10^4	10^7
(iii)	10^4	10^8
(iv)	10^6	10^7
(v)	10^6	10^9

Do not take ionization of hydrogen into account.

TOPIC J

THE CHANDRASEKHAR LIMIT, NEUTRON STARS AND BLACK HOLES

J.1. SOME BASIC QUANTUM MECHANICS PRINCIPLES

For a normal main-sequence star the material behaves for the most part like a perfect gas although at very high temperatures radiation pressure may have an important role in determining the state of equilibrium. However, a body consisting of electrons and positive ions (neutral on the whole) will, at high densities, have its properties constrained by quantum mechanical considerations. In a one-dimensional model if the uncertainty in the position of a particle is Δx, then the corresponding uncertainty in momentum, Δp, has a lower bound given by the Heisenberg uncertainty principle

$$\Delta x \Delta p \geq h \tag{J.1}$$

where h is Planck's constant

Electrons, protons, neutrons and neutrinos are spin-$\frac{1}{2}$ particles, i.e. fermions, and are governed by the Pauli exclusion principle. This means that two electrons in the same quantum-mechanical state cannot occupy the same space and 'space' here is defined in terms of the uncertainty principle. Thus if we define the state of an electron by its position and momentum $(\mathbf{r}, \mathbf{p})$ then the smallest region of the six-dimensional position-momentum space that the particle can occupy is h^3.

These are the basic results that enable us to consider the configurations of stars at very high density.

J.2. DEGENERACY AND WHITE DWARF STARS

Since a fermion occupies a region of position-momentum space h^3, taking the minimum possible value, then the number that can be contained within a sphere of radius r having momentum magnitude in the range p to $p + \mathrm{d}p$ is

$$\mathrm{d}N = \frac{\frac{4}{3}\pi r^3 \times 4\pi p^2\,\mathrm{d}p}{h^3} \times 2. \tag{J.2}$$

The factor 2 in equation (J.2) allows for electrons with opposite spins, and hence different states, to be in the same position-momentum cell. Hence the total number within a sphere of radius r and with momentum magnitude from 0 to P is

$$N = \frac{32\pi^2 r^3}{3h^3} \int_0^P p^2 \, \mathrm{d}p = \frac{32\pi^2 r^3 P^3}{9h^3}. \tag{J.3}$$

In the non-relativistic case the kinetic energy of a particle with mass m and momentum magnitude p is $p^2/2m$. Hence the total kinetic energy for all particles in a sphere of radius r and momentum magnitude between p and $p + \mathrm{d}p$ is given by

$$\mathrm{d}E_\mathrm{K} = \frac{p^2}{2m} \, \mathrm{d}N = \frac{16\pi^2 r^3 p^4}{3h^3 m} \, \mathrm{d}p \tag{J.4}$$

or, for all the particles,

$$E_\mathrm{K} = \int_0^P \frac{16\pi^2 r^3 p^4}{3h^3 m} \, \mathrm{d}p = \frac{16\pi^2 r^3 P^5}{15h^3 m}. \tag{J.5}$$

This kinetic energy creates a pressure (energy per unit volume); because of the way that m occurs in equation (J.5) it is clear that the dominant contributors to the pressure will be electrons. The pressure caused in this way is referred to as *electron degeneracy pressure*. This pressure occurs even at absolute zero temperature and, indeed, it can dominate over kinetic pressure so that the pressure is virtually independent of temperature. Material in such a state is said to be *degenerate*. Substituting for P in terms of N from equation (J.3)

$$E_\mathrm{K} = \frac{16}{15} \left(\frac{9}{32}\right)^{5/3} \frac{h^2 N^{5/3}}{r^2 m\pi^{4/3}}. \tag{J.6}$$

The gravitational potential energy of a uniform sphere of mass M and radius r is

$$E_\mathrm{G} = -\frac{3GM^2}{5r} \tag{J.7}$$

so that the total energy associated with the spherical mass is

$$E_\mathrm{T} = E_\mathrm{K} + E_\mathrm{G} = \frac{16}{15} \left(\frac{9}{32}\right)^{5/3} \frac{h^2 N^{5/3}}{r^2 m\pi^{4/3}} - \frac{3GM^2}{5r}. \tag{J.8}$$

The star is stable, i.e. will neither collapse nor expand, when $\mathrm{d}E_\mathrm{T}/\mathrm{d}r = 0$ or

$$-\frac{32}{15} \left(\frac{9}{32}\right)^{5/3} \frac{h^2 N^{5/3}}{m\pi^{4/3} r^3} + \frac{3GM^2}{5r^2} = 0$$

which gives

$$r = \left(\frac{9}{32}\right)^{2/3} \frac{h^2 N^{5/3}}{\pi^{4/3} GM^2 m}. \tag{J.9}$$

We now apply this result to a white dwarf, a small star which is supported by electron degeneracy pressure. If the average mass of a nucleon is m_N then the total number of nucleons in the star is M/m_N. A white dwarf is the final stage of a well-processed star in which much of the material has been converted into atoms such as carbon and oxygen for which the number of electrons equal to one-half of the number of nucleons. Taking

$$N = \frac{M}{2m_\mathrm{N}} \tag{J.10}$$

then

$$r = \left(\frac{9}{32}\right)^{2/3} \frac{h^2}{\pi^{4/3} G m (2m_N)^{5/3}} \frac{1}{M^{1/3}}. \tag{J.11}$$

Substituting for the various physical constants

$$r = 9.05 \times 10^{16}/M^{1/3}$$

where r is in metres if M is in kilograms. Thus the radius of a white dwarf with the mass of the Sun is 7.18×10^6 m or about 10% more than the radius of the Earth. It will be seen that the relationship gives the interesting result that the radius *decreases* with increasing mass of the white dwarf. This raises the question of whether there is some limiting mass for a white dwarf at which its radius becomes vanishingly small.

J.3.   RELATIVISTIC CONSIDERATIONS

If the density becomes extremely high then this makes the uncertainty in position of the electrons very small. This means that their momenta, and hence kinetic energies, will be so high that we have to abandon classical mechanics and move to relativistic mechanics. The classical kinetic energy expression $p^2/2m$ in equation (J.4) must be replaced by the corresponding relativistic expression $(p^2c^2 + m^2c^4)^{1/2} - mc^2$ so that the kinetic energy of a star of mass M and radius r is

$$_M E_K(r) = \frac{32\pi^2 r^3}{3h^3} \int_0^P \{(p^2c^2 + m^2c^4)^{1/2} - mc^2\} p^2 \, dp \tag{J.12}$$

with P given by equation (J.3). By a change of variable to $q = p/mc$ and from equations (J.3) and (J.10)

$$_M E_K(r) = Ar^3 \int_0^B \{(1 + q^2)^{1/2} - 1\} q^2 \, dq \tag{J.13}$$

where

$$A = \frac{32\pi^2 m^4 c^5}{3h^3} \quad \text{and} \quad B = \frac{h}{mc}\left(\frac{9}{64\pi^2 m_N}\right)^{1/3} \frac{M^{1/3}}{r}.$$

The total energy of the star is thus

$$E_T = {}_M E_K(r) - \frac{3GM^2}{5r}. \tag{J.14}$$

and the condition for stability is $dE_T/dr = 0$, corresponding to a minimum energy. If dE_T/dr is positive for all values of r then the star will shrink without limit. The simplest way to explore the conditions for stability, or its lack, is numerically. The finite-difference approximation

$$\frac{d({}_M E_K(r))}{dr} = \frac{{}_M E_K(r + \delta) - {}_M E_K(r - \delta)}{2\delta}$$

with integrals evaluated by numerical quadrature enables the derivative of the first term on the right-hand side of equation (J.14) to be easily found for any combination of M and r.

   The results of such a calculation are seen in figure J.1. For a mass of $1.0M_\odot$ the radius for stability is 5000 km—somewhat smaller than the value found using the classical expression for kinetic energy. For a mass of $1.7M_\odot$ the stability radius is about 800 km but for a mass of $1.8M_\odot$ it can be seen that dE_T/dr is always positive. The limiting mass at which no stable configuration can

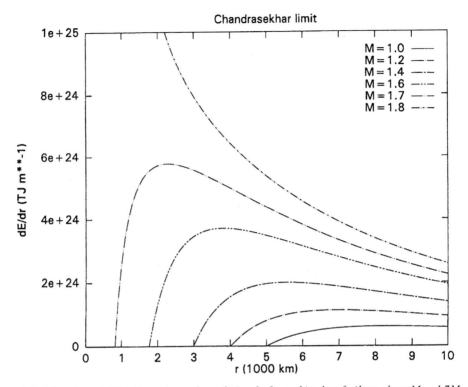

Figure J.1. *The values of* dE/dr *for various values of M and r for a white dwarf. Above about* $M = 1.7M_\odot$ *no radius gives stability. This corresponds to the Chandrasekhar limit.*

exist is called the *Chandrasekhar limit*. Here we have found it to be somewhat over $1.7M_\odot$ but the value usually quoted is $1.44M_\odot$. In his original treatment Chandrasekhar considered factors that are not included here.

J.4.   NEUTRON STARS AND BLACK HOLES

The question is now what happens to a collapsing star consisting of degenerate material if its mass exceeds the Chandrasekhar limit. The first thing that happens is that the electrons in the star combine with the protons in the ions to form neutrons so that the star becomes a *neutron star*. Neutrons are also spin-$\frac{1}{2}$ particles, fermions, so they are able to generate a *neutron degeneracy pressure*. Since they are much more massive than electrons then, according to equation (J.5), they must be much more highly compressed before they exert the pressure necessary to resist gravity and, consequently, neutron stars are much smaller even than white dwarfs. Investigating the size of a neutron star can be done by the same sort of analysis as has been given above with minor modifications. The number of degenerate particles is now equal to the number of nucleons (not one-half as many, as in the neutron case) since all the particles are now neutrons. The other change is that the neutron mass must be used where previously the electron mass appeared. The derivative of equation (J.14) is evaluated to find the stability conditions but now with

$$A = \frac{32\pi^2 m_n^4 c^5}{3h^3} \quad \text{and} \quad B = \frac{h}{mc}\left(\frac{9}{32\pi^2 m_n}\right)^{1/3}\frac{M^{1/3}}{r}$$

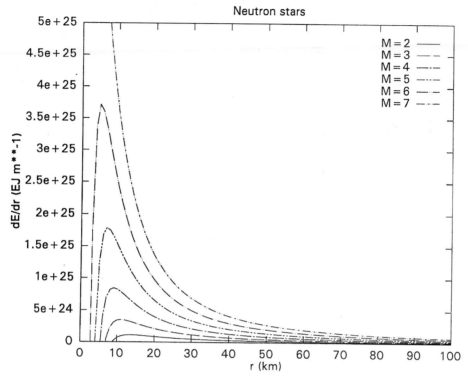

Figure J.2. *The values of dE/dr for various values of M and r for a neutron star. The limiting mass for a neutron star is indicated as about $6M_\odot$.*

where m_n is the mass of the neutron. Calculations for the neutron star are shown in figure J.2. A neutron star of mass $2M_\odot$ has a radius of about 9 km and this falls to less than 3 km for a mass of $6M_\odot$. For a mass of $7M_\odot$ there is no stable configuration and the star will collapse without limit to a *black hole*. Most published estimates of the maximum mass of a neutron star are $3M_\odot$ or so but some are as high as $6M_\odot$, which is suggested by the present analysis.

Problems J

J.1   Using the classical approximation find the radius of a white dwarf of mass $0.8M_\odot$. What is its mean density? Find the relationship between density and mass for a classical white dwarf.

J.2   A neutron star has the density of a neutron, which has a radius of 10^{-15} m. What is the period of a neutron star of mass $2M_\odot$ if the centripetal acceleration at the equator is 0.1 of the gravitational field at the surface?

TOPIC K

PLANETS AROUND OTHER STARS

The first extra-solar planet about a normal star was detected by Mayor and Queloz (1995) and in the next six years more than fifty other normal stars were found to have one or more associated planets. These planets are usually referred to as exoplanets to distinguish them from the planets of the Solar System. Although only the more massive members can be detected at the present time it is clear that these newly-discovered stellar systems are different from our own Solar System and their discovery raises new issues about the way that our own system may have been produced.

K.1. PLANETS AROUND NEUTRON STARS

As mentioned in section 2.7 the first exoplanets to be detected were those about a neutron star, a detection made possible by the extreme constancy of the rate of emission of radio pulses by the star. A neutron star is a product of a cataclysmic event, a supernova, so the problem presents itself of how the planets survived the event, assuming that they were present around the original star. There are at least two possibilities. The first is that they were *not* present before the supernova and were accumulations of material from the debris that was left surrounding the neutron star. The other is that they *were* present but with orbits so large that they could survive the explosion, albeit that they lost most of their outer and more volatile layers. Subsequently their orbits decayed in the medium left around the neutron star. There may be other possibilities but in the light of present knowledge any postulate is just that and cannot be tested.

K.2. EFFECTS OF COMPANIONS ON THE CENTRAL STAR

For normal stars the detection of planets depends on the motion of the star itself relative to the centre of mass of the system of which it is the dominant member. Considering a single companion, the star and planet orbit each other in two linked elliptical paths with a common focus and the same eccentricity, as shown in figure K.1. The shape of each orbit is the same but the linear scale of the orbits, and the speeds of the bodies following them, are inversely proportional to the mass of the body.

Data from the Solar System provide an example of the magnitudes that can be involved. Data for Jupiter, Saturn and Earth, for which the orbits are closely circular, are given in table K.1. The centre of the Sun is displaced from the centre of mass of the system by about 1.1 solar radii ($\sim 7.7 \times 10^8$ m), mainly determined by the most massive planet, Jupiter. Thus the Sun follows an orbit of this small radius about the centre of mass which lies just outside its visible surface. The full solar motion is

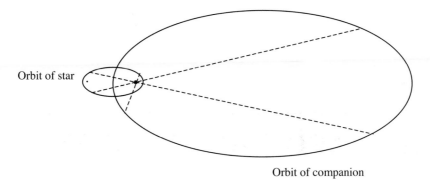

Orbit of star

Orbit of companion

Figure K.1. *Similar orbits of the star and companion. The dashed lines join the positions of the star and companion at various times.*

Table K.1. *The effect of three planets on the motion of the Sun about the centre of mass of the Solar System. The associated solar orbital period is given in column 3, the mean solar orbital radius in column 4 and the mean solar orbital speed in column 5.*

Planet	Eccentricity of orbit	Period (years)	Radius (km)	Speed (m s^{-1})
Jupiter	0.0483	11.86	7.4×10^5	12.4
Saturn	0.0560	29.46	4.1×10^5	2.7
Earth	0.0167	1.00	4.5×10^2	0.09

the superposition of the motions caused by the individual planets but the effect of Jupiter dominates those of all the other planets. It will be seen that the speed of the Sun's motion will be small, but by no means negligible and, as it happens, such speeds can be detected by Doppler-shift measurements.

K.3. FINDING THE SPEED AND MASS OF THE PLANET

We consider a simple case where the line of sight from the Earth is contained in the plane of the orbit of a single planet and the orbit is circular (figure K.2). The maximum speed of the star relative to the Solar System is at points P, with the star receding, and Q, with the star approaching. Relative to the mean speed of the star with respect to the Sun, Doppler-shift measurements give a velocity as shown in figure K.3 where the points P and Q are marked. It is clear that v_*, the speed of the star in its orbit, can be determined in this way. It will be seen from table K.1 that it is necessary to measure speeds of a few metres per second and early measurements were made with spectrometers accurate to about $10\,\mathrm{m\,s^{-1}}$ so the measurements were extremely noisy (figure 2.7). However, there have been improvements in instrumentation and now an accuracy of $3\,\mathrm{m\,s^{-1}}$ is attainable.

To find the mass and orbit of the planet two more pieces of information are required. The first of these is the mass of the star. The stars on which measurements are taken are on the main sequence and, for such stars, there is a well-established relationship between the spectral class (section F.1) and mass. The second requirement is the period of the orbit and this is determined by measurements over time that give the variation of velocity as shown in figures K.3 and 2.7.

Assuming that the mass of the planet, M_p, is very small compared with the mass of the star, M_*,

$$\omega^2 = \frac{4\pi^2}{P^2} = \frac{GM_*}{R^3} \tag{K.1}$$

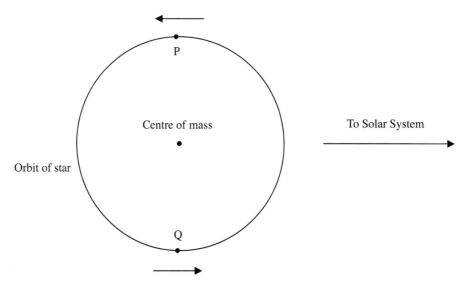

Figure K.2. *The points P and Q corresponding to the greatest speeds relative to the centre of mass as seen by an observer on Earth.*

where P is the period and R the radius of the planetary orbit. Since M_* and P are known then so is the radius of the orbit. From this we find the speed of the planet in its orbit from $v_\mathrm{p} = R\omega$ and hence the mass of the planet from

$$M_\mathrm{p} = M_* \frac{v_*}{v_\mathrm{p}}. \tag{K.2}$$

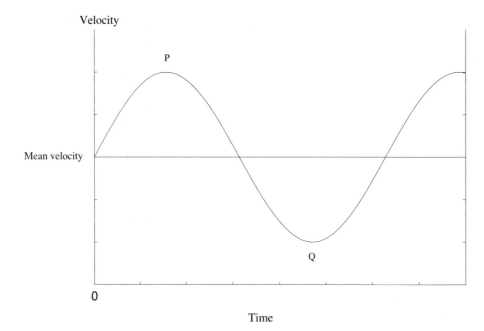

Figure K.3. *The sinusoidal variation of velocity with time for a companion (and star) in a circular orbit. Points P and Q correspond to those in figure K.2*

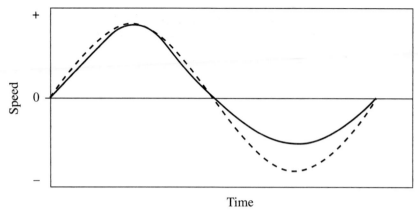

Figure K.4. *The velocity against time curve for an eccentric orbit (full line) and a circular orbit (dashed line).*

Actually it is only the minimum mass of the planet that can be determined since it is not possible to find the inclination of the plane of the orbit to the line of sight. If the normal to the orbital plane to the line of sight, i, is $\pi/2$ then the above analysis will apply. However, if i is other than $\pi/2$ then v_* will be underestimated by a factor $\sin i$ and, from equation (K.2), M_* will be underestimated by the same factor. For this reason any planetary mass estimate is a *minimum* mass and the actual mass could be greater by any factor although, in a statistical sense, the factor will mostly be in the range 1.1 to 3, corresponding to i between $20°$ and $70°$.

Conditions are more complicated if the eccentricity of the orbit is not zero. Now the speed in the orbit will vary and the plot of the velocity with respect to the observer will be as shown in figure K.4. Again, if there are several companion planets then the velocity curve will be the superposition of the separate curves. Decomposing the observed velocity curve into separate periodic components allows the companion motions to be identified and the semi-major axes and eccentricities to be estimated. The mass estimates will still be of minimum masses as explained previously.

If i is very close to $\pi/2$ then, not only will a precise mass be found for the planet but also a transit of the planet across the surface of the star will occur. The fractional loss of brightness of the star as the transit occurs will be the ratio of the area of the planetary disk to that of the stellar disk. Since for main sequence stars the radius is related to spectral class this gives an estimate of the radius of the planet. Such an observation has been made for the star HD209458 (figure K.5; Charbonneau *et al.*, 2000) in which a planet of mass $0.69M_J$ was found to have a radius approximately 1.5 times that of Jupiter. The mean density of about $250\,\mathrm{kg\,m}^{-3}$ suggests that this body is largely hydrogen (presumably within the restriction of the cosmic abundance of elements) and the planet is fluid. The small semi-major axis ($0.045\,\mathrm{AU}$) and zero eccentricity suggests a high surface temperature. This accounts for the large radius and low mean density.

Finding the period is a necessary prerequisite for determining the orbit and mass of an exoplanet. The short-period orbits, of a few days, are the easiest to study because they require the least observing time. Short periods have small semi-major axes. The longer period orbits, involving larger orbital distances from the star, require longer times of observation. In the case of the Sun, an accurate measure of the presence of Jupiter would require measurements over a period of about a decade and detecting Saturn would take even longer. This means that the first studies favour systems with a large companion close in. Only later, when longer observation times have been possible, could

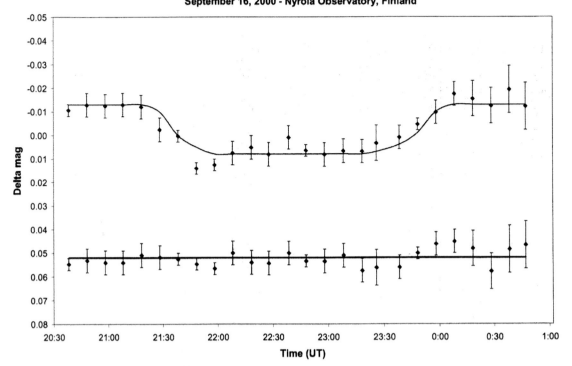

Figure K.5. *Magnitude variation from the mean during an exoplanet transit of HD 209458. The lower line is from a check star. This convincing observation was made by a group of amateur astronomers in Finland (Nyrölä observatory).*

multiple systems be detected accurately. The detection of Earth-mass objects is not possible with present technology (table K.1).

K.4.   THE PRELIMINARY RESULTS OF OBSERVATIONS

The initial exoplanet discovery was of a single companion of 51 Pegasus with a minimum mass $0.47M_J$ orbiting the star in 4.23 days, corresponding to a semi-major axis of 0.05 AU and eccentricity close to zero (the velocity curve is very nearly sinusoidal). The small value of the semi-major axis was unexpected since, by comparison, the closest Solar-System planet, Mercury, orbits the Sun at a distance some eight times greater. The pattern shown by 51 Pegasus is repeated for other orbiting systems found subsequently but other patterns are also found—for example, orbits of greater dimension and of greater eccentricity and also where more than one companion has been identified.

K.4.1.   Mass distributions

The measured minimum masses of companions vary from about 5×10^{26} kg, a quarter of a Jupiter mass, to 1.2×10^{29} kg, or 60 Jupiter masses. The distribution of 48 companion masses is shown in figure K.6 for those up to $12M_J$. The numbers per unit mass range decrease with increasing mass

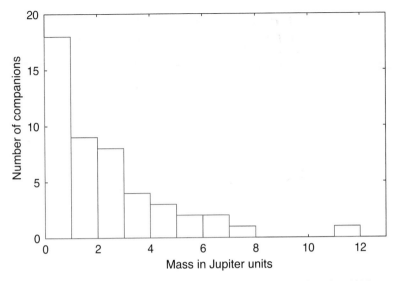

Figure K.6. *Mass distribution of 48 companions with masses less than $13M_J$.*

and the profile can be fitted to an exponential curve. There certainly seems to be a preference for smaller masses but the gap gives witness to the paucity of data.

If the companion is composed of hydrogen then there is a natural divide between masses since above about $13M_J$ deuterium reactions will take place. Above this limit the companion is conventionally called a brown dwarf. Some theoreticians maintain that the formation of planets is an accretion process and so is different from the collapse of gas thought to form brown dwarfs and stars. The largest minimum mass found so far below this apparent divide is $11M_J$ for the companion of HD 114762; the lowest minimum mass above the divide is $13.73M_J$ for the companion of HD 162020. The data for 12 companions with masses in excess of $13M_J$ are collected in table K.2. These brown-dwarf objects may form a long tail to the distribution of figure K.6 if the gap in masses from 11 to $13.73M_J$ is not significant.

Table K.2. *The orbital data for 12 stars with companions of minimum mass greater than $13M_J$.*

Star	Minimum mass (M_J)	Semi-major axis (AU)	Eccentricity
HD 162020	13.73	0.072	0.28
HD 110833	17	0.8(?)	0.69
BD-04 782	21	0.7	0.28
HD 112758	35	0.35	0.16
HD 98230	37	0.06	0.0
HD 18445	39	0.9	0.54
Gl 229	40	—	—
HD 29587	40	2.5	0.37
HD 140913	46	0.54	0.61
HD 283750	50	0.04	0.02
HD 89707	54	—	0.95
HD 217580	60	≈ 1	0.52

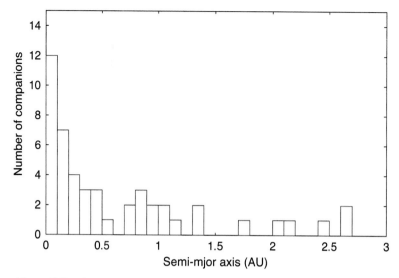

Figure K.7. *The semi-major axes of 48 companions with masses less than 13M$_J$.*

K.4.2.   Characteristics of orbits

Examination of tables 2.1 and K.2 shows semi-major axes varying from 0.04 to 2.5 AU and eccentricities as high as 0.67 for planetary companions and 0.95 for brown dwarf companions. The distribution of semi-major axes and eccentricities for planetary companions are shown in figures K.7 and K.8 respectively.

The lack of orbits with semi-major axes similar in scale to those of the major planets in the Solar System is almost certainly due to selection; as previously mentioned, longer period of observation are required to determine larger orbits. The closeness to the central star of the orbits of some exoplanets is not seen in the Solar System and, indeed, the presence of gas-giant planets so close to stars suggest that they formed farther out and then migrated inwards. How this might have happened is indicated in figure 12.6 and topic AP.

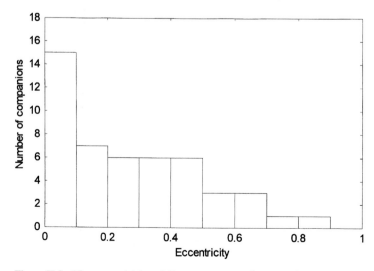

Figure K.8. *The eccentricities of 48 companions with masses less than 13M$_J$.*

K.5.   THE CONSTITUTION OF THE COMPANIONS

The most abundant cosmic chemical element is hydrogen and it is natural to suppose that the companion bodies consist of hydrogen but this need not be so. The stars where companions have been found so far contain a significant proportion of heavy elements (they are said to have high metallicity). The rarer elements can be expected to form silicates, oxides, sulphides and other compounds familiar in the Solar System because these come from a universally-applicable chemistry. Although these compounds are associated with the less massive members of the Solar System this need not be so elsewhere. One can envisage hydrogen bodies with a very substantial silicate/ferrous core—for example, a total mass of $10M_J$ and a core of $5M_J$ or more. Knowledge of the constitution of exoplanets could have important implications for our understanding of planets generally.

The relationship between the mass and radius of a body depends on its constitution and is displayed in figure P.1. The exoplanet of HD 209458 with $\log(M_p) = 26.6$ and $\log(R_p) = 7.78$ fits almost exactly on the line indicating a hydrogen composition. If it was found that, where a radius estimate could be made by transit measurements, the $(\log R, \log M)$ point fell well below the hydrogen line then it might be inferred that considerable amounts of heavier elements were present.

K.6.   ATMOSPHERES

The principles controlling the Earth's atmosphere are considered in Topic O and these apply more widely. The atmosphere is retained by the parent body by gravitational forces whose effects can be weakened both gravitationally (by the proximity of the star) and by the high prevailing temperature. Calculations show that the atmosphere is, in fact, stable for bodies of Jupiter mass provided the semi-major axis is greater than about 0.02 AU (about 5 solar radii). This restriction is met in all cases of companions met so far.

The small semi-major axes and sometimes large eccentricities of the companion orbits means that their atmospheres are under strong thermal and gravitational influence from the central stars. For a planet in a very eccentric orbit the temperature may vary between 1000 K and 100 K. This change of energy content of the atmosphere will be associated with a change of thickness of the atmosphere, the companion having a greater radius when hotter near the star than when at its farthest distance. If the gas is approximately ideal then the radius will vary approximately as $T^{1/3}$ so the thickness of the atmosphere will fluctuate by a factor of ~2 over the few Earth-days of the orbit. It is also likely that a body so near the primary will be in synchronous rotation, having one hemisphere permanently facing the star. This gives a rotation period for the companion of several days making rotation forces quite insignificant. The dynamics of such an atmosphere presents problems not met with in treating planetary atmospheres in the Solar System.

The closeness of the companion to the star will also lead to strong material and magnetic interaction between the two bodies. The surface of the companion will undergo severe particle and radiation bombardment from the star.

K.7.   POSSIBILITIES OF CONDITIONS FOR LIFE

The possibility of life in the universe is considered in Topic AM. A distinction must be drawn between elementary and advanced life forms. Experience from the Earth suggests that elementary life is very hardy and will begin whenever and wherever it can. It survives on Earth in extreme circumstances of hot and cold and under high and low pressures. It probably attempts to start again at a particular location from which it has been removed by chance events. The chemistry of life involves the normal chemical elements common throughout the Galaxy. This leads to the reasonable hypothesis that life may exist elsewhere although there is, as yet, no evidence that this is so.

Advanced life requires more stringent conditions. It is generally agreed that a solid planet is required, located in the broad temperature range given by distances of 0.7 to 1.6 AU from a solar-type star—that is, between Venus and Mars in Solar-System terms. Liquid water is essential if experience on Earth is generally valid. The planetary orbit must be stable and significant surface impacts must be extremely rare. This implies the existence of at least one Jupiter mass planet in exterior orbit to act as a trap for stray bodies. How large the semi-major axis of the protecting planet should be is not too certain but probably at least 2 AU. There are requirements for the solid planet as well. It must be sufficiently massive to retain an atmosphere over cosmic time periods. This will block radiation with photon energy in excess of about 10^{-18} J which is harmful to living tissue (the energy of inter-atomic interactions is about 10^{-21} J). Again, the planet must not be too massive. For example, it is difficult for bone to withstand falls in large gravity fields and the energy required for locomotion also becomes very large.

The question, then, is whether any of the systems observed so far could begin to satisfy these conditions. There are a number of stars with about Jupiter mass or greater planets in orbits above 2 AU that will give the trapping mechanisms for stray objects. However, what have not been detected, because the technology will not allow it, are the smaller solid planets that could be the home of some life form or other.

K.8.   A FINAL COMMENT

The study of exoplanets, which is ongoing, has already provided some interesting differences from what exists in the Solar System. Many planets of large mass orbit very close to the parent star with semi-major axes less than 1 AU. The eccentricities are often large. Comparable observation of the Solar System would give one companion of mass $1M_J$ with semi-major axis 5 AU and eccentricity 0.04, with a second companion of mass $0.3M_J$, a semi-major axis of about 10 AU and eccentricity 0.06. A comparable system to this has not yet been found. It could be that our arrangement of planets might not be a common one, at least for those stars up to a distance of 65 pc that have been investigated so far. It may also turn out that the conditions for advanced life forms to exist are also rare. The study of these matters is called astrobiology

Problem K

K.1   Spectroscopic observation of a star, of known mass $1.15M_*$, shows a sinusoidal variation of Doppler shift with time, with a period of 2.06 years corresponding to a maximum speed of approach or recession of $52\,\mathrm{m\,s^{-1}}$. What is the minimum mass of the accompanying planet and the radius of its orbit?

TOPIC L

SOLAR-SYSTEM STUDIES TO THE BEGINNING OF THE SEVENTEENTH CENTURY

L.1. VIEWS OF THE ANCIENT WORLD

Early *Homo sapiens* must have marvelled at the splendour of the heavens and noticed the patterns of stars and their movements. When civilizations arose, in China and the Middle East, studies of the heavens began from which practical empirical knowledge was acquired and used, for example, by mariners who guided their craft by the stars. Four thousand years ago the Babylonians defined the arrival of spring, when planting began, as when the Sun was seen in the direction of the constellation Aries although, because of a slow precession of the Earth's spin axis (Topic U), the beginning of spring now brings the Sun near the constellation Aquarius.

Most points of light, the stars, form fixed patterns, the *constellations*, but others move around the sky against this backcloth; these are the planets, named from the Greek word for *wanderers*. The early Greeks concluded that the Earth, Sun, Moon and planets form a system separate from the other fixed heavenly bodies. Anaximenes (585–528 BC) took the Earth and the Sun to both be flat but Pythagorus (572–492 BC) proposed that the Earth and the stars were spheres, not for any scientific reason but rather because the sphere was regarded by the Greeks as the 'perfect' figure.

It might be said that scientific studies of the Solar System began with Aristarchus of Samos (310–230 BC). He realized that moonlight was derived from the Sun and that a half-moon is seen when the Sun–Moon–Earth angle is 90° (figure L.1). He attempted to measure the Moon–Earth–Sun angle θ at the time of the half-moon but this angle is too close to 90° to measure accurately and he found the ratio of distances of the Sun and Moon as 19 ± 1 compared with the true value of 390. Aristarchus also proposed that the Sun and the stars were fixed in position and that the Earth moved round the Sun on a circular path—the first heliocentric theory of the solar system. This was to be revived many centuries later by Copernicus and considerably developed.

Another Greek achievement was the measurement of the Earth's radius by the Alexandrian, Eratosthenes (276–195 BC). At mid-day at the beginning of summer the Sun's rays shone straight down a deep well at Syene (modern Aswan) meaning that the Sun was at the *zenith*, or directly overhead. Alexandria is due north of Syene and, from the length of the shadow of a vertical column there at the same time, it could be deduced that the Sun was about 1/50th of a complete rotation ($\theta \approx 7°$) from the zenith (figure L.2). From the distance between Syene and Alexandria, 5000 stadia, Eratosthenes deduced that the circumference of the Earth was $50 \times 5000 = 250\,000$ stadia. The stadium is believed to have been about 180 m giving an Earth radius of 7200 km, or about 13% greater than the actual value of 6367 km.

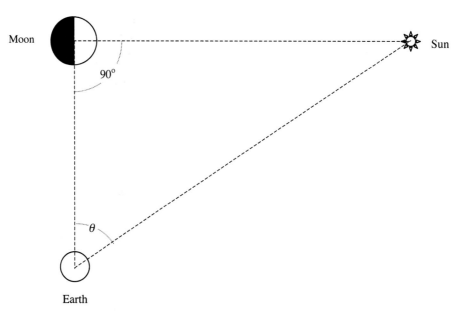

Figure L.1. *A half-illuminated Moon with the Sun–Earth–Moon angle equal to 90°. Aristarchus measured the angle θ to establish the shape of the triangle but it was too close to 90° to measure with sufficient accuracy.*

The idea of a moving Earth was difficult for many to accept since the Earth's motion gives no sensation of movement. It was believed that the Earth must be at rest and that everything else moved round it which, of course, puts mankind at the centre of the universe. An Alexandrian Greek, Ptolemy (c. 150 AD), formally proposed this geocentric model of the Solar System that was to be dominant for 1400 years. A problem in describing the geocentric system is that while the Sun makes a uniform circular orbit of the Earth once a year, and the Moon does likewise once a month, the planets are

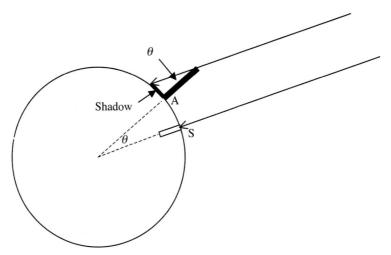

Figure L.2. *The Eratosthenes experiment. The angle subtended by the arc connecting Syene (S) and Alexandria (A) at the centre of the Earth is the same as the angle made by the Sun's rays with a vertical post in Alexandria.*

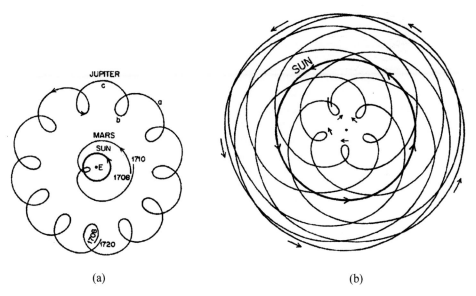

(a)   (b)

Figure L.3. *The path in the sky as seen from Earth of (a) Mars, Jupiter and the Sun and (b) Venus and the Sun.*

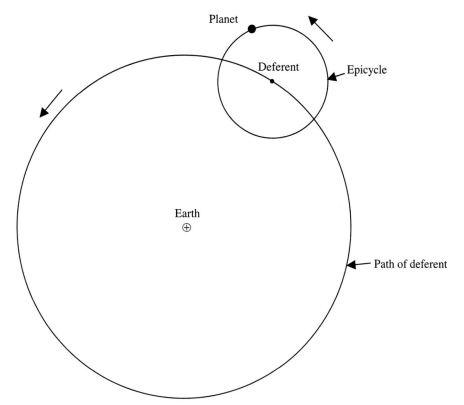

Figure L.4. *According to Ptolemy's model the Earth is fixed while the planet moves in a circular orbit (the epicycle) about the deferent, which itself performs an orbit of period one year about the Earth.*

seen to move in a series of loops in the sky, although predominantly they move in an eastward direction (figure L.3). Ptolemy dealt with this apparent anomaly by describing the motions of the planets in terms of two components—the first motion is a uniform circular orbit with a period of one year of a point called the *deferent* and the second an *epicycle*, uniform motion around a circular path centred on the deferent (figure L.4). The deferents of Mercury and Venus, the *inferior* planets, had always to be on the line connecting the Earth to the Sun because they were always seen at a small angular separation from the Sun. To explain the motion of the *superior* planets, the line joining the deferent to the planet had to remain parallel to the Earth–Sun direction. Although this was a complicated system, with arbitrary rules, it adequately described the motions of the planets relative to the Earth. With no theories available to explain why and how bodies should move it was as good a system as any other and satisfied all the current observational criteria.

L.2. NICOLAUS COPERNICUS

The next major player on the scene was a Polish cleric, Nicolaus Copernicus (1473–1543) who for some time was a professor of mathematics and astronomy in Rome and also practised medicine. His interest in astronomy was almost certainly influenced by Arabic and Greek astronomical thinking, introduced by the influx of scholars who settled in Rome after the collapse of the Holy Roman Empire and the fall of Constantinople in 1453. He became interested in planetary movements and constructed tables describing planetary motions as had Ptolemy before him. He redeveloped the heliocentric theory, but now based on much improved observational data. His model required planetary orbital angular speeds to vary slightly which he regarded as being due to the orbits being circular but not exactly concentric. However, he still needed small epicycles to obtain good agreement of his model with observations. He presented his ideas in a treatise, *De Revolutionibus Orbium Coelestium*, dedicated to Pope Paul III, which was published shortly before he died.

L.3. TYCHO BRAHE

It must not be thought that *De Revolutionibus* brought about a sudden and complete change of belief in astronomical circles. One person who was partly influenced was Tycho Brahe (1546–1601), a Danish nobleman and the most effective observational astronomer of the pre-telescope period. The model he accepted was a compromise between the Ptolemaic and Copernican systems in which the planets, other than the Earth, orbited the Sun but Sun and the Moon orbited the Earth. This model had the characteristic that it gave the correct relative motions of all the bodies in the system. In 1576 Tycho built a splendid observatory on a Baltic island, Hven, with the financial support of the Danish king, Frederick II. His line-of-sight instruments were based on large circles for measuring angles (figure L.5) and his measurements were of much greater accuracy than any made previously.

Tycho Brahe's haughty attitude caused tensions with other inhabitants of Hven and, when his sponsor died, the same arrogance brought him into conflict with the new king, Christian. Eventually he was forced by lack of money to leave his observatory. In 1596, he went to Prague to become the Imperial Mathematician at the court of Rudolph II of Bohemia where he compiled tables of planetary motion assisted by a very able younger man, Johannes Kepler. Tycho had realized that the motion of Mars is unusual and he asked Kepler to study the data related to that planet.

Figure L.5. *Tycho Brahe's quadrant was attached to a wall that had on it a painting of Tycho Brahe himself. The engraving shows the observer, presumably Tycho Brahe, someone keeping an eye on clocks so that times of observations could be recorded and someone at a desk noting the readings.*

L.4. JOHANNES KEPLER

Kepler (1571–1630) had already become interested in the Copernican system when a student at Tübingen and later, while he was a teacher of mathematics at Graz, had written a defence of the system in his first scientific book, *Mysterium Cosmographicum*. He refined Copernicus' idea that the planetary orbits could be represented by a series of non-concentric circles with small epicycles. He felt instinctively that epicycles should not be necessary and that the planetary path is a simple curve. Of the planets known at that time and that could also be well observed, Mars has the most eccentric orbit, which made its motion the most difficult to understand. He deduced that its orbit is not circular, unlike the Earth where the orbit can be taken as circular but with the centre of the circle displaced from the Sun. It took Kepler some time to realize that the curve which could unify the orbits of all the planets is an ellipse with the Sun at one focus.

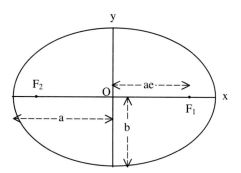

Figure L.6. *The characteristics of an ellipse.*

Actually the Copernican description of circular orbits was reasonably accurate. An ellipse, shown in figure L.6, is given in a Cartesian coordinate system by

$$\frac{x^2}{a^2} + \frac{y^2}{b^2} = 1 \tag{L.1}$$

where a and b are the *semi-major* and *semi-minor* axes respectively. The foci are displaced from the geometrical centre of the ellipse by $\pm ae$ in the x direction, where e is the *eccentricity* of the ellipse. Finally the semi-major axis is related to the semi-minor axis by

$$b = a(1 - e^2)^{1/2}. \tag{L.2}$$

The eccentricities of most planetary orbits are quite small; that of Mars is 0.093. For Mars $b = 0.991a$ and for the remaining planets known to Kepler, except Mercury which was difficult to observe, b would be even closer to a. It took Tycho Brahe's high-quality measurements to distinguish elliptical from circular orbits.

After analysing his results Kepler eventually formulated his three laws of planetary motion:

1.   Planets move on elliptical orbits with the Sun at one focus.
2.   The radius vector sweeps out equal areas in equal time.
3.   The square of the period is proportional to the cube of the mean distance.

The first two laws came early but the third one took more time to discover. The determination of the shape of the orbit of Mars and the first two laws were published by Kepler in *Astronomica Nova* in 1609 and the final law was given in 1619 in his book *Harmonica Mundi*.

L.4.1.   Kepler's determination of orbital shapes

It can be seen from equation (L.2) that, to a good approximation, the orbits of most of the planets known at that time were almost circular. However, Kepler had available Tycho Brahe's accurate and abundant observations of Mars which he used to determine the shape of its orbit. An item of information he required for this exercise was the *sidereal period* of Mars, P_{sid}, which is the time taken for the Sun–Mars vector to return to the same point on the celestial sphere. This cannot be directly determined for a planet by observation from the Earth but it can be derived from the *synodic period*, P_{syn}, which is the time from one Sun–Earth–planet alignment to the next and is easily found. The configuration of the bodies for two successive *oppositions* of Mars is shown in figure L.7, where the Earth is between Mars and the Sun. When Mars has reached position M_2 it has moved an angle $2\pi + \theta$ from M_1 while the Earth at E_2 has moved an angle $4\pi + \theta$ from E_1.

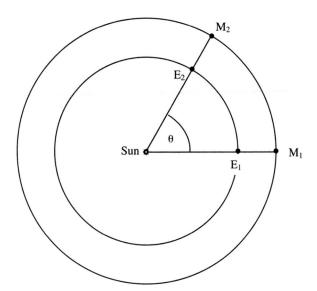

Figure L.7. *Positions of the Earth and Mars for two successive oppositions of Mars.*

Since one synodic period has ensued between positions E_1 and E_2 we may write

$$\frac{P_{\text{syn}}}{P_{\text{E}}} = \frac{4\pi + \theta}{2\pi} \tag{L.3}$$

where P_{E} is the sidereal period of the Earth, i.e. one year. Since Mars moves through an angle 2π during one sidereal period we have

$$\frac{P_{\text{syn}}}{P_{\text{sid}}} = \frac{2\pi + \theta}{2\pi} \tag{L.4}$$

and combining equations (L.3) and (L.4) we find

$$\frac{1}{P_{\text{sid}}} = \frac{1}{P_{\text{E}}} - \frac{1}{P_{\text{syn}}}. \tag{L.5}$$

From the synodic period of Mars, 2.135 years, the sidereal period is found to be 1.881 years.

The first step taken by Kepler was to find an estimate of the Earth's orbit better than that of being circular and centred on the Sun. To this end the direction of Mars from the Earth, noted on the background celestial sphere, was found first at opposition and then at the end of a period equal to a sidereal period of Mars. The position of Mars, M_1, shown in figure L.8, is the same for both readings but the Earth will have moved from E_1 to E_2 an angle of about 677°; this value is based on a constant angular velocity of the Earth about the Sun but is good enough to determine a better orbit for the Earth. Since the two measured directions of Mars give the angle ψ in figure L.8 the distance SE_2 may be found in units of the distance SM_1. By the time another Mars sidereal period has passed the Earth is at position E_3 and SE_3 may be found in the same units of distance, SM_1. Continuing this process gives different Earth–Sun distances which enables the shape of the orbit to be found, but not its absolute size, and also its relationship to the Sun. Kepler found:

(i)   the Earth's orbit was indistinguishable from a circle,
(ii)   the Sun is not at the centre of the circle,
(iii)   the Earth's angular speed around the Sun is not constant.

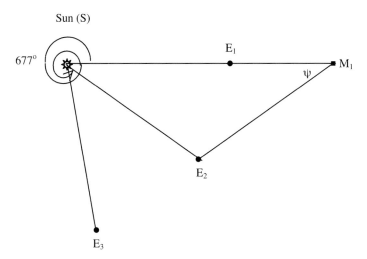

Figure L.8. *E_1 and E_2 are Earth positions at the beginning and end of one sidereal period of Mars. The directions of Mars from Earth in these two positions give ψ and hence the geometry of the triangle Sun–E_2–M_1, since the angle E_1–Sun–E_2 is known.*

With the Earth's orbit established the process was applied in the reverse direction to find the orbit of Mars. If now, in figure L.8, SE_2 is considered known then SM_1 can be found. By pairs of measurements of directions of Mars from the Earth, with Mars in various positions in its orbit, e.g. M_2, M_3 etc., then the Mars–Sun distances may be determined in terms of Earth–Sun distances which are known relative to each other.

Kepler found that the orbit of Mars is not a circle, although the difference from a circle is not great. It was surprising that Kepler, an expert geometrician, spent several years in matching his orbit to an ellipse after trying a great variety of other curves. However, eventually he did so and with the aid of Tycho Brahe's very precise data, with angular errors less than $6'$ of arc, he was able to expound his three laws.

Tycho had begun, but did not live to finish, the compilation of what he called the *Rudolphine Tables* of planetary positions and Kepler completed them, incorporating his own discoveries. This provided a means of calculating the planetary positions for any date in the future or the past. Logarithms were involved which he developed independently of Napier. Kepler used his Tables to predict the transits of Mercury and Venus across the face of the Sun and the predicted transit of Mercury was actually observed in 1631, a year after he died. His methods of analysis were included in his book *Stereometrica Doliorum* and later formed the basis of the integral calculus. It is hard to overestimate the achievements of Kepler. He had deduced three exact empirical laws describing nature on the basis of observational data which included errors of measurement—in a sense he had corrected the observations. This is now expected of scientific analyses and there are sophisticated rules for approaching the matter but this was the first time it had actually been achieved.

Although Kepler's name is nowadays irrevocably linked with the laws of planetary motion, in fact he made a wider contribution to our understanding of the world. He was the first to realize that the tides are due to the action of the Moon, although his great contemporary Galileo thought he was wrong. He also realized that the Sun spins on its axis as he explained in *Astronomia Nova*. He investigated the formation of optical images in a physical system including the mechanism of vision itself (originally to discover *how* a telescope works). In practical terms this gave rules for designing eyeglasses for long and short sight. This work was published in his books *Dioptrice* (a name he invented and which is still used today in connection with the refraction of light) and in *Astronomia Pars Optica*. These studies laid the foundations of what was later to become geometrical optics. He understood the way that both eyes together gave depth to perception and extended this idea of parallax to attempting to measure the distances of stars using the Earth's orbit as a baseline. He was unsuccessful because his instrumentation was not good enough but the principle was correct and has been used in more recent times with great success.

L.5. GALILEO GALILEI

The development of the study of planets from the purely empirical into the theoretical arena was the achievement of Galileo Galilei (1562–1642), a contemporary of Kepler. Galileo had been appointed professor of mathematics in Pisa in 1589, at the age of 25, and also taught at Padua university for 18 years. It can be said that Galileo laid the foundations of much of modern science, in particular in the field of mechanics, in which he realized that the acceleration of a body was as important as its velocity in describing its motion and was also the first to formulate the concept of relative motion. He was interested in the motion of bodies under gravity and also in astronomy—in particular in the motions of the planets. Kepler sent him a copy of his *Epitome of the Copernican Astronomy*, which described all his scientific work, including that on planetary motion and in his response Galileo revealed that he too favoured the Copernican system.

Soon after its invention by a Dutch spectacle-maker in 1608, Galileo made a telescope and used it for astronomical observations. He saw lunar mountains, of which he made height estimates from the shadows they cast, and also the large satellites of Jupiter (Io, Europa, Ganymede and Callisto) now known as the Galilean satellites. In 1610 Galileo observed that Saturn had a pair of appendages that looked like ears; his telescope could not resolve Saturn's rings and he assumed that what he saw were two companion bodies. Two years later, in 1612, he was astonished to find that the 'companions' had disappeared; the rings were presenting themselves edge-on towards the Earth and so were invisible.

Galileo also observed the phases of Venus. The Ptolemaic model required the deferent of an inferior planet always to be on the Earth–Sun line so that only part of the illuminated hemisphere away from the Earth could be seen. Venus should thus only ever be observed in a crescent phase (figure L.9a). On the other hand, for the Copernican model, all phases would be observed (figure L.9b). A full Venus will be seen when Venus is in *superior conjunction*, with the Sun between the Earth and Venus, and a crescent Venus when it is close to *inferior conjunction*, i.e. directly between the Earth and the Sun. Hence the angular size of the disk will be greater in the crescent phase, when Venus is closer, than in the full phase. This observation irrefutably supported the Copernican prediction.

With such evidence it was not surprising that both Kepler and Galileo agreed that the Jovian system is mechanically a miniature version of the Solar System with Jupiter playing the role of the Sun. The orbits of the satellites must surely also obey Kepler's empirical laws of motion. Galileo recognized the correctness of the Copernican model and he attempted to persuade others by his *Dialogue on the Two Chief World Systems*, in which two individuals dispassionately discuss the relative merits of the geocentric and heliocentric models. The heliocentic theory came over as quite clearly the superior system. Galileo thought that his professional and personal prestige would support him in this venture but he was mistaken. At the time of the publication of *Revolutionibus* the Church had not raised any objections to it—indeed Pope Clement VII encouraged Copernicus and the publication was paid for by a cardinal. However, over the course of the next 60 years or so the attitude of the Church changed. An event which triggered the Church into decisive action was

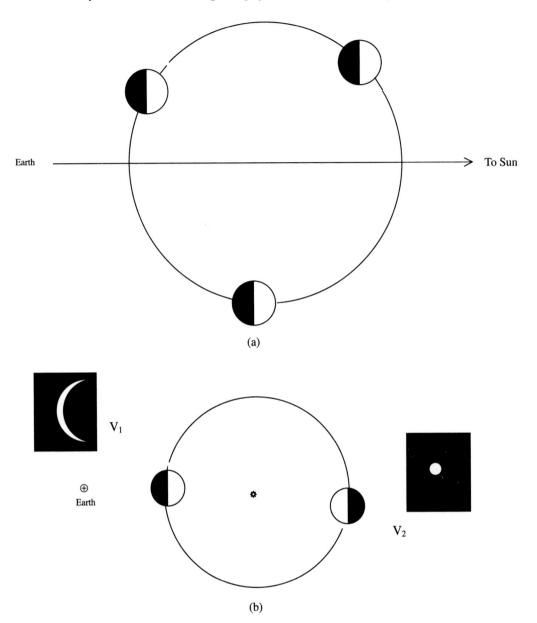

Figure L.9. *(a) According to Ptolemy's model Venus moves on the shown epicycle and is always seen in a crescent phase. (b) According to the heliocentric model Venus is seen as a large crescent at V_1 and a smaller complete disk at V_2.*

an idea put forward by an Italian Renaissance philosopher and Dominican monk, Giordano Bruno, who argued that all other stars were like the Sun and would have planets around them inhabited by other races of men. This challenged the religious concept of the central role of humankind and its relationship to the deity. Bruno refused to retract his views and, in 1600, paid with his life for his obstinacy. In 1616 *Revolutionibus* was added to the *Index Librorum Prohibitorum*, the list of forbidden books—as were Kepler's works.

Kepler lived in the Protestant north of Europe and so was little affected by what went on in Rome. The effect on Galileo, living in the Catholic south, was more profound. He was interviewed by the inquisition and humiliated by being required to make a public rejection of the Copernican world view. Due to his eminence he was neither ill-treated nor put in prison but his freedom was restricted and he was effectively placed under house arrest for the last years of his life. It is a very sad story but was probably the last time that the new scientific method of argument was rejected on matters of material experiment and observation.

Problems L

L.1   Aberdeen is due north of Bournemouth at a distance estimated to be about 750 km. When the Sun is at its zenith on a particular day a vertical stick of length 1 m throws a shadow 150.8 cm long in Aberdeen and 119.0 cm long in Bournemouth. Estimate the radius of the Earth.

L.2   Tycho Brahe's quadrant had a radius of 2.1 m. If the sixteenth century engravers could work to an accuracy of 0.10 mm then to what angular precision was Brahe able to measure?

L.3   Calculate the angular size of Venus when seen (a) in full phase and (b) in extreme crescent phase.

TOPIC M

NEWTON, KEPLER'S LAWS AND SOLAR-SYSTEM DYNAMICS

M.1. ISAAC NEWTON, KEPLER AND THE INVERSE-SQUARE LAW

In the year that Galileo died the greatest scientist of all time, Isaac Newton (1642–1727), was born. He formulated 'Newton's laws of motion', studied light and various aspects of hydrostatics and hydrodynamics and developed calculus. His greatest contribution to the understanding of the Solar System was his proof that the inverse-square law of gravitational attraction followed from Kepler's third law which can be expressed in the form

$$\frac{P_1^2}{P_2^2} = \frac{a_1^3}{a_2^3} \tag{M.1}$$

where P is the period and a the semi-major axis of an orbit. Newton considered the force law between bodies to be of the form

$$F = C\phi(r)$$

where C is a constant. Since the centripetal acceleration is proportional to the force then, for a circular motion

$$\frac{F_1}{F_2} = \frac{\phi(r_1)}{\phi(r_2)} = \frac{v_1^2/r_1}{v_2^2/r_2} \tag{M.2}$$

where v is the velocity in a circular orbit of radius r. For circular motion $v = 2\pi r/P$ so that

$$\frac{F_1}{F_2} = \frac{r_1/P_1^2}{r_2/P_2^2} = \frac{1/r_1^2}{1/r_2^2}. \tag{M.3}$$

From equation (M.3) the force is proportional to the inverse square of the distance. Since it was known that the acceleration of bodies falling under gravity is independent of mass it follows that the gravitational force is proportional to the mass of the falling body. By symmetry, since bodies attract each other, it had to depend on the product of the attracting masses. The overall force law was found by Newton to be

$$F = G\frac{m_1 m_2}{r^2} \tag{M.4}$$

in which the gravitational constant, G, was subsequently found by laboratory measurement to be $6.67 \times 10^{-11}\,\mathrm{m^3\,kg^{-1}\,s^{-2}}$.

Newton described his theory of gravitation and much else in a famous publication, the *Principia*, which appeared in 1687 but took 15 years to write. His importance as a scientist cannot be overstated. His work represents the greatest ever watershed in science, particularly in the study of the Solar System. From Newton onwards the essential structure of the Solar System was clearly established and clearly understood.

    Newton was somewhat irascible and quarrelsome but he also had firm friends, one of whom was Edmond Halley (1656–1742); it was Halley who persuaded Newton to publish the *Principia* and he paid for the cost of the publication. Halley is best remembered for the comet that bears his name. In 1705 he suggested that the comet of 1682 was a reappearance of a comet observed in 1456, 1531 and 1607, with a period just under 76 years. He correctly predicted its return in 1758, although he did not live to see it. He established comets as part of the Solar-System family, although in such eccentric orbits that they can only be seen when they approach the Sun.

M.2. GENERAL ORBITS

Kepler's empirical determination of the shapes of planetary orbits and his three empirical laws were described in sections 3.2 and L.4 and we have shown that this implied an inverse-square law for the

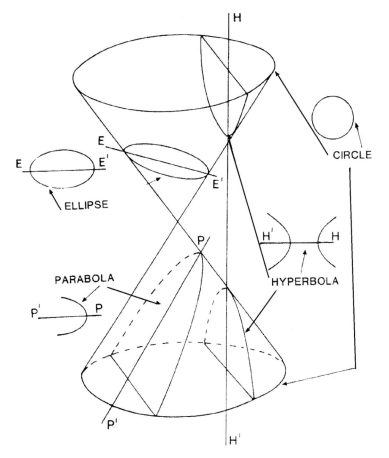

Figure M.1. *The sections of a cone giving a circle, ellipse, parabola and hyperbola.*

gravitational force. Here we shall explore the detailed mechanics of orbits using the theoretical tools provided by Newton. To do this we require the description of an elliptical orbit in terms of polar coordinates with the centre of mass of the system (the massive body if one of the bodies has negligible mass) at the origin as given in section 3.2.

We must recognize that the ellipse represents just one type of trajectory produced by the gravitational interaction of a pair of bodies. If $e = 0$ then equation (3.2) becomes $r = a$, so that a circular orbit may be regarded as the special case of an elliptical orbit with zero eccentricity. The circle and ellipse are just two types of *conic sections*, so-called because they relate to plane sections that can be taken through a cone. In figure M.1 the sections that give the circle and ellipse are shown. Another section gives a hyperbola, a conic section that is not closed but goes to infinity. The section which marks the interface between ellipses and hyperbolae gives the parabola which is also a non-closed conic section.

One way of describing a parabola is to consider the ellipse in figure 3.2 and imagine that the perihelion distance is kept fixed but that e steadily increases. The ratio of the aphelion distance to the perihelion distance is $(1 + e)/(1 - e)$ so that when $e = 1$ the aphelion has retreated to infinity and the curve has become a parabola. Another way of describing a parabola is shown in figure M.2(a) where it is the locus of points which are the same distance from the focus, F, and the straight line AB. Now in figure M.2(b) we show a curve where the distance from the focus is twice the distance from the straight line. This is a hyperbola of eccentricity $e = 2$.

The importance of the conic sections is that they represent all the possible orbits of a body moving under a central inverse-square-law force. We shall now show that such a force leads to an elliptical orbit and and also gives the other two Kepler laws.

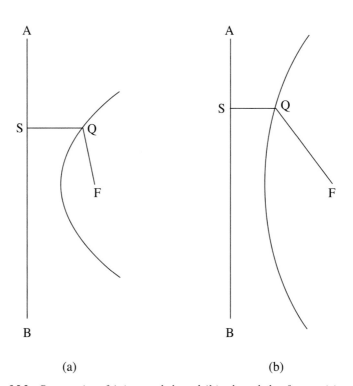

(a)   (b)

Figure M.2. *Construction of (a) a parabola and (b) a hyperbola of eccentricity $e = 2$.*

M.3. KEPLER'S LAWS FROM THE INVERSE-SQUARE-LAW FORCE

We now consider the motion of a body P, of negligible mass, moving under the gravitational attraction of a body, Q, of mass M (figure M.3). According to Newton the gravitational field at P is in the direction of Q and has magnitude GM/r^2 where r is the distance between P and Q. In a frame of reference centred on Q and rotating with P the radial acceleration is given by

$$\frac{d^2r}{dt^2} = -\frac{GM}{r^2} + r\left(\frac{d\theta}{dt}\right)^2 \tag{M.5}$$

where G is the gravitational constant. From the conservation of angular momentum we write

$$r^2\frac{d\theta}{dt} = H \tag{M.6}$$

where the constant H is the *intrinsic angular momentum* or angular momentum per unit mass.

For the first part of our analysis we modify equation (M.5) by changing the independent variable from t to θ and we use equation (M.6) in this process. First we write

$$\frac{dr}{dt} = \frac{dr}{d\theta}\frac{d\theta}{dt} = \frac{H}{r^2}\frac{dr}{d\theta}. \tag{M.7}$$

Then

$$\frac{d^2r}{dt^2} = \frac{d\theta}{dt}\frac{d}{d\theta}\left(\frac{H}{r^2}\frac{dr}{d\theta}\right) = \frac{H}{r^2}\left[-2\frac{H}{r^3}\left(\frac{dr}{d\theta}\right)^2 + \frac{H}{r^2}\frac{d^2r}{d\theta^2}\right]. \tag{M.8}$$

The dependent variable is now changed to $u = 1/r$ giving

$$\frac{dr}{d\theta} = \frac{d}{d\theta}\left(\frac{1}{u}\right) = -\frac{1}{u^2}\frac{du}{d\theta} \tag{M.9}$$

and

$$\frac{d^2r}{d\theta^2} = \frac{2}{u^3}\left(\frac{du}{d\theta}\right)^2 - \frac{1}{u^2}\frac{d^2u}{d\theta^2}. \tag{M.10}$$

Inserting results from equations (M.8), (M.9) and (M.10) in equation (M.5) we find

$$\frac{d^2u}{d\theta^2} = -u + \frac{GM}{H^2} \tag{M.11}$$

the solution of which is

$$u = A\cos\theta + \frac{GM}{H^2} \tag{M.12}$$

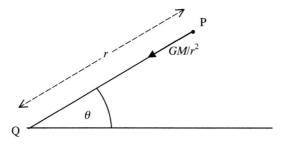

Figure M.3. *The acceleration at P due to a mass at Q.*

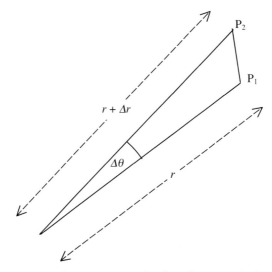

Figure M.4. *The area swept out by the radius vector in time* Δt.

where A is an arbitrary constant. Converting back to r as the dependent variable

$$r = \frac{H^2/GM}{1 + (AH^2/GM)\cos\theta}.$$
(M.13)

Comparing equation (M.13) with equation (3.2) the reader will see that it is of the same general form with

$$a(1 - e^2) = \frac{H^2}{GM} \quad \text{and} \quad e = \frac{AH^2}{GM}.$$

Hence equation (M.13) represents an ellipse, showing that the inverse-square-law gravitational field leads to the first of Kepler's laws.

The second law is actually inherent in equation (M.6) which is a statement of the conservation of angular momentum. In figure M.4 the positions of a planet are shown at the beginning and end of a time interval Δt. The radius vector has rotated through a small angle $\Delta\theta$ and the area swept out is

$$\tfrac{1}{2}r \times (r + \Delta r) \times \sin(\Delta\theta) \approx \tfrac{1}{2}r^2\,\Delta\theta.$$

The rate that area is swept out is thus

$$\left[\frac{1}{2}r^2\frac{\Delta\theta}{\Delta t}\right]_{\Delta t \to 0} = \frac{1}{2}r^2\frac{\mathrm{d}\theta}{\mathrm{d}t} = \frac{1}{2}H,$$
(M.14)

which is a constant. This is a statement of Kepler's second law.

To verify the third law we equate the numerators of equations (3.2) and (M.13) to give

$$H = [GMa(1 - e^2)]^{1/2}$$
(M15a)

or, from equation (L.2),

$$H = (GMb^2/a)^{1/2}.$$
(M.15b)

The period, P, is the total area of the elliptical orbit, πab, divided by the rate at which area is swept out by the radius vector. From equations (M.14) and (M.15b)

$$P = \frac{\pi ab}{H/2} = 2\pi \left(\frac{a^3}{GM} \right)^{1/2}. \tag{M.16}$$

This shows that $P^2 \propto a^3$, which is Kepler's third law.

M.4.   ESTABLISHING A SOLAR-SYSTEM DISTANCE SCALE

Although the laws of planetary motion given by Kepler and Newton enabled relative distances to be found they did not allow an absolute distance scale to be determined. Reasonable estimates of the Earth–Sun distance were available from the time of Newton but they were not very precise. Transit measurements, especially of the planet Venus, were attempted to improve the estimate and, indeed, a major objective of one of Captain Cook's expeditions to the Pacific was to observe the transit of 1769. However, the image of Venus against the Sun's disk was difficult to place accurately and the fuzzy edges of both bodies, due to their atmospheres, made the transit time difficult to measure.

Since the relative distances of orbiting bodies can be determined quite accurately it only requires an absolute measurement of any one distance to fix the absolute scale. This was done with some precision in 1931 using parallax methods when the asteroid Eros approached the Earth at a distance of 23 million kilometres. Since the Space Age the distance scale has been confirmed by timing radio signals to and from spacecraft. In the case of the Moon retro-reflecting mirrors have been left on its surface and very precise distances can be found by timing laser pulses reflected from them. It is assumed that the vacuum speed of light is constant throughout the double journey so the accuracy in measuring the distance is limited only by the accuracy of the best available clocks—which are atomic clocks. Any corrections which involve an intervening medium (for instance an atmosphere or ionosphere) can be used to investigate the medium itself.

M.5.   THE DYNAMICS OF ELLIPTICAL ORBITS

In equation (M.15a) we found that the magnitude of the intrinsic angular momentum depends on the combination $a(1 - e^2)$ which is the *semi-latus rectum* of the ellipse, the length of the radius vector when $\theta = \pm \pi/2$. However, angular momentum is a vector quantity and we should therefore have some means of specifying its direction. To do this requires the establishment of a coordinate system and we saw how to do this in section 3.2.

All planets orbit the Sun in the same sense, that is counter-clockwise looking towards the ecliptic from the north, and this sense of rotation is called *prograde* or sometimes *direct*. Any orbital or spin motion in the opposite sense is referred to as *retrograde*. Prograde motion occurs if the component along z of

$$\mathbf{H} = \mathbf{r} \times \mathbf{v} \tag{M.17}$$

is positive. For example, if with the Sun at the origin the orbiting body is on the positive X axis so that $\mathbf{r} = (x, 0, 0)$ and it is moving in the positive Y direction so that $\mathbf{v} = (0, \dot{y}, 0)$ then

$$\mathbf{H} = \begin{pmatrix} \hat{\mathbf{i}} & \hat{\mathbf{j}} & \hat{\mathbf{k}} \\ x & 0 & 0 \\ 0 & \dot{y} & 0 \end{pmatrix} = x\dot{y}\hat{\mathbf{k}} \tag{M.18}$$

where $\hat{\mathbf{i}}$, $\hat{\mathbf{j}}$ and $\hat{\mathbf{k}}$ are the unit vectors in the principal directions. The motion is prograde and the magnitude of the Z component is positive which agrees with the convention we have established.

In general with an orbiting body at position $\mathbf{r} = (x, y, z)$ with velocity $\mathbf{v} = (\dot{x}, \dot{y}, \dot{z})$ the angular momentum is

$$\mathbf{H} = H_x\hat{\mathbf{i}} + H_y\hat{\mathbf{j}} + H_z\hat{\mathbf{k}}$$

where

$$H_x = y\dot{z} - z\dot{y}, \qquad H_y = z\dot{x} - x\dot{z}, \qquad H_z = x\dot{y} - y\dot{x}, \tag{M.19}$$

with the sign of H_z indicating whether the orbit is prograde or retrograde. The angle between the plane of the orbiting body and the ecliptic is called the inclination. Since H_z is the component of $\mathbf{H}$ along z, the normal to the ecliptic, the inclination, i, is given by

$$\cos i = H_z/H. \tag{M.20}$$

Another conserved quantity of interest for the orbiting body is the *intrinsic total energy* or the total energy per unit mass. This is the sum of the intrinsic potential and kinetic energies at any point of the orbit and so is

$$E = -\frac{GM_\odot}{r} + \frac{1}{2}v^2. \tag{M.21}$$

The perihelion and aphelion are points in the orbit where the motion is orthogonal to the radius vector so that at these points it is particularly easy to find the speed. At perihelion the intrinsic angular momentum is

$$rv = a(1 - e)v_P = [GM_\odot a(1 - e^2)]^{1/2}$$

which gives

$$v_P^2 = \frac{GM_\odot(1 + e)}{a(1 - e)}. \tag{M.22}$$

Then from equation (M.21) it is found that

$$E = -\frac{GM_\odot}{a(1 - e)} + \frac{1}{2}\left(\frac{GM_\odot(1 + e)}{a(1 - e)}\right)$$

which gives

$$E = -\frac{GM_\odot}{2a}. \tag{M.23}$$

If E and H are known then from equations (M.15a) and (M.23) it is possible to find both a and e. From equation (M.23)

$$a = -\frac{GM_\odot}{2E} \tag{M.24}$$

and substituting this in equation (M.15a) gives

$$e = \left(1 + \frac{2EH^2}{G^2M_\odot^2}\right)^{1/2}. \tag{M.25}$$

It will be seen from equation (M.25) that to have an elliptical (i.e. bound) orbit with $e < 1$ it requires the intrinsic total energy, E, to be negative. If $E = 0$ then $e = 1$ and we have the special case of a parabolic orbit. With $E > 0$ then $e > 1$ and the orbit is hyperbolic.

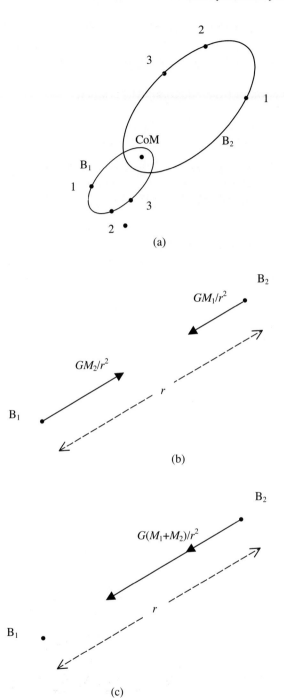

Figure M.5. *(a) Corresponding points on the orbits of two bodies about their centre of mass. (b) The accelerations of the bodies B_1 and B_2. (c) The acceleration of B_2 when an overall acceleration is applied to the system to bring B_1 to rest.*

We have already seen that if the position and velocity of an orbiting body are known then we can find, from equation (M.19)

$$H^2 = H_x^2 + H_y^2 + H_z^2.$$ (M.26)

The value of E can also be found from equation (M.21) where

$$E = -\frac{GM_\odot}{(x^2 + y^2 + z^2)^{1/2}} + \frac{1}{2}[(\dot{x})^2 + (\dot{y})^2 + (\dot{z})^2].$$ (M.27)

It is evident that a and e can be found directly from the position and velocity of the body.

The values of a, e and i do not define the orbit completely. It is also necessary to define some orientation angles and a time reference; the way of doing this was described in section 3.2 and illustrated in figure 3.3.

Although the discussion of orbits has been couched in terms of orbits around the Sun—which applies to planets, asteroids and comets—the results also apply to satellites, both natural and artificial, moving around planets. At the beginning of section M.3 the condition was imposed that the orbiting body should effectively be massless. If the mass of the orbiting body is not negligible compared with the central mass then the various formulae need some modification. In figure M.5a we show two bodies, B_1 and B_2, of masses M_1 and M_2 respectively, orbiting around each other; the point about which they both move is their centre of mass. They follow orbits with the same period, the same shape but different sizes and corresponding points of the two bodies are shown in figure M.5a.

The accelerations of the two bodies relative to the centre of mass are indicated in figure M.5b. The acceleration of B_2 relative to B_1, and hence the motion of B_2, relative to B_1 can be found by applying an acceleration to the whole system which cancels that of B_1. This is shown in figure M.5c and the effect is equivalent to making B_2 massless and transferring its mass to B_1. To describe the motion of the most massive planet, Jupiter, around the Sun, ignoring the effects of the other planets, involves replacing $M_\odot$ by $M_\odot + M_{Jup}$ in all the equations describing the motion. Since the mass of Jupiter is one-thousandth that of the Sun, the effect is small but by no means negligible.

M.6.   SOME SPECIAL ORBITAL SITUATIONS

We show here how the mechanics described in section M.3 can be applied to two different situations, one totally confined to motion close to the surface of the Earth and the other concerned with the manoeuvre of spacecraft.

M.6.1.   *Parabolic paths of projectiles*

It was first recognized by Galileo that a projectile on Earth, e.g. a cannon ball from a gun, follows a parabolic path, although this is only strictly true if air resistance is ignored. If the shell is projected at a speed V at angle α to the ground then its initial vertical speed is $V \sin \alpha$ while its horizontal speed, $V \cos \alpha$, is constant. The path of the projectile in relation to the Earth is shown in figure M.6; at the apex of its trajectory, at height h, at most a few kilometres above the Earth's surface, it is travelling horizontally and is at the apogee of an elliptical orbit with the centre of the Earth at a focus. There is an equation, similar to equation (M.22), which gives the speed at apogee as

$$V_a = \left(\frac{GM_\oplus(1 - e)}{a(1 + e)}\right)^{1/2}.$$ (M.28)

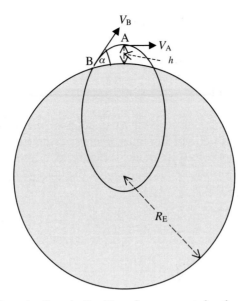

Figure M.6. *The motion of a projectile at the Earth's surface represented as though in orbit about the centre of the Earth.*

The apogee distance is $a(1 + e) = R_\oplus + h$ where $R_\oplus$ is the Earth's radius. Hence, from equation (M.28) we find

$$e = 1 - \frac{V_a^2 (R_\oplus + h)}{GM_\oplus}. \tag{M.29}$$

Making $V_a = 300\,\text{m s}^{-1}$, approximately the speed of sound, and ignoring h compared with $R_\oplus$, we find with approximate values $R_\oplus = 6.4 \times 10^6\,\text{m}$ and $M_\oplus = 6 \times 10^{24}\,\text{kg}$ that $e = 0.9986$—very close to unity, the value for a parabola.

Another illustration of the way that the orbit description of the motion links with the projectile description comes from equating the intrinsic energies of the projectile at points B and A (see figure M.6). This gives

$$\frac{1}{2}V_B^2 - \frac{GM_\oplus}{R_\oplus} = \frac{1}{2}V_A^2 - \frac{GM_\oplus}{R_\oplus + h}. \tag{M.30}$$

Since $h \ll R_\oplus$, application of the binomial theorem gives

$$\frac{1}{2}(V_B^2 - V_A^2) = \frac{GM_\oplus}{R_\oplus} - \frac{GM_\oplus}{R_\oplus}\frac{1}{1 + h/R_\oplus} = \frac{GM_\oplus h}{R_\oplus^2} = hg \tag{M.31}$$

where g is the acceleration due to gravity at the Earth surface. Considering just the region of the orbit outside the Earth it is clear that $V_B^2 - V_A^2$ is the square of the vertical component of the velocity of the projectile, V_P^2, so that

$$h = \frac{V_P^2}{2g} \tag{M.32}$$

which is exactly what would be expected for a projectile launched from the surface of the Earth with a constant acceleration due to gravity.

M.6.2.   Transfer orbits between planets

Consider a spacecraft moving in the same orbit as the Earth but free from the Earth's influence. Its speed, V_E, is given by

$$V_E^2 = \frac{GM_\odot}{r_E} \tag{M.33}$$

where r_E is the radius of the Earth's orbit. Now let us suppose that it is required to move the orbit inwards to end up in the orbit of one of the inner planets, assumed circular. The method of doing this will be to decelerate the spacecraft at point T (figure M.7) so that perihelion distance of the new orbit will be at P, where r_P is the orbital radius for the inner planet. It is clear from figure M.7 that the major axis of the new orbit is given by

$$2a = r_E + r_P \tag{M.34}$$

so that if the speed after deceleration at T is V_D then, by equating intrinsic energies,

$$E = -\frac{GM_\odot}{r_E + r_P} = \frac{1}{2}V_D^2 - \frac{GM_\odot}{r_E} \quad \text{or} \quad V_D^2 = \frac{2GM_\odot r_P}{r_E(r_E + r_P)}. \tag{M.35}$$

Since $2r_P/(r_E + r_P) < 1$ then $V_D^2 < V_E^2$.

The change of speed required in the deceleration stage is $V_E - V_D$. On reaching the point P the spacecraft will be moving in the direction of the inner planetary orbit but with a speed, V_F, given by

$$\frac{1}{2}V_F^2 - \frac{GM_\odot}{r_P} = -\frac{GM_\odot}{r_E + r_P} \quad \text{or} \quad V_F^2 = \frac{2GM_\odot r_E}{r_P(r_E + r_P)}. \tag{M.36}$$

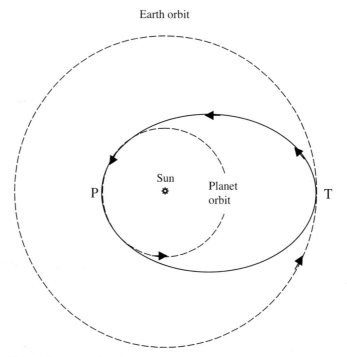

Earth orbit

Sun

Planet orbit

P

T

Figure M.7. *Transfer of a spacecraft from Earth orbit to a planet orbit.*

To go into a circular orbit at P the speed of the spacecraft must be changed to V_P where

$$V_P^2 = \frac{GM_\odot}{r_P} \tag{M.37}$$

and comparing equations (M.36) and (M.37) it is found that $V_F > V_P$, which means that the spacecraft must be decelerated again to go into the required orbit.

This has been a rather simplified account of orbit transfer; in practice the orbits of the planets are elliptical and the planets may slightly perturb the spacecraft. However, the essential features of orbit transfer are illustrated by this simple example.

Problems M

M.1 The perihelion distance of an orbiting body is 2.3 AU. When it has travelled an angular distance of 60° from perihelion its distance from the Sun is 3.0 AU. What are its semi-major axis and eccentricity?

M.2 A satellite is moving in a circular orbit of radius 1 AU well apart from the Earth. Retro-rockets are fired to slow it down so that its perihelion is at the distance of Venus. At that point retrorockets are fired again to bring it into a circular Venus orbit. What are the changes of speed at the two retro-rocket firings?

M.3 An asteroid moving in the ecliptic is seen with coordinates $x = 1.2$ AU, $y = 0.8$ AU moving along a direction making an anticlockwise angle of 55° with the x-axis at a speed of $20\,\text{km s}^{-1}$. What are the semi-major axis and eccentricity of its orbit? Is it an Earth-crossing asteroid?

TOPIC N

THE FORMATION OF COMMENSURATE ORBITS

In section 7.3.5 the mechanism suggested by Yoder (1979) was described by which commensurabilities for the Galilean satellites was established. This involves tidal interactions between Jupiter and the various satellites. Despite the larger masses of the primary and secondary bodies (the Sun and the major planets) the much greater distances make the establishment of commensurability between planetary orbits by a tidal mechanism involving the Sun quite implausible. Here we describe an alternative mechanism (Melita and Woolfson, 1996). The scenario is appropriate to the early Solar System in which planets have been formed in the presence of a resisting medium. The effect of the resisting medium on an isolated planet is to cause its orbit to decay and round off, if it was originally eccentric, and to reduce its inclination, measured relative to the mean plane of the flattened medium.

For one of the computational models studied by Melita and Woolfson two bodies were placed in orbit, the inner one with the mass of Jupiter and the outer with the mass of Saturn. The initial eccentricities and inclinations of the orbits were 0.1 and 0.06 radians (3.4°) respectively for both orbits. The initial ratio of the orbital periods was set at 2.5, close to the present ratio for Jupiter:Saturn which is 2.48. Three runs of the computation were made with different initial relative positions in their respective orbits. Figure N.1 shows the variation with time of the semi-major axis, eccentricity and inclination of the orbits of the two bodies. Both the semi-major axis and the inclination fall monotonically but the eccentricity falls at first but then rises again for both bodies. What is not readily seen in the figure is that the ratio of the two periods departs from 2.5 and settles down close to 2.0, actually oscillating about 2.02.

As with many numerical studies of complex systems the results are easier to interpret in terms of what is happening in the computation than to predict. When the computation is begun the first effect is that the eccentricities fall quite quickly, as is seen in figure N.1c and d. A fall in eccentricity reduces the speed of a planet relative to the resisting medium, which reduces the rate of energy dissipation and hence the rate of change of orbital parameters. The differential decay changes the ratio of the periods of the orbits until it becomes close to 2.0—which means that, in a relative sense, Saturn's orbit has been decaying more rapidly than that of Jupiter. The relative rate of decay of the bodies will depend both on their masses and the distribution of the resisting medium.

At this stage a new process comes into play. We know that Kirkwood gaps in the asteroid belt correspond to periods commensurate with that of Jupiter (section 8.3) and that the gaps in Saturn's rings correspond to simple commensurabilities with the periods of Mimas and Tethys (section 7.8.1). The reason for this is that the main orbiting body, Jupiter, around the Sun or a close satellite around Saturn, *removes* energy from *interior* bodies in commensurate orbits. Conversely, although

288

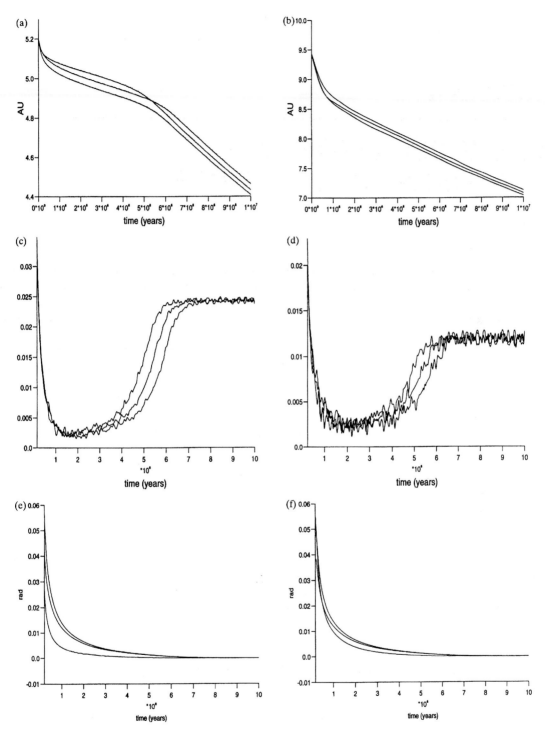

Figure N.1. *Semi-major axes, eccentricities and inclinations for Jupiter–Saturn systems started near the 2 : 5 resonance in three different relative positions. (a) Semi-major axis: Jupiter. (b) Semi-major axis; Saturn. (c) Eccentricity: Jupiter. (d) Eccentricity: Saturn. (e) Inclination: Jupiter. (f) Inclination: Saturn.*

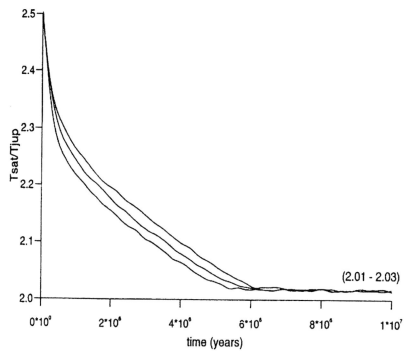

Figure N.2. *Ratio of periods for Jupiter–Saturn systems starting near the* 2:5 *resonance. Each curve represents a different initial relative position.*

there is no obvious demonstration of this in the Solar System, energy is *added* to *exterior* bodies in commensurate orbits.

Another effect of commensurability is that a resonance effect is set up such that the major mutual perturbations, corresponding to when the bodies are closest together, occur repeatedly at the same points of the orbits and amplify the effects of the perturbations. This increases the eccentricity which, in its turn, increases the rate of dissipation and hence the rate of decline in the semi-major axis. This latter effect is seen for Jupiter in figure N.1a by a change of slope of da/dt but it is not seen for Saturn. The reason for this is that the increasing dissipation due to increasing eccentricity of Saturn's orbit is balanced by a gain of energy due to the 2:1 resonance. Thus the rate of change of Saturn's semi-major axis is little affected while that of Jupiter is greatly increased. A non-obvious result of this is that the ratio of the periods, which had fallen from 5:2 to 2:1 now becomes locked at 2:1 although both orbits are still decaying (figure N.2).

The numerical experiments were repeated for the Jupiter–Saturn system with different initial eccentricities—the nine combinations with each eccentricity having the possible values 0.1, 0.2 and 0.3—but always with an initial orbital period ratio 2.5. The results are shown in figure N.3. Five of the combinations end up with ratio close to 2.0 and four stay at the original ratio 2.5.

Other trials gave a Uranus–Neptune system with a ratio close to 2.0 starting with the present observed ratio of 1.96 (figure N.4). A factor that has not been taken into account in the calculations is the dissipation of the resisting medium that would take place in 10^7 years or less (section 12.6). As the medium is removed so the rate of change of orbital parameters declines and will cease once the medium is removed completely. For this reason some pairs of orbital periods may become

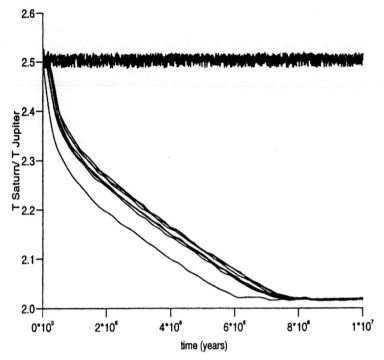

Figure N.3. *The Jupiter–Saturn system for different pairs of initial eccentricities. The value of the eccentricity, once the resonance is stabilized, does not depend on the initial value.*

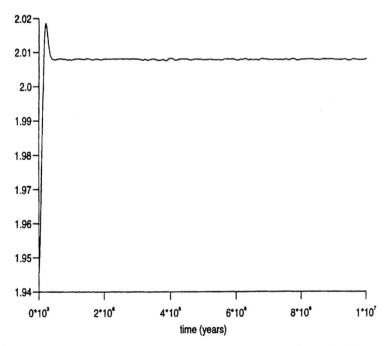

Figure N.4. *The Uranus–Neptune system started near the present ratio of periods. The system reaches a resonant configuration with a ratio of periods slightly greater than 2.*

stranded away from the commensurabilities towards which they were evolving. Thus if the orbit of Uranus was decaying more slowly than that of Saturn then the ratio of the periods would gradually increase. The present ratio of the periods is 2.85. That could indicate that it was going towards 3.0 but was terminated by the removal of the medium.

TOPIC O

THE ATMOSPHERE OF THE EARTH

The planets are of two basic kinds, the gas giants consisting of volatile outer materials with a solid core and the terrestrial planets that have iron cores and silicate mantles. All the giants have atmospheres that consist of the outer parts of their main structures where the material is diffuse enough to have the properties we associate with gases. The major difficulty with these planets is to define just where the atmosphere begins, for there is no sharp boundary in going from outside towards the centre. In the case of the terrestrial planets there is no such difficulty. Three of them, Venus, the Earth and Mars, have atmospheres consisting of material that is quite different from that of the bulk planet. The fourth terrestrial planet, Mercury, has no detectable atmosphere. The most familiar atmosphere is that of the Earth and we shall discuss this here in some detail. As it turns out the structure of the Earth's atmosphere is more complex than that of any of the other planets and it acts as an ideal comparison model when considering other atmospheres.

O.1. A SIMPLE ISOTHERMAL ATMOSPHERE

The simplest atmospheric model is one that is isothermal. In figure O.1 the density of an isothermal atmosphere at temperature T is ρ_0 at the surface and ρ_h at height h above the surface. We shall now find the relationship between ρ_h and h assuming that between the surface and height h the acceleration due to gravity, g, is a constant. Considering the small slab of the atmosphere at height x of unit cross section and thickness dx then the forces on it are a pressure P acting upwards, a pressure $P + dP$ acting downwards and its weight $\rho_x g \, dx$ also acting downwards. Equating these forces for equilibrium of the slab gives

$$P + dP + \rho_x g \, dx + P = 0 \qquad \text{or} \qquad \frac{dP}{dx} = -g\rho_x. \tag{O.1}$$

For an ideal gas the pressure is expressed in terms of density by

$$P = \frac{\rho k T}{\mu} \tag{O.2}$$

where k is Boltzmann's constant and μ is the mean mass of the molecules constituting the atmosphere. From this we find

$$\frac{dP}{dx} = \frac{kT}{\mu} \frac{d\rho}{dx} \tag{O.3}$$

293

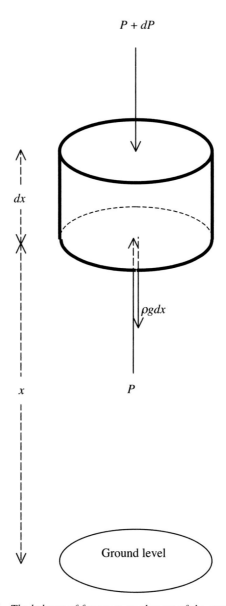

Figure O.1. *The balance of forces on an element of the atmosphere.*

that, with equation (O.1), gives

$$\frac{\mathrm{d}\rho}{\mathrm{d}x} = -\frac{\mu g \rho}{kT}. \tag{O.4}$$

Integrating equation (O.4), putting in boundary conditions at $x = 0$ and $x = h$ gives

$$\rho_h = \rho_0 \exp\left(-\frac{\mu g h}{kT}\right). \tag{O.5a}$$

Since for an isothermal atmosphere pressure is proportional to density an alternative form of equation (O.5a) is

$$P_h = P_0 \exp\left(-\frac{\mu g h}{kT}\right).$$ (O.5b)

A useful concept in atmospheric studies is the *scale height*, that is the height through which the pressure falls by a factor $1/e$. From equation (O.5b) this is

$$H = \frac{kT}{\mu g}.$$ (O.6)

Assuming that such an isothermal atmosphere could exist and was of infinite extent then the total mass of atmosphere per unit area of the surface is called the *column mass* of the atmosphere. If the density of the atmosphere did not decline rapidly with height then it would be necessary to take into account all the material in a flared region going out to a large distance (figure O.2). In that case equation (O.5a) would not be applicable since g would vary with distance from the surface. In practice most of the mass *is* very close to the surface so that the column mass, c_m, can be calculated as

$$c_m = \int_0^\infty \rho \, dx = \rho_0 \int_0^\infty \exp\left(-\frac{\mu g}{kT}x\right) dx = \frac{\rho_0 kT}{\mu g} = \frac{P_0}{g}.$$ (O.7)

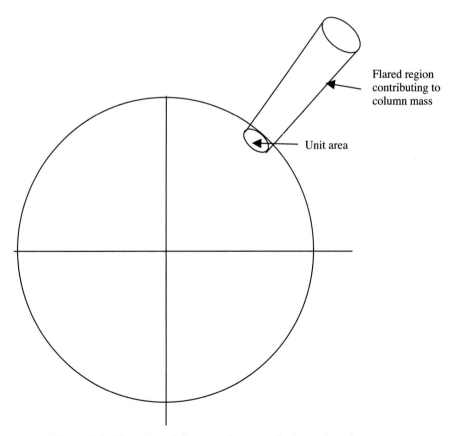

Flared region contributing to column mass

Unit area

Figure O.2. *The region of the atmosphere contributing to the column mass.*

The structures of real atmospheres are much more complicated than is given by equation (O.5) because they depart greatly from being isothermal. However, since the expression for c_m does not involve temperature it is valid for a non-isothermal atmosphere as long as g is appreciably constant. For the Earth the surface pressure is 1.014 bar (1 bar $= 10^5 \, N \, m^{-2}$) and $g = 9.81 \, m \, s^{-2}$. From equation (O.7) we find the column mass to be $1.034 \times 10^4 \, kg \, m^{-2}$. Multiplying this column mass by the surface area of the Earth which is $4\pi(6.38 \times 10^6)^2 \, m^2$ gives the total mass of the atmosphere as 5.3×10^{18} kg or just under one-millionth of the mass of the Earth.

O.2.   THE STRUCTURE OF THE EARTH'S ATMOSPHERE

The Earth's atmosphere has a very complex structure dictated by its composition and the way in which radiation from outside the Earth interacts with it. First we shall simply note the variation of temperature with height as recorded by instruments. After that we shall see what are the factors that govern the temperature profile, first considering the higher reaches of the atmosphere where the major influence comes from outside the Earth itself and then the lower reaches where the presence of the solid Earth itself is having an influence.

O.2.1. The variation of temperature with height

The composition of the Earth's atmosphere is unique in the Solar System in that it is rich in oxygen, the essential ingredient for life on Earth. However, as seen in table O.1, oxygen is not the major component and accounts for just 21% of the atmosphere, most of the remainder being nitrogen (78%). The inert gas argon occurs at the 1% level and also found are carbon dioxide (0.033%), hydrogen (0.01%) and traces of helium, neon and other gases. These constituents and their proportions exclude water vapour which occurs at levels of 0.1–2.8%. As we shall see, the composition of the atmosphere has an important role in controlling the passage of radiation through it and consequently its temperature profile.

It is well known that, in general, temperature decreases with height and the general rule that used to be quoted, in traditional units, was 'one degree Fahrenheit for every 300 feet'. That is why Mt. Kenya, which is on the equator, is snow-capped and why snow persists in shady areas on Alpine

Table O.1. *The composition of the Earth's atmosphere excluding water vapour.*

Constituent	Formula	Proportion
Nitrogen	N_2	0.77
Oxygen	O_2	0.21
Argon	Ar	0.009
Carbon dioxide	CO_2	3.3×10^{-4}
Neon	Ne	1.8×10^{-5}
Helium	He	5.2×10^{-6}
Methane	CH_4	1.5×10^{-6}
Krypton	Kr	1.1×10^{-6}
Hydrogen	H_2	5×10^{-7}
Ozone	O_3	4×10^{-7}
Nitrous oxide	N_2O	3×10^{-7}
Carbon monoxide	CO	1×10^{-7}
Ammonia	NH_3	1×10^{-8}

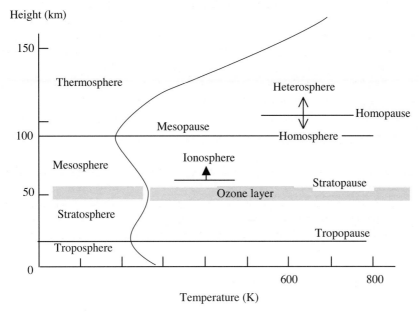

Figure O.3. *The structure of the Earth's atmosphere.*

passes in the summer. The Earth's atmosphere has been monitored by several means, by balloons up to tens of kilometres and spacecraft beyond that. The pattern of temperature that is found is shown in figure O.3. The initial fall-off with altitude, which is our every-day experience, is seen to be maintained up to 18 km or so but then, over a fairly small region, the temperature gradient is reversed and temperature increases up to about 50 km. The temperature then falls again up to a height of 90 km falling to about 175 K and thereafter rises to very high temperatures of 2000 K or so. It should be pointed out that bodies moving in the very high temperature region would not be greatly heated by the atmosphere. These temperatures are *kinetic temperatures*, a measurement of the mean random speed of the gas molecules, but the number density of these molecules is so low that the heat content of the atmosphere per unit volume is very small indeed.

O.2.2. The upper reaches of the atmosphere

O.2.2.1. The exosphere

To explain the pattern of temperature variation with height we imagine that we start at a great distance from the Earth and gradually move inwards towards the surface. The first important region we reach is the *exosphere*, about 500 km from the surface, which is where the atmosphere is so tenuous that the mean-free-path of a gas molecule is virtually infinite. To put it simply, starting from the exosphere a molecule moving in an outward direction through a distance equal to one mean-free-path would reach a region where the atmosphere would have disappeared or simply merged with the interplanetary medium. At this height what governs the ability of a molecule (or atom) to escape is whether or not its speed exceeds the escape speed, v_{esc}, from the Earth. This is given by

$$v_{esc}^2 = \frac{2GM_\oplus}{R_{ex}}$$ (O.8)

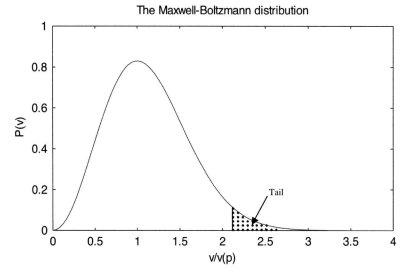

Figure O.4. *The Maxwell–Boltzmann distribution showing the tail which extends to infinity. The velocity is made non-dimensional by dividing by $v(p) = \sqrt{2kT/\mu}$.*

in which $M_\oplus$ is the mass of the Earth and R_{ex} is the radius of the exosphere. At a height of 500 km above the Earth's surface this is about $10.8 \, \mathrm{km \, s^{-1}}$. For any gas of molecular mass μ in equilibrium at a particular temperature T the probability distribution of speeds is governed by the Maxwell–Boltzmann distribution

$$p(v) = \left(\frac{2\mu^3}{\pi k^3 T^3}\right)^{1/2} v^2 \exp\left(-\frac{\mu v^2}{2kT}\right). \tag{O.9}$$

Thus the proportion of molecules with speed greater than the escape speed at any time will be

$$f = \int_{v_{esc}}^{\infty} p(v) \, \mathrm{d}v. \tag{O.10}$$

The form of the Maxwell–Boltzmann distribution is shown in figure O.4 and, in a particular situation, the shaded area in the tail of the distribution will represent the proportion of molecules able to escape.

We can carry out an analysis to estimate the rate at which molecules of a particular mass escape from the exosphere if we make some simplifying assumptions. Let us consider molecules of mass μ and speed v and number density n in the region of the exosphere. If v is greater than the escape speed then one half of them will be moving outwards and hence will escape. To estimate the flux of such particles we must take into account their different directions of motion. A fraction $\frac{1}{2}\sin\alpha \, \mathrm{d}\alpha$ of those moving outwards do so at an angle between α and $\alpha + \mathrm{d}\alpha$ to the normal to the exosphere (figure O.5). Their component of velocity perpendicular to the exosphere is $v\cos\alpha$ so they will contribute an outward flux

$$\mathrm{d}F = \tfrac{1}{2}nv\sin\alpha\cos\alpha \, \mathrm{d}\alpha. \tag{O.11}$$

The total flux, i.e. the number crossing the exosphere per unit area, will thus be

$$F = \tfrac{1}{2}nv\int_0^{\pi/2}\sin\alpha\cos\alpha \, \mathrm{d}\alpha = \tfrac{1}{4}nv. \tag{O.12}$$

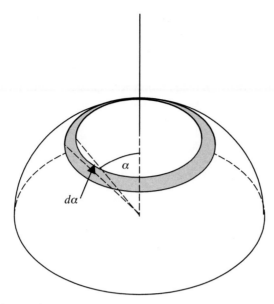

Figure O.5. *Molecules leaving the exosphere at angles between α and $\alpha + \mathrm{d}\alpha$ to the normal pass through the shaded area.*

If we now take into account that the molecules move at different speeds then the number with speed between v and $v + \mathrm{d}v$ is $n_{\mathrm{ex}}p(v)\,\mathrm{d}v$, in which n_{ex} is the total number density in the exosphere region. They will contribute a flux $n_{\mathrm{ex}}p(v)\,\mathrm{d}v/4$ so that the total flux of escaping particles is

$$F_{\mathrm{p}} = \frac{1}{4}n_{\mathrm{ex}}\left(\frac{2\mu^3}{\pi k^3 T^3}\right)^{1/2}\int_{v_{\mathrm{esc}}}^{\infty} v^3 \exp\left(-\frac{\mu v^2}{2kT}\right)\mathrm{d}v. \qquad (\mathrm{O}.13)$$

This integral may readily be evaluated. Writing

$$\phi = \frac{\mu v_{\mathrm{esc}}^2}{2kT} = \frac{3v_{\mathrm{esc}}^2}{2v_T^2} \qquad (\mathrm{O}.14)$$

where $v_T^2\,(= 3kT/\mu)$ is the mean-square thermal speed we find

$$F_{\mathrm{p}} = \frac{1}{3\pi^{1/2}}n_{\mathrm{ex}}\frac{1}{\phi^{1/2}}v_{\mathrm{esc}}(1+\phi)\,\mathrm{e}^{-\phi}. \qquad (\mathrm{O}.15)$$

This is the rate of loss of molecules per unit area of the exosphere so the total rate of loss from the body, planet or satellite, is

$$F_{\mathrm{T}} = 4\pi R_{\mathrm{ex}}^2 F_{\mathrm{p}} = \frac{4}{3}\pi^{1/2}n_{\mathrm{ex}}v_{\mathrm{esc}}R_{\mathrm{ex}}^2\frac{1}{\phi^{1/2}}(1+\phi)\,\mathrm{e}^{-\phi}. \qquad (\mathrm{O}.16)$$

The probability that a molecule moving directly outwards would collide with another molecule in moving a distance $\mathrm{d}x$ is $\mathrm{d}p_c = \mathrm{d}x/\lambda$ where λ is the mean free path. The mean free path is related to the number density of molecules, n, and the collision cross section, A, by $\lambda = (nA)^{-1}$. To find n_{ex} we use the arbitrary, but reasonable, condition that a molecule leaving the exosphere along the normal will make, on average, one collision in travelling to infinity. In moving through a distance $\mathrm{d}x$ the ratio of the number of collisions to the total number of molecules is

$$\mathrm{d}p = nA\,\mathrm{d}x.$$

Starting with $x = 0$ at the exosphere we write

$$n = n_{ex} \exp\left(-\frac{\mu g x}{kT}\right)$$

assuming that g does not vary greatly in the region where n is significant. Hence the ratio of the number of collisions to the total number of molecules in the journey to infinity is

$$p = n_{ex} A \int_0^\infty \exp\left(-\frac{\mu g x}{kT}\right) dx = \frac{kT}{\mu g} n_{ex} A. \tag{O.17}$$

Making $p = 1$ gives $n_{ex} = \mu g / kTA$. From $g = v_{esc}^2 / 2R_{ex}$ and equation (O.14), $n_{ex} = \phi / AR_{ex}$ so that

$$F_T = \frac{4\pi^{1/2}}{3A} v_{esc} R_{ex} \phi^{1/2} (1 + \phi) e^{-\phi} \tag{O.18a}$$

Eliminating v_{esc}

$$F_T = \frac{4}{3A} (2\pi G M_p R_{ex} \phi)^{1/2} (1 + \phi) e^{-\phi} \tag{O.18b}$$

where M_p is the mass of the planet or other body.

For the Earth the temperature to take, at the level of the exosphere, is certainly more than 1000 K. For this temperature with $R_{ex} = 6900$ km, $A = 10^{-18}$ m² and $\mu = 4.67 \times 10^{-26}$ kg, the rate of loss is 3.3×10^{-53} molecules per second which is effectively zero. Even if the temperature is taken as 2000 K the rate of loss is still only 5.5×10^{-11} molecules per second. It is clear that the Earth's dominantly nitrogen atmosphere is very stable.

If the above analysis is repeated for a hypothetical hydrogen component of the atmosphere then there is the complication of the fact that it would exist alongside the main components. However, assuming that a pure hydrogen-molecule atmosphere once existed on Earth we could find the loss rate for various temperatures assuming the exosphere was at the same height. In table O.2 there are shown the molecular loss rates for various combinations of temperature and atmospheric constituent for the Earth and the Moon. At 2000 K a primitive H_2 terrestrial atmosphere consisting of 10^{44} molecules (the number in the present atmosphere) would have had a lifetime of about 1.4×10^9 years. For a temperature of 1000 K the lifetime would have exceeded the age of the Solar System. For the Moon, assuming a temperature of 1000 K at the exosphere and an initial atmosphere of 10^{43} molecules, the lifetime of an N_2 atmosphere would have been about 4.6×10^9 years and of an H_2 atmosphere about 6.7×10^7 years.

Table O.2. *Loss rates of various atmospheres from the Earth and the Moon. The exosphere radii are taken as 6900 km for the Earth and 2000 km for the Moon. It is assumed that initial Earth atmospheres contained 10^{44} molecules and initial Moon atmospheres contained 10^{43} molecules.*

Body	Gas	Temperature (K)	Loss rate (molecules s^{-1})	Estimated lifetime (years)
Earth	N_2	1000	1.5×10^{-53}	∞
Earth	N_2	2000	2.4×10^{-11}	∞
Earth	H_2	2000	2.2×10^{27}	1.4×10^9
Moon	N_2	1000	4.6×10^{25}	6.9×10^9
Moon	H_2	1000	4.8×10^{27}	6.7×10^7

The actual atmospheric-loss situation is much more complicated than our simple model implies. The loss rate is very sensitive to temperature and the appropriate temperatures for Solar-System bodies immediately after their formation are unknown. A high temperature in the upper reaches of the atmosphere might imply a rapid rate of loss of a particular constituent. A lower temperature closer to the surface may mean that the lost constituent may not be readily replaced from below, so that its rate of loss from the atmosphere as a whole will actually be low. Again, if the temperature is sufficiently high then molecules may completely or partially dissociate so that one ought to be considering the escape of atoms instead of, or together with, molecules. Another difficulty is when there is a mixture of atmospheric constituents which influences the mean-free path of the molecules. However, our analysis gives a good general picture of the way that atmospheres are retained or lost.

O.2.2.2. The thermosphere

The region within which the exosphere is encountered, where temperature increases with increasing height, is the *thermosphere*. In this region ultraviolet radiation is interacting with atoms and molecules and producing heat. The energy of a photon is given by

$$E = h\nu = hc/\lambda \tag{O.19}$$

in which ν and λ are respectively the frequency and wavelength of the radiation and h is Planck's constant. For ultraviolet radiation of wavelength 10^{-7} m the photon energy is 1.99×10^{-18} J or about $12.4\,\text{eV}$. This is only just less than the ionization energy for hydrogen, $13.6\,\text{eV}$, but the outer electrons of heavier atoms are less tightly bound so that the energies of photons of wavelengths around 10^{-7} m are capable of ionizing many types of atom. For longer wavelengths, greater than about 3×10^{-7} m, the photon energies are then large enough to break bonds between atoms in molecules, a process called *photodissociation*. The processes of ionization and photodissociation represent the absorption of energy and so act as a source of heating within the thermosphere. As the density of the radiation, particularly of the most energetic radiation, diminishes as it gets absorbed so its heating power also diminishes and the temperature decreases with decreasing height.

O.2.2.3. The homopause

If on the journey through the thermosphere the atmospheric constituents were sampled it would be noticed that, at first, the mean molecular weight of the atmospheric gas increased with decreasing distance from the Earth. This indicates that there is a tendency for the atmosphere to become layered with lesser-molecular-weight gases being farther out. This separation tendency diminishes with decreasing height and eventually a level is reached below which, because of frequent collisions, all constituents are intimately mixed regardless of height. This level, at a height of about 100 km, is called the *homopause*; the region above, where separation occurs, is the *heterosphere* and the region below is the *homosphere*.

O.2.3. The lower reaches of the atmosphere

O.2.3.1. The mesosphere

Below a height of about 90 km we cross the *mesopause* and enter the *mesosphere* below which there is still some, but comparatively little, remaining very short wavelength ionizing radiation. Below a height of about 60 km there is virtually no ionization and the region above this level, where there are appreciable numbers of ions and free electrons, is called the *ionosphere*. The ionosphere is important in that it reflects radio waves and hence plays an important part in radio communication. However, a high proportion of the radiation with wavelengths between 10^{-7} m and 3×10^{-7} m is unabsorbed by the thermosphere and is capable of dissociating molecules within the mesosphere. An important molecule in this respect is O_2 that, when dissociated, produces atomic oxygen which then readily combines with another O_2 molecule to produce ozone, O_3. This triatomic molecule has the property that it is a very powerful absorber of ultraviolet radiation and most absorption of such radiation in the atmosphere is by ozone. As the ultraviolet radiation travels towards the Earth so its ability to dissociate molecules gets reduced by absorption reducing its intensity, but increased by the fact that it has a denser atmosphere on which to act. The net effect of this is that absorption increases with decreasing height, until at a height of about 50 km there is a runaway absorption effect where absorption gives dissociation of O_2, dissociation gives ozone and ozone gives even more absorption. Life on Earth depends for its protection on the *ozone layer* and it is a matter of great concern that scientists have discovered a hole developing in the ozone layer above the Antarctic region. An important contribution to ozone reduction is the industrial and domestic use of various aerosol substances, in particular chlorofluorocarbons, which drift into the upper atmosphere and break down the O_3 produced there. Steps are now being taken to reduce the production of these harmful substances but, unfortunately, once they are present in the atmosphere it takes about one hundred years for them to break down and become innocuous.

O.2.3.2. The stratosphere and troposphere

The ozone layer is one of high absorption and therefore one of high temperature and this explains the profile of temperature in the mesosphere region. Below the ozone layer one crosses the *stratopause* and enters the *stratosphere*, the lower reaches of which are regularly entered by supersonic aircraft. In this region there is little ultraviolet radiation to be absorbed and so the temperature starts falling again with decreasing height. This continues until, at a height of about 16 km, one encounters the *tropopause* and enters the *troposphere*, the part of the atmosphere that supports life.

The troposphere is heated from below by radiation from the Earth's surface. All solar radiation striking the Earth's surface is absorbed, heating the surface to a global-average temperature of 288 K, well above that at the tropopause. Heated air at the surface rises and so the upper atmosphere receives heat by convection and a temperature gradient is established as shown in figure O.3. This temperature gradient is called the *adiabatic lapse rate* and its value depends on the nature of the atmospheric gases. The process of convection that leads to the adiabatic lapse rate can be explained in terms of what happens to packets of air moving upwards or downwards in the atmosphere. The explanation involves the concept of *adiabatic* expansions and contractions of a gas, implying that during the change of volume the gas is thermally insulated and exchanges no energy with its surroundings.

In figure O.6 a small packet of air, A, with pressure P, volume V and temperature T, moves a small distance upwards to B. Consider that it expands adiabatically and cools. If its temperature is higher than the air at the new level then it will heat up the surrounding air thus reducing the temperature gradient. Another package of air, C, will have been displaced in a downward direction, cooling the air near the surface at D and also contributing to a reduction in the temperature gradient. This convective condition is extremely efficient and will quickly bring the temperatures at the levels B and D to the point where the temperatures of the packets, if they moved and expanded or contracted, would always exactly match that of their environment. At that stage convection ceases and the adiabatic lapse rate is established.

We can now follow this analytically. For an adiabatic change the relationship between pressure and volume is given by

$$PV^\gamma = K \tag{O.20a}$$

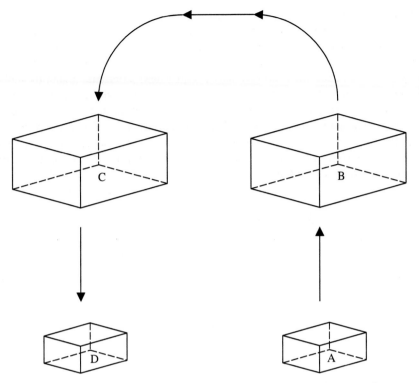

Figure O.6. *A packet of air moving from A to B while an equivalent packet moves from C to D.*

in which K is a constant and γ is the ratio of the specific heat of the gas at constant pressure to that at constant volume. Given that density ρ is inversely proportional to V this can also be written as

$$P = K'\rho^{\gamma}. \tag{O.20b}$$

Substituting for P from the perfect gas law (equation O.2) we find

$$T = C\rho^{\gamma-1} \tag{O.21}$$

where C is a constant. From equation (O.21) we find that

$$\frac{\mathrm{d}T}{\mathrm{d}\rho} = C(\gamma - 1)\rho^{\gamma-2}. \tag{O.22}$$

Combining this result with equation (O.4) gives

$$\frac{\mathrm{d}T}{\mathrm{d}x} = \frac{\mathrm{d}T}{\mathrm{d}\rho}\frac{\mathrm{d}\rho}{\mathrm{d}x} = -(\gamma - 1)C\frac{g\mu}{kT}\rho^{\gamma-1}$$

and using equation (O.21) the adiabatic lapse rate is

$$\frac{\mathrm{d}T}{\mathrm{d}x} = -(\gamma - 1)\frac{g\mu}{k}. \tag{O.23}$$

We notice that this is independent of temperature and so should be constant over a region of the atmosphere where g is appreciably constant. At the surface of the Earth with μ for N_2 and with

$\gamma = 1.4$, the value for a diatomic gas, we find the adiabatic lapse rate as $0.0133\,\mathrm{K\,m^{-1}}$ or $13.3\,\mathrm{K\,km^{-1}}$. Measured lapse rates are somewhat less, varying from $10\,\mathrm{K\,km^{-1}}$ to $5\,\mathrm{K\,km^{-1}}$ and decreasing with increasing humidity. This is because the condensation of water vapour is a contributor to the transfer of heat. When heated air rises and cools so water will condense and give up latent heat so that the packet of air will cool more slowly. This is equivalent to having a smaller value of γ (note: for an isothermal gas where $\gamma = 1$ there will be no cooling at all) and hence, from equation (O.23), a lower value of the lapse rate. It is interesting to note that the quoted rule 'one degree Fahrenheit for each 300 feet' corresponds to $5.74\,\mathrm{K\,km^{-1}}$, an appropriate value for a damp climate, as is that of the UK.

This concludes our description of the composition and structure of the Earth's atmosphere and we shall now consider some aspects of its dynamics.

O.3.   THE DYNAMICS OF THE ATMOSPHERE

In considering atmospheric dynamics we are mostly concerned with the generation of winds due to the thermal effects of solar heating and the rotation of the Earth. We shall be concerned only with general large-scale effects since, on a small scale, wind structures are affected by local topography and are sometimes not simple to explain or understand. In figure O.7(a) we illustrate the formation of a circulation pattern in the atmosphere due to what are called *Hadley cells*. Solar heating is strongest close to the equator and the surface temperature becomes high there. A pattern of circulation can be imagined where air rises at the equator, then flows horizontally in higher regions of the troposphere towards the poles, sinks at the poles and then flows closer to the surface back to the equator. In fact a simple Hadley cell pattern of this type would not form because it would be unstable. This is due to the *Coriolis force* that applies to moving material in a rotating system. The air rising from the equatorial region will be at rest relative to the surface and have a speed v_e ($\sim 460\,\mathrm{m\,s^{-1}}$, the Earth's equatorial speed) in an easterly direction relative to the centre of the Earth. On its journey at altitude towards the pole it moves closer to the Earth's spin axis and hence must speed up to conserve angular momentum. At a latitude θ the reduction in distance is proportional to $\cos\theta$ and so the speed relative to the centre of the Earth is $v_e/\cos\theta$. At the same time the speed of the surface below the moving air has been reduced by a factor $\cos\theta$ so the relative speed of the air and the ground below, which is a west wind speed (i.e. from the west), is

$$v_w = v_e\left(\frac{1}{\cos\theta} - \cos\theta\right). \tag{O.24}$$

With air flowing towards the equator at ground level, similar reasoning leads to the prediction of an east wind but much smaller in magnitude since the falling air near the poles would have had little initial motion relative to the centre of the Earth.

Equation (O.24) gives an infinite value of v_w for $\theta = \pi/2$ so that a breakdown in the pattern must occur. What actually happens is that three cells form in each hemisphere, as shown in figure O.7(b); the circulation in the mid-latitudes cell is opposed to the temperature variation at its extremes and is dictated by the requirement to fit in with the polar and equatorial cells. The motions are seen to give surface east winds (trade winds) in the equatorial region, west winds between about $30°$ and $60°$ latitude and east winds again in the polar regions. At the equator this pattern gives no wind and this region is referred to as the *doldrums*, where sailing ships had little help in the past. Similarly, at latitudes of $\pm 30°$ and $\pm 60°$ where the air currents are making a transition from one level to another or where air in the middle-latitude cell is rising to accommodate the polar cells, there are other regions of slack winds known as the *horse latitudes*.

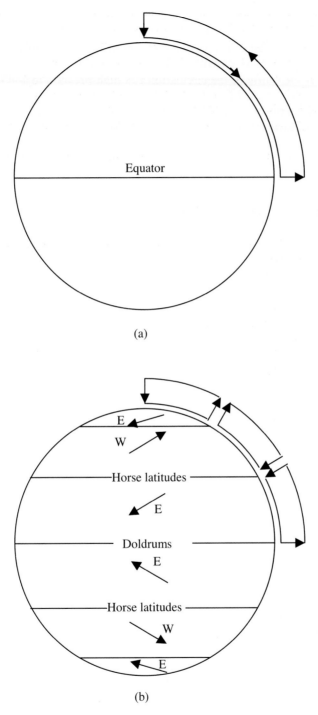

(a)

(b)

Figure O.7. *(a) A hypothetical, but impossible, Hadley cell extending from the equator to the pole. (b) The actual pattern of Hadley cells. The cell between about latitude 30° and 60° is driven against the natural direction by the neighbouring cells. Winds are marked according to the direction from which they come.*

These general wind patterns are often overwhelmed by movements caused by instabilities in the atmosphere giving rise to circulations of air several thousands of kilometres in extent, known as cyclones and anticyclones. If the centre of the circulation is a low-pressure region then the winds around it are cyclonic and are clockwise in the southern hemisphere and anti-clockwise in the northern hemisphere. If the centre of the circulation is a high-pressure region then the winds around it are anticyclonic and are anticlockwise in the southern hemisphere and clockwise in the northern hemisphere. In tropical regions very energetic cyclones, called hurricanes, can occur generating wind speeds that can do great structural damage and cause loss of life. These circulation patterns have lifetimes measured in days but there are similar structures on other planets that last a much longer time (section 5.1.3).

Problems O

O.1 Assuming an isothermal atmosphere ($\theta = 260$ K) then what is the pressure at the top of Mount Everest (height 8 km) given a pressure of 10^5 N m^{-2} at sea level? What would be the pressure at a height of 20 km?

O.2 (i) Pressure at sea level is 10^5 N m^{-2} and the temperature 300 K. Taking water-vapour effects into account the effective value of γ is 1.25. What is the temperature and pressure at the top of Mount Everest?

(ii) Assuming that Venus has an atmosphere of pure CO_2, for which $\gamma = 1.28$, then calculate the adiabatic lapse rate. Compare the value you find with that indicated by figure 4.9. If the ground temperature and pressure are 10^7 N m^{-2} and 730 K then find the temperature and pressure at a height of 50 km.

O.3 Given the masses, exosphere radii and temperature of various bodies in the following table, find the rates of loss for the given possible atmospheric constituents from equation (O.18b). (Assume $A = 10^{-18}$ m^2 in all cases.)

Body	Mass (kg)	R_{ex} (km)	Temperature (K)	Constituent
Jupiter	1.90×10^{27}	75 000	300	H_2
Mars	6.42×10^{23}	3800	500	CO_2
Mars	6.42×10^{23}	3800	500	H_2O
Moon	7.35×10^{22}	2000	1000	O_2
Io	8.92×10^{22}	2100	500	SO_2
Titan	1.43×10^{23}	2800	200	N_2

TOPIC P

THE PHYSICS OF PLANETARY INTERIORS

P.1. INTRODUCTION

It has been found that a general way of studying the overall structure of a planet is by treating it as a collection of atoms in equilibrium. Several different energies are involved and the relations between them allow the planets to be understood as a single class of object. The relationship between a planet and other bodies in the Universe is made clear. Unlike for the Sun and main-sequence stars, the composition may range from being solid to almost completely gaseous. The treatment here gives the essential physics but greater detail about the interiors of planets has been given by Cole (1984, 1986).

P.2. APPLYING THE VIRIAL THEOREM

The total energy of the body is the sum of the kinetic and potential energy contributions, the former accounting for the translational motion and the latter for the interactions restraining the motions. A central relationship between the kinetic and potential energies of a group of particles in an equilibrium configuration is expressed by the *Virial Theorem* (Topic C). Writing E_k for the total translational kinetic energy and E_v for the total potential energy, the Virial Theorem states that

$$2E_k = -E_v. \tag{P.1}$$

For planets, the particles are the atoms, composed of a positively charged nucleus and negatively charged electrons. This means the kinetic and potential energies can be described in terms of the magnitudes of the electric charges, the particle masses, the total mass, M_P, and the radius, R_P, of the planet as a whole.

P.3. THE ENERGIES INVOLVED

It is necessary to account explicitly for the kinetic and potential energies. The kinetic energy can be supposed to come entirely from the motion of the electrons. The motions of the positive nuclei are negligible by comparison. To see this consider the hydrogen atom. Relative to the centre of mass the momenta of the proton and electron have equal magnitude, p. Since kinetic energy is $p^2/(2m)$ it is clear that the kinetic energy of the proton can be neglected because of its relatively high mass. The potential energy has two contributions, one due to electrostatics and the other due to gravity. It is necessary to express these various energies in terms of the mass and radius of the body.

P.3.1.   The kinetic (degeneracy) energy

The planetary body of mass M_P is composed of N_p atoms. If A is the average mass number of the atoms then $N_p = M_P/Am_p$ where the nucleon mass $m_p = 1.67 \times 10^{-27}$ kg. As usual, no distinction needs be made between the masses of the proton and neutron. Each atom will have Z extra-nuclear electrons giving ZN_p electrons in the body. Electrons are fermions (section J.2) and so obey the Pauli principle of all having a different quantum mechanical state. The number of such states in the body is then equal numerically to the number of electrons there. If the volume of the planet is V_p then the volume available to each electron in the body is $v_p = V_p/ZN_p = 4\pi R_P^3/3ZN_p$. Assuming that the 'cell' occupied by the electron is spherical, its radius, d, is given by the expression

$$d = \left(\frac{3v_p}{4\pi}\right)^{1/3} = \left(\frac{1}{ZN_p}\right)^{1/3} R_P = \left(\frac{Am_p}{ZM_P}\right)^{1/3} R_P \tag{P.2}$$

The electron kinetic energy is composed of two parts. The first part is the motion of the electron within its cell. The second part is the kinetic energy of motion of the cell, which is negligible in comparison with the first. The kinetic energy can then be expressed in terms of the dimensions of the cell alone. This energy is independent of the temperature and so is called *degenerate*. In Topic J an estimate of the degenerate energy was found from the Heisenberg uncertainty principle. Here we use another approach.

The kinetic energy of the electron is written $E_k = p^2/(2m_e)$, where p is the momentum, and $m_e = 9.1 \times 10^{-31}$ kg is the mass of the electron. In quantum mechanics, momentum is related to the electron de Broglie wavelength, λ, by the relationship $p\lambda = h$ where $h = 6.63 \times 10^{-34}$ J s is the Planck quantum of action. The wavelength λ can have a range of values (related as harmonics) and the longest will be one wavelength spread around the circumference of the cell. Therefore we can set to sufficient approximation $\lambda = 2\pi d$ and the lowest degenerate energy, E_k, per electron cell is

$$\mathsf{E}_k = \frac{h^2}{2m_e} \frac{1}{4\pi^2 d^2}. \tag{P.3}$$

The total degenerate energy is thus

$$E_K = \mathsf{E}_k ZN_p. \tag{P.4}$$

Combining equations (P.2) and (P.3) with equation (P.4) gives

$$E_K = \gamma_K \frac{M_P^{5/3} Z^{5/3}}{A^{5/3} R_P^2} \tag{P.5}$$

with

$$\gamma_K \approx \frac{h^2}{8\pi^2 m_e m_p^{5/3}} = 2.60 \times 10^6 \ \mathrm{kg^{-2/3} \, m^4 \, s^{-2}}.$$

More precise theory gives $\gamma_K = 9.8 \times 10^6 \ \mathrm{kg^{-2/3} \, m^4 \, s^{-2}}$, a value that we shall use hereafter. Equation (P.5) is similar to equation (J.6) except for the numerical factor, which differs because of different approximations that have been made.

P.3.2.   The electrostatic energy

The problem of summing the contributions of all electrons and nuclei to the electrostatic energy is rather complicated but we can appeal to an approximate argument for our present purpose. Under the density conditions of normal planets the speeds of the charges are non-relativistic so the interaction is adequately described by electrostatics. The electrostatic energy for a single cell, E_e, is written, to a good approximation, as the interaction of the two charges, one positive and the other negative, separated by the size of a cell. Then

$$E_e \approx -a \frac{1}{4\pi\varepsilon_0} \frac{Ze^2}{d} = -a \frac{Z^{4/3}e^2}{4\pi\varepsilon_0} \frac{M_P^{1/3}}{A^{1/3}m_p^{1/3}R_P},$$

where e is the electronic charge, ε_0 is the permitivity of free space ($8.854 \times 10^{-12}\,\mathrm{F\,m^{-1}}$) and a is a numerical constant close to unity. The total electrostatic energy is given by summing over all the cells: $E_e = E_e N_p Z$, that is

$$E_e \approx -a\gamma_e \frac{M_P^{4/3} Z^{7/3}}{A^{4/3} R_P}$$

with

$$\gamma_e = \frac{e^2}{4\pi\varepsilon_0 m_p^{4/3}} = 1.16 \times 10^8\,\mathrm{kg^{-1/3}\,m^3\,s^{-2}}. \tag{P.6}$$

We set $a = 1$ although a precise calculation gives $a = 0.9$.

P.3.3. The gravitational energy

The gravitational energy E_g is readily calculated. With b a numerical constant of order unity

$$E_g \approx -b \frac{GM_P^2}{R_P} = -\gamma_g \frac{M_P^2}{R_P}. \tag{P.7}$$

The value of b will depend on the degree of compression in the body but an average value is 0.9 giving $\gamma_g \equiv bG = 6.0 \times 10^{-11}\,\mathrm{m^3\,kg^{-1}\,s^{-2}}$.

P.3.4. The energies combined

The kinetic energy has the single contribution, equation (P.5). The total potential energy is the sum of the expressions (P.6) and (P.7). Explicitly, the virial expression (P.1) becomes

$$2\gamma_K \frac{M_P^{5/3} Z^{5/3}}{A^{5/3} R_P^2} \approx \gamma_e \frac{M_P^{4/3} Z^{7/3}}{A^{4/3} R_P} + \gamma_g \frac{M_P^2}{R_P}$$

which is easily rearranged to become a relation between R_P and M_P.

$$\frac{1}{R_P} \approx \frac{\gamma_e}{2\gamma_K} \frac{A^{1/3} Z^{2/3}}{M_P^{1/3}} + \frac{\gamma_g}{2\gamma_K} \frac{M_P^{1/3} A^{5/3}}{Z^{5/3}}. \tag{P.8}$$

Also $\gamma_e/2\gamma_K = 5.9\,\mathrm{kg^{1/3}\,m^{-1}}$ and $\gamma_g/2\gamma_K = 3.0 \times 10^{-18}\,\mathrm{kg^{-1/3}\,m^{-1}\,s^2}$.

It is seen from equation (P.8) that for small masses the first term on the right hand side is dominant so that $M_P^{1/3}/R_P = $ constant. If we make the assumption that in a solid body the number of atoms per unit volume is approximately constant then the density is proportional to A and we find

$$M_P \propto \rho R_P^3$$

that is the normal relationship between mass and radius for a constant density. Physically there is equilibrium between the degeneracy and the electric (chemical) forces. On the other hand, for very large masses the second term on the right hand side of equation (P.8) becomes dominant and the different relation $M_P^{1/3} R_P = $ constant then applies. The radius now decreases with increasing mass.

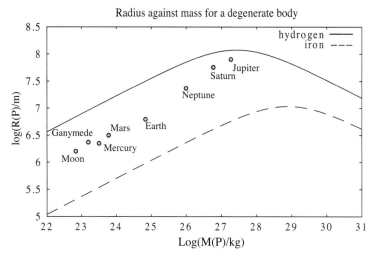

Figure P.1. *The radius–mass relationship for bodies made of hydrogen and of iron. According to their compositions various Solar-System bodies fall in the region between the two lines.*

The plot of R_P against M_P is shown in figure P.1. The upper curve is for a pure hydrogen composition ($A = 1$, $Z = 1$) while the lower curve is for a pure iron composition ($A = 56$, $Z = 26$). It is seen that there is a maximum radius in each curve corresponding to a particular mass. The radius increases at first with mass but then decreases as the compression of the material increases with increasing gravitational force to more than compensate for the extra material.

P.4.   MAXIMUM RADIUS

The mass for the maximum radius, $M_{R_{\max}}$, can be found from equation (P.8) by differentiation, setting dR_P/dM_P equal to zero, to give the condition for the maximum radius. Because of the form of the relationship for small M_P and large M_P we know that the extremum must be a maximum. It is found that $dR_P/dM_P = 0$ if

$$M_{R_{\max}} \approx \left(\frac{\gamma_e}{\gamma_g}\frac{Z^{7/3}}{A^{4/3}}\right)^{3/2}. \tag{P.9}$$

Inserting this value for M_P in equation (P.8) gives the expression for the maximum radius as

$$R_{\max} = \frac{\gamma_K}{\sqrt{\gamma_e\gamma_g}}\frac{Z^{1/2}}{A}. \tag{P.10}$$

Since $A \geq Z$ it follows that $R_{\max}$ decreases with increasing A so that the maximum radius is smaller for a silicate body than for a hydrogen body. The value of $M_{R_{\max}}$ depends on the actual values of A and Z. For a silicate body with, say, $Z = 15$ and $A = 30$ the value is some fifteen times higher than that for a hydrogen body. On the other hand for helium ($Z = 2$, $A = 4$) the maximum mass is smaller than that for hydrogen. Inserting the various numbers

$$R_{\max} \approx 1.19 \times 10^8 Z^{1/2}/A \quad \text{(m)}, \qquad M_{R_{\max}} \approx 2.69 \times 10^{27} Z^{7/2}/A^2 \quad \text{(kg)}. \tag{P.11}$$

The maximum radii and masses for a hydrogen body ($A = 1$, $Z = 1$) and or a helium body ($A = 4$, $Z = 2$) follow immediately from equation (P.11). These are compared with the observed data for

Jupiter, the largest planet in the Solar System, and for Saturn, the second largest, as follows:

$$R_{max}(H) \approx 1.19 \times 10^8 \, m \qquad M_{R_{max}}(H) \approx 2.69 \times 10^{27} \, kg$$

$$R_{Jupiter} = 7.15 \times 10^7 \, m \qquad M_{Jupiter} = 1.90 \times 10^{27} \, kg$$

$$R_{Saturn} = 6.03 \times 10^7 \, m \qquad M_{Saturn} = 5.69 \times 10^{26} \, kg$$

$$R_{max}(He) \approx 4.21 \times 10^7 \, m \qquad M_{R_{max}}(He) \approx 1.90 \times 10^{27} \, kg$$

These data repeat the conclusion to be drawn from figure P.1, namely that both Jupiter and Saturn are essentially hydrogen bodies but with some helium—although the mass of Saturn falls somewhat outside the range.

P.5. CONDITIONS WITHIN A PLANET OF MAXIMUM RADIUS AND MASS

The mass and radius data for some planets and principle satellites are also plotted on figure P.1. It is seen that these data fit the curve well and all points lie between the hydrogen and iron lines. The farther from the hydrogen line, the greater the mean A for the body. It is interesting that the satellite data fit just as well suggesting that, in structural terms, there is no fundamental difference between satellites and planets. The data for Jupiter are closely those for the maximum radius and the data for Saturn are close by. The question arises, what are the physical criteria for a maximum planetary size?

Using equation (P.11) it follows that the number of nuclei, $N_{R_{max}}$, for a body with the maximum radius is given by the expression

$$N_{R_{max}} \approx M_{R_{max}}/Am_p = 1.61 \times 10^{54} Z^{7/2}/A^3.$$

The gravitational energy is

$$E_{R_{max}} \approx -\frac{GM^2_{R_{max}}}{R_{max}} = -4.1 \times 10^{36} Z^{13/2}/A^3 \quad (J).$$

The gravitational energy per particle is thus

$$\mathsf{E}_{R_{max}} = E_{R_{max}}/N_{R_{max}} \approx -2.5 \times 10^{-18} Z^3 \quad (J).$$

The hypothetical largest planetary body has a hydrogen composition ($A = Z = 1$), so that $\mathsf{E}_{R_{max}}(H) \approx -2.5 \times 10^{-18}$ J, the magnitude of which is very closely the ionization energy of the hydrogen atom ($13.6\,eV \approx 2.2 \times 10^{-18}$ J). It can be concluded, within our admitted approximate arguments, that the constituent atoms are on the verge of being ionized by the gravitational pressure. Any greater mass will begin the ionization process. The central pressure, P_c, is, as to orders of magnitude,

$$P_c \approx \frac{GM^2_{R_{max}}}{R^4_{max}} = 2.43 \times 10^{12} \, N\,m^{-2} \approx 25 \text{ million atmospheres}.$$

This is, in fact, similar to the magnitude of the degeneracy pressure opposing the electrostatic attraction of the electron and the proton in the atom. A greater pressure within the planetary body will cause the atomic forces to become unbalanced and so cause ionization.

As M_P increases well beyond the maximum then, according to equation (P.8), R_P decreases and the ionization energy is certainly exceeded. The same conclusions follow for heavier atomic compositions. It would seem that the maximum radius marks the transition from a body where ionization is not important to one where ionization progressively increases. The fact that atoms are ionized does not mean that thermonuclear process begin. The mass where such processes become dominant and a wide range of nuclear reactions occur is about $0.075 M_\odot$.

The usual definition of a *brown dwarf* is that it is a body within which deuterium-based nuclear reactions, but no others, take place. The lower mass limit for this to occur is about 13 Jupiter masses or $0.013 M_\odot$. However, we have identified another natural division in the progression of masses—where ionization begins to take place in the planetary interior. There is no distinctive name for bodies in the mass range from the beginning of ionization to the beginning of deuterium reactions. One possibility is to extend the description of a brown dwarf so as to include this state—although this is not the usual definition.

P.6.   SPECIFYING A PLANET: THE PLANETARY BODY

For bodies with less than the maximum radius, which includes planets and satellites within the Solar System, there are three defining criteria:

(a)   *It has a mean surface figure of symmetry.* All the planets are essentially spherical or prolate spheroids due to rotation.

(b)   *The material inside is not ionized.* This means the composition is of atoms and molecules and does not involve a plasma.

(c)   *The radius increases with the mass.* This is essentially a consequence of (b). The constant of proportionality is a function of the material density.

The dominating force is that of gravity and there are several attributes that are *not* involved in planetary equilibria. For instance:

(d)   *There has been no reference to the internal temperature.* The kinetic energy is degenerate (temperature independent) and the potential energies depend on the temperature only indirectly through changes of volume. This makes it very difficult to determine the internal temperature distribution.

(e)   *There has been no mention of magnetic energy.* The energy associated with intrinsic magnetic phenomena is negligible in relation to the gravitational energy.

(f)   *There has been no mention of the precise composition.* Planets can be made of many materials.

(g)   *There has been no mention of rotation.* The state of rotation is not relevant provided the rotation is not so fast that it breaks the stability of the inner structure. This is obviously not the case for a stable body.

(h)   *There is no requirement that the body should orbit the Sun.* The arguments apply to any body where gravity is the dominant force and there is no distinction to be drawn in these terms between a planet and a principal satellite. The theory, therefore, covers planets and satellites equally and these bodies can be called collectively *planetary bodies*.

P.7.   THE MINIMUM MASS FOR A PLANETARY BODY

We have defined a planetary body as one that takes up a spherical form, or approximately so. It is known that small asteroids and some smaller satellites have irregular shapes and so would not come within the definition we have been using. It is intuitively obvious that there cannot exist a large planetary body, say as big as the Moon, in the form of a cube. The configuration of minimum energy is a sphere and the forces operating to produce that configuration would overwhelm the strength of the material of the body. To find the minimum size of a body that would just take up a spherical form we first have to understand the nature of the forces that hold solids together.

P.7.1. The rigidity of a solid body

The atoms within the silicates that form rocks and the ices that form parts of some planets and satellites are held together by bonds that can be of various kinds. One type of bond, occurring in alkali halides, is the ionic bond where neighbouring atoms, one positively charged and the other negatively charged, are attracted to each other by electrostatic forces. More commonly in minerals, neighbouring atoms are bound together by covalent forces produced by the sharing of some of their outer valence electrons. This is how the oxygen and silicon atoms would be bound in a silicate material. Finally there are weaker van der Waals forces that act between molecules. In some silicates, sheets of atoms strongly bound by covalent bonds are bound together by van der Waals forces between the sheets. Such silicates are easily cleaved since the weak van der Waals forces are comparatively easy to break.

To disrupt a solid so that it is no longer a rigid collection of atoms but one where individual atoms, or perhaps rather stable entities such as SiO_2, can easily move relative to each other requires an input of energy. The energy associated with the lattice structure, the *lattice energy*, is negative since it corresponds to forming a bound system. It can be expressed in terms of the energy per mole, e.g. $500\,kJ\,mole^{-1}$ or, as we prefer here, as average energy per unit mass of the mineral. It is quite straightforward to estimate this. When a solid is heated then the individual atoms of which it is formed take up thermal energy $3kT/2$ and they oscillate about some mean position with this amount of mean kinetic energy. As the temperature is raised so the amplitude of oscillation increases until eventually the bonds are broken, the material loses its rigidity and becomes a liquid. The lattice energy per atom can then be approximated as $3kT_m/2$ where T_m is the melting temperature. Thus for an average olivine with a melting point of 1500 K the lattice energy per atom is 3×10^{-20} J. Dividing this by the average atomic mass, say $20m_p$ for a silicate, gives the lattice energy per unit mass as $9.0 \times 10^5\,J\,kg^{-1}$. If the body is a fluid then, even if it is a small body so that the gravitational energy per atom is tiny compared with the lattice energy, it will still take up a spherical form. It is possible that even very small bodies could be spherical if they were formed from molten material.

When it comes to bodies that have adopted a spherical shape because of gravitational forces it is not the lattice energy that is relevant. Planetary material, like most natural material, is not in the form of a perfect crystal of infinite extent. It consists of small regions of perfection, that might be crystalline mineral grains, bound together by forces of cohesion that are much weaker than the covalent bond forces that keep a crystal together. Consequently the material is much less rigid and less strong than would be suggested by a consideration of lattice energy. For example, carbon fibre is a synthetic perfect material and has the strength of steel. However, we know how easy it is to snap a pencil lead that is made of the same chemical substance. To estimate the rigidity of a natural material we need to know its crushing strength, the pressure required to make it crumble. The orthopyroxene mineral, bronzite $(Mg,Fe)SiO_3$, has density $3.66 \times 10^3\,kg\,m^{-3}$, melting point 1650 K and crushing strength $1.627 \times 10^8\,N\,m^{-2}$. The crushing strength is a pressure, that can be considered as energy per unit volume, so that dividing it by density gives energy per unit mass. For bronzite this is $4.4 \times 10^4\,J\,kg^{-1}$. For this material the corresponding lattice energy is $1.0 \times 10^6\,J\,kg^{-1}$, more than 20 times greater.

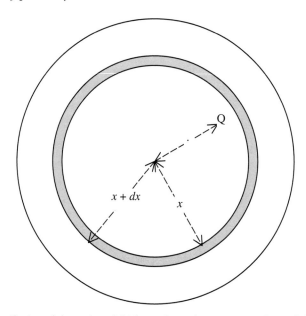

Figure P.2. *The contribution of the region of thickness* dx *to the pressure at* Q *is* $\rho g(x)\,dx$ *where* $g(x)$ *is the gravitational field at distance* x *from the centre.*

If a solid body forms then it will take up a configuration of minimum energy. It requires work to crush it into a spherical form and if the difference in the gravitational energy between the original configuration and a spherical form is less than this amount of work then it will retain its original shape. The ratio of the energy required to the energy available is a complicated matter, depending as it does on the exact form of the original configuration. We can make an approximate estimate of the minimum size of a body that will take up a spherical shape from the condition that the mean pressure within it should exceed the crushing strength of its material. Figure P.2 shows a uniform spherical body. The pressure at the point Q, distance r from the centre, is due to the weight of all the material above it. The pressure contribution of the material between x and $x + dx$ is $\rho g(x)\,dx$ or

$$dP = \rho \frac{\frac{4}{3}\pi x^3 \rho G}{x^2}\,dx = \frac{4}{3}\pi G \rho^2 x\,dx.$$

Integrating between $x = r$ and $x = R$, the radius of the body, gives

$$P(r) = \tfrac{2}{3}\pi G \rho^2 (R^2 - r^2). \tag{P.12}$$

A representative pressure is found by taking $r = \frac{1}{2}R$ and is $\frac{1}{2}\pi G \rho^2 R^2$. Equating this to the crushing strength, S, gives

$$R = \left(\frac{2S}{\pi G \rho^2} \right)^{1/2}. \tag{P.13}$$

For bronzite, with the figures previously given, $R = 340\,\text{km}$. However bronzite is a dense strong material and crushing strengths for rocks can vary between $6 \times 10^6\,\text{N}\,\text{m}^{-2}$ and $2 \times 10^9\,\text{N}\,\text{m}^{-2}$. The radii of the smallest satellites of Saturn and Uranus with figures of symmetry are $R_{\text{Mimas}} = 195\,\text{km}$ and $R_{\text{Miranda}} = 242\,\text{km}$. The densities of these bodies indicate that they have a large ice component and the effective ratio S/ρ^2, that occurs in equation (P.13), for these bodies is probably at the low end of the silicate range.

P.8.   THE INTERNAL STRUCTURE OF A PLANETARY BODY

It is now possible to make several deductions about the general conditions inside a representative solid planetary body of the type found in the Solar System. Examples are the terrestrial planets, the satellites and the largest asteroids.

P.8.1.   The crust

Larger solid planetary bodies have a crust, an outer shell of material of lower density and more rigidity than the underlying mantle. The crust moves in solid masses under the forces of plate tectonics. The lower boundary of the crust may be taken as where the pressure equals the crushing strength of the crust material. The crust is comparatively thin, so it is reasonable to take the acceleration due to gravity as constant within its thickness. Hence, the thickness of the crust, z, is given by

$$\rho g z = S \quad \text{or} \quad z = \frac{S}{\rho g}. \tag{P.14}$$

The igneous rocks forming the Earth's surface are quite strong; taking crust material with $S \approx 10^9 \, \mathrm{N \, m^{-2}}$ and $\rho = 2.8 \times 10^3 \, \mathrm{kg \, m^{-3}}$ gives $z = 36 \, \mathrm{km}$. The crust-mantle boundary is given by seismic measurements (Topic R) and is known as the *Mohorovicic discontinuity* or *Moho*. Beneath the surface of continents this is in the range 20–90 km below the surface, which contains the estimate we have found. However, the Moho is only 5–10 km deep below the ocean crust, indicating that the rocks that form in these regions have smaller crushing strength than those in continental regions.

From equation (P.14) we see that, for material of similar composition, the thickness of the crust varies as $1/g$. Thus for Mars, with $g = 3.7 \, \mathrm{m \, s^{-2}}$, the crust would be almost three times as thick as on Earth.

P.8.2.   The maximum height of surface elevations

A mountain of height h consisting of material of uniform density ρ will give a pressure $\rho g h$ at the level of its base. If that pressure is greater than the crushing strength of the mountain material then the material at the base will crumble and the mountain will slump to give a height such that $\rho g h = S$. Without knowledge of the precise material forming a mountain—that will not be uniform in any case—a maximum height cannot be estimated. By taking $S = 2.5 \times 10^8 \, \mathrm{N \, m^{-2}}$ and $\rho = 2800 \, \mathrm{kg \, m^{-3}}$ the maximum height is found to be 8.9 km, a little more than the height of Mount Everest. However, these physical properties were selected to give the result, so there is no significance in this match of values.

What can be seen is that for the same kind of material the maximum height of a mountain is proportional to $1/g$. For Earth the product of the height of Mount Everest and g is $8 \times 10^5 \, \mathrm{m^2 \, s^{-2}}$. The tallest mountain on Venus is Maxwell Montes, some 11–12 km high, and the product hg is $\sim 10^6 \, \mathrm{m^2 \, s^{-2}}$. On Mars Olympus Mons is 25 km high and the product hg is $\sim 9.2 \times 10^5 \, \mathrm{m^2 \, s^{-2}}$.

P.8.3.   Hydrostatic equilibrium

The gravitational force in the interior regions below the crust is dominant and this allows the material to exhibit self creep under the directed action of gravity. The normal fluid state arises because the interaction energy between the atoms becomes smaller than the thermal energy beyond a particular temperature. The force of self-gravity within a planetary body replaces the thermal energy in determining the normal conditions on the surface. The consequence is that heavier atoms will tend to sink within the volume, forcing lighter atoms upwards as if the material were indeed a very viscous liquid. The time scale for such a chemical separation of the constituents to be complete may be long if the material were well mixed originally. Of course, the effect is enhanced by any source of energy that raises the temperature farther. The material behaves on the whole as a high density fluid and can be considered as being under the condition of *hydrostatic equilibrium* over sufficiently long time scales. One consequence is that the bulk of the body assumes a shape of symmetry.

P.8.4.   Mantle and core

The bodies of the Solar System have distinctive chemical compositions. The solar abundance for the chemical elements (see table 1.1) leads to four components dominating others, these are:

(i)   a hydrogen/helium mixture accompanied by hydrogen compounds in small proportions;
(ii) silicates of various kinds but mainly of magnesium and aluminium;
(iii) ferrous materials, Fe, Co, Ni; and
(iv) the combination of the first and second most abundant active elements to form water (H_2O).

Of particular interest here is the distinctions between the densities of the various materials. Hydrostatic equilibrium will lead to the denser components moving to the central regions forcing the less dense materials upwards. The separation between the different materials can be expected to be sharp because the densities are so different $\rho(\text{ferrous}) \approx 2\rho(\text{silicates}) \approx 4\rho(\text{ice}) \geq 8\rho(H, He)$. There can, of course, be secondary divisions within each region reflecting variations of the density, partly due to structural phase changes resulting from increasing pressure.

Whether the material behaves like a solid or liquid on the short time scale depends on the local melting points of the particular minerals. It should be realized, of course, that the distinction between the two is not sharp. No material is ideally rigid (and therefore perfectly solid) and no fluid material is entirely devoid of rigidity (and therefore totally incapable of sustaining any shear like the ideal liquid). The time scale of events is often of the greatest importance in planetary science.

The interior is, then, quite generally divided into a central *core*, a less dense *mantle* and an even less dense *crust*. There may, of course, be divisions within these regions. The degree of concentration of density towards the centre of the body determines the moment of inertia. Measurements of the moment of inertia factor (Topic S) allow estimates to be made of the relative sizes of the core and mantle in any particular case.

P.8.5.   Variation of pressure and density with depth

The condition of hydrostatic equilibrium is where the pressure at any level is equal to the weight of material above. Consider a spherical surface distance r from the centre of a spherical body where the pressure is P and a second surface at radius $r + dr$ (and so farther from the centre) where the pressure is $P + dP$ and dr is small. For hydrostatic equilibrium $dP = -\rho(r)g(r)\,dr$ and $g(r) = GM(r)/r^2$, where $M(r)$ is the mass within the distance r from the centre. The pressure gradient can be deduced if the density gradient is known and the density gradient can be deduced if the bulk modulus of elasticity, $K = \rho(dP/d\rho)$, is known. For then

$$\frac{\partial P}{\partial r} = \frac{\partial P}{\partial \rho}\frac{\partial \rho}{\partial r} = \frac{K}{\rho}\frac{\partial \rho}{\partial r} = -\rho g \quad \text{or} \quad \frac{\partial \rho}{\partial r} = -\frac{\rho^2 g}{K}. \tag{P.16}$$

It is sometimes more convenient to measure the distance downwards, into the Earth, and for this purpose we write $z = R_P - r$ so that $z = R_P$ when $r = 0$ and $z = 0$ when $r = R_P$. Then $\partial \rho/\partial r = -\partial \rho/\partial z$ and equation (P.16) becomes alternatively

$$\frac{\partial \rho}{\partial z} = \frac{\rho^2 g}{K}.$$

P.8.6. *Specifying K*

Equation (P.16) can be used to find the variation of density with depth, by iteration from the surface, if the bulk modulus is known as function of the depth.

(i)   One possibility is to invoke an empirical equation of state such as the Murnaghan equation (see Problem P.2) deduced from observations of Earth material. There $K/\rho^B = A$, a constant, so that equation (P.16) becomes

$$\frac{\partial \rho}{\partial r} = -\frac{g}{A\rho^{B-2}}$$

For silicates it is usual to take $B = 4$. The equation takes a particular form when $B = 2$, a value often useful for discussions involving hydrogen and helium.

(ii) Seismology is a powerful practical second alternative for exploring the physical conditions with depth in a planet, and particularly the Earth and the Moon on which seismometers have been placed. This is explained in some detail in Topic S.

Problems P

P.1   Calculate from equation (P.8) the radius of a cold body made of carbon if its mass is 10^{28} kg.

P.2   The bulk modulus of a material may be expressed as $K = \rho(dP/d\rho)$. The dependence of K on P may be written expressed as $K = K_0 + BP + CP^2 + DP^3 \ldots$ where K_0 is the bulk modulus at zero pressure and coefficients $B, C, D, \ldots$ depend on the composition.

  (i)   For low pressures all the coefficients may be taken as zero so that $K = K_0$. Show that this gives the equation of state $\rho = \rho_0 \exp(P/K_0)$ where ρ_0 is the density at zero pressure.

  (ii) For higher pressures $K = K_0 + BP$. Show that this leads to an equation of state for the material

$$\rho = \rho_0 \left(1 + B\frac{P}{K_0}\right)^{1/B}.$$

This is the equation of Murnaghan. Show that it is equivalent to $K/\rho^B = $ constant. In applications to silicate materials B is in the range $3 \le B \le 5$.

  (iii) Using the quadratic form $K = K_0 + BP + CP^2$ show that, if $K_0C - B^2/4 > 0$, then

$$\rho = \rho_0 \exp\left\{\frac{1}{\sqrt{K_0C - B^2/4}}\left[\tan^{-1}\left(\frac{PC + B/2}{\sqrt{K_0C - B^2/4}}\right) - \tan^{-1}\left(\frac{B/2}{\sqrt{K_0C - B^2/4}}\right)\right]\right\}.$$

TOPIC Q

THE TRANSFER OF HEAT

The internal temperature does not affect the overall long-term equilibrium of the body but temperature variations can affect the local dynamics of the interior and the surface. It is necessary, therefore, to consider the internal temperature in order to understand the *dynamics* of the internal structure. This presents some difficulties in practice since internal temperatures cannot be measured directly. The only firm measurements are the surface heat flow and mean near-surface temperature. This is, incidentally, directed outwards for each of the planets.

A planetary body has a hot interior for three reasons. One is the rise of temperature due to compression within the volume. The second is the heat remaining from the formation of the body. The third is the heating due to radioactive components. The consequence of these different sources of heat is a temperature that decreases through the body from the central region to the outer crust. There will be a mean outward heat flow from the heat of formation and from the radioactive sources but the temperature differences due to compression will not give rise to a heat flow. The temperature gradient here is consequently called the *adiabatic gradient* (section O.2.3.2). Its representation follows from thermodynamics and involves the local coefficient of isothermal compression. Heat transfer in planetary interiors is essentially either by *convection* or by *conduction* depending on the circumstances. The third form of energy transfer is by radiation but the temperatures in planets will not be great enough for the transfer by radiation to be important, although it is the basic heat transfer mechanism for most of the volume of the Sun (Topic G). It is possible that radiative transfer could play a minor role in restricted regions of the major planets but that is by no means certain.

Q.1. CONDUCTION OF HEAT IN A SOLID

If a temperature difference is maintained across a region between two surfaces, heat will flow from the hotter to the colder surface. The quantity of heat that flows is found empirically to be proportional to the local gradient of the temperature that is maintained. This is *Fourier's law*. Suppose we consider the flow of heat in the x direction (the one-dimensional problem). The quantity of heat flowing in the x direction through unit area of surface normal to the x direction (that is in the y–z plane) per unit time, denoted by $q(x)$, is called the heat flux in the x direction. According to Fourier's empirical law we write

$$q(x) = -\zeta \frac{\partial T}{\partial x} \qquad (Q.1)$$

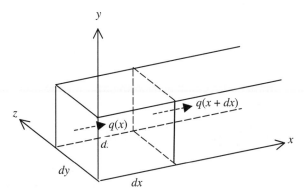

Figure Q.1. *Heat flow in a small element.*

where ζ is the thermal conductivity (a characteristic of the material and with dimensions $\mathrm{W\,m^{-1}\,K^{-1}}$) and T is the *absolute* temperature. The minus sign accounts for the flow of heat from hot to cold. The magnitude of ζ for materials of planetary interest is often of the order $10^2\,\mathrm{W\,m^{-1}\,K^{-1}}$. The passage of heat in heat conduction proceeds without any movement of the medium and so may occur in solids. Heat transfer is a diffusion process and is *irreversible*. By this we mean that once heat has passed from a hotter to a colder body it will not spontaneously move in the opposite direction. Heat flows until the temperature is uniform everywhere.

Q.1.1.   The equation of heat conduction in a solid

The empirical law (equation Q.1) forms the basis of the quantitative theory. A mathematical description of heat conduction can be constructed by considering the balance of heat in a rectangular volume formed by equal two flat parallel plates of area $\mathrm{d}y\,\mathrm{d}z$ separated by the distance $\mathrm{d}x$ (figure Q.1). The heat flow is in the x direction. The heat energy entering the volume through the face with coordinate x is written $q(x)\,\mathrm{d}y\,\mathrm{d}z$ while that leaving through the face with coordinate $x + \mathrm{d}x$ is $q(x + \mathrm{d}x)\,\mathrm{d}y\,\mathrm{d}z$. In addition there may be heat generated within the volume, or removed from it, at a rate h per unit volume per unit time. The rate of change of heat within the volume, Q, is the combination of these two contributions.

$$Q = q(x)\,\mathrm{d}y\,\mathrm{d}z - q(x + \mathrm{d}x)\,\mathrm{d}y\,\mathrm{d}z + h\,\mathrm{d}x\,\mathrm{d}y\,\mathrm{d}z$$

$$= q(x)\,\mathrm{d}y\,\mathrm{d}z - \left[q(x) + \frac{\partial q}{\partial x}\,\mathrm{d}x\right]\mathrm{d}y\,\mathrm{d}z + h\,\mathrm{d}x\,\mathrm{d}y\,\mathrm{d}z$$

$$= -\frac{\partial q}{\partial x}\,\mathrm{d}x\,\mathrm{d}y\,\mathrm{d}z + h\,\mathrm{d}x\,\mathrm{d}y\,\mathrm{d}z. \tag{Q.2}$$

The accumulation or loss of heat energy within the volume will cause the temperature to rise or fall and the rate of change of heat energy can be expressed at the rate of change of temperature, T, according to

$$Q = \frac{\partial}{\partial t}(\rho C_\mathrm{p} T)\,\mathrm{d}x\,\mathrm{d}y\,\mathrm{d}z = \rho C_\mathrm{p}\frac{\partial T}{\partial t}\,\mathrm{d}x\,\mathrm{d}y\,\mathrm{d}z$$

where ρ is the density of the material and C_p is the specific heat capacity at constant pressure. These can usually be supposed constants of the heat flow in planetary problems so, combining this expression with equations (Q.1) and (Q.2), gives

$$\frac{\partial T}{\partial t} = \frac{\zeta}{\rho C_p} \frac{\partial^2 T}{\partial x^2} + \frac{h}{\rho C_p}.$$

This accounts only for flow in the x direction: accounting also for the flows in the y and z directions gives

$$\frac{\partial T}{\partial t} = \alpha \nabla^2 T + H \qquad (Q.3)$$

with $\nabla^2 = (\partial^2/\partial x^2) + (\partial^2/\partial y^2) + (\partial^2/\partial z^2)$ (the Laplacian operator), $\alpha = \zeta/\rho C_p$ (called the thermometric conductivity) and $H = h/\rho C_p$. The solution of equation (Q.3) must be subject to the restrictions of *boundary conditions*. In planetary problems these are usually taken to be the temperature and the heat flow (the gradient of the temperature) at a boundary.

While equation (Q.3) has been derived for the Cartesian (box) coordinates, it can equally well be applied to other, including spherical geometries. This is done by writing ∇^2 in terms of the appropriate coordinates. For a spherical polar coordinates (r, θ, φ) the Laplace operator takes the form

$$\nabla^2 = \frac{1}{r^2} \frac{\partial}{\partial r} \left(r^2 \frac{\partial}{\partial r} \right) + \frac{1}{r^2 \sin \theta} \left(\frac{\partial}{\partial \theta} \sin \theta \frac{\partial}{\partial \theta} \right) + \frac{1}{r^2 \sin^2 \theta} \frac{\partial^2}{\partial \varphi^2}.$$

If there is complete spherical symmetry (so that neither θ nor φ appear) this reduces to the form

$$\nabla^2 = \frac{\partial^2}{\partial r^2} + \frac{2}{r} \frac{\partial}{\partial r} = \frac{1}{r^2} \frac{\partial}{\partial r} \left(r^2 \frac{\partial}{\partial r} \right).$$

Other geometries, for example cylindrical, have appropriate alternative forms.

Q.2.   COMMENTS ON THE DESCRIPTION OF FLUID FLOWS

The passage of heat in fluids is of importance in planetary science, either in gases (as in an atmosphere) which have no rigidity or in the interior of a planet where the material has limited rigidity. Planetary material, although appearing to be a solid on short time scales, can behave like a very viscous fluid over sufficiently long time scales. Convection involves the motion of the medium to carry the heat energy from one location to another and is usually the main mechanism for the passage of heat in a fluid. The driving force is natural buoyancy under gravity, due to the reduced density of material relative to that above. The increased heat content of a small region is dissipated by conduction to the contiguous fluid. As an example, this is the action of the domestic convection heater. Convection, bringing heat from one region to another, can be due to the action of other forces independently of the action of gravity. These alternative modes of heat transport, where convection can be regarded as 'forced' in some sense, are particularly important where the fluid density and specific heat capacity are high and where the effects of viscous resistance are not insignificant. An example is motion of the hot magma extruded on to the surface of the Earth from a volcano due to interior pressure. Such a forced movement of heat from one region to another is often called *advection*. In this case heat energy is advected to the surface from below. Advection usually involves the passage of a greater quantity of heat energy than does convection. In the case of magma at the Earth's surface, cooling certainly involves conduction with the surrounding material but radiation is also an important cause of energy loss by the sample. On the other hand, radiation plays no significant part in the cooling of magma extruded under water such as at mid-ocean ridges.

There is no general theory of convective motion to apply to *any* fluid and planetary magma is a particularly complicated material. It has, nevertheless, proved possible to understand in general terms the role of natural convection in the thermal dynamics of a planetary body by studying the motion of simple fluids, such as water and air. These are the so-called linear or Newtonian fluids, named after Isaac Newton who first studied their properties. This simplification of the discussion is possible because the dynamics of both simple and complex fluids are investigated by invoking the conservation statements of mass and of momentum (sufficient for an isothermal flow), perhaps augmented by the conservation of energy (for thermal flows). The full analysis is very complicated but it is sufficient for our present purposes to use the dimensional arguments to describe the behaviour. Several variables are needed to describe a particular situation but if these are arranged into a dimensionless form it is possible to compare the situations in different cases which are geometrically and dynamically similar. This means that the cases differ essentially by a simple scaling and have comparable flow conditions (for instance turbulent or not). The convective fluid motion is studied by comparing the values of the different dimensionless groupings under different circumstances.

Q.2.1. The fluid parameters

A fluid is a continuum but the description of its motion must involve the specifications of conditions at particular points in the fluid. The conditions are described by a set of variables that vary from one point to another. The variables are:

(i)   The characteristic size of the region, written ℓ.
(ii)   The density at the point, denoted by ρ.
(iii)   A characteristic speed, U, describing the motion of the fluid.
(iv)   The viscous (friction) force within the fluid that opposes differences in velocity between neigh-bouring planes of the fluid. If $\Delta U/\ell$ is the local velocity gradient associated with the stress Γ (force per unit area in the plane) then $\mu \equiv \Gamma/(\Delta U/\ell)$ defines the coefficient of shear viscosity. The shear viscosity divided by the density is often of interest: this is called the kinematic viscosity $\nu = \mu/\rho$. For Newtonian fluids, μ is independent of the flow conditions but this is not the case for complex fluids.
(v)   The rate of rotation of the region expressed through a mean angular speed, Ω.
(vi)   Thermal conditions are described by introducing the thermometric conductivity, α. This is defined as follows: the flow of heat energy is proportional to the thermal conductivity, ζ; the heat content of a fluid volume per unit temperature rise is ρC_p where C_p is the specific heat capacity; the ratio of these two quantities is $\alpha = \zeta/\rho C_p$.
(vii)   Different temperatures across a small region gives rise to a thermal force which causes the fluid to carry heat from the hotter to the colder regions. This arises from the difference in density between the hot and cold ends which produces a buoyancy force in a gravitational field. If ΔT is the difference of temperature between the regions and $\beta = (1/\rho)(\Delta\rho/\Delta T)$ is the coefficient of thermal expansion, the buoyancy force per unit mass $F_B = g\rho\beta\Delta T$. Here g is the mean accelera-tion due to gravity and ρ is a mean density taken as standard.
(viii) Conditions can change at a given point so time, t, is the last variable.

These parameters will be found sufficient to describe a general thermal fluid flow.

Q.2.2. The dimensionless parameters

The parameters identified in section Q.2.1 are now combined into dimensionless groupings. We invoke the *Buckingham Pi theorem*. This theorem tells us that if there are n independent variables described using m dimensions then $n - m$ relationships can be constructed. In our case there are eight independent variables (ℓ, ρ, U, Ω, μ, α, F_B, t) and we will describe them using the three dimensions of length, L, mass, M, and time, t. Therefore, according to the Pi theorem, we can construct $8 - 3 = 5$ independent collections of dimensionless variables. The three dimensions (L, M, t) are contained in the three variables ℓ, U, ρ. The five independent groupings of variables are then

$$\Pi_1 = \Pi_1(\ell, U, \rho, \mu), \qquad \Pi_2 = \Pi_2(\ell, U, \rho, \Omega), \qquad \Pi_3 = \Pi_3(\ell, U, \rho, \alpha),$$

$$\Pi_4 = \Pi_4(\ell, U, \rho, F_B), \qquad \Pi_5 = \Pi_5(\ell, U, \rho, t)$$

From dimensional arguments we easily find the dimensionless groupings

$$\Pi_1 \Rightarrow \rho U \ell / \mu = U \ell / \nu \equiv Re, \quad \text{the Reynolds number}$$

where $\nu = \mu / \rho$ is the kinematic viscosity defined previously;

$$\Pi_2 \Rightarrow \ell \Omega / U = Ro, \quad \text{the Rossby number}$$

$$\Pi_3 \Rightarrow \ell U / \alpha = Pe, \quad \text{the Pèclet number}$$

$$\Pi_4 \Rightarrow g \ell \beta \Delta T / U^2.$$

$$\Pi_5 = U t / \ell = St, \quad \text{the Strouhal number}.$$

Q.2.3. *Physical interpretation: rearrangements*

The various dimensionless numbers are, in fact, the ratios of the magnitudes of forces so each number has a physical interpretation. Thus

$$Re = \text{inertia force/viscous force}$$

$$Ro = \text{rotation force/inertia force}$$

$$\Pi_4 = \text{buoyancy force/inertia force}$$

$$St = \text{time dependent inertia force/convection inertia force}$$

$$Pe = \text{heat transfer by advection/heat transfer by conduction}.$$

These numbers are rearranged to form others. For instance, the rotation force can be compared with the viscous friction force by making the rearrangement

$$Ro = (\ell \Omega / U) = (\rho \ell^2 \Omega / \mu)(\mu / \rho U \ell) = Ta / Re$$

where Ta is a new dimensionless number, the Taylor number.

The Grashof number can be defined from Π_4 by introducing the viscous force through the Reynolds number. Then

$$\Pi_4 = g \beta \ell \Delta T / U^2 = [g \beta \ell^3 \Delta T / \nu^2](\nu^2 / \ell^2 U^2)$$

$$Gr = g \beta \ell^3 \Delta T / \nu^2 \text{ is the Grashof number}$$

so that $\Pi_4 = Gr / Re^2$. The Grashof number can be rearranged further to provide the Rayleigh number, Ra, by introducing the Prandtl number, Pr. Explicitly

$$Ra = g \ell^3 \beta \Delta T / \nu \alpha$$

and the Prandtl number

$$Pr = \nu/\alpha$$

so that

$$\Pi_4 = Gr/Re^2 = Ra/Pr \times Re^2.$$

It can be noticed that the Prandtl number is the only dimensionless grouping that is concerned only with the properties of the fluid and not with the flow.

The parameter Π_5 describes the time dependent features. Then

$$St = Ut/L = (UL/\nu)(\nu/\alpha)(\alpha t/L^2) = Re \times Pr \times (\alpha t/L^2).$$

This means the Strouhal number involving the speed has been replaced by an equivalent number involving only the thermal properties and the size.

This number allows the time of decay of heat in a body of linear dimension L to be estimated. For, as far as orders of magnitude

$$\alpha t/L^2 \approx 1.$$

This enables us to define a *characteristic time*, $\tau \approx L^2/\alpha$, where τ can be interpreted as the time for the initial quantity of heat to decrease by the factor $1/e \approx 0.367$. It is seen that this time is proportional to L^2 so that, for the same material compositions, larger bodies cool more slowly than smaller ones. The time to lose initial heat can be very long for cosmic bodies. As an example for Earth, $\alpha \approx 10^{-5}\,\mathrm{m^2\,s^{-1}}$, $R \approx 6.4 \times 10^6\,\mathrm{m}$, making $\tau \approx 3.6 \times 10^{18}\,\mathrm{s} \approx 1 \times 10^{11}$ years for that case. This is much greater than its present age so that most initial heat energy is still retained inside. The same conclusion applies to Mars. This will be true also for Venus but the hot surface conditions there will affect the temperature distribution inside and lengthen the decay time for that planet. A larger planet such as Jupiter will also have an extremely long decay time. The presence of radioactive heat sources adds to the heat content of the body. For sources of reasonable strength the planet might remain liquid throughout most of its volume but the situation could be more complicated for the smaller bodies. According to the Strouhal number estimate the Moon should have retained a large proportion of the heat with which it began. This is consistent with it having a partially molten core (section 6.5.4).

For silicates under planetary conditions $Pr \approx 10^{22}$, due to the very large value of the shear viscosity, $\eta \approx 10^{21}\,\mathrm{kg\,m^{-1}\,s^{-1}}$. This very large value distinguishes planetary conditions from those of the laboratory where $Pr \approx 1$. For planetary problems $Re \approx 10^{-19}$ so that $Pe \approx 10^3$ ($\gg 1$) and the advection of heat is greater than its convection. Such applications strain the usual concepts of the Newtonian viscosity.

Problems Q

Q.1 A roughly-spherical iron asteroid of radius 10 km is formed at the melting temperature of iron, 1811 K. Given its density, $7850\,\mathrm{kg\,m^{-3}}$, thermal conductivity $80\,\mathrm{W\,m^{-1}\,K^{-1}}$ and specific heat capacity $500\,\mathrm{J\,kg^{-1}\,K^{-1}}$ then estimate the time it will take to become a cool body in thermal equilibrium with its environment.

Q.2 The near-surface material of a planetary body has the following physical properties: density $3.0 \times 10^3\,\mathrm{kg\,m^{-3}}$, thermal conductivity $20\,\mathrm{W\,m^{-1}\,K^{-1}}$ and specific heat capacity $800\,\mathrm{J\,kg^{-1}\,K^{-1}}$. The surface temperature gradient is $1.83 \times 10^{-3}\,\mathrm{K\,m^{-1}}$ and at a depth of 10 km it is $1.71 \times 10^{-3}\,\mathrm{K\,km^{-1}}$. The near surface material contains radioactive elements generating $3 \times 10^{-7}\,\mathrm{W\,m^3\,s^{-1}}$. Estimate the rate at which the material is heating or cooling at a depth of 5 km.

TOPIC R

SEISMOLOGY—THE INTERIOR OF THE EARTH

Planetary material behaves like an elastic solid on a short time scale but as a viscous fluid on a long time scale. In describing this behaviour the simplifications are made that the elasticity is perfect and that the fluid behaves in a simple way. This duality allows both fluid statics and elastic vibration theory to be combined to provide unique and useful formulae for exploring the interior of an Earth-like planet. The applications require a knowledge of the elastic properties of the planetary interior and these are derived from the studies of earthquakes. The approach has great practical value and was first used by Adams and Williamson as a means of determining the distribution of density within the Earth. There are two assumptions, one concerned with the long-term behaviour of the material and the other with the short term. We take the short-term behaviour first.

R.1. THE BEHAVIOUR OF PLANETARY MATERIAL FOR AN IMPULSIVE RELEASE OF ENERGY

Suppose the release to be confined to a small volume within the body. This acts as the source of energy for a system of elastic waves that propagate throughout the volume. The amplitudes of the waves decrease in a way determined by the dissipative properties of the medium carrying them. There are differences between a medium of effectively infinite volume and one that is finite.

R.1.1. Waves without a boundary

If the source is deep within the Earth and boundaries are not relevant (at least to good first approximation) two types of elastic wave emerge within the body, one longitudinal (called the P-wave) and the other transverse (called the S-wave). The P-wave is the analogue of the sound wave in a fluid and involves alternately the pure compression and the pure dilatation of a volume. The S-wave involves instead the transverse vibration of an element at constant volume. These are shown in figure R.1a and b.

The elastic properties that are involved in these waves are the bulk modulus (incompressibility), κ, and the shear modulus, μ. All elastic moduli are described in terms of stress/strain, where stress has the dimensions of pressure and strain is a dimensionless quantity describing the fractional distortion of the body. The bulk modulus describes the fractional change in volume when the body is subjected to an additional pressure. If the body with *specific volume* V (volume per unit mass) changes in volume by dV when subjected to an increase of pressure dP then

$$\kappa = -\frac{dP}{dV/V} = \rho \frac{dP}{d\rho} \tag{R.1}$$

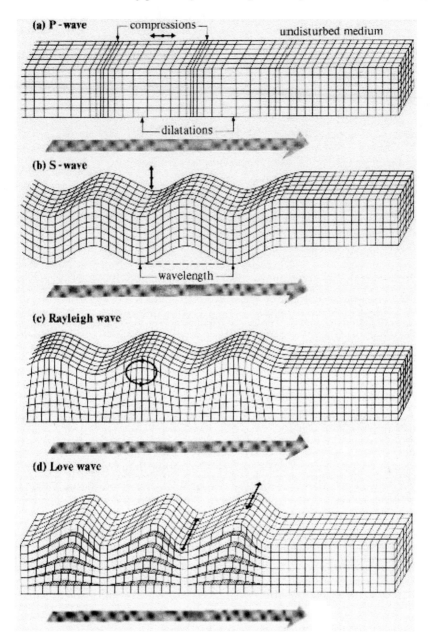

Figure R.1. *(a) A longitudinal (push-pull) P-wave. (b) A transverse S-wave. (c) A Rayleigh wave. (d) A Love wave.*

since $\rho = 1/V$. The rigidity modulus can be envisaged by considering a stress applied as a shear force per unit area as shown in figure R.2. In this case the stress, with the dimensions of pressure, is F/A and the strain is dl/l, equal to θ if the distortion is small. Hence

$$\mu = \frac{F}{A\theta}. \qquad \text{(R.2)}$$

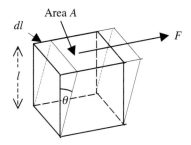

Figure R.2. *Shearing a block of material by a tangential force F over the area A. There must be an equal and opposite force on the lower face to give a shearing torque.*

The longitudinal wave speed V_P depends on both κ and μ. The transverse wave speed, V_S, is determined by the shear modulus, μ, alone. Explicitly:

$$\rho V_P^2 = \kappa + \tfrac{4}{3}\mu, \qquad \rho V_S^2 = \mu. \tag{R.3}$$

It follows that $V_P > V_S$ and hence the P-wave arrives from a common source before the S-wave. It is for this reason that the terminology P, i.e. principal, wave and S, i.e. secondary, wave was adopted.

If the material density is known and the wave speeds can be measured then the elastic constants can be inferred. The shear modulus is given directly by the S-wave speed but the coefficient of bulk modulus requires a knowledge of both wave speeds,

$$\kappa(r) = \rho[V_P^2(r) - \tfrac{4}{3}V_S^2(r)] \tag{R.4}$$

where r is the distance from the centre of the Earth.

R.1.2.   *Waves near a boundary surface*

The simple P- and S-body wave pattern does not follow if the waves encounter a surface. The energy is now distributed over the two dimensions of the surface rather than the three dimensions of the main body. The surface is a 'plane' of zero stress and the wave pattern must accommodate itself to comply with this requirement. The result is a wave, first discovered by Rayleigh and named after him, which is entirely analogous to the surface wave in a fluid. It travels with a speed $V_R \approx 0.9V_S$. The material motion contains both horizontal and vertical components but constrained to a vertical plane containing the direction of propagation of the wave. Like the surface water wave, the particle motion in this free Rayleigh wave is elliptical and with decreasing amplitude downwards—the motion is retrograde. The major axis of the elliptical path is normal to the surface.

It was found by Love that horizontally polarized shear waves could also be propagated through a layered surface if the S-wave speed for the surface layer containing the wave is lower than that for the sub-stratum. The speed of these Love waves, V_L, must then satisfy $V_S(\text{layer}) < V_L < V_S(\text{sub})$. The precise speed of the Love waves depends upon the thickness of the layer and the wavelength of the wave—Love waves are thus dispersive. It is found that a range of waves of different wavelength can be associated with the same wave speed so the wave can be propagated in different modes. These waves can be regarded as resulting from the constructive interference between plane waves reflected successively from the upper and lower surfaces of the layer. Rayleigh and Love wave modes are shown in figure R.1c and d.

Other wave modes are possible such as waves 'guided' along an interface between two layers, the so-called Stoneley waves. These various surface wave modes are of great importance to exploration geologists but play essentially no part in planetary seismology.

R.1.3. *Full-body waves*

The waves discussed so far have local features but do not involve an oscillation of the body as a whole. In fact, a complete study of the equations of elasticity shows solutions that represent coordinated oscillations involving the entire sphere. For such oscillations the Earth rings like a bell or like a glass tumbler. The theoretical existence of such oscillations had been known since the work of Lamb at the beginning of the twentieth century but it was only in May 1960 (as a result of a major earthquake in Chile) that these collective oscillations were observed for the first time.

There are two general wave modes, longitudinal and transverse. Vibrations of the first class are simple torsional oscillations without radial motion and without dilatation. For a uniform sphere with a radius the same as for the Earth, the longest period oscillation has a period of about 19 minutes. Vibrations of the second kind are mixed oscillations involving both radial displacements and torsional oscillations. The longest period oscillation for the Earth in this case has a period of about 16 minutes. These are shown schematically in figure R.3.

The interpretation of full body oscillations is highly complicated and not appropriate for us to explore now. However, it is interesting to note that the analysis of collective oscillations is now used

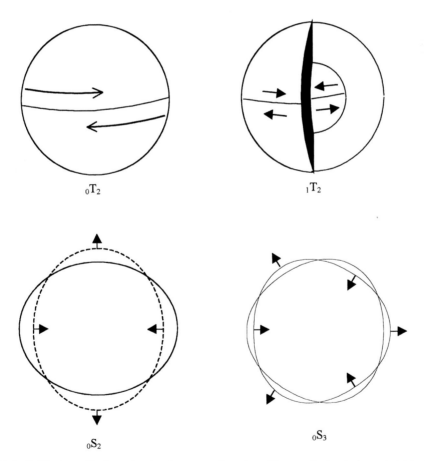

Figure R.3. *Full-body oscillations showing four normal modes:* $_0T_2$ *opposite hemispheres twist in opposite directions;* $_1T_2$ *inner and outer spheres twist as in* $_0T_1$ *but* π *out of phase;* $_0S_2$ *and* $_0S_3$ *breathing modes with 2 or 3 undulations around the circumference.*

to explore the interior of the Sun and is even beginning to be applied to the study of the interiors of distant stars.

R.2. ATTENUATION OF SEISMIC WAVES

Seismic waves would pass through material without loss of energy in a perfectly elastic medium but planetary material is never perfect. The amount of energy lost as the wave passes a particular point is expressed in terms of the *quality factor* Q for the medium. Q is defined as the ratio

$$Q = -2\pi \frac{\text{elastic energy stored in the wave}}{\text{energy lost in one cycle}}.$$

For a perfectly elastic material $Q = \infty$ because there is no energy lost in a cycle. For a perfectly inelastic system (totally dissipative) $Q = 0$ because no energy is stored in the wave. A highly attenuating region of a body is often said to be a region of low Q. If E is the energy in the seismic wave, τ its period and t the time, the definition of Q gives

$$Q = -\frac{2\pi E}{\tau(\mathrm{d}E/\mathrm{d}t)} \tag{R.5}$$

where $\mathrm{d}E/\mathrm{d}t$ is the rate of loss of energy averaged over one cycle. This gives

$$\frac{\mathrm{d}E}{\mathrm{d}t} = -\frac{2\pi E}{Q\tau}$$

or, by integration,

$$E = E(0)\exp\left(-\frac{2\pi t}{Q\tau}\right) \tag{R.6}$$

where $E(0)$ is the energy at time $t = 0$. If A is the amplitude of the seismic wave we know that $A \propto \sqrt{E}$ so that for the decrease of the amplitude with time

$$A = A(0)\exp\left(-\frac{\pi t}{Q\tau}\right).$$

The angular frequency of the wave is $\omega = 2\pi/\tau$ and the wavelength $\lambda = V/\tau$. This gives alternatively

$$A = A(0)\exp\left(-\frac{\omega t}{2Q}\right).$$

The distance, x, travelled by the wave in time t is $x = Vt$. But $V\tau = \lambda$ so that

$$\frac{\omega t}{2Q} = \left(\frac{2\pi}{\tau}\right)\left(\frac{x}{V}\right)\left(\frac{1}{2}Q\right) = \frac{\pi x}{Q\lambda} \quad \text{and} \quad A = A(0)\exp\left(-\frac{\pi x}{Q\lambda}\right).$$

These formulae allow an estimation to be made of Q. Estimates of Q for various regions of the Earth are in the range 100–1000. It is found that $Q(\text{P-wave}) \approx 2Q(\text{S-wave})$.

R.3. SEISMOMETERS AND SEISMOGRAPHS

Seismic waves result from the explosive local release of energy in an earthquake. The resultant waves (both longitudinal and transverse) spread throughout the body of the planet and through its surface and can be observed at points on the surface as a complicated periodic motion. The instrument designed to detect these motions is the seismometer which, when linked to a recording system, becomes a seismograph. In its most basic form a seismometer consists of a frame, rigidly mounted so that it will follow the local motion of the Earth, from which is suspended a very massive object. When the Earth, and hence the frame, moves the inertia of the massive object prevents it from following the motion. The relative motion of the frame and the massive object is then a measure of the Earth's motion and allows the true motion of the surface to be deduced. The instrument is designed to provide the maximum accuracy of observation and this depends on the waves being observed. For P- and S-waves and surface waves the required frequency is of the order of seconds. For the full-body waves it is of the order of hours. The design of seismometers has developed enormously over the hundred years that they have been in use and they are now highly sensitive and sophisticated instruments.

R.3.1.   Travel times and seismic speeds

The observations at seismic stations consist of the arrival of the different P- and S-waves comprising the total wave package. The P-waves arrive first, followed by the slower S-waves. To find the travel times and trajectories, the time of arrival of the waves must be correlated with the time of the earthquake and with its location underground (called the *focus*—the point on the surface directly above the focus is called the *epicentre*). It is not usually easy to locate the source of the earthquake and its time of occurrence without a prior knowledge of the travel times but if the location is known then mean speeds of the waves can be deduced. Underground nuclear explosions are preferable to natural earthquakes on this count because the source can be supposed to be at the surface and the time of the explosion is known precisely. Again, the source is very much a point source, whereas natural earthquake sources spread over a volume which need not be small. However, the strengths of nuclear sources are orders of magnitude lower than those of stronger earthquake disturbances.

Assuming a general model for the interior, and especially isotropic conditions, the behaviour of wave trajectories inside the body can be found and this knowledge converted to most likely wave behaviour if the travel times for a large number of surface stations are available. The method began in a small way during the early years of the twentieth century. Wierchert and Zoppritz (1907) devised elementary travel time curves for earthquakes near the surface and near the point of observation. Later, the method was developed by Jeffereys, starting about 1931, who both improved the accuracy of the method and extended it to include distant and deep earthquakes. Together with his student Bullen he provided definitive travel times for earthquakes on a worldwide basis. With the advent of computers and an increasing number of observatory sites around the world the method has achieved now a high degree of sophistication. Wave speeds throughout the body have been inferred so the bulk modulus and shear modulus of the material are known. Comparison with laboratory measurements, under the same conditions of pressure and temperature, give some indication of the physical form of the material in various locations. It should be noted that the S-wave has a zero speed if the material is fluid (so that $\mu = 0$) but the P-wave remains with a finite speed. The method can, therefore, be used to detect solid and liquid interfaces within the planetary body. Although developed for the Earth, the method is available for application to all the solid bodies of the Solar System when suitable equipment can be installed.

R.3.2.   Reflection and refraction across a boundary

The reflection and refraction of seismic waves across a surface of material discontinuity follows closely that of the optical case across a discontinuous surface between media of different refractive indices. The polarization of the waves is the essential characteristic. The pure P-wave is polarized in the direction of motion while the pure S-wave is polarized at right angles to it. For either wave approaching a plane

Table R.1. *The name system for wave paths from focus to seismometer.*

Wave name	Wave description
P	a P-wave in the mantle
S	an S-wave in the mantle
K	a P-wave through the outer core
J	an S-wave through the inner core
I	a P-wave through the inner core
c	a reflection from the mantle—outer core boundary
i	a reflection from the outer core—inner core boundary
s	an S-wave reflected from the Earth's surface close to an earthquake focus
p	a P-wave reflected from the Earth's surface close to an earthquake focus
PKP	a P-wave through the mantle and the outer core and on through the mantle
PKIKP	a P-wave though the mantle, outer core, inner core, outer core and mantle
SP	a wave which travelled through the mantle as an S-wave, was reflected at the surface and travelled on through the mantle as a P-wave
LR	Rayleigh wave
LQ	Love wave

interface at an angle to the normal the single polarization of the initial wave is divided into a wave polarized in the direction of motion *and* one polarized at right angles to it. This means that in general a pure P-wave is converted into a mixture of P-waves and S-waves as is an initially pure S-wave. This circumstance considerably complicates the interpretation of received seismic waves following an earthquake.

The only application of detailed seismic analysis so far has been to the Earth and a nomenclature has been developed to describe seismic waves (table R.1).

The general inner structure of the Earth is described in section 4.3.6, where it is seen that there is a central core (with a separate inner core and outer core) encased by a mantle and the whole enclosed in the crust. We might notice that the outer core is liquid and so no S-waves can traverse it. Both P- and S-waves are found in the inner core (solid) and this is only possible by the details of the refraction process outlined above whereby P-waves, traversing the outer core, are converted to both P- and S-waves in the inner core. Some types of paths within the Earth are shown in figure R.4.

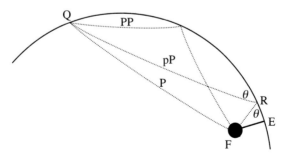

Figure R.4. *Waves within the Earth due to an earthquake with focus at F corresponding to epicentre E. The first wave detected by the seismometer at Q is the direct one P. The first reflected wave, via R near the epicentre, is coded pP (or sS). Later the reflected wave PP (or SS) arrives. The difference in the arrival time of P and pP indicates the focal depth. (The diagram is not to scale.)*

R.4. SEISMIC TOMOGRAPHY

With the accumulation of world wide data and the availability of powerful electronic computers it is natural to attempt more refined analysis of seismic information. One such analysis has been that of isolating local variations of wave speed to provide a detailed map of wave behaviour. This is called seismic tomography and is roughly an equivalent of the medical full body scan. The results have been used as an indicator of particular conditions locally and especially conditions of temperature. The wave speeds decrease with increasing temperature so a region where the wave speed is lower than the surroundings has been interpreted as being hotter than the surroundings and vice versa. Clearly, the interpretation is not unique because changes in the wave speed could also follow from compositional changes in the material carrying the wave. In practice, the variation of wave speeds locally can be represented only against a standard model so the method depends on the reliability of a zero-order model interior.

R.5. LONG-TERM HYDROSTATIC EQUILIBRIUM OF PLANETARY MATERIAL

An important assumption in seismology is that the material of the planet is in long-term hydrostatic equilibrium. According to the rules of hydrostatic equilibrium, the local pressure $P(r)$ at a distance r from the centre of the planet is related to the local material density $\rho(r)$ and the local acceleration of gravity $g(r)$ through the gradient of the pressure

$$\frac{dP(r)}{dr} = -\rho(r)g(r) = -\frac{G\rho(r)m(r)}{r^2} \tag{R.7}$$

where $m(r)$ is the mass contained within the radius r. The negative sign shows that the pressure increases with depth (that is with decreasing distance from the centre). This expression can be used to determine the variation of pressure with depth only if $m(r)$ is known throughout the body— which amounts to knowing the density distribution. An expression to provide the variation of density with depth can be deduced by noting that

$$\frac{dP}{dr} = \left(\frac{d\rho}{dr}\right)\left(\frac{dP}{d\rho}\right). \tag{R.8}$$

Combining the formulae (R.1), (R.7) and (R.8) leads to an expression for the dependence of the density on the distance from the centre

$$\frac{d\rho(r)}{dr} = -\frac{Gm(r)\rho^2(r)}{\kappa(r)r^2}. \tag{R.9}$$

In terms of the density, the mass of a thin spherical shell of radius r and thickness dr is given by

$$dm(r) = 4\pi\rho(r)r^2\,dr \quad \text{or} \quad \frac{dm(r)}{dr} = 4\pi\rho(r)r^2. \tag{R.10}$$

The pair of coupled differential equations (R.9) and (R.10) enable the variation of density with depth to be found, but only if $\kappa(r)$ is known. While the properties of the material are not known *a priori* for planetary bodies, seismic measurements do give the required information.

R.6. THE ADAMS–WILLIAMSON METHOD USING EARTHQUAKE DATA

Starting with surface values $r = R$, and known values of $m(R)$ and $\rho(R)$ as boundary conditions, the solution in terms of values of $m(r)$ and $\rho(r)$ can be followed towards the Earth's centre. From equations

(R.4) and (R.9)

$$\frac{d\rho(r)}{dr} = -\frac{Gm(r)\rho(r)}{[V_P^2(r) - \frac{4}{3}V_S^2(r)]r^2}.$$

(R.11)

This expression is central to the classical application of seismology to Earth studies using earthquakes. It is presumed that $V_P(r)$ and $V_S(r)$ are known throughout the body. Starting near the surface, $r = R$ with the density ρR and mass $m(R)$, equations (R.10) and (R.11) are integrated inwards as a pair of coupled equations. Accumulated errors in measurements will generally not provide the necessary condition $m(0) = 0$ and best fit solutions will be required to achieve this.

It is possible to eliminate specific reference to $\rho(r)$ and to produce a non-linear second-order differential equation involving only $m(r)$. Differentiating equation (R.10) with respect to r gives

$$\frac{d^2m(r)}{dr^2} = 8\pi\rho(r)r + 4\pi r^2 \frac{d\rho(r)}{dr}.$$

Substituting for $d\rho(r)/dr$ from equation (R.11) and $\rho(r)$ from equation (R.10) gives

$$\frac{d^2m(r)}{dr^2} - \frac{2}{r}\frac{dm(r)}{dr} - \frac{\beta}{r^2}m(r)\frac{dm(r)}{dr} = 0$$

(R.12)

with $\beta = G/(V_P^2 - \frac{4}{3}V_S^2)$.

The starting boundary conditions are $m(R) = M$, the planet's mass, and $[dm(r)/dr]_{r=R}$, determined from the surface density. If when integrated inwards the equation does not give $m(0) = 0$ then the internal properties of the planet must be modified as little as possible, within the uncertainties indicated by the seismic measurements, until the essential central boundary condition is obtained.

R.7. MOMENT OF INERTIA CONSIDERATIONS

A further criterion to restrict the calculation of the density distribution within a planet is the moment of inertia (Topic S). For a spherically-symmetric mass M with radius R the moment of inertia I for rotation about a diameter is written

$$I = \alpha MR^2$$

(R.13)

where α is called the *moment-of-inertia factor*. For a uniform sphere $\alpha = 2/5$ and for a centrally-condensed body α has a lower value than this. At the limit, for a point mass, α has the value zero. A given value of I for a particular body restricts severely the possible density distributions compatible with equation (R.12) since it is now necessary not only to achieve $m(0) = 0$ but also the correct value of I.

All the above has been based on the idea of a spherically-symmetric body but, rather curiously, the ability to be able to determine the moment of inertia of the Earth, and other bodies, depend on the fact that they are only *approximately* spherical. Because of the spin of planets they take on the form of oblate spheroids (section 4.3.1) with a polar radius, R_p, and a larger equatorial radius, R_e. The polar axis is one of cylindrical symmetry and the moment of inertia about the polar axis may be written

$$C = \alpha_p MR_e^2.$$

(R.14)

There is no cylindrical symmetry for a diameter in the equatorial plane but we may write the moment of inertia for spin about such an axis as A, although no such physical spin actually occurs. For a body like the Earth, where the flattening is small, A and C will be similar in magnitude.

A consequence of the distortion of a planetary body from spherical form is that the gravitational field around it also lacks spherical symmetry. In Topic T we discuss the gravitational field due to a distorted planet and show that the gravitational potential can be represented by a series in which the leading term is that for a point mass and subsequent terms are successively weaker in their influence. The second term involves a numerical quantity J_2 that can be estimated by observing the motion of a satellite in the vicinity of the distorted body. The value of J_2 is well determined for the Earth and is 1.0826×10^{-3}. Its importance here is that there is a theoretical relationship between J_2 and the moments of inertia

$$C - A = J_2 M R_p^2. \tag{R.15}$$

Since everything on the right-hand side of equation (R.15) is known the difference of the two principal moments of inertia are also known.

In Topic U the precession of the Earth's spin axis is discussed. This is due to the gravitational effect of the Moon acting on the equatorial bulge of the Earth due to its spin. Measurement of the rate of precession, plus other known parameters of the Earth and the Moon, enables an estimate to be made of the ratio C/A. The two relationships between C and A enable them both to be determined. For bodies other than the Earth where precession of the spin axis cannot be measured, the flattening of the planet is another, but less accurate, way of determining C/A. This uses the assumption that the planet behaves like a perfectly fluid body and that its external shape is controlled by gravity and spin without the constraint of any rigidity.

The determined value of C, or something intermediate between C and A, may be used in the modelling as a constraint, assuming that the Earth is spherical. The uncertainties in the modelling assumptions are much more serious than any involved in the value found for the moment of inertia.

Problem R

R.1 (i) Over a small region at the centre of a planet the density is approximately constant. Why is this?

(ii) The density at the centre of the Earth is $1.29 \times 10^4 \, \mathrm{kg \, m^{-3}}$ and varies very little throughout the inner core. Seismic measurement indicate $V_P = 11.2 \, \mathrm{km \, s^{-1}}$ and $V_S = 3.6 \, \mathrm{km \, s^{-1}}$ throughout the inner core. What is the approximate gradient of density at $1000 \, \mathrm{km}$ from the centre. Assuming that this gradient is the average over the whole inner core, that has a radius of $1300 \, \mathrm{km}$, then what is the difference of density between the centre of the Earth and the inner-core boundary?

TOPIC S

MOMENTS OF INERTIA

Many bodies in the Solar System are spherical, or approximately so, and spin about a diametrical axis. It is sometimes possible to estimate the moment of inertia of a body and this can give important information about its internal structure. Here we shall be concerned with bodies that are spheres with a spherically symmetric density distribution or with spheroids that are not too far from spherical form.

S.1. THE MOMENT OF INERTIA OF A UNIFORM SPHERE ABOUT A DIAMETER

The basic equation that defines the moment of inertia is

$$I = \sum_{i=1}^{N} m_i r_i^2 \tag{S.1}$$

where the system consists of N bodies, the ith of which has mass m_i and is at distance r_i from the rotation axis. Where the system is a continuous distribution of matter then a corresponding integral form for I must be used.

We first consider the moment of inertia of a uniform disk, illustrated in figure S.1, about an axis through its centre and normal to its plane. The radius of the disk is a and its areal density is σ. The

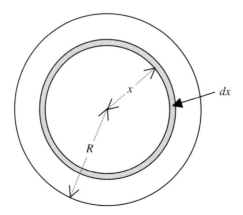

Figure S.1. *A ring in a disk.*

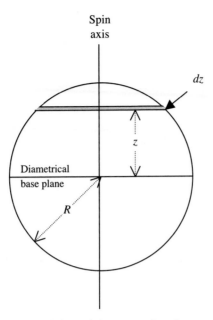

Figure S.2. *A disk section of a sphere.*

shaded area has mass $2\pi x \sigma\,\mathrm{d}x$ and is all at the same distance x from the rotation axis. Its contribution to the moment of inertia is thus

$$\mathrm{d}I = 2\pi x^3 \sigma\,\mathrm{d}x$$

and the total moment of inertia is

$$I = 2\pi\sigma \int_0^R x^3\,\mathrm{d}x = \tfrac{1}{2}\pi\sigma R^4 = \tfrac{1}{2}MR^2 \tag{S.2}$$

where M is the total mass of the disk.

We can now move to a uniform sphere of density ρ by considering it as a combination of disks, as illustrated in figure S.2. The figure shows an edge view of a disk of thickness $\mathrm{d}z$ at distance z from a diametrical base-plane perpendicular to the rotation axis. From equation (S.2) the contribution of the disk to the moment of inertia of the sphere is

$$\mathrm{d}I = \tfrac{1}{2}\pi(R^2 - z^2)^2 \rho\,\mathrm{d}z.$$

The moment of inertia of the sphere is thus

$$I_{\mathrm{sp}} = \tfrac{1}{2}\pi\rho \int_{-R}^{R} (R^2 - z^2)^2\,\mathrm{d}z = \tfrac{8}{15}\pi\rho R^5 = \tfrac{2}{5}M_{\mathrm{s}}R^2. \tag{S.3}$$

From equation (S.3) it is straightforward to find the moment of inertia of a uniform shell. From equation (S.3)

$$\frac{\mathrm{d}I}{\mathrm{d}R} = \frac{8}{3}\pi\rho R^4. \tag{S.4}$$

The moment of inertia of a thin shell of radius R and of small thickness ΔR is the difference between the moments of inertia of uniform spheres of radius $R + \Delta R$ and R. From equation (S.4) this is

$$I_{sh} = \tfrac{8}{3}\pi\rho R^4 \, \Delta R = \tfrac{2}{3} M_{sh} R^2 \tag{S.5}$$

where M_{sh} is the mass of the shell.

S.2. THE MOMENT OF INERTIA OF A SPHERICALLY SYMMETRIC DISTRIBUTION

We now consider a sphere where the density is of the form $\rho(r)$ where r is the distance from the centre. The sphere may be decomposed into a set of thin uniform shells for each of which the moment of inertia is given by equation (S.5). If the radius of the sphere is R then its moment of inertia is

$$I = \tfrac{8}{3}\pi \int_0^R \rho(r) r^4 \, dr. \tag{S.6}$$

A variation of density which is linear with distance from the centre

$$\rho(r) = \rho_0 \left(1 - \frac{r}{R}\right)$$

gives

$$I = \tfrac{4}{45}\pi\rho_0 R^5.$$

A similar decomposition into shells gives the mass of the sphere as

$$M = 4\pi\rho_0 \int_0^R \left(1 - \frac{r}{R}\right) r^2 \, dr = \tfrac{1}{3}\pi\rho_0 R^3.$$

Inserting this into the expression for moment of inertia gives

$$I = \tfrac{4}{15} M R^2. \tag{S.7}$$

It can be seen that equation (S.7) is of the same general form as equation (S.3), i.e.

$$I = \alpha M R^2 \tag{S.8}$$

and α is referred to as the *moment-of-inertia factor*. It is smaller for the centrally condensed non-uniform body because the average distance of the constituent mass from the rotation axis is smaller. In fact, from dimensional considerations, any moment of inertia can be put in the form of equation (S.8) if R is taken as some characteristic dimension of a body of defined form. For example, for a straight uniform rod with length, l, much greater than any cross-sectional dimension the moment of inertia is $Ml^2/12$.

S.3. THE MOMENT OF INERTIA OF A SPHEROID ABOUT THE SYMMETRY AXIS

The way that planets are distorted by their spin was described in section 3.4. They flatten along the spin axis and go into the form of an oblate spheroid with equatorial radius greater than the polar radius. A non-spinning planet would be very closely spherically symmetric but the flattened planet would not retain spherical symmetry. However, it *would* retain axial symmetry about the rotation axis and the moment of inertia can still be written in the form $\alpha M R^2$. If the oblate spheroid were uniform then α would be 0.4, just as for the uniform sphere, as long as R was taken as the equatorial radius. This can be seen by imagining that every particle of a uniform sphere of radius a changes its distance from the diametrical base-plane by a factor c/a. The sphere has now been transformed into a uniform oblate spheroid but, since all particles are at the same distance from the rotation axis, the moment of inertia is unchanged. The same would be true of a non-uniform but spherically symmetric planet if the distortion could be regarded as a simple uniform compression along the rotation axis. Actually the effect of spin is to increase the equatorial radius as well as decrease the polar radius. If the volume remains constant then the reduction in polar radius is twice the increase in equatorial radius, assuming that all changes of radius are small.

One way of changing a sphere of radius R to an oblate spheroid is to imagine that each particle changes its distance from the base plane by a factor c/R and moves out from the rotation axis so that its distance changes by a factor a/R. If the interior does change in this way and the moment of inertia of the undistorted sphere was $\alpha M R^2$ then the moment of inertia of the oblate spheroid would be

$$C = \alpha M a^2 \tag{S.9}$$

with no change of α.

It can be shown that the moment of inertia of a uniform spheroid about an axis contained in the diametrical base plane is

$$I_a = \tfrac{1}{5} M (a^2 + c^2) \tag{S.10}$$

that becomes (S.3) with $a = c = R$. In a similar way, with a change of shape of a spherically symmetric sphere to a spheroid as described above, the moment of inertia is given by

$$A = \tfrac{1}{2}\alpha M (a^2 + c^2) \tag{S.11}$$

with the same moment-of-inertia factor as for the undistorted sphere.

From a measurement of the gravitational coefficient J_2 it is possible to estimate the difference $D = C - A$ (Topic T). From equations (S.9) and (S.11)

$$D = \alpha M a^2 - \tfrac{1}{2}\alpha M (a^2 + c^2) = \tfrac{1}{2}\alpha M (a^2 - c^2) \tag{S.12}$$

and since a and c can be found from observation it is possible to estimate α. Knowledge of α can then be used to test various models for the internal structure of the body. Actually a measurement of the precession rate of the spin axis of the body gives the value of C/A and a better estimate of C and A individually, but precession rates are generally difficult to measure.

It is interesting to note that without spin of the body there would be no distortion and no way of estimating α and hence learning about the interior of the body. Estimates of the moment-of-inertia factors of planets, other than Neptune and Pluto, and of the Moon are given in table S.1.

Table S.1. *The moment-of-inertia factors of some planets and the Moon.*

Body	α
Mercury	0.33
Venus	0.33
Earth	0.331
Moon	0.392
Mars	0.366
Jupiter	0.254
Saturn	0.210
Uranus	0.225

Problem S

S.1 One way of modelling the density distribution in the Sun is as

$$\rho = \rho_0 \left(1 - \frac{r}{R} \right)^n$$

where r is the distance from the centre and R is the Sun's radius. Show that the value $n = 7.57$ gives the required moment-of-inertia factor, $\alpha = 0.055$.

[Note: you are expected to derive an expression for α in terms of n and then show that $n = 7.57$ gives the required result.]

TOPIC T

THE GRAVITATIONAL FIELD OF A DISTORTED PLANET

To a first approximation a planet is a spherical body with a spherically-symmetric distribution of density within it. For such an ideal body the gravitational potential at an external point would be the same as that if all its mass was concentrated at its centre and there would be no information available about the actual internal distribution of mass. Real planets are different. On a small scale we can see departures from spherical symmetry—continents and oceans, mountain ranges and polar ice. These are comparatively minor departures from spherical symmetry. Measurements of the shape of solid bodies often reveal significant departures from spherical symmetry; thus the Moon is very slightly pear-shaped with the pointed end of the pear facing the Earth. Another feature shown by gravity measurements is that the centre of mass of the Moon is displaced by about 2.5 km from its *centre of figure*. The centre of figure is where the centre of mass would be situated if the Moon had the same shape but was of uniform density. There is a COM-COF offset of about the same magnitude for Mars. An important departure from spherical symmetry comes from the spin of a planet. This distorts it into the shape of an oblate spheroid (section 3.4), or at least approximately so.

Gravitational surveys in the neighbourhood of planets show measurable departures from spherical symmetry and high precision measurements show great complexity in the pattern of gravitational potential. However, for planets with appreciable spin, the signature of the distortion due to spin is a dominant one and that is what we shall be concerned with here.

T.1. THE GRAVITATIONAL POTENTIAL OF A SPINNING PLANET

For a general distribution of mass, defined in terms of spherical polar coordinates, the gravitational potential will need to be defined in terms of all three of the coordinates (r, θ, ϕ). However, for a mass distribution with axial symmetry the potential can be represented in terms of only two variables, r and θ, and this is the case for an ideal spinning planet when θ is measured from the spin axis. The potential at external points can be expressed as

$$\Phi(r, \theta) = -\frac{GM}{r} \left[1 - \sum_{n=1}^{\infty} \left(\frac{r_e}{r} \right)^{2n} J_{2n} P_{2n}(\cos \theta) \right] \tag{T.1}$$

where M is the mass of the planet, r_e its equatorial radius and $P_{2n}(\cos \theta)$ is the *Legendre polynomial* of order $2n$. In section V.4 there is further discussion on series of this kind. The coefficients J_{2n} are the *gravitational moments* and the determination of these is a major objective when the gravitational

Table T.1. J_2 *coefficients for the planets.*

Planet	J_2
Mercury	6.0×10^{-5}
Venus	4.5×10^{-6}
Earth	1.082×10^{-2}
Mars	1.960×10^{-3}
Jupiter	1.474×10^{-2}
Saturn	1.630×10^{-2}
Uranus	3.343×10^{-3}
Neptune	3.411×10^{-3}

potential is mapped around a planetary body. The Legendre polynomials of lowest order in the summation are:

$$P_2(\cos\theta) = \tfrac{1}{2}(3\cos^2\theta - 1) \quad \text{and} \quad P_4(\cos\theta) = \tfrac{1}{8}(35\cos^4\theta - 30\cos^2\theta + 3).$$

The gravitational moments are approximately proportional to ω^2, where ω is the spin angular speed of the planet, and they also depend on the material of the planet as does the flattening shown in table 3.3. It is obvious that solid planets, which most resist distortion through their mechanical strength, would show the least departure from spherical symmetry and hence have smaller values for the gravitational moments. Conversely, the gas giants would distort comparatively easily and so have larger J values. In fact the values of J_2 are much larger than the coefficients of higher order and only for Jupiter and Saturn are there estimated values of J_4. The values of J_2 are given in table T.1.

It can be shown theoretically that the value of J_2 is given by

$$J_2 = \frac{C - A}{Mr_p^2} \tag{T.2}$$

where C is the moment of inertia about the spin axis, A is the moment of inertia about a diametrical axis in the equatorial plane and r_p is the polar radius. We saw in equation (S.12) that a knowledge of the flattening, essentially the values of the polar and equatorial radii, enable $C - A$ to be found in terms of the moment of inertia factor as the only unknown. Actually it is better to find some observationally-based way of determining C and A individually. For the Earth this can be done by measuring the precession of the spin axis (Topic U) that gives the ratio C/A that can then be combined with $C - A$ from J_2. The moment of inertia factor, once found, is an important constraint on models of planetary interiors.

Problems T

T.1   Given the value of J_2 in table T.1, the mass and equatorial radius of the Earth in table 3.2 and the flattening of the Earth in table 3.3, estimate the Earth's moment-of-inertia factor. Compare this with the value in table S.1.

T.2   A satellite is in circular orbit 100 km above the equator. By what fraction is the gravitational field modified by the J_2 term in equation (T.1)?

TOPIC U

PRECESSION OF THE EARTH'S SPIN AXIS

U.1. THE BASIC MECHANISM

It was noted in section 4.3.1 that, because of its spin, the Earth takes up the form of an oblate spheroid with polar radius less than the equatorial radius. A schematic cross-section is illustrated in figure U.1; the distorted Earth may be regarded as the sum of a spherical portion with radius, c, equal to the polar radius and an additional part, shown shaded. The shaded part has a maximum thickness $a - c$ at the equator, where a is the Earth's equatorial radius.

Also shown in figure U.1 is a distant body B, of mass M, in a direction making an angle θ with the equatorial plane. To a first approximation the spherical region in figure U.1 may be regarded as spherically symmetric so, due to this portion, B exerts a force on the Earth acting at its centre.

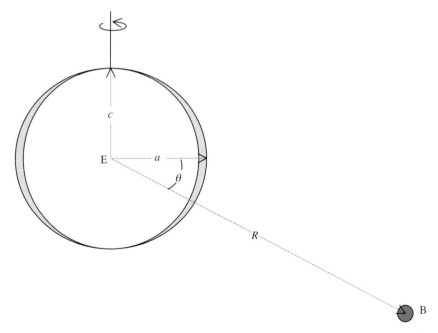

Figure U.1. *The Earth distorted by spin represented as a sphere plus the shaded material.*

However, the shaded bulge region does not have spherical symmetry, only axial symmetry, and for this reason the force due to B on this part of the Earth is a combination of a force acting at its centre plus a torque. Since the Earth is a spinning object, if it experiences a torque with a component perpendicular to the spin axis then the spin axis will precess. This is commonly seen when a toy gyroscope or top spins with its axis inclined to the vertical. For a top the torque is due to the Earth's gravitational field acting at the centre of mass and the reaction at the point of contact with the ground.

In the case of the Earth the body B may be either the Moon or the Sun—both contribute to the precession. The force on the Earth due to the Sun is in the plane of the ecliptic and the *time-averaged* force of the Moon on the Earth is also in the ecliptic. The Moon's orbit makes an angle of just over $5°$ with the ecliptic: sometimes it is to the north of the ecliptic and sometimes to the south. The precession of the spin axis of the top described above is about the vertical, the line for the spin axis that would make the torque equal zero. Similarly the Earth's spin axis precesses around the normal to the ecliptic, the line for the spin axis that would, on average, make the torque equal zero.

Here we shall consider a simple configuration for the Earth and a single outside body and show that, from the precession rate, one can in principle determine the ratio A/C where A and C are defined in section T.1.

U.2. THE SIMPLE CONFIGURATION

In figure U.1 we take the spin axis as the z axis and the body B in the x–z plane so that B has coordinates $(R\cos\theta, 0, R\sin\theta)$. We consider a small element of volume dV and mass $\rho\,dV$ situated at $\mathbf{r} \equiv (x, y, z)$. The force on the element due to B, expressed in vector form, is

$$\Delta\mathbf{F} = \frac{GM\rho\,dV}{D^3}(R\cos\theta - x, -y, R\sin\theta - z) \tag{U.1}$$

where

$$D = [(R\cos\theta - x)^2 + y^2 + (R\sin\theta - z)^2]^{1/2}. \tag{U.2}$$

The torque exerted by this force on the Earth is

$$\Delta\mathbf{T} = \Delta\mathbf{F} \times \mathbf{r} = \frac{GM\rho\,dV}{D^3}\begin{vmatrix} \hat{i} & \hat{j} & \hat{k} \\ R\cos\theta - x & -y & R\sin\theta - z \\ x & y & z \end{vmatrix}. \tag{U.3}$$

where $\hat{i}$, $\hat{j}$ and $\hat{k}$ are the unit vectors along x, y and z.

From considerations of symmetry the torque due to B on the whole Earth spheroid will be about the y axis so, taking that component,

$$\Delta T_y = \frac{GM\rho\,dV}{D^3}[(R\sin\theta - z)x - (R\cos\theta - x)z] = \frac{GMR\rho\,dV}{D^3}(x\sin\theta - z\cos\theta). \tag{U.4}$$

We now expand D^{-3} using the binomial theorem including terms x/R, y/R and z/R but excluding higher powers. This gives

$$D^{-3} = [(R\cos\theta - x)^2 + y^2 + (R\sin\theta - z)^2]^{-3/2} = R^{-3}\left[\left(\cos\theta - \frac{x}{R}\right)^2 + \left(\frac{y}{R}\right)^2 + \left(\sin\theta - \frac{z}{R}\right)^2\right]^{-3/2}$$

$$= R^{-3}\left(1 - 2\frac{x}{R}\cos\theta - 2\frac{z}{R}\sin\theta\right)^{-3/2} = R^{-3}\left(1 + 3\frac{x}{R}\cos\theta + 3\frac{z}{R}\sin\theta\right). \tag{U.5}$$

From equations (U.4) and (U.5) with the same level of approximation

$$\Delta T_y = \frac{GM\rho\, dV}{R^3} [Rx\sin\theta - Rz\cos\theta + \tfrac{3}{2}\sin 2\theta(x^2 - z^2) - 3xz\cos 2\theta]. \tag{U.6}$$

The symmetry of the spheroid gives points $(x, y, z), (x, y, -z), (-x, y, z)$ and $(-x, y, -z)$ all with the same density and by adding their contributions to T_y it is found that only the third term in the bracket of equation (U.6) remains. Thus to find T_y we integrate this term over the whole spheroid, thus

$$
\begin{aligned}
T_y &= \frac{3GM\sin 2\theta}{2R^3} \int_{\text{ellipsoid}} \rho(x^2 - z^2)\, dV \\
&= \frac{3GM\sin 2\theta}{2R^3} \left(\int_{\text{ellipsoid}} \rho(x^2 + y^2)\, dV - \int_{\text{ellipsoid}} \rho(y^2 + z^2)\, dV \right) \\
&= \frac{3GM\sin 2\theta}{2R^3} (C - A).
\end{aligned}
\tag{U.7}
$$

From gyroscopic theory the spin axis will precess with an angular speed equal to the torque divided by the spin angular momentum or

$$\Omega = \frac{T_y}{C\omega} = \frac{3GM\sin 2\theta}{2R^3\omega}\left(1 - \frac{A}{C}\right) \tag{U.8}$$

where ω is the spin angular speed, corresponding to a period of 1 day (or more precisely 23 h 56 m since in 24 hours, in going from one noon to the next, because of the Earth's motion round the Sun its axial spin is more than 2π).

   If there were just one body involved in causing the precession then equation (U.8) would give the instantaneous precession rate. However, the Moon is orbiting the Earth so the angle θ is constantly changing, as is R because of the Moon's quite eccentric orbit ($e = 0.056$). In addition the Earth is orbiting the Sun so the θ involved in this interaction is also changing with time, albeit more slowly, and there is an eccentricity effect here also, although smaller than that for the Moon. A smaller contribution, at about the 1% level, is perturbation due to the other planets. Consequently, to apply equation (U.8) in a quantitatively precise way to find the precession period (25 725 years) requires a detailed analysis of the positions of the Moon and the Sun relative to the Earth over extended periods and this is beyond the scope of our treatment. What has been demonstrated is the principle that from the rate of precession, if all the complicating factors can be dealt with, a value of A/C can be estimated. A measurement of J_2 then gives $C - A$ and hence both C and A can be determined. We also know that $C = \alpha Mr_e^2$ and since the mass, M, and the equatorial radius, r_e, of the Earth are known this immediately gives the moment-of-inertia factor.

   Because of the variations in the torque as the Moon and the Sun move relative to the Earth's equatorial plane (changing the θs) the Earth's spin axis wobbles slightly, a motion known as *nutation*. This has components with periods of one month and one year, associated with relative motions of the Moon and Sun. Another important component is due to precession of the Moon's orbit caused by the gravitational effect of the Sun on the Earth–Moon system. This causes the line of nodes, the line in which the Moon's orbit intersects the ecliptic, to rotate in a retrograde sense with a period of 18.6 years.

Problem U

U.1   At the time of an equinox the Sun is crossing the equator and exerts no torque on the Earth. At one such time the Moon is at a distance of 365 000 km and the Earth–Moon direction makes an angle of 63° with the Earth's polar axis. Over a 50 hour period, spanning the equinox, careful measurement shows precession of the spin axis through 0.204″. Given that for the Earth $J_2 = 1.08 \times 10^{-3}$, then estimate the moment-of-inertia factor for the Earth, ignoring influences other than the Moon. (Equations (S.9) and (T.2) will be relevant.)

TOPIC V

INTRINSIC PLANETARY MAGNETISM

The Earth has an intrinsic magnetism that has been explored for more than three centuries. Exploration of the Solar System using space probes has shown clearly that intrinsic magnetism is not confined to the Earth. It would seem, from what observations there are, that the Earth's magnetic behaviour is likely to be typical of that of other bodies, which gives special importance to our knowledge of the Earth's field. For this reason the study of the intrinsic magnetism of the Solar System can properly begin with a study of the terrestrial field. The magnetic properties are detected at the most elementary level by suspending a small piece of lodestone on a string (forming an elementary compass). It is found that the lodestone aligns itself at every location very roughly, though by no means exactly, along the geographical north–south line. This behaviour shows the existence of magnetic properties near the surface of the Earth, the region being permeated by a magnetic field. Careful observations show it to vary from place to place on the surface and to change with time over periods of decades or longer. Lower amplitude shorter period events also occur associated with the interaction between the atmosphere and the Sun. These secondary, though often important, phenomena are considered in Topic W.

V.1. MAGNETIC POLES

On Earth the geographical north and south poles define the spin axis. The direction of the geographical axis is defined from the spin by the 'corkscrew rule'. This means that the positive direction is that in which a corkscrew would travel if turned in the sense of the spin. Looking from above the northern hemisphere the spin is anti-clockwise so the positive direction of the geographical axis is from the south pole to the north pole.

The term *pole* is also used in relation to magnets and the terms *north* and *south* are also used to describe the two different kinds of pole. A simple magnet may be regarded as a separated north pole and south pole, the resulting magnetic field being referred to as a *dipole field*. As in the case of electrostatics opposites attract, so the north pole of one magnet will be attracted by the south pole of another.

The Earth acts like a magnet, the magnetic axis of which varies with time and at present makes an angle of about 11.5° with the geographical axis. In the way that magnetic poles are defined it is the *south* pole of the terrestrial magnet that is in the northern hemisphere and vice-versa. Hence the direction of the magnetic axis (defined as from south magnetic pole to north magnetic pole) is anti-parallel to the geographical axis (table V.1). This means that it is the north pole of a magnetized compass needle that points to the north.

As well as being tilted away from the geographical axis the magnetic axis is also slightly displaced from the centre of the Earth. Such displacements, on a much larger scale, are common for planets (see figure 5.14).

A magnetized needle generally does not hang horizontally even when suspended from its centre of mass. In the northern hemisphere it hangs with its north pole pointing downwards below the horizon. In the southern hemisphere it is the south pole that hangs downwards. There are, however, two points and one line where the magnetic field does something different. The two points are places where the needle hangs vertically. These are the intrinsic magnetic poles. The line, in contrast, connects points where the needle hangs horizontally even though it is magnetized. This defines the magnetic equator and circumnavigates the equatorial regions of the Earth, separating the northern magnetic hemisphere from the southern one. The magnetic equator does not coincide with the geographical equator.

Recent accurate studies of the north magnetic pole, located at the present time in the islands off the northern coast of Canada, show it to be a complicated region. The pole is not a point but rather extends over a region of several kilometres. It is found to move northwards, on average, by 40 m each day. It also executes a daily elliptical path. It is presumed that the south magnetic pole will show similar behaviour. This implies that the magnetic axis is not a fixed line within the Earth.

V.2. MAGNETIC ELEMENTS: ISOMAGNETIC CHARTS

The quantitative study of the strength and orientation of the magnet field at a particular location is made by introducing *magnetic elements*. Three axes are defined, X pointing to the geographical north, Y to the east and Z pointing vertically downwards as shown in figure V.1. The horizontal plane is defined by the X and Y axes. The suspended magnet will generally make an angle with the horizontal called the magnetic dip angle, I, or sometimes the magnetic inclination. The horizontal component of the magnetic field is denoted by H which makes the angle D with the

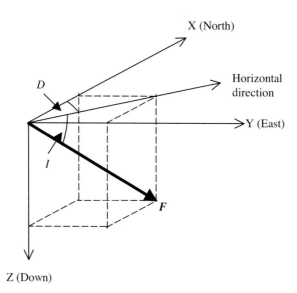

Figure V.1. *The magnetic elements set against the geographical axes, north and east, and vertically downwards.*

north direction, called the angle of magnetic declination. The total magnetic field strength is denoted by **F** with magnitude F. The five quantities F, H, D, I, Z $(=F_Z)$ form the set of magnetic elements.

The magnetic elements are related in several ways. Thus

$$H = F \cos I \qquad Z = F \sin I = H \tan I \qquad F_X = H \cos D \qquad F_Y = H \sin D$$

$$F_X^2 + F_Y^2 = H^2 \qquad F_X^2 + F_Y^2 + F_Z^2 = H^2 + F_Z^2 = F^2 \qquad (V.1)$$

It is convenient to display the magnetic field as a global plot with contours joining points of equal value (direction or magnitude). These plots are called isomagnetic charts, a name and procedure first introduced by Edmund Halley. The charts of some elements have an alternative name. Plots of the magnitude of the total magnetic field strength, F, or of the horizontal component, H, are called isodynamic, those for D are called isogonic and for I isoclinic.

The ages of western exploration, which started in the fifteenth century, allowed global magnetic measurements to be made for the first time, including those at sea. By the seventeenth century quite reliable world magnetic maps were available, among the earliest being those constructed by Edmund Halley showing the magnetic conditions in the North and South Atlantic oceans. Magnetic measurements of increasing accuracy were also being made at certain magnetic observatories on land (especially at Kew, then outside London, and at Potsdam, near Berlin) so that accurate data for Europe are available from the sixteenth century onwards. Detailed data for the southern hemisphere are relatively recent. Developing technology has allowed the magnetic field at any location in the northern or southern hemisphere to be measured with ever-increasing precision. Space vehicles in Earth orbit have extended the measurements out into space. The result is that now very fine details of the field are measured regularly.

The strength of the field at the equator is similar to that of a small magnet used at home for picking up pins—something near 0.5 gauss (the traditional unit for the geomagnetic field is the gauss where 1 gauss $= 10^{-4}$ T). Variations of the field must sometimes be measured in gamma (γ) units ($1\gamma = 10^{-5}$ gauss $= 10^{-9}$ T). Since the determination of the field of the Earth and its variations, and of the fields associated with other planets, involve the very accurate measurement of small magnetic fields magnetometers must be shielded to eliminate stray magnetic fields.

V.3. THE FORM OF THE FIELD

In his 1601 treatise on magnetism, William Gilbert describes the magnetic field of the Earth as having the same form as that of a uniformly magnetized sphere. This is equivalent to a simple bar magnet which is the source of a dipole field, shown in figure 3.6. The directions of the field lines from north (positive) to south (negative) is conventional. This is still a useful approximation to the actual form. To see this we refer to the magnetic charts.

The isodynamic world map of the total magnetic field for the epoch 1980 is shown in figure V.2, and for the epoch 1922 in figure V.3. The isodynamic world map of the horizontal force for the epoch 1922 is shown in figure V.4. Figure V.5 is an isoclinic map for the magnetic inclination while figure V.6 is a map of the magnetic declination for the same epoch. Finally, the lines of equal declination and inclination for the northern hemisphere are shown in figure V.7 for the year 1835. The difference between the geographical and the magnetic coordinates is clearly visible as is the distortion of the magnetic elements from the form of a simple dipole.

The form of the representation should be mentioned. This is normally the Mercator projection, as seen in figure V.3 and other figures, which gives great distortion in the polar regions. Maps of

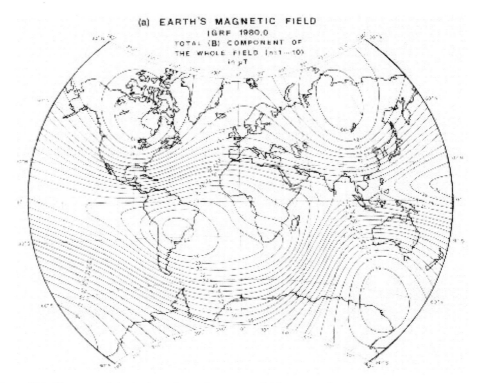

Figure V.2. *The map of the isodynamic total magnetic field intensity for the epoch 1980. The units are microteslas (after Nevanlinna, Pesonen and Blomster).*

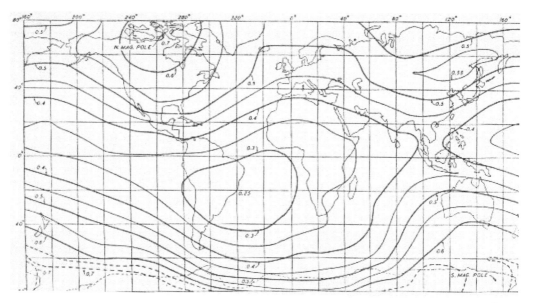

Figure V.3. *The isodynamic chart of lines of equal magnetic intensity for the epoch 1922. The units are 10^2 μT (from British Admiralty Charts).*

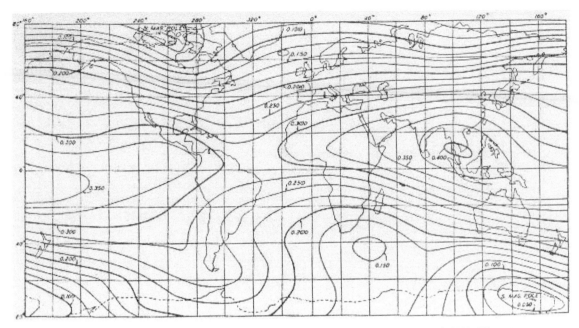

Figure V.4. *The isodynamic map of the lines of equal horizontal intensity for the epoch 1922. The units are* $10^2 \, \mu T$ *(from British Admiralty Charts).*

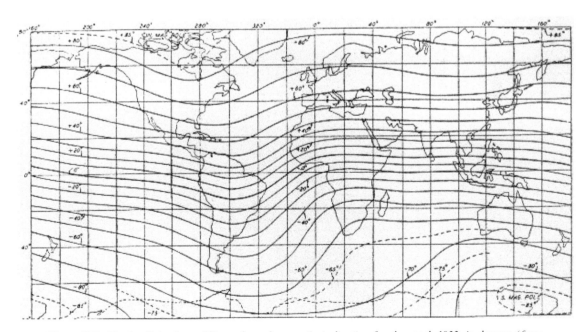

Figure V.5. *The isoclinic chart of lines of equal magnetic inclination for the epoch 1922, in degrees (from British Admiralty Charts).*

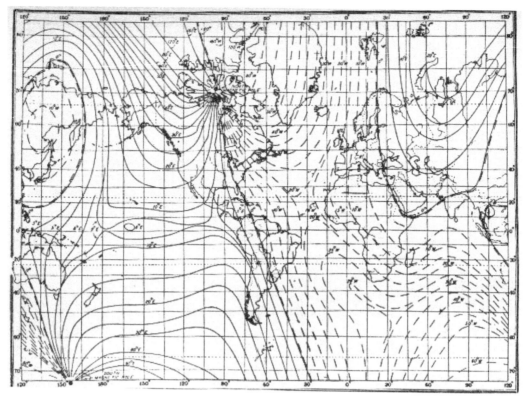

Figure V.6. *The isogonic chart for lines of equal magnetic declination for the epoch 1935, in degrees (U.S. Hydrographic Office).*

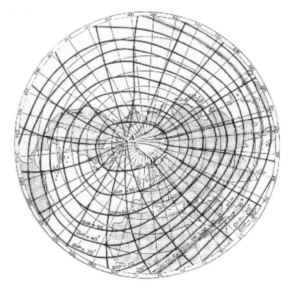

Figure V.7. *Lines of equal inclination and of declination for the northern hemisphere for the epoch 1835. The units are degrees (after Airy).*

equal area can be constructed but they are generally not convenient to use. The Mercator projections have in the past been taken as the most convenient for practical purposes and data are usually plotted in this way.

V.4.   ANALYSING THE FIELD

The resemblance between the observed magnetic field to that of a dipole allows the total field to be represented by that of a dipole supplemented by a non-dipole component. This representation has a theoretical basis first recognized by Gauss in the eighteenth century. The method of representation is based on a study of the potential of the magnetic field, U. The components of the magnetic field are obtained by differentiating the expression for the magnetic potential with respect to the separate space coordinates. The Earth is closely spherical so it is natural to use spherical polar coordinates (r, θ, ϕ) for this purpose. Then

$$\mathbf{F} = -\text{grad}\, U(r, \theta, \phi).$$

If there are no electric currents, U is a solution of the Laplace equation

$$\nabla^2 U(r, \theta, \phi) = 0.$$

Gauss showed that there is an external and internal solution provided there are no electric currents in the surface, which is true to good approximation. The data available to him were not good enough to explore such detail but he was able to show that the main external magnetic field is intrinsic and is of internal origin. The solution of the Laplace equation has the form of an expansion in spherical harmonic functions

$$U(r, \theta, \phi) = -\sum_{l=1}^{\infty} \sum_{l \geq m} C_{lm} \left(\frac{1}{r}\right)^{l+1} P_l^m(\cos\theta)[a_m \cos m\phi + b_m \sin m\phi] \tag{V.2}$$

where C_{lm}, a_m and b_m are sets of coefficients to be determined. The $P_l^m(\cos\theta)$ are *associated Legendre polynomials*. If, as is often approximately the case, the magnetic field does not depend on the longitude then we have the simpler expression

$$U(r, \theta) = -\sum_{l=1}^{\infty} C_l \left(\frac{1}{r}\right)^{l+1} P_l(\cos\theta). \tag{V.3}$$

The coefficients C_l must be determined from observations. The factors $P_l(\cos\theta)$ are the *Legendre polynomials*. They can be calculated in several ways including the use of the Rodrigues formula

$$P_n(x) = \frac{1}{2^n n!} \left(\frac{d}{dx}\right)^n (x^2 - 1)^n$$

where, for convenience, we have set $\cos\theta = x$. Using this formula it is readily found that the polynomials up to P_4 are given by the expressions

$$P_0(x) = 1, \qquad P_1(x) = x, \qquad P_2(x) = \tfrac{1}{2}(3x^2 - 1),$$

$$P_3(x) = \frac{x}{4}(5x^2 - 1), \qquad P_4(x) = \tfrac{1}{8}(35x^4 - 30x^2 + 3).$$

It is seen that the even polynomials are even functions of $\cos\theta$ while the odd polynomials are odd functions. The magnetic potential due to a dipole is anti-symmetric, meaning to say that if the direction of the dipole was reversed then the role of the north and south poles would reverse leading to a change of sign of the potential. This is equivalent to saying that $U(r, \theta) = -U(r, \pi - \theta)$. For this to be so even for non-dipole components, which is approximately true, the representation of the magnetic potential must involve only odd functions so the even polynomials are removed from the description.

Thus, with only the first two odd terms

$$U(r, \theta) = C_1 \left(\frac{1}{r}\right)^2 \cos \theta + \frac{1}{4} C_3 \left(\frac{1}{r}\right)^4 (5 \cos^3 \theta - \cos \theta). \tag{V.4}$$

The corresponding expressions for the orthogonal components of the field, one in the direction of increasing r and the other in the direction of increasing θ, are

$$F_r = -\frac{\partial U}{\partial r} = 2C_1 \left(\frac{1}{r}\right)^3 \cos \theta + C_3 \left(\frac{1}{r}\right)^5 (5 \cos^3 \theta - \cos \theta) \tag{V.5a}$$

$$F_\theta = -\frac{1}{r}\frac{\partial U}{\partial \theta} = C_1 \left(\frac{1}{r}\right)^3 \sin \theta + \frac{1}{4} C_3 \left(\frac{1}{r}\right)^5 (15 \cos^2 \theta \sin \theta - \sin \theta). \tag{V.5b}$$

The total field has magnitude

$$F = \sqrt{F_r^2 + F_\theta^2} \tag{V.6a}$$

and makes an angle ψ with the radius vector in a clockwise direction where

$$\tan \psi = \frac{F_\theta}{F_r}. \tag{V.6b}$$

and the phase ambiguity is resolved by noting that $\sin \psi$ has the sign of F_θ.

The hierarchy of terms in equation (V.3) has a specific physical interpretation. The leading term, with $l = 1$, proportional to $P_1(\cos \theta)/r^2$, represents a magnetic dipole potential. The next term, proportional to $P_3(\cos \theta)/r^4$, represents a quadrupole contribution, and so on. The source of a planetary magnetic field has many components of different types with different dependence on the distance. The effect is that, as the distance from the source increases, the higher-order terms vanish more quickly than the lower-order ones. Viewed at large distances, the potential has a closely dipole form, increasingly so as the distance becomes greater. Any magnetic source appears as a dipole at sufficiently large distance.

The various coefficients in equation (V.3) are found in practice by inserting observational values into the equation. The number of terms in the polynomial that can be determined depends on the number of sets of accurate observational data that are available. Early applications had only restricted data with poor computational aids and were rarely able to go beyond the term P_2. Modern analyses are supported by a substantial database that gives a better fit between observation and the theoretical representation. The latest analyses include as many as eight terms in the expansion. The effect is to achieve better accuracy for each component of the expansion and especially for the dipole component.

V.5.   THE RESULT FOR THE EARTH

These analyses have been applied to the Earth and the details of the intrinsic field have been established.

V.5.1. The dipole approximation

It is found from the analysis that, close to the Earth, the dipole component accounts for some 80% of the measured field, the non-dipole part accounting for the rest. The dipole that comes from analysis is called the 'best fit' dipole field for the measured data. This defines the International Geomagnetic Reference Field (IGRF) and is agreed internationally. This field is often denoted by the symbol F_1. The data are updated every few years partly to account for improved measurements but also because the dipole itself changes slowly. A vector representation of the IGRF is shown in figure V.8. The corresponding declination lines are shown in figure V.9.

The axis of the IGRF cuts the surface of the Earth at two *geomagnetic poles*, as opposed to the magnetic poles referring to the actual measured pole positions as defined in section V.1. In the year 2001 the geomagnetic poles were located approximately at the positions 79°N, 71°W and 79°S, 109°E. The axis makes an angle of about 11.5° with the geographical spin axis and is displaced from the centre of the Earth by a distance of 640 km. The geomagnetic equator is a line circumventing the Earth in the plane perpendicular to the dipole axis and mid-way between the geomagnetic poles. It also makes an angle of 11.5° with the geographical equator. These relationships are shown in figure V.10. The difference between the magnetic poles and the geomagnetic poles is clear, as is the difference between the geomagnetic equator (the hypothetical plane which cuts the Earth's surface perpendicular to the axis of F_1) and the magnetic equator (the line for which $I = 0$).

V.5.2. The non-dipole component of the magnetic field

The non-dipole component of the magnetic field is the field that remains when the IGRF is subtracted from the observed total field. It is often described as $[F - F_1]$ and described as the secular magnetic

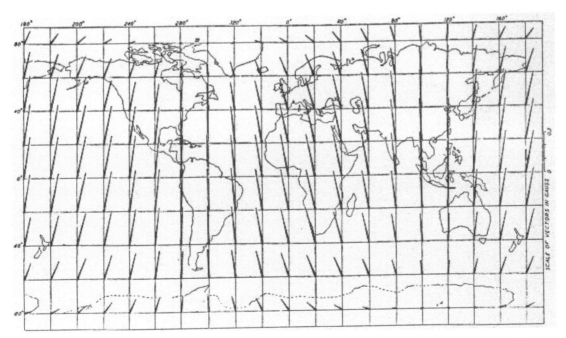

Figure V.8. *Vectors representing the magnitude and direction of the horizontal component of the magnetic field corresponding to the best fit dipole or the field of uniform magnetization of the sphere that best fits the measured field of the Earth (Chapman and Bartels).*

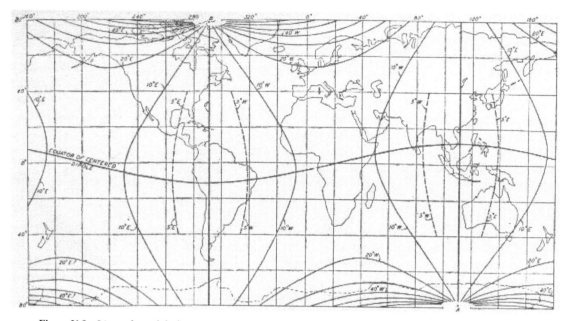

Figure V.9. *Lines of equal declination associated with the field* **F₁** *of the sphere of uniform magnetization that fits most closely to the Earth's field (Chapman and Bartels).*

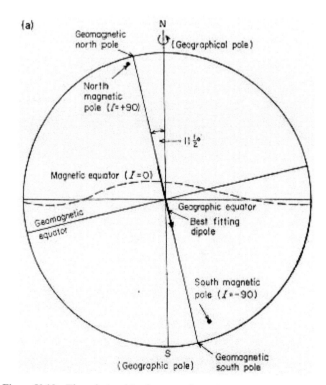

Figure V.10. *The relationships between the various magnetic quantities.*

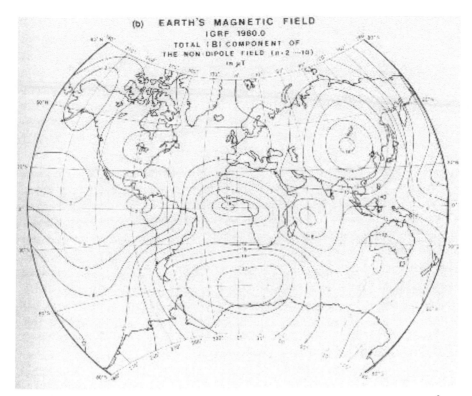

Figure V.11. *Showing the non-dipole magnetic field* [**F** − **F₁**] *for the epoch 1980. The units are* 10^2 μT *(after Nevanlinna, Pesonen and Blomster).*

field. As an example, the non-dipole magnetic field for the year 1980 is shown in figure V.11. The amplitude of [**F** − **F₁**] is some 20% of the total or $\frac{1}{4}$ that of the dipole field. The relative strength of the dipole component suggests, from equation (V.3), that the source of the total magnetic field is some distance below the surface of the Earth. The details of the interior of the Earth are considered in section 4.3.6 and Topic R.

V.6.   TIME DEPENDENCIES OF THE MAGNETIC FIELD

The Earth's intrinsic magnetic field is found to change slowly with time. Some indication that this is so can be gathered from the comparison between figures V.2 and V.3 for the epochs 1922 and 1980. There are two separate components of change.

V.6.1.   The dipole field

Data for the magnetic field date back to 1836 for a limited number of observatories but it was 1886 before a wider range of observatories became available to provide data for analysis. With this proviso it is possible to explore the field back more than 150 years and to draw some conclusions. The most obvious one is that the dipole moment has fallen over the period of data collection. A plot of dipole moment against epoch is shown in figure V.12 for the period 1840 to 1960. It is clear that the moment has decreased continuously by some 8% over this period and it seems that this decrease will continue into the future. It is generally agreed that the geomagnetic poles (referring to

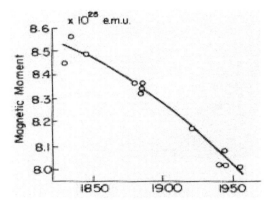

Figure V.12. *The falling magnitude of the magnetic dipole moment for the best fit dipole field of the Earth over the period 1840 to 1960.*

the best dipole) have always remained close to the geographical poles although this may not have been the case for the magnetic poles (referring to the total field). The problem of disentangling the possible changes in the dipole field from those of the total field is fraught with difficulty. It is generally accepted that the dipole field changes its magnitude, though not its direction, over substantial periods of time.

V.6.2.   The non-dipole secular field—the secular variation

It has long been realized that the total field changes with time. This change is called the secular variation. As an example, figure V.13 shows the magnetic declination as measured at Kew during the period 1560 to 1980. The change is not periodic. More detailed data are available from more recent surveys. The secular variations for the two epochs 1922.5 and 1942.5, a period of 20 years, are shown for the vertical component, Z, in figure V.14. The field patterns are very similar for the

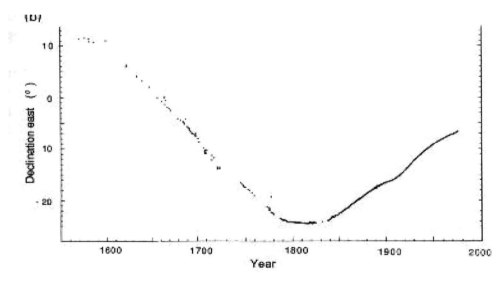

Figure V.13. *The magnetic declination measured at Kew over the period 1560 to 1980 (after Malin and Bullard). The change is not periodic.*

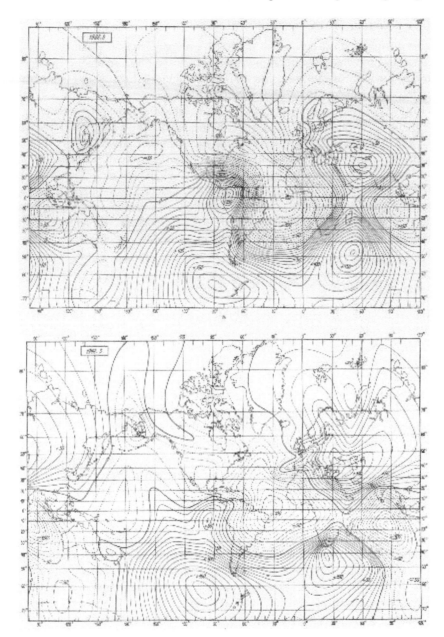

Figure V.14. *The secular variation in the Earth's magnetic field for the epochs 1922.5 and 1942.5 for the vertical component Z. The changes are in* 10^{-9} T *(gamma) (after Vestine).*

two years but are not the same. A westward drift of the pattern is discernible with general magnitude of a few degrees of longitude per year. The westward movement is more marked in the west than in the east, suggesting the cause is not a simple rotation of the magnetic field.

The general feature of the change involves regions of continental size though not coincident with the continents themselves. This suggests an origin within the Earth below that where continental/oceanic features play a role.

V.6.3.   Reversals of the direction of magnetization

Perhaps the most intriguing discovery about the Earth's magnetism is its polarity reversals in the past. Sedimentary and magma materials contain small particles of ferromagnetic material that can be magnetized in the direction of the prevailing magnetic field. As the material is compressed to form sedimentary rock or cools below the Curie point (section 6.6) to form volcanic rock, the orientation of the magnetization of the particles is frozen into the rock sample. By measuring this remnant magnetization of rocks the magnetic conditions of the Earth, and particularly the direction of the total magnetic field, can be inferred for the time the rock was laid down.

If the orientation of the rock is carefully recorded then the study of *rock magnetism* enables both the strength of the palaeomagnetic field and the locations of the magnetic poles at past times to be determined. If a remnant magetization M_1 is found to be induced in the rock in the laboratory by a field of known strength B_1, then, by applying a principle of proportionality the strength of the palaeofield can be estimated. This is because $M \propto B$ and the constant of proportionality depends on the magnetic susceptibility of the material, which is essentially independent of the temperature. Consequently, we can write

$$B_0 = \frac{M_0}{M_1} B_1$$

where M_0 is the remnant magnetization and B_0 the palaeofield. Although the technique is simple to define it is made very difficult to perform accurately due to smallness of the fields to be measured—often of the order of gammas.

Two separate pieces of evidence for conditions in the early Earth have been found in this way. First is the variation of the polarity of the dipole field. An example of this covering the past 160 million years is shown in figure V.15. Black represents the normal polarity found now while white refers to the reversed polarity. The present polarity has been present for a longer total period than the reversed polarity. Although the polarity reverses after a period of roughly 10^6 years, there is no periodicity. It is seen that there was a substantial period from about 125 million years ago to about 83 million years ago when the polarity remained steady and in its present direction. The palaeomagnetic records actually go back farther than this but the results are not too reliable. It seems, in fact, that the Earth has been the source of intrinsic magnetism from the very earliest times and that reversals have probably occurred from the earliest times.

The reversal of polarity is always complete, the antiparallel alignment being replaced by a parallel alignment and vice versa. The magnetic poles retain the same positions, the north magnetic pole becoming a south pole and the south polarity changing to north. The pole strength seems to decrease to zero and changes sign to essentially the same strength over a period of about 10^4 years. The behaviour of the pole strength shown in figure V.11 fits these observations and suggests the dipole field is decreasing at the present time towards a possible future reversal.

There is recent evidence that the total Earth's field undergoes very short period changes called *jerks*, where a reversal seems to begin but is aborted over a period of a month or more, the field direction remaining unchanged. It would seem that the magnetic field behaviour is most complicated on a range of time scales and it is not certain that all its possible characteristics have yet been recognized and catalogued.

V.6.4.   Pole wander

One result of the study of remnant magnetism is the location of the positions of the geomagnetic poles at different ages, obtained by obtaining data from rocks of different ages. A remarkable discovery was made when data were compared from different continents. An example is shown in figure V.16a

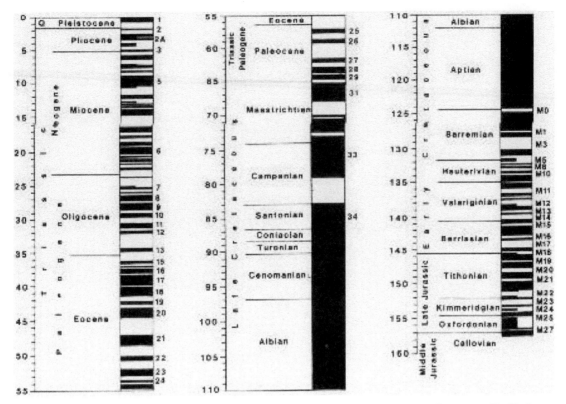

Figure V.15. *The pattern of reversals of the Earth's magnetic field over the past 180 million years. The black regions are normal polarity while the white regions are reversed polarity. There are no intermediate regions (after Harland, Llewellyn, Pickton, Smith and Walters).*

showing data covering 550 million years for the position of the north magnetic pole. The data from each continent showed that the position of the north magnetic pole has moved over considerable distance, moving over both land and sea. The shapes of the paths are the same but the predicted positions of the poles are very different. They are most at odds for the earlier times and come closer into coincidence as the present time is approached. The two curves are brought into almost complete coincidence if allowance is made for changes in the relative positions of Europe and North America. This is known to have arisen from the opening of the Atlantic Ocean over the past 500 million years. When allowance is made for this the two curves fit closely together as is seen from figure V.16b. This is convincing evidence for continental drift, first proposed from a study of the flora of Indonesia by Alfred Russell Wallace in the mid-1800s and independently suggested by Wegener from studies of the rock formations apparently common to Africa and South America.

V.6.5. Sea floor spreading

The movement of continents (continental drift) results from the continuous ejection of crustal magma from active mid-ocean ridges. The surface area of the Earth is unchanging so there must be a corresponding absorption into the crust again, at subduction zones. When it is extruded from the mid-ocean ridge the material is at a temperature as high as 1300 K, well above its Curie point, and is permeated by the local magnetic field. At this point the encasing seawater cools the magma very

(a)

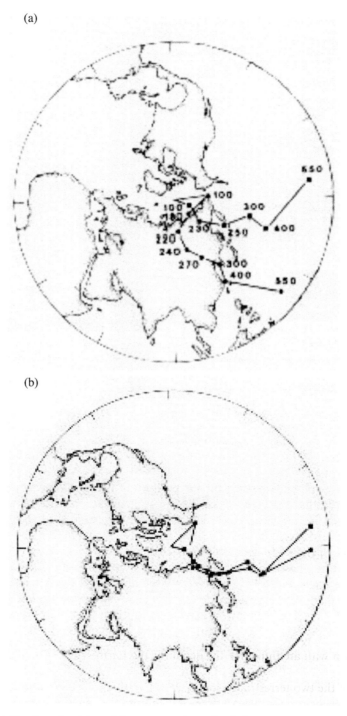

(b)

Figure V.16. *(a) The plots of positions of the geomagnetic north pole for remnant magnetic data from Europe (squares) and N. America (circles). (b) The polar wander curves for Europe (Squares) and N. America (circles) over the past 550 million years when account is taken of the opening of the Atlantic Ocean. The fit is remarkably good (after McElhinney).*

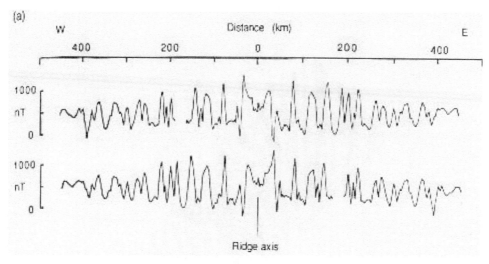

Figure V.17. *The anomalous profile associated with the expansion of magma from the Pacific–Antarctic ridge. The symmetry across the ridge is made clear by comparing the actual trace with its mirror image.*

rapidly to a temperature below the Curie point, the magma then acquiring remnant magnetism. The consequence is a magnetic profile, expressed as deviation from the local mean field, developing outwards perpendicular to the ridge in both directions. Such a magnetic profile is shown in figure V.17 straddling the Pacific–Antarctic ridge. The plot contains a mirror image profile to make the symmetry about the central ridge clear. The ridge axis runs perpendicular to the page and the extruded magma moves slowly both to the right and to the left. The symmetry of the profile is very clear. The magnetic profile can be compared with alternative magnetic profiles of known age so that the spreading material can be dated and the rate of spreading can be found. This is found to be about 4.5 cm per year. The amplitudes of the anomalies are about ± 500 nT, $\sim 1\%$ of the total intrinsic field strength. This is a typical amplitude for other ridge developments around the globe.

Studies of this kind were important for establishing ocean-floor spreading as a universal mechanism for continental plate movement. The analysis is often most complex. This can be appreciated from figure V.18 for the total field in the Pacific Ocean south-west of Vancouver Island. This was the first published map showing the details of the magnetic anomalies associated with an active mid-ocean ridge. The regions of positive magnetic anomaly are shown black and of negative anomaly in white.

V.7. MAGNETISM OF OTHER SOLAR-SYSTEM PLANETS

Other members of the Solar System have also been found to possess intrinsic magnetism although the details are not known with anything like the detail for the Earth's field. The data available so far are listed in table V.1.

The polarity of the two terrestrial planets is anti-parallel while those of the major planets are all parallel. The significance of this is difficult to assess, especially when it is remembered that the Earth's field undergoes polarity reversals. For the planets other than the Earth, the details have been deduced from space vehicle measurements and assume that the field is essentially of dipole form. The state of the polarity can, nevertheless, be found reliably. This also involves the interaction between the planet and the solar wind, details of which are treated in Topic W. It is seen that the strengths of the fields at the

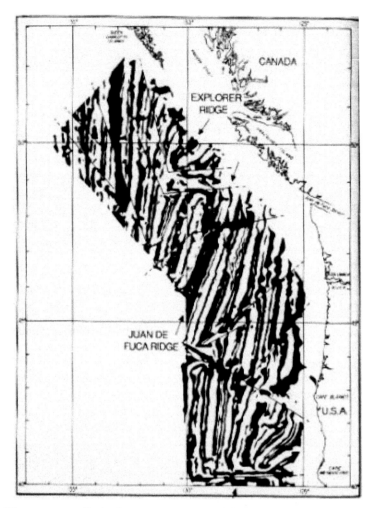

Figure V.18. *The magnetic profile for the active ocean ridge off the coast of Canada and the USA. Black refers to the current polarity and white to the reverse polarity (after Raff and Mason).*

Table V.1. *The intrinsic magnetism of the planets. The angles in the last column refer to the rotation and dipole axes The dipole moment for the Earth $= 7.9 \times 10^{15} \, \text{T m}^3$.*

Planet	Rotation period (days)	Dipole moment (Earth = 1)	Equatorial field (10^{-4} T)	Polarity	Angle between axes (degrees)
Mercury	58.65	7×10^{-4}	0.002	anti-parallel	~10
Venus	243.01 (R)	$<4 \times 10^{-4}$	<0.0003	?	?
Earth	1.00	1.00	0.305	anti-parallel	11.5
Mars	1.03	$<2 \times 10^{-4}$	0.0004 (?)	?	?
Jupiter	0.41	2×10^4	4.2	parallel	9.5
Saturn	0.44	6×10^2	0.20	parallel	<1
Uranus	0.72 (R)	50	0.23	parallel	58.6
Neptune	0.67	25	0.006–1.2	parallel	46.8

(R) in the table means a retrograde motion.

equator vary enormously from one planet to another. The dipole moment, depending as it does on the volume of the planet, is not a reliable guide to the strength of the magnetism. The measurement of the fields is at some distance from the planet and, according to equation (V.3), the dipole field will be dominant. More details of the field of Jupiter were found during the Galileo Space Mission. One feature is a higher proportion of quadrupole field in the total, suggesting that the source is closer to the surface than for the Earth.

Mars does not show a perceptible intrinsic magnetic field but there are magnetic signatures on its southern surface. One such is shown in plate 8 taken by the Mars Global Surveyor in 1999. The orbit of the surveyor provides north–south (green) strips with blue and red banded magnetic field profiles in the east–west direction. These have opposite polarities, each band being about 180 km wide and 950 km long. These patterns are similar to the magnetic patterns associated with the material extruded by mid-ocean ridges on Earth. The full implications of these measurements are yet to be established but they might imply the early presence of plate motions on Mars, moving in an intrinsic magnetic field. It seems that the conditions on Mars could have been very different early in its history than they are now.

V.8.   INTRINSIC MAGNETISM OF NON-SOLAR PLANETS

The magnetic conditions in the recently discovered companions to solar-type stars (Topic K) are as yet unknown but, from our knowledge of Solar-System bodies it is almost certain that they will be a source of magnetic fields. The magnetic properties of Jupiter are likely to be a good approximation to the conditions in the companions discovered so far. A small semi-major axis for a companion, as is often observed, implies a strong interaction between the magnetic fields of the star and the companion. Such interactions are described for Solar-System bodies in Topic W.

Problem V

V.1   A spacecraft is in a polar elliptical orbit around a planet. When it is over the pole at distance 8×10^4 km from the planet's centre the measured magnetic field is 2.00×10^{-4} T in the outward direction. When it is over the equator at 1×10^5 km from the planets centre the field is 4.00×10^{-5} T in the tangential direction. If there are only dipole and quadrupole components then what are the values of the coefficients C_1 and C_3 in equation (V.5)? What proportion of the field is due to the quadrupole component on the surface at the pole if the polar radius of the planet is 5×10^4 km?

TOPIC W

MAGNETIC INTERACTIONS BETWEEN PLANET AND STAR

The magnetic environment of a planet orbiting a star will interact with the magnetic environment of the star itself. The interaction will be quite complicated, especially if the semi-major axis of the orbit is small. Then the companion will lie close to the star either for most of its orbit or for some of it. Nothing is yet known directly of conditions to be found in such cases.

In the Solar System planets are so distant from the Sun that direct linkage of planetary fields with that of the Sun is of little importance. The dipole moment of the Sun, $\sim 8 \times 10^{22}\,\mathrm{T\,m}^3$, gives a field of 0.02γ ($1\gamma = 10^{-9}\,\mathrm{T}$) in the vicinity of the Earth, which is trivial compared with the Earth's own field. However, the magnetic environment of the Earth *is* greatly influenced by the Sun through the solar wind, the charged particles from the Sun that pervade the whole Solar System. Space vehicles, sampling both magnetic fields and charged particle conditions both close to and distant from planets, have enabled a detailed picture of these Sun–planet interactions to emerge.

We begin by considering the Sun–Earth interaction in some detail and then move on to the other Solar-System planets.

W.1. TRANSIENT MAGNETIC COMPONENTS

Although the major component of the magnetic field at any location on the Earth's surface is the intrinsic field (Topic V), very careful measurements of the magnetic elements show there are also a secondary magnetic field component of much smaller magnitude. These are variable and have no long-term effect on the mean field. The magnitude of the accompanying field is at most 10% of the intrinsic field and often less, which is no greater than a few times $10^{-6}\,\mathrm{T}$.

The secondary field is variable over a range of time scales, all less than 11 years and many as short as hours or minutes. The effects vary from one surface location to another and show a dependence on local solar and lunar times. One component, called the diurnal variation **S**, changes in a broadly cyclical way with a period of about 24 hours. There is a difference between day and night conditions, that is whether the part of the Earth in question is facing the Sun or otherwise. The diurnal variations follow solar time. There is another component which varies with a period of about 28 days and is based on lunar time: this is the lunar variation **L**. There is also an annual effect. These are not all. The profile of disturbances is also found to vary over the 11 year period of the solar sun-spot cycle, the effects being stronger the greater the solar activity. The Sun's 11 year cycle of activity coincides with the variation of the density of sunspots on its surface (figure W.1). Sunspots are regions, slightly cooler that their surroundings so they appear dark, that are

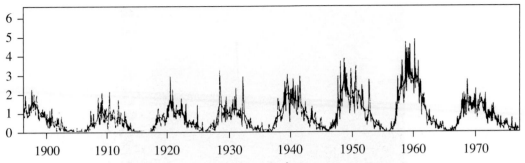

Figure W.1. *Sunspot coverage (in units of 10^{-3} of disk) from 1900 to 1970 (RGO).*

associated with stronger local magnetic fields. When sunspot density is higher then there tends to be greater solar activity. One form of activity is represented by *prominences*, emissions of solar material, often in the form of a loop extending for tens of thousands of kilometres above the Sun's surface. The solar wind is a constant feature of the Sun's activity but it can be greatly enhanced by *flares*, transient generation of energy in the form of electromagnetic radiation in a region of the Sun's surface. The solar wind is augmented by a burst of particles of much greater speed and energy and the effects on Earth can be quite dramatic. In 1989 a massive flare disrupted radio communications on Earth for some time and power surges blacked out a great part of Quebec province in Canada.

Each magnetic component shows variations which seem random and change continually. Some days show a minimum effect and the magnetic profiles are smooth. These are magnetically quiet days. Other days show substantial irregularities and are termed magnetically disturbed, or magnetically active. The magnetic elements for disturbed conditions are sometimes given the symbol **D**. If the disturbance is particularly strong and/or protracted the Earth is said to undergo a magnetic storm. Extremely disturbed conditions may be associated with the appearance of an aurora display in the sky of green and ruby curtains of light. These are the aurora borealis in the north and aurora australis in the south, produced by the de-excitation of atoms in the upper atmosphere previously excited by solar-wind particles. These displays usually form a ring around the magnetic poles. Although normally restricted to the polar regions, the auroral displays can be seen in mid-latitude regions for strong magnetic disturbances or even, very unusually, nearer the equator during extreme disturbances. Pictures of the aurora borealis and of the aurora australis are shown in plate 1. The display seen from below at the Earth's surface is shown in (a) while that seen from above, looking down from Earth orbit, is shown in (b). The green emission is the 5577 Å line which generally occurs at heights between 110 and 250 km. The ruby emission is the pair of lines 6300 Å and 6364 Å which appear at lower gas densities, and so at greater heights, perhaps up to 800 km or even more.

The study of these various magnetic phenomena from the surface of the Earth was, before about 1990, very difficult and gave confusing conclusions. This was partly because data at any time were available only for restricted parts of the Earth and also because it was not possible earlier on to take direct account of all influences, including the Sun. The study was revolutionized by the introduction of orbiting space vehicles. Of special importance has been the ability to survey points all over the Earth's surface effectively simultaneously (certainly within 100 minutes for a single satellite but over shorter times for more than one craft). One lingering question, answered immediately, was the very basic one of whether the aurorae occur simultaneously at the north and south poles—that they do is seen in figure W.2 taken from orbit by the Dynamics Explorer 1. That this was not previously certain indicates the inadequacy of the earlier techniques.

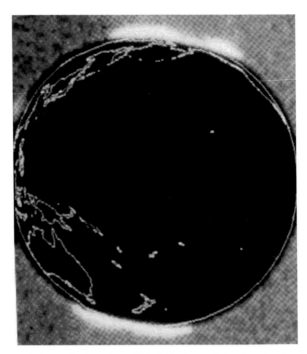

Figure W.2. *A photograph from the American Global Explorer 1 showing the simultaneous occurrence of aurorae at the north and south poles (NASA).*

W.2. THE ORIGIN OF THE ATMOSPHERIC FIELDS

The links between the transient magnetic activity on the surface of the Earth with the influences of the Sun and Moon suggest that these effects have their origins outside the surface of the Earth. That this is so can be established using the arguments first developed by Gauss (section V.4). Provided there are no electric currents crossing the Earth's surface, the potential, U_e, of the magnetic field external to the surface satisfies the equation

$$\nabla^2 U_e = 0 \tag{W.1}$$

where ∇^2 is the Laplacian operator. With the source of the field *inside* the Earth, and assuming axial symmetry (θ but not ϕ dependence in spherical polar coordinates) the potential can be expressed in the form of equation (V.3) and is seen to fall off with increasing distance from the Earth's surface. In the circumstance where there is a source of axially-symmetric magnetic field outside the Earth, the source being within a well defined region, then, between the source and the Earth's surface, the form of the potential is

$$U_e(r, \theta) = \sum_i E_e (r/R_\oplus)^i P_i(\cos \theta) \tag{W.2}$$

where $R_\oplus$ is the radius of the Earth. It will be seen that within this region the potential *increases* with increasing distance from the Earth's surface. However, farther from the Earth than this source region, the source now becomes an internal one and the potential it creates again falls with increasing distance from the Earth. Analysis of observations around the Earth demonstrate uniquely that the external source of transient fields is within the Earth's atmosphere.

The atmosphere of the Earth (Topic O) is composed primarily of nitrogen and oxygen, either as free atoms near the top of the atmosphere or as molecules lower down. These atoms are ionized by solar photons of sufficiently high energy. The proportion of atoms ionized is proportional both to the number of atoms capable of being ionized and the number of photons with sufficient energy to ionize them. The outer regions of the atmosphere have low material density and so few atoms but the radiation from the Sun has its full initial number of high-energy photons. The number of atoms increases with depth but the number of high-energy photons decreases inwards. The rate of ionization, therefore, must increase at first and then subsequently decrease, with the result that a relatively thin spherical shell of ionized gas is maintained by the Sun high up in the atmosphere. This ionized shell is called the *ionosphere* and has a mean height of about 120 km above the Earth's surface. It extends out to ~1000 km although there its density is very low. The ionosphere is an efficient conductor of electricity.

The dynamics of the ionosphere is complicated. Its thickness and degree of ionization depends on the intensity of radiation from the Sun and this is known to vary with conditions on the solar surface. The solar intensity varies with the time so the degree of ionization in the ionosphere must also vary correspondingly. There is a difference at any location between daytime, when the atmosphere receives radiation, and at night, when it does not. The effects of thermal expansion make the daytime atmosphere somewhat more extensive than at night. There are other effects. Charged particles enter the atmosphere from outside thus enhancing the local electrical conductivity. The atmosphere is also under the separate gravitational influences of the Moon and of the Sun which cause atmospheric tides similar in origin to the sea and solid-earth tides (Topic Y) but of much greater magnitude.

An electric current is induced within an electrical conductor moving relative to a permeating magnetic field. This induced current has an associated magnetic field enclosing it. The ionosphere is an electrical conductor that moves relative to the intrinsic magnetic field of the Earth; both the Earth and the ionosphere have spin but they do not move as a rigid system. An atmospheric current flows in the ionosphere as a consequence. It is the small magnetic field associated with this current that is measured at the surface of the Earth. It is the gravitational effects of the Moon and Sun, and the thermal effects of the Sun, that cause movements of the ionosphere and hence changes in the ionospheric currents the associated external magnetic field. The strength of the normal electric currents in the ionosphere can be augmented by the reception of charged particles from the Sun. In this case the electric current strength is increased considerably and the magnitude of the induced magnetic field can become much greater. The most severe conditions are magnetic storms. Because the ionosphere is used as a reflecting conductor for electromagnetic communications, terrestrial communications can be modified or even disrupted by these unpredictable events on the Sun.

W.3. THE SOLAR WIND

The remarkable activity of the Sun was first recognized in the 1950s. The recognition began in an inauspicious way. Comet tails are observed to point consistently away from the Sun and in 1951 Ludwig Biermann suggested that this was due to the effects of a stream of particles coming from the Sun with sufficient momentum to affect the particles of the tail. This led to the idea of the continuous emission of plasma from the Sun, called the *solar wind*. It had been known from past experience that features on the solar surface such as sunspots and solar flares give rise to magnetic storm conditions. Since the time delay between the feature and the Earth disturbance was days rather than minutes it was also known that the link is not electromagnetic. The novelty in the solar-wind concept was that

Figure W.3. *The Sun emits fast particles from the polar regions and slow particles from the equatorial regions (Ulysses craft, NASA).*

the solar emission is continuous and not occasional. This was confirmed dramatically later by measurements from space vehicles. Astronomical measurements have indicated winds from other stars, many of them much more dense and vigorous than that from the Sun.

The solar wind consists of particles of very high energy, primarily electrons and protons. These particles have been observed to leave the Sun with a range of speeds up to $1000\,km\,s^{-1}$. The emission is not uniform over the solar surface. Measurements from SOHO, a spacecraft observing the Sun, show that the fastest particles are emitted from the poles and the slower particles from the equator, as in figure W.3. The origin of the wind is still not fully understood but some of the details of its emission have been found. The solar surface is covered by a series of large convection cells (figure W.4) some of which emit particles while others have an inflow of particles. The emission of fast particles has been found to be associated with coronal holes, one of which is shown in figure W.5. Matter enters a coronal hole and then is ejected into the Solar System. It moves away from the rotating Sun with constant angular momentum so that in angular terms it lags behind the spinning Sun. Thus a stream of particles seems to have a spiral form, often pictured by analogy as water from a rotating hose pipe. Solar-wind speeds are very variable. At the distance of the Earth the speed can be between 300 and $700\,km\,s^{-1}$, typically 400–$500\,km\,s^{-1}$, while at the distance of Jupiter it is perhaps $300\,km\,s^{-1}$. The density of particles is of the order of a few million particles per cubic metre in the Earth's vicinity.

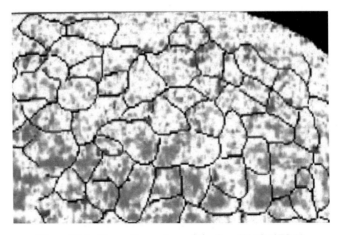

Figure W.4. *Convective regions of the Sun (NASA/ESA).*

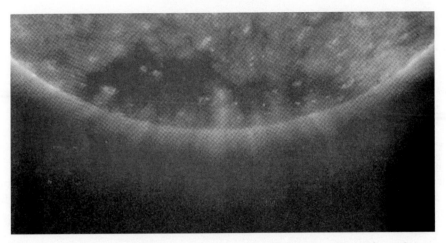

Figure W.5. *The dark patch is the south polar coronal hole from which the fast solar wind particles are emitted, pictured at the wavelength 171 Å (far ultraviolet). The surface temperature is 10^6 K. (SOHO, NASA)*

W.4. COUPLING BETWEEN PLASMA STREAMS AND MAGNETIC FIELDS

The main features of the way that the Earth's magnetic environment is affected by the solar wind can be understood in terms of the way that streams of charged particles, including a plasma stream, and magnetic fields become coupled together. In figure W.6a we show a charged particle, of mass m, charge e and velocity **V** within the region of a magnetic field of strength **B**. The force experienced by the charged particle, the *Lorentz force*, is

$$\mathbf{F} = e\mathbf{V} \times \mathbf{B} \qquad\qquad (W.3)$$

and hence is perpendicular to the plane defined by **V** and **B**. In the particular case that **V** and **B** are perpendicular to each other (figure W.6b) the force will result in circular motion in a plane of

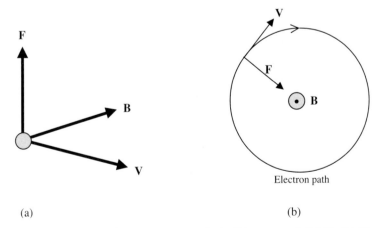

(a)   (b)

Figure W.6. *(a) The force, **F**, on an electron moving at velocity **V** in a magnetic field **B**. (b) Circular motion in a plane.*

radius r where

$$\frac{mV^2}{r} = eVB \quad \text{or} \quad r = \frac{mV}{eB}. \tag{W.4}$$

Since $V = r\omega$ and the period $P = 2\pi/\omega$ the period is found to be

$$P = \frac{2\pi m}{eB} \tag{W.5}$$

and hence is independent of V.

In the case that **V** makes an angle α with **B** the result is that the particle travels on a helical path with a constant component of velocity $V\cos\alpha$ along the direction of the field line with a period of spin around the field line given by equation (W.5). The radius of the motion around the field line is given by

$$r = \frac{mV\sin\alpha}{eB}. \tag{W.6}$$

The strength of the coupling is indicated by r; the smaller is r the stronger is the coupling. This shows, for example, that electrons will be more strongly coupled than positive ions—perhaps an obvious result since they have less mass to react to the force—and that slower particles will also be more strongly bound. When magnetic field lines follow curved paths then strongly bound particles will follow the field but those with large momentum may detach themselves.

A converse situation is that the magnetic field lines may also be strongly bound to the plasma, just considered now as a conducting fluid. For example, we consider a highly conducting body within a uniform magnetic field corresponding to having evenly spaced field lines. If now the body collapses under the influence of gravity then the field lines can be frozen into the conducting fluid to give the result shown in figure W.7. In this case the magnetic field becomes stronger within the collapsed body, as indicated by the higher concentration of field lines within it.

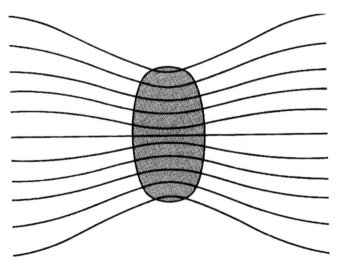

Figure W.7. *The concentration of magnetic field lines frozen into a collapsing conducting body.*

W.5. EFFECTS OF THE SOLAR WIND

The solar wind of charged particles also carries lines of magnetic field. The plasma/field combination acts like a compressible fluid with energy and momentum. This plasma fluid exerts a pressure on an obstacle in its path and flows around it. If the obstacle has plasticity, its shape will adjust itself to ensure the equality of the pressures on each side of the fluid/obstacle interface. This process applies when the solar wind encounters a planet.

W.5.1. *The effect on the Earth's field*

The solar wind compresses the Earth's intrinsic field at the higher latitudes on the sunlight side but draws it out on the night side. The effect is the formation of a magnetic tail, the *magnetotail*, extending several thousand kilometres downstream. The features are shown schematically in figure W.8. The structure was observed in ultraviolet light by the NASA space craft IMAGE and shown in figure W.9 looking downwards to the north pole. The Sun is beyond the lower right corner. The direction of the shadow, on the night side, is towards the upper left and contains the magnetotail. The circle on the Earth is an aurora display around the north magnetic pole.

At large distances from the Earth the solar wind is hardly affected by the Earth's magnetic field but as it approaches the Earth it experiences greater deflection forces, as described in section W.4. Then, at a certain distance from the Earth the deflection forces become so great that the charged particles are reflected backwards. The location where this begins to occur is referred to as the *bow shock*. The region within which the particles decelerate before they move backwards is a rather turbulent region known as the *magnetosheath*. Starting from the Earth there is a region within which the Earth's field behaves in a reasonably continuous way, gradually decreasing in strength

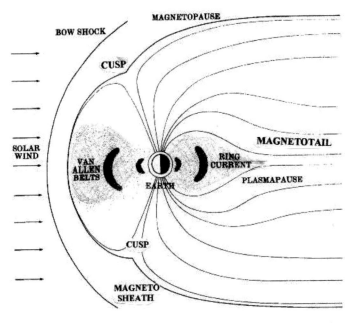

Figure W.8. *The magnetosphere of the Earth. The magnetic field is compressed on the day side and extended on the night side.*

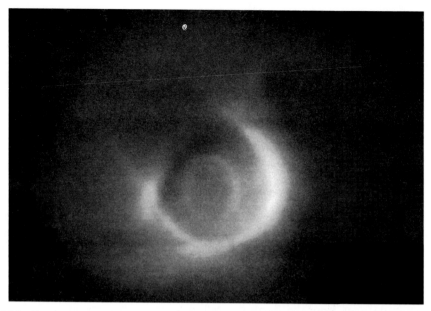

Figure W.9. *The observed plasmasphere of the Earth in ultraviolet light looking down over the north pole. The direction of the solar wind is from the bottom right hand corner producing the bow shock while the upper left shows the magnetotail. The central ring is an aurora borealis (northern light) (Image Space Craft NASA).*

with distance from the Earth. However, there is a sudden discontinuity in the field, the *magnetopause*, corresponding to the boundary of the dominion of the Earth's field and this marks the inner boundary of the magnetosheath.

The distance of the bow shock from the Earth can be estimated from the characteristics of the Earth's field and of the solar wind expressed in terms of pressures, or energy densities. The magnetic pressure associated with a field of strength B is given by

$$P_m = \frac{B^2}{2\mu_0}$$

where μ_0 is the permeability of free space, $4\pi \times 10^{-7}\,\mathrm{H\,m^{-1}}$. If the dipole moment of the Earth is D then at distance r in the magnetic equator the field is D/r^3 so that

$$P_m = \frac{D^2}{2\mu_0 r^6}. \tag{W.7}$$

The solar wind is a plasma stream and the particles that will be the most resistant to deflection are the ions, mostly protons. The relevant pressure in this case is the ram pressure of the stream (Topic AC) given by

$$P_{sw} = nmV^2 \tag{W.8}$$

where n is the density of protons, m the proton mass and V the solar wind speed. The condition for the distance of the bow shock is the equality of the pressures given by equations (W.7) and (W.8) or

$$r = \left(\frac{D^2}{2\mu_0 nmV^2}\right)^{1/6}. \tag{W.9}$$

The dipole moment of the Earth is $7.9 \times 10^{15}\,\mathrm{T\,m^3}$ and taking $n = 5 \times 10^6\,\mathrm{m^{-3}}$ and $V = 4 \times 10^5\,\mathrm{m\,s^{-1}}$ we find $r = 5.1 \times 10^7\,\mathrm{m}$ or about $8R_{\oplus}$.

The magnetic poles play a special role. The magnetic lines of force are open, extending downstream in the *magnetotail*. The division between the lines blown downstream on the night side and those compressed on the day side has the form of a cusp, allowing access to the low atmosphere in the polar regions. It is this limited region that allows charged particles, electrons especially, to move to the auroral zones and gives the link between magnetically disturbed days on Earth and events on the solar surface.

W.5.2. The trapped particles

The intrinsic magnetic field, which has a form very close to that of a dipole, is largely unaffected by the solar wind at the lower latitudes and forms closed loops as for a static field. In 1958, measurements using radiation detectors on the Explorer 1 and 3 satellites showed the existence of a large reservoir of trapped charged particles in this region held by the magnetic field. These were analysed by van Allen and his students and two separate belts were recognized, now called the van Allen radiation belts. They do not, in fact, radiate and the term is historical. They are shown in figure W.8. The inner belt contains protons with energies in excess of $10^7\,\mathrm{eV}$ and electrons with energies greater than $5 \times 10^5\,\mathrm{eV}$. The axis of the dipole field is displaced from the centre of the Earth by about $550\,\mathrm{km}$ so the array of particles is not symmetric about the geographical axis. In fact, the particle belt comes close to the surface in the South Atlantic, providing an electrical anomaly there. Protons and electrons also populate the outer belt but the proton energies are lower, less than $1.5 \times 10^6\,\mathrm{eV}$.

The origin of the belts is, at first sight, unexpected. Energetic particles from the Sun and galactic cosmic rays enter the Earth's atmosphere and a small proportion collide with the oxygen and nitrogen atoms producing neutrons, among other particles. Most neutrons are lost to space but a few suffer decay in the atmosphere, transforming to protons and electrons. These charged particles are trapped by the magnetic field for substantial periods of time extending into decades. Even though the source of the particles is very weak the rate of loss is small and the reservoir remains full of particles. Some contribution may also come directly from solar wind particles that leak through the magnetic barrier.

As previously described, the charged particles trapped by the Earth's magnetic field move in a spiral path along the magnetic lines of force. The field strength is greater near the Earth than farther away and the particles are reflected by the stronger fields near the ends of their trajectories (figure W.10). The times for such motions are in the range of seconds and minutes. Apart from moving in magnetic latitude they also move, but more slowly, in longitude, the electrons drifting eastwards and the protons westwards. The drift around the Earth has a time scale of hours.

W.5.3. Whistlers

Lightning strikes, either between clouds or between a cloud and the ground, cause ionization of the air and the resulting charged particles also move along the magnetic lines of force. Since they are accelerating they emit radiation which can be detected as a whistle with changing frequencies. The sound is characteristic. The study of the travel times allows information to be inferred about the charged particle concentrations.

W.5.4. The plasma tail

The conditions downstream in the magnetic tail are dictated by the conditions on the solar surface. The particles that enter the outer magnetosphere are of low energy (less than $10^4\,\mathrm{eV}$) but influence the

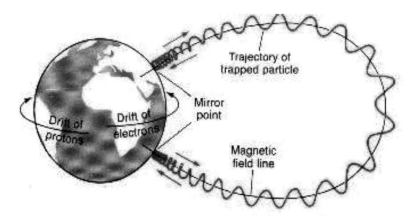

Figure W.10. *The trajectory of a trapped charged particle in the Earth's intrinsic magnetic field. It spirals around a line of magnetic force.*

magnetosphere quite strongly. There are two different regions. Close to the Earth the ions, protons and electrons are trapped by the magnetic field and corotate with the Earth and the magnetic field. This region contains the ring currents that cause magnetic storms. It extends over 4 to 8 Earth radii. Farther away, the magnetosphere does not corotate and can be strongly affected by the solar wind. The plasma here has its origin entirely in the main solar wind. Moving downstream, the plasma is continually restricted towards the axis, forming a plasma sheet of substantial particle density. The solar wind is continually changing and this affects the plasma sheet in ways that are seen in figure W.11. It could be said that the wind shows variable weather with storms driven by the Sun. For a steady solar wind the plasma sheet is stable as is shown in (a) of the figure. For more blustery conditions the plasma sheet becomes unstable. A magnetic instability can form between 6 and 20

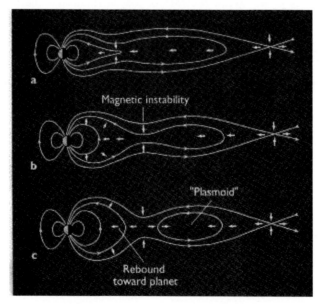

Figure W.11. *Three conditions of the Earth's magnetotail: (a) quiet conditions, (b) disturbed conditions and (c) storm conditions (van Allen and Bagenal).*

Earth radii downstream, as seen in (b), and this is recognized on Earth as a magnetic sub-storm. For stronger effects from the solar wind the plasma sheet can be divided by a strong pinch. The electrons on the Earth side pass into the polar regions to form aurorae. On the opposite side the electrons form a restricted region called the plasmoid, constrained between the pinched region and the end parts of the tail, illustrated in (c).

W.6.   THE MAGNETOSPHERES OF OTHER PLANETS

The major planets are accompanied by substantial magnetospheres. Those for Jupiter and Saturn are especially large but by 2001 only that for Jupiter has been studied in any detail. The details of Saturn will be revealed in due course by the Cassini mission due to begin in situ measurements in 2004.

W.6.1.   *The major planets*

Jupiter

The strength of the intrinsic magnetic field of Jupiter, plus a lesser intensity of solar wind, leads to a very large magnetosphere in comparison with that of the Earth. The effects are enhanced by charged particles emitted by the major satellites due to the bombardment of their surfaces by the solar wind. Io particularly, with its volcanic surface, acts as a substantial reservoir of charged particles, especially ions of sulphur and sodium.

   The first recognition of Jupiter as a magnetic planet was made in 1956 when non-thermal variable deca*metric* radiation was detected, with a frequency of 22.2 MHz. A correlation was soon established between the radiation and the longitude of Io in its orbit about Jupiter. Alerted to the possibility of radiation, a constant deci*metric* radiation was next detected with frequencies between 300 and 3000 MHz. This was interpreted as synchrotron radiation from relativistic trapped particles held by a intrinsic Jovian magnetic field, and analogous to the van Allen belts around the Earth. The symmetry of the radiation shows the magnetic axis making an angle of about $9.5°$ to the rotation axis. These conclusions were confirmed by measurements made 20 years later by the Pioneer 10 and 11 space vehicles, which found that there is, indeed, a substantial intrinsic magnetic field and that the particle density in the radiation belts is orders of magnitude greater than for Earth. The presence of a Jovian magnetic field is independently shown by the observation of polar aurora, as is seen in figure W.12.

   In general terms the Jovian magnetosphere shows a close similarity to that of the Earth. The solar wind is only 4% as strong at the distance of Jupiter as it is at the distance of the Earth. The effect of the intrinsic field is, consequently, substantially greater. Added to this, the magnetic moment of the Jovian field is some 20 000 times that of the Earth. One effect is the magnetotail extending a great distance downstream—to some 6.5×10^8 km, which is beyond the orbit of Saturn. A representation of the magnetosphere is shown schematically in figure W.13.

   The region corotates with the planet, which means a rotation period of 9 hours 59 minutes. The associated Coriolis force is strong and draws the equatorial plasma out along the equatorial plane to form a plasma disc along the magnetic equatorial plane, The concentration of charged particles in the Jovian system is strongly dependent on the satellite Io as a source. In particular, the sodium ion is of great importance, which is deposited into the Jovian ionosphere due to volcanic activity on Io. Sulphur ions have, in fact, been observed all around the Io orbit. The interaction between Jupiter and Io is complex. Electric currents flow in Io, linking it with Jupiter where the current flows in the ionosphere. Other current flow in the doughnut of Io's orbit.

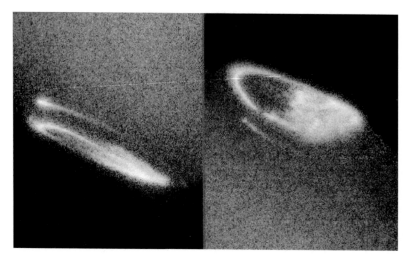

Figure W.12. *Aurorae at the poles of Jupiter, north on the left and south on the right (HSTI).*

The corotation of the plasma and the planet cannot be maintained throughout the magnetotail and the downwind structure of the magnetosphere remains unclear. Many problems remain in understanding the system and this will require further study of Jupiter to elucidate.

Saturn

The magnetism of Saturn was discovered in 1979 by measurements from Pioneer 11. There had been no previous indications from observations on the Earth of magnetic properties of Saturn. Aurorae have been observed at the magnetic poles using the Hubble space telescope (figure W.14), confirming both the existence of intrinsic magnetism and its unexpectedly exact coincidence with the rotation axis. Such measurements do not, however, allow details of the magnetosphere to be observed. Preliminary data were returned by the Pioneer 11 spacecraft during its flyby.

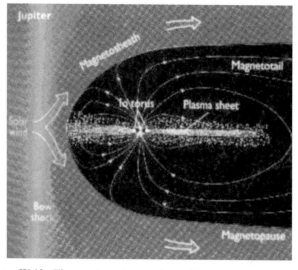

Figure W.13. *The massive magnetosphere of Jupiter (after van Allen).*

Figure W.14. *Showing the aurorae on Saturn (HSTI).*

A bow shock was detected 24 radii upstream (that is about 1.5×10^6 km). The magnetotail spreads downstream as for the Earth and Jupiter. The plasma sheet of charged particles, familiar in Earth and Jupiter, is replaced by an agglomeration of particles with a major component of neutral particles. This different structure for Saturn is caused by the presence of ice particles in its ring system. These absorb local charged particles very effectively. The satellites again act as a source of charged and uncharged particles for the magnetosphere. The details of the magnetosphere will be explored in greater detail by the Cassini mission which reaches the neighbourhood of Saturn during the year 2004.

Uranus and Neptune

The asymmetric geometries of the intrinsic magnetic fields of these two planets sets them apart from Jupiter and Saturn. Added to this, the weakness of the solar wind at their orbital distances allows their fields to have a stronger effect. The magnetospheres were observed during the flyby of Voyager 2.

The case of Uranus is the more complicated because the magnetic field makes an angle of 59° with the rotation axis, which itself makes an angle of 98° with the plane of the orbit. Consequently, the angle between the magnetic axis and the solar wind changes between 41° and 149°. The orientation changes throughout the orbital period of 84 years, leading to a substantial change in the configuration of the field during this period. The plasma tail is in the plane of the magnetic equator near to the planet but relaxes to lie along the direction of the solar wind some 250 000 km downstream. The whole tail structure rotates about the axis formed by the line joining Uranus to the Sun. Detailed analyses will be possible in the future when a detailed survey of the environment of the planet has been undertaken.

Neptune also has a magnetosphere but with a peculiar property due to the large angle between the magnetic and rotation axes. Again the details are lacking in the absence of data but it does seem that the structure of the magnetosphere shows significant changes over the rotation period of 16 hours.

W.6.2. Examples of other planetary bodies

An intrinsic magnetism without an atmosphere: Mercury

In this case the magnetic field will be compressed on the day side and extended on the night side, as for the Earth, and the field will hold the solar wind away from the surface. There will be a bow shock and a

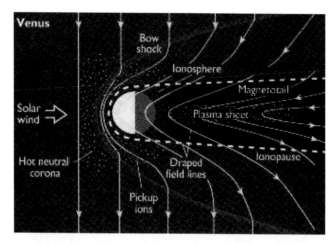

Figure W.15. *The magnetosphere of Venus—an atmosphere but no intrinsic magnetism (van Allen and Bagenal).*

magnetotail as for the Earth. The distance of the bow shock is determined by the strength of the intrinsic planetary field. The magnetotail has a restricted form because there is no atmosphere to act as a source of additional charged particles.

No intrinsic magnetism but with an atmosphere: Venus

The solar radiation induces an ionosphere with its own weak magnetism. This opposes, albeit perhaps only weakly, the solar wind plasma, preventing it from actually reaching the surface. The magnetopause is entirely symmetrical about the direction of the solar wind. The planet acts as a blunt body in the wind flow. The case of Venus is shown in figure W.15. There is a plasma sheet downstream.

No intrinsic magnetism and no atmosphere: the Moon.

The solar wind now penetrates to the surface. This is the case for the Moon, shown in figure W.16. The surface is bombarded by charged particles, causing radiation damage and fragmenting ('gardening')

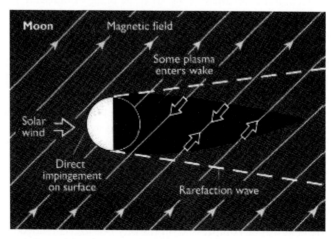

Figure W.16. *The magnetosphere of the Moon—no atmosphere and no intrinsic magnetism (van Allen and Bagenal).*

the surface to produce micro-particles. There is a shadow zone behind the main body leading to a region with few, if any, solar wind particles.

W.7. MOTION THROUGH THE INTERSTELLAR MEDIUM

The Solar System moves as a unit through the interstellar medium, which is permeated by a weak interstellar magnetic field. The solar magnetic field acts as an obstacle to the interstellar field leading to a bow shock and magnetopause, called now the *heliopause*, as for a planet. There is a shock wave ahead of the Sun and a tail behind that extends until the magnetic pressure is the same as that of the interstellar field. This point may be some 150 AU ($\sim$2.3 $\times$ 10^{10} km or four times the Sun–Pluto distance) downstream. There the heliopause melts into the interstellar field. Charged particles can enter the Solar System at this end of the heliopause. The interstellar field varies locally with time providing 'weather'. This leads to changing conditions within the interplanetary regions of the Solar System, affecting the dynamics of the solar wind to a greater or lesser extent.

The Explorer and Voyager spacecraft have moved through the Solar System making close encounters with the major planets and have now reached out to about 5 $\times$ 10^9 km. This is close to the estimated distance of the heliopause. There is some communication with the craft though the communication time is about 10 days. The craft have been operating for some 20 years, working on electrical systems fuelled by the radioactive decay of plutonium atoms (providing heat energy).

W.8. COMPANIONS TO OTHER STARS

There is no direct evidence of the magnetism of companion bodies to other stars but from Solar System studies it would seem safe to assume that the companions possess intrinsic magnetic fields with a dominant dipole component. It can also be supposed that the star emits a stellar wind. The orientations of the rotation and magnetic axes are unknown so little can be said about the details of the magnetic interactions with the stellar winds. For those systems with very small semi-major axes or periastron distances, direct magnetic interactions will be important but we have no parallels to this within the Solar System.

Problem W

W.1 The solar wind in the vicinity of Jupiter has a proton particle density 2 $\times$ 10^5 m^{-3} and a speed of 300 km s^{-1}. Given the dipole moment of Jupiter as 1.6 $\times$ 10^{20} T m^3, calculate the distance of the bow shock from the centre of the planet.

TOPIC X

PLANETARY ALBEDOES

It is clear from spacecraft photographs that some bodies have a bright appearance, particularly those that are covered by light-coloured clouds, while others are dark in appearance. No Solar-System bodies, except the Sun, are self-luminous so their brightness just depends on the amount of received sunlight they scatter. This is determined by the *visual albedo*, A_V, defined as the ratio of the amount of visible light scattered by a surface to that falling on it, and clearly $0 \leq A_V \leq 1$. The visual albedoes of the planets are given in table X.1.

Table X.1. *The visual and Bond albedoes of the planets.*

Planet	Visual albedo	Bond albedo
Mercury	0.106	0.119
Venus	0.650	0.750
Earth	0.367	0.306
Mars	0.150	0.250
Jupiter	0.520	0.343
Saturn	0.470	0.342
Uranus	0.510	0.300
Neptune	0.410	0.290
Pluto	0.300	0.367

The visual albedo, which determines the brightness of a planet or other Solar-System object, is only concerned with the *visible* radiation that is scattered and is not the only albedo of interest. There is also the *Bond albedo*, which gives the fraction of all radiation, both visible and non-visible, that is scattered. If one wishes to know the equilibrium temperature of a body exposed to solar radiation at a particular distance from the Sun then it is the Bond albedo that is involved. It will be seen from table X.1 that the Bond albedo can be either larger or smaller than the visual albedo, depending on the nature of the surface that is scattering.

X.1. THE BRIGHTNESS OF SOLAR-SYSTEM BODIES SEEN FROM EARTH

The observed brightness of a Solar-System body will depend on the intensity of the light arriving at Earth by scattering from the body in question. If the body, of radius R_B, is at a distance r_{BS} from the Sun then it intercepts energy at a rate

$$S = L_{SV} \frac{R_B^2}{4r_{BS}^2} \qquad (X.1)$$

where L_{SV} is the luminosity of the Sun in the visible range of wavelengths. A fraction A_V of this light is scattered but the scattering is only over one hemisphere, or a solid angle of 2π steradians. If the Earth is at a distance r_{EB} from the body then the intensity (energy per unit area per unit time) of the light arriving at Earth from the body, which is its brightness, is

$$I = \frac{A_V S}{2\pi r_{EB}^2} = \frac{A_V L_{SV} R_B^2}{8\pi r_{EB}^2 r_{BS}^2}. \qquad (X.2)$$

Equation (X.2) assumes that the Earth can see a completely illuminated hemisphere of the planet. This is only true for planets in superior conjunction, where the planet, Sun and Earth are in line with the planet on the far side of the Sun, or in opposition, where the Earth is between the Sun and the planet.

X.2. THE EQUILIBRIUM TEMPERATURE OF THE PLANETS

The temperature of a planetary surface depends on several factors—the internal energy generated by the planet itself, the Bond albedo, which determines how much of the incident solar radiation falls on the surface and the greenhouse effect (section 4.2.2). Here we shall just be concerned with the equilibrium temperature as determined by the Bond albedo of a body and its distance from the Sun.

The energy per unit time, coming from the Sun falling on a spherical body, is

$$S = L_\odot \frac{R_B^2}{4r_{BS}^2} \qquad (X.3)$$

which is similar to equation (X.1) except that the total luminosity of the Sun is now involved. If the Bond albedo is A_B then the fraction of it absorbed by the body is $(1 - A_B)$. We now assume that the body is spinning so that the heating effect is uniformly distributed over the whole spherical surface of the body. This is rather a severe assumption. We know, for example that there is a very large difference of temperature between the hemisphere of Mercury facing the Sun and the opposite hemisphere. However, if the body has an atmosphere then heat is transported around to give some semblance of uniform heating.

Equilibrium is established when the heat absorbed by the body from the Sun equals the heat radiated by the body. It is a common assumption that, although the body *absorbs* a fraction $(1 - A_B)$ of the radiation that falls on it, it *radiates* energy as though it were a black body. This would assume that the surface is really a black body and the albedo is due to some external factor, such as clouds. It also assumes that these same clouds will not block the emitted radiation. In this case we may equate the radiation emitted to that received to find an equilibrium temperature, T_B, from

$$4\pi\sigma R_B^2 T_B^4 = L_\odot \frac{R_B^2}{4r_{BS}^2} (1 - A_B)$$

where σ is Stefan's constant, or

$$T_B = \frac{1}{2} \left(\frac{L_\odot (1 - A_B)}{\pi\sigma r_{BS}^2} \right)^{1/4}. \qquad (X.4)$$

Inserting the values for the Earth, $A_B = 0.306$ and $r_{BS} = 1.50 \times 10^{11}$ m, we find the equilibrium temperature for the Earth as 255 K. If the Earth temperature was that average temperature ($-18°C$) then there would be no liquid water, and probably no life. The difference between the actual temperature and the theoretical temperature is due to a natural greenhouse effect.

For a simple absorbing body, without an atmosphere or any other complicating feature, the equilibrium temperature in Earth orbit is found by inserting $A_B = 0$ in equation (X.4) and is 279 K. It would not matter whether it was a black body or not since the *absorption coefficient* of a body, the fraction it absorbs of the incident radiation, equals it *emissivity*, the power it radiates as a fraction of that from a black body.

Problems X

X.1. Compare the brightness of Venus at quadrature (half the disk illuminated as seen from Earth) with that of Jupiter when it is closest to Earth.

X.2. Assuming that the asteroid Icarus (table 8.1) is always in thermal equilibrium then find its minimum and maximum temperatures during one orbit. Take the luminosity of the Sun as 4×10^{26} W.

TOPIC Y

THE PHYSICS OF TIDES

A naïve view of tides as they affect the Earth's oceans is that when the Earth's spin brings a particular region to a position facing the Moon then the water is drawn up on that side so producing a high tide. If this were the total story then there would be one tide per day but the actual situation is that there are, in general, two tides per day at any location, so clearly a closer examination of tides as a phenomenon is required.

The action of a tide is to modify the surface profile of the affected body. In the case of the Earth this primarily affects the oceans, because they are fluid and will, under the influence of tidal forces, move towards a configuration that gives an equipotential surface. Solid bodies are also distorted by tides but, because they so strongly resist changes of shape due to elastic forces, the tides are much smaller in magnitude. It is usually quoted that in mid-ocean, where the interaction of water with coastlines is unimportant, the high tide level is ~1 m above the average sea level. There are also land tides but these are much smaller—a few centimetres on the Earth.

Although here we shall be mainly concerned with Earth–Moon tidal interactions, tides are of importance in many other Solar-System situations. As examples of this we have the commensurabilities of the orbital periods of the Galilean satellites due to tidal effects between the satellites and Jupiter (section 7.3.5) and also the tidal heating of Io (Topic AB).

In dealing with tides in an analytical way it is useful to discount the distortion of the tidally affected body due to its spin. Spin creates a distortion that is axially symmetric and constant at any place and hence it does not give a tide, i.e. a periodic rise and fall of the surface. Both spin effects and tidal effects are small in magnitude and it is legitimate to apply a superposition principle and to regard the behaviour of the surface at any point as a sum of the two effects taken separately. For this reason in what follows, unless stated otherwise, we regard the Earth as a non-spinning body.

Y.1. THE BASICS OF THE TIDE-RAISING MECHANISM

The Earth and the Moon are both in orbit around the centre of mass of the system. Since the ratio of the masses of the two bodies is 81.3, the distance of the centre of mass from the centre of the Earth is 1/82.3 times the Earth–Moon distance. This is 4680 km, which puts the centre of mass within the Earth, but in figure Y.1 we have shown it at C, outside the Earth. The figure is not to scale in any case. For a non-spinning Earth the line joining the centre of the Earth, E, to the nearpoint of the Earth to the Moon, N, always points in the same direction and both E and N follow similar circular paths. For that reason the centrifugal acceleration at E and N is always the same and there

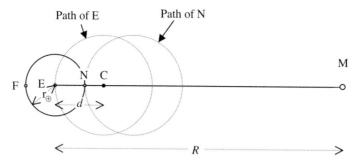

Figure Y.1. *The path of the centre of the Earth,* E, *and the nearside point,* N, *for a non-spinning Earth. The Moon is at* M.

is no relative acceleration (force per unit mass) at these points due to this cause. The only relative acceleration is due to the fact that they are at different distances from the Moon and is

$$A_{NE} = \frac{GM_M}{(R - r_\oplus)^2} - \frac{GM_M}{R^2} \tag{Y.1}$$

where R is the Earth–Moon distance, $r_\oplus$ is the radius of the Earth and M_M is the mass of the Moon.

Since $r_\oplus \ll R$ we can use the binomial expansion to give

$$A_{NE} = \frac{GM_M}{R^2}\left(1 + 2\frac{r_\oplus}{R}\right) - \frac{GM_M}{R^2} = \frac{2GM_M r_\oplus}{R^3}. \tag{Y.2}$$

This acceleration at N relative to E acts towards the Moon. By a similar analysis, again involving a binomial expansion, there will be an acceleration at F, the farside point, of equal magnitude and pointing away from the Moon. These two accelerations, shown in figure Y.3, give forces stretching the Earth along the Earth–Moon direction.

There are also accompanying compressive accelerations acting in the cross section perpendicular to the Earth–Moon direction and the point S in figure Y.2 is at the surface of this cross section. The acceleration at S towards M, the Moon, is

$$A_S = \frac{GM_M}{R^2 + r_\oplus^2}.$$

The component of this acceleration in the Earth–Moon direction is

$$A_{EM} = \frac{GM_M}{R^2 + r_\oplus^2}\frac{R}{(R^2 + r_\oplus^2)^{1/2}}.$$

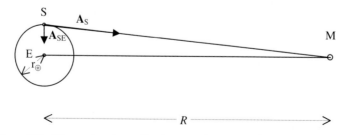

Figure Y.2. *The acceleration at* S *relative to the centre of the Earth,* E, *is* $\mathbf{A}_{SE}$.

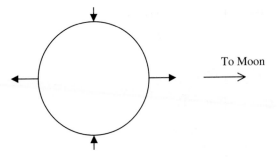

Figure Y.3. *The pattern of forces on the Earth along the Earth–Moon line and in the diametrical plane perpendicular to the Earth–Moon line.*

If in the binomial expansion we ignore terms involving $r_\oplus^2/R^2$ compared with unity then we find

$$A_{EM} = \frac{GM_M}{R^2}, \qquad (Y.3)$$

which means that there is no acceleration of S relative to E in the direction EM. However in direction SE the relative acceleration is

$$A_{SE} = \frac{GM_M}{R^2 + r_\oplus^2} \frac{r_\oplus}{(R^2 + r_\oplus^2)^{1/2}} \approx \frac{GM_M r_\oplus}{R^3}, \qquad (Y.4)$$

that is one half of the value found in equation (Y.2). These accelerations are illustrated in figure Y.3. There are, of course, accelerations relative to E at all points on the surface but those at the points shown indicate the stretching along the Earth–Moon direction and the compression in perpendicular directions.

As the Earth spins so the oceans on different parts of its surface are exposed to maximum tidal effects. However, due to inertia the water at high tide takes a little time to recede after it passes the position of highest tidal forces. At the same time the water exposed to highest tidal forces takes a little time to rise to the maximum height. For this reason the position of the high tide is ahead of the direction of the Moon. It is as though the spin of the Earth has dragged forward the position of the high tide and, indeed, as one would expect with drag, there are frictional forces between the solid Earth and the oceans. This advance of the tide has important long-term effects on the Earth–Moon relationship (section Y.3). For each complete rotation of the Earth each point arrives twice at the location of a tidal bulge and these will be the times of high tide.

It is well known that successive high tides may be of very different magnitude and the reason for this is illustrated in figure Y.4. This difference is due to the fact that the Moon can be far from the Earth's equatorial plane. The combination of the tilt of the Earth's spin axis to the ecliptic, $23\frac{1}{2}°$, and the Moon's orbit being at about 5° to the ecliptic, can make the Earth–Moon line inclined at an angle up to about 28° to the equatorial plane. Figure Y.4 is in the plane containing the high tide—that is, the plane containing the Moon. The high tide at A is clearly a small one but about 12 hours later when point A has moved around to A' the tide is much higher. This inequality of the *diurnal tides* can be so marked that there is effectively only one high tide per day.

The effects of coastal features can also be quite strong and high tides of more than 10 m can occur, for example at various locations bordering the English Channel. Another curious phenomenon is that the port of Southampton may effectively have four high tides per day because of a backwash effect due to the Isle of Wight. It is not possible to deal in a general way with local

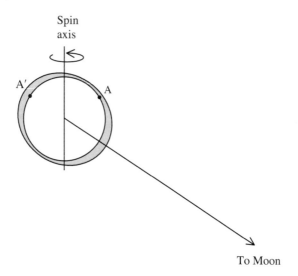

Figure Y.4. *The tides at 12 hour intervals at a point on the Earth's surface can be at very different heights.*

special effects but we shall be finding an expression to give an estimate of the magnitude of a high tide in mid-ocean in section Y.4.

Y.2. SPRING TIDES AND NEAP TIDES

The times when high tides occur would suggest that the only body causing tides is the Moon but the Sun also has a significant influence. From equations (Y.2) and (Y.4) the strength of the tidal effect depends on M/R^3 where M is the mass of the tide-raising body and R its distance. The ratio of this quantity for the Sun and the Moon in relation to the Earth is

$$\frac{1.99 \times 10^{30}}{(1.496 \times 10^{11})^3} \frac{(3.84 \times 10^8)^3}{7.35 \times 10^{22}} = 0.46.$$

Although the Moon's contribution is certainly dominant, the contribution of the Sun is by no means negligible and the magnitudes of high tides can be greatly affected by the way in which the contributions of the Moon and the Sun combine.

If the Sun, Moon and Earth are collinear in projection looking down on to the ecliptic, which means either at full Moon or new Moon, then the tidal effects of the Moon and the Sun reinforce each other to give maximum high tides. This configuration, with the tidal accelerations of the two bodies illustrated, is shown in figure Y.5a. Such strong high tides are called *spring tides*, where *spring* is derived from the German *springen* meaning *to leap up* and must not be confused with the season. Spring tides can occur throughout the year.

Another extreme situation is seen when the Moon is at quadrature, which means the position when the Moon is just one-half illuminated as seen from Earth. This configuration is shown in figure Y.5b. In this case the compression and extension effects of the Moon and the Sun are opposed to each other and the magnitude of the tide is at a minimum. Such tides are called *neap tides*.

The actual heights of the tides depend on other factors as well, in particular where the Moon and the Earth are in their respective orbits around the Earth and the Sun. The orbital eccentricity of the Moon, 0.056, gives a considerable difference in tidal effect between apogee and perigee; the smaller

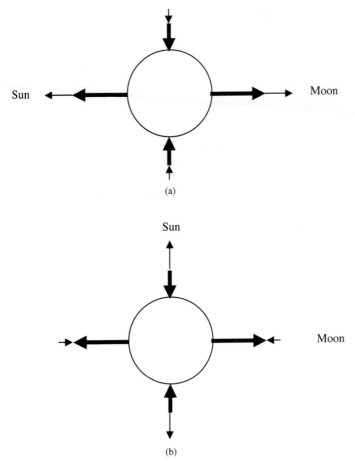

Figure Y.5. *The tidal contributions of the Moon (thick line) and Sun (thin line). (a) Spring tides when the Sun, Earth and Moon are collinear. (b) Neap tides when the Moon is at quadrature.*

eccentricity of the Earth's orbit, 0.017, gives a smaller, but still significant, difference, due to the Sun, between aphelion and perihelion. These points are illustrated in Problem Y.1.

Y.3. THE RECESSION OF THE MOON FROM THE EARTH

If the Earth and the Moon were tidally locked, so that the spin periods of the Moon and the Earth both equalled their orbital period, and with the Sun's influence ignored, then the surface of a water-covered Earth would be an equipotential surface. This is not true for the real Earth for two main reasons. The first reason can be understood by considering a non-spinning Earth. As the Moon orbits so the equipotential surface rotates at the same angular speed. However, due to inertia the oceans could not react immediately to the ever-changing equipotential surface and there would be some lag in the high tide. However, since the Moon's orbital period is quite large the water surface has time to adjust, so the lag in the tide would be small and quite negligible.

Of greater importance is the effect of the Earth's comparatively rapid spin. Due to friction this pulls in a forward direction so that the high tide direction is actually ahead of the Moon. Once

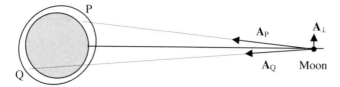

Figure Y.6. *The accelerations on the Moon due to the nearside bulge and the farside bulge. with resultant* **A**$_\perp$.

again the effect is small, and very difficult to analyse because it depends on the way that friction operates and this could depend on the detailed topology of the continents that channel the water in various ways. Although the advance of the tide is small it has important long-term effects. Figure Y.6 shows the advanced high tide (greatly exaggerated for clarity) and the tidal bulges at P and Q. The force on the Moon can be approximately represented by the central force due to a spherically symmetric Earth plus components due to single masses in the vicinity of P and Q. The accelerations that these give on the Moon are shown as **A**$_P$ and **A**$_Q$. Since P is closer to the Moon than Q, **A**$_P$ has a larger magnitude than **A**$_Q$ and there is thus a net component of acceleration due to the bulges, **A**$_\perp$, in the direction of motion of the Moon. The torque due to **A**$_\perp$ acts in the same sense as the Moon's motion and hence increases the angular momentum associated with its orbit around the Earth. This causes it to retreat from the Earth; another effect is to increase the eccentricity of its orbit but we shall not deal with that.

Another effect that can be seen from figure Y.6 is that forces on the Earth will be acting on the bulges at P and Q that are equal and opposite to those shown, which act on the Moon. The net effect of these is to oppose the Earth's spin and to slow it down. This interaction, which increases the angular momentum in the Moon's orbit and also reduces that in the Earth's spin, continues until the two periods become equal at something like 50 present Earth days (Problem Y.2). At this stage the two bodies are completely tidally locked and the whole system rotates like a solid body. This account excludes the complicating factor of the Sun's influence that makes the whole mechanism much more difficult to envisage, although the general pattern ending with a tidally-locked Earth–Moon system is as described.

We can estimate the rate of lunar recession using ancient astronomical records. The Babylonians, for example, made records of when eclipses of the Sun took place and, using the current length of the day, the time of day when they took place can be calculated. Based on such calculations they were observing eclipses of the Sun in the middle of the night, which cannot be true. The discrepancies can be accounted for if the day has lengthened over the past few thousand years at a rate of 1.6 ms per century. Thus over 4000 years, say, the length of the day has changed by 64 ms or, over the 4000 year period, it has been on average 32 ms shorter than now. Thus the time slippage over 4000 years is $0.032 \times 4000 \times 365\,\mathrm{s} = 13$ hours, a value that explains the ancient records.

Neglecting any effect of the Sun we can assume that the angular momentum of the Earth–Moon system has remained constant. This is

$$H = \alpha_\oplus M_\oplus r_\oplus^2 \omega_\oplus + \alpha_M M_M r_M^2 \omega_M + (GM_\oplus R)^{1/2} M_M \tag{Y.5}$$

where the α are the moment of inertia factors (Topic S), the ω are the spin periods of the Earth and Moon respectively and the other quantities have previously been defined. The first term is the angular momentum in the Earth's spin, the second term that in the Moon's spin and the final term that of the Moon in its orbit (see equation M.15a). It is easily verified that the second term on the right-hand side is very small compared with the other two terms and so can be ignored. Replacing $\omega_\oplus$ by $2\pi/P_\oplus$, where $P_\oplus$ is the Earth's spin period, and differentiating equation (Y.5) with respect to time

$$-\frac{2\pi\alpha_\oplus M_\oplus r_\oplus^2}{P_\oplus^2}\frac{\mathrm{d}P_\oplus}{\mathrm{d}t}+\frac{1}{2}\left(\frac{GM_\oplus}{R}\right)^{1/2}M_\mathrm{M}\frac{\mathrm{d}R}{\mathrm{d}t}=0$$

or

$$\frac{\mathrm{d}R}{\mathrm{d}t}=\frac{4\pi\alpha_\oplus r_\oplus^2}{P_\oplus^2 M_\mathrm{M}}\left(\frac{RM_\oplus}{G}\right)^{1/2}\frac{\mathrm{d}P_\oplus}{\mathrm{d}t}. \tag{Y.6}$$

The value of $\mathrm{d}P_\oplus/\mathrm{d}t$ is 0.0016 s per century or 5.07×10^{-13}. This gives $\mathrm{d}R/\mathrm{d}t=0.029\,\mathrm{m\,yr^{-1}}$, which seems quite little but is significant over Solar-System timescales. Thus in 1000 million years the Moon will have retreated by 29 000 km and, since its angular diameter as seen from Earth would then be 7% smaller, total eclipses of the Sun would not be possible.

Another effect of this Earth–Moon interaction is that energy is dissipated by friction between the moving oceans and the solid Earth. This happens particularly in places like the Bering Straits between Siberia and Alaska and the Irish Sea where the seas are channelled through narrow passages. The rate of energy loss can be estimated from the rate of slowdown of the Earth's spin and the rate of recession of the Moon. The combined energy in the Earth's spin and the Moon's orbit is

$$E=\tfrac{1}{2}\alpha_\oplus M_\oplus r_\oplus^2\omega_\oplus^2-\frac{GM_\oplus M_\mathrm{M}}{R}=\frac{2\pi^2\alpha_\oplus M_\oplus r_\oplus^2}{P_\oplus^2}-\frac{GM_\oplus M_\mathrm{M}}{R}$$

which gives

$$\frac{\mathrm{d}E}{\mathrm{d}t}=-\frac{4\pi^2\alpha_\oplus M_\oplus r_\oplus^2}{P_\oplus^3}\frac{\mathrm{d}P_\oplus}{\mathrm{d}t}+\frac{GM_\oplus M_\mathrm{M}}{R^2}\frac{\mathrm{d}R}{\mathrm{d}t}. \tag{Y.7}$$

The first term in equation (Y.7) is dominant and the value is $-2.3\times10^{12}\,\mathrm{W}$. Part of the search for renewable energy sources is in ways of exploiting this energy.

Y.4. THE MAGNITUDE OF THE MID-OCEAN TIDE

While the details of the tides depend on the detailed topography of the Earth it is possible to make a model that indicates the magnitude of a high tide in mid-ocean. In figure Y.7a is shown a model Earth which, in the absence of tide-raising bodies, is a sphere, the outermost shell of which is an ocean covering the whole globe. For the purposes of analysing the tidal effect of the Moon (we ignore the Sun) it is considered as a superposition of two spheres—a hypothetical water sphere of radius $r_\oplus$ and mass M_W ($1.09\times10^{24}\,\mathrm{kg}$) plus a smaller solid sphere of mass M_S ($4.91\times10^{24}\,\mathrm{kg}$) and radius less than $r_\oplus$. The mass of the solid sphere is the difference between that of the Earth and that of the complete water sphere. The effect of the Moon is to distort the water sphere while leaving the solid sphere unaffected. We take it that the water sphere is distorted into a prolate spheroid which, to preserve its volume, has a unique axis, pointing towards the Moon, of length $r_\oplus+\varepsilon$ and the other two principal axes of length $r_\oplus-\tfrac{1}{2}\varepsilon$. This configuration is shown in figure Y.7b.

One possible approach would be to use the condition that the water surface is equipotential but since the system is a dynamic one it is unlikely that this will be so. It is clear that as the high-tide position is approached the net force per unit mass acting on water at the surface is such as to draw it towards the Moon relative to the centre of the Earth. Later, in the post-high-tide situation, the net force is moving the surface water towards the centre of the Earth. It is clear that at high tide the forces are in balance and we shall use this balance to estimate the height of a high tide. The effect of the Moon at the water surface is easily found since the Moon can be considered as a point mass. To

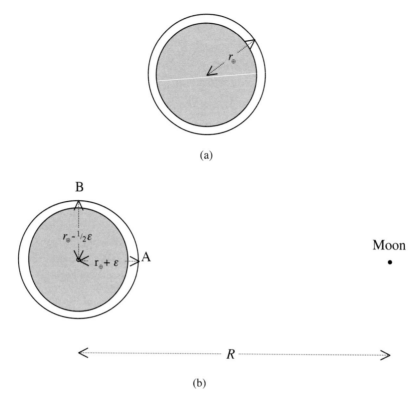

(a)

(b)

Figure Y.7. *(a) A model non-spinning Earth in the absence of the Moon decomposed into a solid sphere, with density equal to the difference between that of the Earth and of water, plus a water sphere. (b) In the presence of the Moon the water sphere distorts into a spheroid while the solid sphere is unaffected.*

find the force per unit mass on surface material at high tide we shall use a known result on the oscillations of spherical liquid spheres under gravity.

Y.4.1. Oscillations of fluid spheres

The configuration of the water sphere at any time is a prolate spheroid with the major axis pointing towards the Moon. In a frame of reference in which the Earth is not spinning the major axis of this prolate spheroid rotates with a period of about one day, thus giving every place on the Earth two high tides per day. At any point on the Earth's surface what is seen is a periodic rise and fall of the water surface, as would be seen if the hypothetical water sphere was in a state of oscillation. The period of oscillation of a fluid sphere of density ρ in its simplest mode is given by

$$P = \frac{\pi}{\sqrt{G\rho}}, \tag{Y.8}$$

and each surface element executes simple harmonic motion. This means that the force per unit mass on a surface element is proportional to its displacement from the undistorted spherical surface and equals ω^2 where ω is the angular frequency. Thus for a displacement ε the force per unit mass is

$$F_{\mathrm{W}} = -\omega^2 \varepsilon = -\frac{4\pi^2}{P^2}\varepsilon = -4G\rho\varepsilon = -\frac{3GM_{\mathrm{W}}}{\pi r_\oplus^3}\varepsilon \tag{Y.9}$$

where the negative sign indicates that it acts towards the centre when ε is positive. This force is due to the distorted hypothetical water sphere of mass M_W but there will also be an inward force due to the central undistorted solid sphere. Clearly this will act inwards to restore the surface element to its equilibrium position at distance $r_\oplus$ from the centre. To determine the form of this force we express the potential at distance r from the centre of a sphere of mass M as

$$\Omega = -\frac{GM}{r}.$$

Hence the potential difference between points at distances $r + \varepsilon$ and r is

$$\Delta\Omega = -\frac{GM}{r+\varepsilon} + \frac{GM}{r} = \frac{GM}{r^2}\varepsilon.$$

And the force at $r + \varepsilon$ towards r is

$$F = -\frac{d(\Delta\Omega)}{dr} = \frac{2GM}{r^3}\varepsilon.$$

Putting $r = r_\oplus$, $M = M_S$ and using a negative sign to indicate inwardly acting the restoring force due to the solid sphere is

$$F_S = -\frac{2GM_S}{r_\oplus^3}\varepsilon. \tag{Y.10}$$

The forces F_W and F_S are both relative to a point on a surface at a fixed distance, $r_\oplus$ from the centre of the body. For this reason both forces are also relative to the centre of the body; when they are negative they give an acceleration towards the body centre and when they are positive the acceleration is away from the body centre.

The outwardly acting force per unit mass relative to the body centre due to the Moon on a surface element is, making the usual approximations, as given by equation (Y.2),

$$F_M = \frac{2GM_M}{R^3}(r_\oplus + \varepsilon) \approx \frac{2GM_M}{R^3}r_\oplus, \tag{Y.11}$$

since $r_\oplus \gg \varepsilon$. To find the height of the tide we now use the balance of forces to give

$$F_W + F_S + F_M = 0$$

or

$$-\frac{3GM_W}{\pi r_\oplus^3}\varepsilon - \frac{2GM_S}{r_\oplus^3}\varepsilon + \frac{2GM_M}{R^3}r_\oplus = 0.$$

This gives

$$\varepsilon = \frac{2M_M r_\oplus^4}{R^3[(3M_W/\pi) + 2M_S]}. \tag{Y.12}$$

Inserting values $M_W = 1.09 \times 10^{24}$ kg, $M_S = 4.91 \times 10^{24}$ kg, $M_M = 7.35 \times 10^{22}$ kg, $r_\oplus = 6.4 \times 10^6$ m and $R = 3.84 \times 10^8$ m we find

$$\varepsilon = 0.40 \text{ m}.$$

A more elaborate analysis gives a lower value—about 0.24 m.

Problems Y

Y.1 Taking into account the eccentricities of the Moon's orbit around the Earth (0.056) and the Earth's orbit around the Sun (0.017) find the ratio of the maximum to the minimum tidal forces at the Earth's surface.

Y.2 Show that when the day and the month become equal they will both be equal to approximately 50 present Earth days. The moment-of-inertia factor of the Earth is given in table S.1.

TOPIC Z

DARWIN'S THEORY OF LUNAR ORIGIN

In 1878 George Darwin suggested that when the Earth was formed it was spinning so quickly that it became rotationally unstable and broke up to give the Earth–Moon system. In section Y.3 it was shown that the tidal interaction between the Earth and the Moon causes the Moon gradually to recede from the Earth. As it does so it gains orbital angular momentum and to compensate for this the Earth spins more slowly. Although the tidal influence of the Sun is a disturbing influence we can, for present purposes, take the angular momentum of the Earth–Moon system as remaining approximately constant. It is possible, from this conservation principle alone, to deduce the closest distance that the Moon could ever have been from the Earth. This would be when the orbital period of the Moon–Earth orbit (around the centre of mass of the system) equals the spin period of the Earth. The Moon would be locked with one face towards the Earth so that Earth plus Moon would be spinning as a rigid system (figure Z.1). It turns out that this period is about $5\frac{1}{2}$ hours. We now deduce this.

One quantity required is the angular momentum associated with the orbits of the Earth and the Moon about their centre of mass. This is

$$M_\oplus \left(\frac{M_M R}{M_\oplus + M_M} \right)^2 \omega_o + M_M \left(\frac{M_\oplus R}{M_\oplus + M_M} \right)^2 \omega_o = \frac{M_\oplus M_M}{M_\oplus + M_M} R^2 \omega_o$$

where M and r are the mass and radius of a body, R the radius of the Earth–Moon orbit and subscripts $\oplus$, m and o represent the Earth, Moon and orbit respectively. Hence the total present angular momentum of the Earth–Moon system is

$$H = \alpha_\oplus M_\oplus r_\oplus^2 \omega_\oplus + \frac{M_\oplus M_m}{M_\oplus + M_m} R^2 \omega_o + \alpha_m M_m r_m^2 \omega_o \tag{Z.1}$$

in which α represents the moment-of-inertia factor (Topic S). For the initial state $\omega_\oplus = \omega_o = \omega_I$ and it is convenient to eliminate the Earth–Moon distance by equating the gravitational force to the mass times centripetal acceleration of either the Earth or the Moon. This gives

$$\frac{GM_\oplus M_m}{R^2} = \frac{M_\oplus M_m}{M_\oplus + M_m} R\omega_I^2. \tag{Z.2}$$

For the nearest approach, from equations (Z.1) and (Z.2)

$$(\alpha_\oplus M_\oplus r_\oplus^2 + \alpha_m M_m r_m^2)\omega_I + \frac{G^{2/3} M_\oplus M_m}{(M_\oplus + M_m)^{1/3} \omega_I^{1/3}} = H. \tag{Z.3}$$

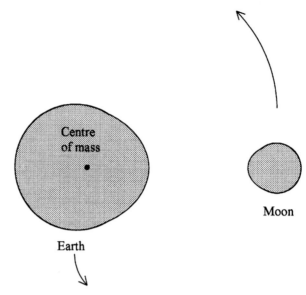

Figure Z.1. *The early Earth–Moon system according to Darwin. The spin of both bodies is locked to the orbital rate as both bodies orbit the centre of mass.*

The value of H can be derived from equation (Z.1) with the present parameters of the Earth–Moon system inserted. Using the accepted value of $\alpha_{\oplus} = 0.33$, $\alpha_{\mathrm{m}} = 0.40$ and the sidereal month as 27.32 days, $H = 3.44 \times 10^{34}\,\mathrm{kg\,m^2\,s^{-1}}$. The value of ω_{I} satisfying equation (Z.3) is then found to be equivalent to an orbital period of 4 h 35 m, somewhat different from Darwin's estimate of 5 h 36 m. However, if one takes $\alpha_{\oplus} = 0.4$ then Darwin's result is obtained. Darwin's result corresponds to a centre-to-centre separation of the two bodies of about 16 400 km or about 2.6 Earth radii, whereas the shorter period derived here corresponds to 2.2 Earth radii.

TOPIC AA

THE ROCHE LIMIT AND SATELLITE DISRUPTION

In Topic Y the distortion of a body due to tidal influences was considered. The distortions, such as those between planets and satellites, are generally small although they can have important effects, such as causing the recession of the Moon from the Earth. In the lifetime of the Solar System there may have been occasions when bodies approached so closely that the tidally produced distortions became very large—in fact so large that one of the bodies was completely disrupted. That is what we are concerned with here.

AA.1. THE ROCHE LIMIT FOR FLUID BODIES

The only fluid bodies in the Solar System are the major planets and there are no realistic circumstances when they would be disrupted by interaction with another body. Nevertheless, just for completeness, we give here some indication of what occurs when a fluid body is tidally disrupted.

In modelling astrophysical situations involving interactions between stars one often uses the so-called *Roche model star*. This model star has a finite and constant volume but all the mass is concentrated at the central point. This has the advantage for theoretical work that the star may greatly distort while its gravitational field remains unchanged—being that due to a point mass. This is the model used by Jeans in his tidal model of the origin of the Solar System (section AH.1). In this case the increasing distortion of the star due to the approach of another star eventually leads to it occupying the largest closed equipotential surface that can contain its volume. Any closer approach of the two stars leads to disruption of the Roche model star.

The Roche model is an extreme case of a body with central condensation. At the other extreme one can consider a uniform incompressible body—for example, a hypothetical liquid body. The condition still exists that an equilibrium configuration of the liquid body is an equipotential surface. However, in this case its contribution to the local gravitational potential changes when it distorts. Determining what its shape will be in the presence of another body is a self-consistency problem. The shape it takes up must contribute to the overall gravitational potential in such a way that its surface is an equipotential. This problem was considered by Darwin in 1906 when he considered the distortion of two supposedly-fluid stars in a binary system.

A result that comes from detailed analysis of the disruption of a liquid body by a point mass is that the liquid body, of density ρ_S, will disrupt when its distance from the disrupting body, of density ρ_P and radius a_P, is

$$R = 2.46 \left(\frac{\rho_P}{\rho_S}\right)^{1/3} a_P. \tag{AA.1}$$

This equation gives the *Roche limit* and, since the factor involving densities is usually close to unity, it is often stated that no satellite or other body can resist disruption within a distance 2.5 planetary radii from a planetary centre.

AA.2. THE ROCHE LIMIT FOR A SOLID BODY

We shall now consider the more realistic case of the tidal disruption of a solid satellite body. We know that solid bodies are elastic to some extent and can distort without disruption, e.g. by flattening due to their spin. However, distortion is resisted by the mechanical strength of the solid material and if either compression or tension exceed some critical limit then the body will break up. Here we shall assume that the satellite will break up before it departs too severely from a spherical form so that we can make the approximation that, before it disrupts, its total mass acts as though it was all at the centre. This approximation is similar to that used for the Roche model but here it applies just to a spherically symmetric body and not to one with a very high central condensation.

Another condition we introduce is that the satellite is tidally locked to the planet so that it always presents one face to the planet—much as the Moon does towards the Earth. This condition simply means that the spin period of the satellite equals its orbital period around the planet. If the satellite gradually spirals inwards towards the planet due to some dissipation of energy then its tidal distortion will increase until eventually it breaks up. Such a situation will probably occur with Triton, Neptune's large satellite. Because of its retrograde orbit, tidal interaction with Neptune is causing its orbit to decay. A retrograde orbit gives the opposite effect to what was described in section Y.3 which showed that the Moon is *receding* from the Earth. Eventually Triton will pass within the Roche limit and then, at some stage, it will be disrupted. Depending on the form of the disruption it will either break up into a few smaller satellites or perhaps, eventually, into a myriad of smaller bodies that will reinforce the ring system.

Before we consider the actual disruption of a satellite we shall first derive the condition under which it begins to come under tension. Figure AA.1 shows a satellite, S, of radius a_S, in a circular orbit of radius R around a planet, P, of radius a_P. The acceleration (force per unit mass) at point Q towards P is

$$A_Q = \frac{GM_P}{(R - a_S)^2} - \frac{GM_S}{a_S^2} - (R - a_S)\omega^2. \tag{AA.2}$$

The first term on the right hand side is due to the mass of the planet, M_P, the second due to the mass of the satellite, M_S, and the final term the centrifugal acceleration corresponding to the orbital angular speed, ω. Similarly the acceleration at point C, the centre of the satellite, towards P is

$$A_C = \frac{GM_P}{R^2} - R\omega^2. \tag{AA.3}$$

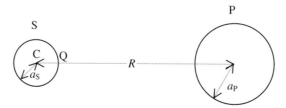

Figure AA.1. *A satellite,* S, *and primary body,* P.

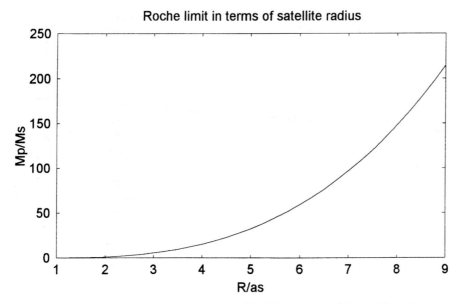

Figure AA.2. *The Roche limit as a function of M_P/M_S in terms of the satellite radius.*

Hence the acceleration of Q relative to C is

$$A_{QC} = A_Q - A_C = \frac{GM_P}{(R - a_S)^2} - \frac{GM_S}{a_S^2} + a_S\omega^2 - \frac{GM_P}{R^2}. \tag{AA.4}$$

If A_{QC} is zero then there is no net tension or compression between the points C and Q and the corresponding value of R, R_L, is the normally defined Roche limit. If we now substitute

$$\omega^2 = \frac{G(M_P + M_S)}{R_L^3}, \qquad \frac{M_P}{M_S} = \phi, \qquad \frac{R_L}{a_S} = \delta$$

then we find

$$\phi = \left(1 - \frac{1}{\delta^3}\right)\left(\frac{1}{(\delta - 1)^2} + \frac{1}{\delta^3} - \frac{1}{\delta^2}\right)^{-1}. \tag{AA.5}$$

This gives the ratio of masses corresponding to various Roche limits expressed in units of the satellite radius. This relationship is shown in figure AA.2. To convert this to give the Roche limit in terms of the planetary radius we use

$$\frac{R_L}{a_P} = \frac{R_L}{a_S}\frac{a_S}{a_P} = \frac{R_L}{a_S}\left(\frac{1}{\phi}\frac{\rho_P}{\rho_S}\right)^{1/3}. \tag{AA.6}$$

Modifying the results in figure AA.2 as indicated by equation (AA.6) gives figure AA.3, which shows the Roche limit for various mass ratios and density ratios. For any density ratio it gives the apparently paradoxical result that the Roche limit, expressed in planetary units, gets *less* as the mass ratio becomes greater. This is not actually a paradox; as ϕ increases so R_L increases, as one would expect. For fixed densities if a_P is changed by a factor ϕ then M_P increases by a factor ϕ^3. However, from figure AA.2 it is found that R_L increases less rapidly than $M_P^{1/3}$, or ϕ. Hence R_L increases less rapidly than α and so R_L/a_P is a declining function of ϕ and hence of M_P.

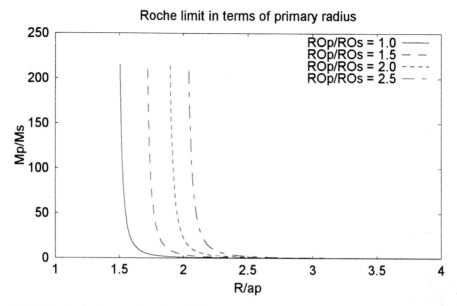

Figure AA.3. *The Roche limit as a function of M_P/M_S in terms of the planet radius for various density ratios.*

We have now found the Roche limit, defined as the distance at which the body just begins to experience tensional forces. Instinctively we understand that the ability of a body to stand up to a stretching tidal field will depend on its size. Although Saturn's rings were probably formed by a large body straying within the Roche limit the boulder-sized bodies that constitute the ring are not themselves being torn apart—and neither are the shepherd satellites that control the motion of these smaller bodies.

AA.3.   THE DISRUPTION OF A SOLID SATELLITE

We now consider the acceleration of a point X (figure AA.4) contained within the satellite and situated on the line joining the centre of the satellite to the centre of the planet. Assuming that the satellite has a uniform density, by similar reasoning that led to equation (AA.4), the acceleration of X with respect to C is

$$A_{XC} = \frac{GM_P}{(R-x)^2} - \frac{4}{3}\pi G\rho_S x + \frac{GM_P x}{R^3} - \frac{GM_P}{R^2}. \tag{AA.7}$$

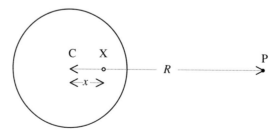

Figure AA.4. *Calculation of the acceleration of* X *relative to* C.

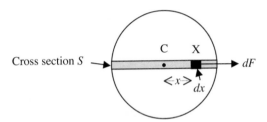

Figure AA.5. *The force acting on a small segment of a thin rod of material straddling the satellite diameter in the direction of the tidal force.*

We now assume that the radius of the satellite is much less than R so that we can use a binomial theorem expansion of the first term on the right-hand side giving

$$A_{\mathrm{XC}} = \frac{GM_\mathrm{P}}{R^2}\left(1 + \frac{2x}{R}\right) - \frac{4}{3}\pi G\rho_\mathrm{S}x + \frac{GM_\mathrm{P}x}{R^3} - \frac{GM_\mathrm{P}}{R^2} = \frac{3GM_\mathrm{P}x}{R^3} - \frac{4}{3}\pi G\rho_\mathrm{S}x. \qquad \text{(AA.8)}$$

We now consider a thin uniform rod of material, of cross-sectional area S through a diameter containing X as illustrated in figure AA.5. A section of the rod, of length dx, has a force acting on it towards the planet and relative to the point C

$$\mathrm{d}F = GS\rho_\mathrm{S}\left(\frac{3M_\mathrm{P}}{R^3} - \frac{4}{3}\pi\rho_\mathrm{S}\right)x\,\mathrm{d}x.$$

Thus the total force relative to C on that part of the rod on the planet side of C, acting towards the planet, is

$$F = GS\rho_\mathrm{S}\left(\frac{3M_\mathrm{P}}{R^3} - \frac{4}{3}\pi\rho_\mathrm{S}\right)\int_0^{a_\mathrm{S}} x\,\mathrm{d}x = \frac{1}{2}GS\rho_\mathrm{S}a_\mathrm{S}^2\left(\frac{3M_\mathrm{P}}{R^3} - \frac{4}{3}\pi\rho_\mathrm{S}\right). \qquad \text{(AA.9)}$$

There will be an equal and opposite force relative to C acting on that half of the rod away from the planet and these forces give a stress at C equal to F/S. If the stress at C is greater than the tensile strength of the satellite material, T, then the satellite will break up. The disruption distance R_D then comes from

$$T = \frac{F}{S} = \frac{1}{2}G\rho_\mathrm{S}a_\mathrm{S}^2\left(\frac{3M_\mathrm{P}}{R_\mathrm{D}^3} - \frac{4}{3}\pi\rho_\mathrm{S}\right)$$

or

$$R_\mathrm{D} = \left(\frac{3M_\mathrm{P}}{(2T/G\rho_\mathrm{S}a_\mathrm{S}^2) + \frac{4}{3}\pi\rho_\mathrm{S}}\right)^{1/3}. \qquad \text{(AA.10)}$$

The strength of the material of small bodies in the Solar System is very variable in character ranging from very hard, dense and strong minerals to low density and rather frangible material. We represent this by taking three values of $T - 10^6\,\mathrm{N\,m^{-2}}$, $10^7\,\mathrm{N\,m^{-2}}$ and $10^8\,\mathrm{N\,m^{-2}}$. We take M_P as $2.0 \times 10^{27}\,\mathrm{kg}$ and $\rho_\mathrm{S} = 2 \times 10^3\,\mathrm{kg\,m^{-3}}$. Figure AA.6 shows the variation of R_D with a_S for each value of the tensile strength. For large bodies the first term in the divisor of the right-hand side of equation (AA.10) becomes small compared with the other term and so R_D goes towards a limiting value. On the other hand for very small a_S, $R_\mathrm{D} \propto a_\mathrm{S}^{2/3}$. We have taken the mass equal to that of Jupiter, which has a radius of $70\,000\,\mathrm{km}$. From the figure it may be seen that, for the greatest tensile strength, a body with radius more than $400\,\mathrm{km}$ could survive in orbit just above the surface of Jupiter. On the other hand for the lowest tensile strength that radius would be about $40\,\mathrm{km}$.

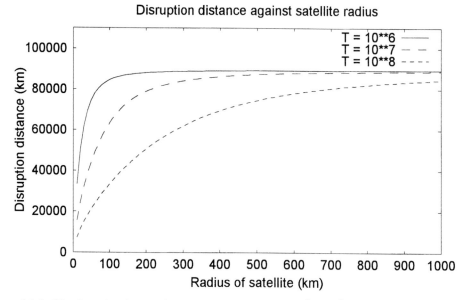

Figure AA.6. *The disruption distance for a satellite of density $2 \times 10^3 \, kg \, m^{-3}$ as a function of its radius for various tensile strengths in the presence of a Jupiter-mass planet.*

There are some assumptions in the analysis that would justify being cautious about the conclusions. However, the general trend of the results is undoubtedly correct, especially as it indicates the influence of the strength of material on the distance for disruption.

AA.4.   THE SPHERE OF INFLUENCE

The magnitude of the acceleration of the Moon due to the Sun is $5.9 \times 10^{-3} \, m \, s^{-2}$ whereas the corresponding figure for the acceleration due to Earth is $2.7 \times 10^{-3} \, m \, s^{-2}$. Although the attraction of the Sun is more than twice as large as that of the Earth, it is the Earth that mainly governs the Moon's motion.

The situation is similar to that of disrupting a satellite where the Earth–Moon system is the 'satellite' and the Sun is the primary body. Figure AA.7 shows the Sun, S, a planet, P, and a body,

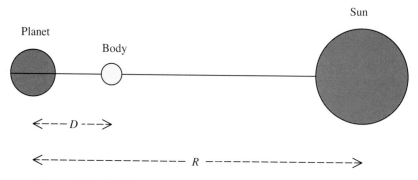

Figure AA.7. *The Sun, planet and a third body in a collinear arrangement.*

Table AA.1. *The spheres of influence of the planets.*

Planet	Sphere of influence (10^3 km)
Mercury	253
Venus	1158
Earth	1714
Mars	1242
Jupiter	64 360
Saturn	74 680
Uranus	80 270
Neptune	132 900
Pluto	8987

B, in orbit around P with the three bodies collinear. The net acceleration on B towards the Sun relative to P is

$$F = \frac{GM_\odot}{(R-D)^2} - \frac{GM_P}{D^2} - \frac{GM_\odot}{R^2}. \tag{AA.11}$$

In the limit when $F = 0$ the body is just about to be removed from the dominant influence of the planet and the value of D corresponding to this condition, D_S, is referred to as the *sphere of influence* of the planet. Assuming that $D \ll R$ and using the binomial approximation (AA.11) gives

$$D_S = \left(\frac{M_P}{2M_\odot}\right)^{1/3} R. \tag{AA.12}$$

Any body within this distance of the planet will have its motion primarily controlled by the planet, notwithstanding that there may be considerable perturbation by the Sun. Taking R as the mean orbital radius of the planet gives the spheres of influence in Table AA.1.

To illustrate the significance of these values, the outermost satellite of Saturn, Phoebe, with an orbital radius of $12\,970 \times 10^3$ km is only at one-sixth of the distance corresponding to the sphere of influence.

Problems AA

AA.1 A satellite, with the same density as the primary body, is in a circular orbit. Assuming that it is too large for its tensile strength to resist disruption then what is the smallest orbit, expressed in units of the radius of the primary body, that will enable it to remain whole?

AA.2 A satellite of density 3×10^3 kg m^{-3} and tensile strength 5×10^7 N m^{-2} is in a circular orbit of radius 10^7 m around a primary body of mass 10^{25} kg. What is the maximum radius it can have if it is to resist disruption?

TOPIC AB

TIDAL HEATING OF IO

AB.1. ELASTIC HYSTERESIS AND Q VALUES

If a uniform rod of a perfectly elastic material is stretched, then the amount by which it is stretched, the *displacement*, is proportional to the stretching force, as illustrated in figure AB.1a. The work done in the stretching process is $\frac{1}{2}F_{max}e_{max}$ where F_{max} and e_{max} are the maximum force and extension respectively and this is the shaded area under the line OP. If the force is gradually reduced back to zero then the state of the rod is again represented by points on the line OP. However, most real materials are not perfectly elastic and when stretched and allowed to contract again the path on the (F, e) diagram is as shown in figure AB.1b. The work done *on* the material is the area under the curve OAP and the work done *by* the material in returning to the unstressed state is the area under the curve OBP. The difference, represented by the shaded area, appears as heat within the material. What has happened physically is that internal friction between different grains within the material has generated heat and, from the principle of conservation of energy, it cannot do as much work in contraction as was done on it in expansion. This phenomenon is known as hysteresis and a similar phenomenon occurs with magnetic materials subjected to an alternating magnetizing field.

The extent of the hysteresis is described by the quality factor, Q, as described in section R.2. Highly compressed materials within large planetary bodies might have Q values as large as 1000,

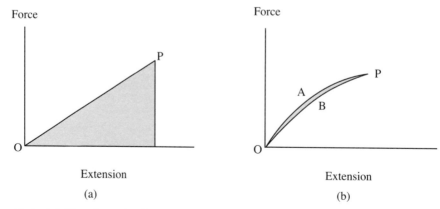

Figure AB.1. *(a) The relationship between force and extension for a perfectly elastic rod. (b) Extension followed by contraction of an imperfectly elastic rod. The hysteresis heating corresponds to the shaded area.*

indicating that less than 1% of the energy stored when it goes through a stress cycle appears as heat. By contrast, material in a less compressed state, close to the surface of a large body or within a smaller body, might have a Q value of 100 so that about 6% of the stored energy appears as heat.

AB.2.   TIDAL STRESSING IN IO

We can make a reasonable order-of-magnitude estimate of the heating of Io by a rough calculation. A representation of Io, with radius a, is shown in figure AB.2 divided by a diametrical plane perpendicular to the orbital radius vector. The tidal force due to Jupiter stretches the satellite, with each hemisphere experiencing forces away from the centre of the satellite. To estimate the total stretching force we calculate the acceleration (force per unit mass) at P relative to O where P is distant $\frac{1}{2}a$ from O. We then assume that this acceleration applies to all the mass in that hemisphere. The acceleration at P is

$$A = \frac{GM_J}{(R - \frac{1}{2}a)^2} - \frac{GM_J}{R^2} - (R - \frac{1}{2}a)\omega^2 + R\omega^2 \tag{AB.1}$$

where M_J is the mass of Jupiter, R the Jupiter–Io distance and ω the orbital speed. Eliminating ω^2 by

$$\omega^2 = \frac{GM_J}{R^3}$$

and with the usual binomial theorem approximations

$$A = \frac{3GM_J a}{2R^3}. \tag{AB.2}$$

Assuming that this is the average force per unit mass for the nearer hemisphere, the total force on it towards Jupiter is

$$F = \frac{3GM_1 M_J a}{4R^3} \tag{AB.3}$$

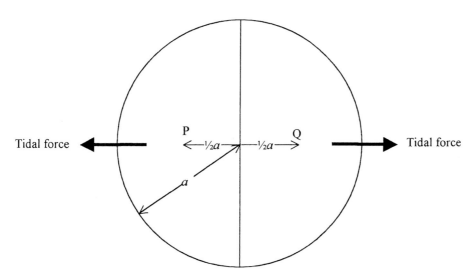

Figure AB.2. *Simulation of the periodic tidal force acting on Io divided into two hemispheres.*

where M_I is the mass of Io and the force on the far hemisphere is equal and opposite. As the next approximation we consider the satellite as a cube of side $2a$ being stretched along one principal direction. If Young's modulus for the satellite material is Y then the extension is

$$\varepsilon = \frac{F \times \text{length}}{Y \times \text{area}} = \frac{F}{2aY}$$

and the energy associated with the stretching is

$$\Psi = \tfrac{1}{2}F\varepsilon = \left(\frac{3GM_IM_J}{4}\right)^2 \frac{a}{4Y} \times \frac{1}{R^6}. \tag{AB.4}$$

The distance R varies in a cyclic fashion between $R_I(1+e)$ and $R_I(1-e)$ where R_I is the mean orbital radius of Io and e the eccentricity of its orbit. Hence the total change in energy between the extreme positions is

$$\Delta\Psi = E = \left(\frac{3GM_JM_I}{4}\right)^2 \frac{a}{4Y} \times \frac{6}{R^7}\,\Delta R = -\frac{27}{16}(GM_IM_J)^2 \frac{ae}{R_I^6 Y}, \tag{AB.5}$$

since $\Delta R = 2Re$.

The energy dissipated within the satellite per orbital period, P, is given by equation (R.5) as $2\pi E/Q$ and hence the rate of energy dissipation is found to be

$$W = -\frac{2\pi E}{QP} = (GM_IM_J)^2 \frac{27\pi ae}{8R_I^6 QPY}. \tag{AB.6}$$

We now consider the values of e, Q and Y to be used in equation (AB.6). The usually quoted value of e is 0.0001 although the fractional uncertainty in that value must be high. We take a $Y = 5 \times 10^{10}\,\mathrm{N\,m^{-2}}$, a typical figure for meteorite material, and $Q = 200$ on the grounds that satellite material is not highly compressed and may resemble crustal or upper mantle material on Earth. With these value we find $W = 2.8 \times 10^{13}\,\mathrm{W}$, which, given the uncertainties in the model, is reasonably close to the accepted figure.

Problems AB

AB.1   Assuming that the Moon is made of similar material to Io, estimate the mean rate of energy generation within it due to tidal flexing in its eccentric orbit around the Earth.

AB.2   Assuming that Mimas is made of similar material to Io, estimate the mean rate of energy generation within it due to tidal flexing in its eccentric orbit around Saturn.

TOPIC AC

THE RAM PRESSURE OF A GAS STREAM

Normal gas pressure, due to the random thermal motion of the gas molecules, acts in all directions and is quite isotropic. However, pressure is just 'force per unit area' and since force is a vector quantity it is possible to have a 'vector pressure' that acts just in one direction. A typical example of this is the pressure due to the wind blowing on to any surface, for example the sail of a yacht. This pressure, due to mass motion of the air, is different from the normal air pressure due to the random thermal motion of the air molecules. The root-mean-square speed of air molecules at 300 K is $365 \, \mathrm{m \, s^{-1}}$ or $1300 \, \mathrm{km \, h^{-1}}$ and no wind speed could be more than a small fraction of that. Nevertheless, even a wind speed of $50 \, \mathrm{km \, h^{-1}}$ is important in the world of sailing. In a similar way vector pressure, referred to as *ram pressure*, can be important in a number of astronomical contexts.

We imagine a uniform stream of particles of mean density ρ, moving at speed u, impinging normally on to an absorbing surface. By absorbing in this sense we mean that all the momentum of the impinging particles is taken up by the surface. The momentum density (momentum per unit volume) in the stream is ρu and in unit time the total momentum absorbed by the surface is that contained within a region of cross section unity and length u, i.e. with volume u. Force is rate of change of momentum and for the absorbing surface this is the absorbed momentum per unit time. Hence the ram pressure is given by

$$P_{\mathrm{ram}} = \rho u^2. \tag{AC.1}$$

As an example we now calculate the ram pressure due to the solar wind in the vicinity of the Earth. The density of protons, n, is variable but we can take $5 \times 10^6 \, \mathrm{m^{-3}}$ as a common value with $u = 4 \times 10^5 \, \mathrm{m \, s^{-1}}$. In this case the ram pressure is

$$P_{\mathrm{ram}} = n\mu u^2 \tag{AC.2}$$

where μ is the proton mass, and is found to be $1.3 \times 10^{-9} \, \mathrm{N \, m^{-2}}$. To put this in context we compare this with radiation pressure in the Earth's vicinity, given by

$$P_{\mathrm{rad}} = \frac{L_\odot}{4\pi R^2 c} \tag{AC.3}$$

with $R = 1 \, \mathrm{AU}$ (Topic AG). This comes to $4.6 \times 10^{-6} \, \mathrm{N \, m^{-2}}$, more than 3000 times greater than the solar-wind ram pressure.

Problem AC

AC.1   A small spherical grain of radius a and density 2×10^3 kg m^{-3} is close to the Earth's orbit. In its vicinity the solar wind has density 5×10^6 protons m^{-3} and speed 400 km s^{-1}. For what value of a is the outward force of the solar wind equal to the inward force of the Sun's gravity?

TOPIC AD

THE TROJAN ASTEROIDS

In Topic K the orbital motions of two bodies of finite mass under the influence of mutual gravitational attraction was described. In figure AD.1(a) S and J represent the Sun and Jupiter of masses $M_\odot$ and M_J respectively. They each orbit around the centre-of-mass C and, assuming circular orbits, the angular velocity will be given by

$$\omega^2 = \frac{G(M_\odot + M_J)}{R^3} \tag{AD.1}$$

where R is the Sun–Jupiter distance. We now introduce a Trojan asteroid, T, of negligible mass such that SJT forms an equilateral triangle. For the asteroid to orbit in equilibrium, it too must move in a circular orbit about C with the angular velocity indicated by equation (AD.1).

First we show that the combined gravitational fields of the Sun and Jupiter at T points in the direction TC. In the triangle STC, $ST = R$ and $SC = M_J R / (M_J + M_\odot)$. Hence we can find a relationship involving the angle α (in degrees) as

$$\frac{SC}{\sin \alpha} = \frac{ST}{\sin(120 - \alpha)}$$

or

$$\frac{M_J}{\sin \alpha} = \frac{M_\odot + M_J}{\sin(120 - \alpha)}.$$

This reduces to

$$M_J[\sin(120 - \alpha) - \sin \alpha] = M_J \sin(60 - \alpha) = M_\odot \sin \alpha. \tag{AD.2}$$

In figure AD.1(b) the fields at T due to the Sun and Jupiter are shown and, since the asteroid is equidistant from the other two bodies, these fields are proportional to $M_\odot$ and M_J respectively. The resultant field makes an angle θ with the direction TS given by

$$\frac{M_J}{\sin \theta} = \frac{M_\odot}{\sin(60 - \theta)}. \tag{AD.3}$$

Comparing equations (AD.2) and (AD.3) it is clear that $\alpha = \theta$ so that the net field at T points towards the centre of mass of the system. The magnitude of the field is also obtained from figure AD.1(b) as

$$F_T = \frac{G}{R^2}(M_\odot^2 + M_J^2 + M_\odot M_J)^{1/2}. \tag{AD.4}$$

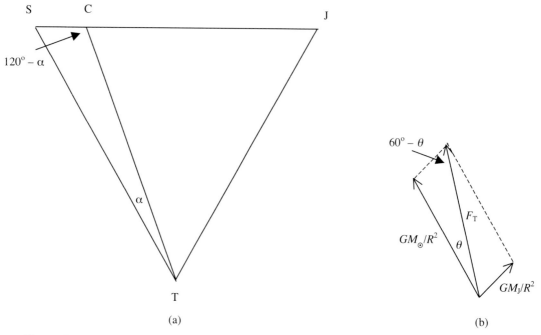

Figure AD.1. *(a) The relative positions of the Sun (S), Jupiter (J), the Trojan asteroid (T) and the centre of mass (C). (b) The forces on the asteroid due to the Sun and Jupiter and the resultant F_T.*

The distance TC ($= a$) is found from the triangle SCT as

$$a^2 = R^2 + \frac{M_J^2}{\left(M_\odot + M_J\right)^2} R^2 - \frac{M_J}{M_\odot + M_J} R^2 = R^2 \frac{M_\odot^2 + M_J^2 + M_\odot M_J}{\left(M_\odot + M_J\right)^2}. \tag{AD.5}$$

For a circular orbit at distance a with the angular velocity given by equation (AD.1) the required field is

$$a\omega^2 = R\frac{(M_\odot^2 + M_J^2 + M_\odot M_J)^{1/2}}{M_\odot + M_J} \frac{G(M_\odot + M_J)}{R^3} = \frac{G}{R^2}(M_\odot^2 + M_J^2 + M_\odot M_J)^{1/2} \tag{AD.6}$$

that is the field given in equation (AD.4).

What has been shown is that the system of three bodies rotating about the centre of mass all at the same angular velocity is in a state of equilibrium but not necessarily in a state of *stable* equilibrium. To

Table AD.1 *The initial parameters for TROJAN.FOR. The velocity components of the asteroids are heavily rounded off to demonstrate the stability by oscillation around a mean position.*

	Sun	Jupiter	Trojan 1	Trojan 2
Mass	1	0.001	0	0
x	0	0	−4.503	4.503
y	0	5.2	2.6	2.6
z	0	0	0	0
V_x	0	−2.756 74	−1.38	−1.38
V_y	0	0	−2.39	2.39
V_z	0	0	0	0

Motions of Trojan asteroids over 300 years

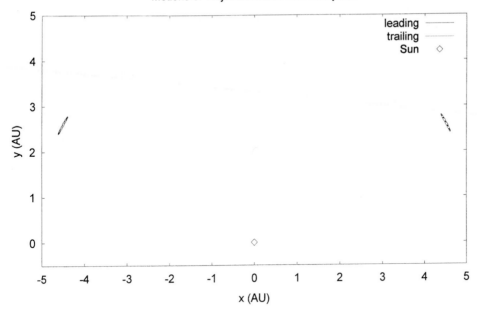

Motions of Trojan asteroids over 300 years

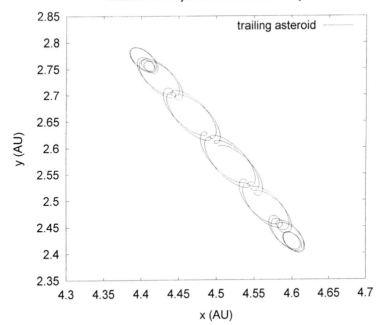

Figure AD.2. *(a) Motions of the leading and trailing Trojan asteroids relative to the Sun and Jupiter. (b) Details of the motion of the trailing asteroid.*

show stability of the three-body system involves a straightforward but lengthy analysis but we can show the stability numerically. Appendix I lists a FORTRAN program TROJANS which calculates the motion relative to the Sun and Jupiter. The calculation is based on keeping the Sun at the origin which is achieved by subtracting from every body in the system the forces exerted by all other bodies on the Sun. By a suitable transformation the position of the asteroid is found relative to the Sun–Jupiter line as the y axis. For calculations involving the Solar System it is convenient to use a system of units where the unit of mass is the solar mass, the unit of distance is the astronomical unit (AU) and the unit of time is one year. In such a system the gravitational constant is $4\pi^2$. Table AD.1 shows some input data which is requested by the program. The Sun, of unit mass is stationary at the origin. Jupiter, with mass 0.001 units is at distance 5.2 AU. on the y axis moving in the negative x direction with a speed 2.756 74 AU yr^{-1}. The coordinates and velocity components for the trailing and leading asteroids can be derived from geometry. In fact, to illustrate the stability of the system the asteroid velocities are not precisely those required for motion in a fixed position. The actual motions of both the leading and trailing asteroids are shown in figure AD.2 for a simulated period of 300 years, which is about 25 Jupiter orbits. It will be seen that they wander around the equilibrium position, illustrating clearly the stability of the system. The observed Trojan asteroids occupy a considerable region around the 60° leading and trailing positions and their motions relative to Jupiter must resemble those shown in figure AD.2.

Problem AD

AD.1   This problem requires the FORTRAN program TROJANS given in Appendix I and a graphics package that will plot a curve from a data file containing consecutive values of (x, y).
The Saturn satellite Tethys has mass 6.26×10^{20} kg and is in a circular orbit of radius 295 000 km. Two tiny satellites, Telesto and Calypso are in the same orbit, one approximately 60° ahead of Tethys and the other 60° behind.
(i)   What is the speed in AU yr^{-1} of Tethys in its orbit around Saturn?
(ii)   Placing Saturn at the origin and Tethys on the y axis then what are the (x, y) coordinates of Telesto and Calypso in AU?
(iii)   What are the velocity components in AU yr^{-1} of Telesto and Calypso at these points?
(iv)   With the information from (i), (ii) and (iii) and the mass of Saturn from table 3.2 run the program TROJANS. Remember to use solar-system units. The following parameters are recommended:

Initial timestep 0.01 (years)
Total simulation time 1000 (years)
Tolerance 10^{-8} (AU)

The output files for the leading and trailing satellites are in FAST.DAT and LAST.DAT respectively.

TOPIC AE

HEATING BY ACCRETION

AE.1. MODELS FOR THE ACCRETION OF PLANETS AND SATELLITES

There are two possible ways for a solid body, such as a terrestrial planet, major planet core or large satellite to form. The first is by the accretion of planetesimals as described in section 12.6.2.3 and analysed in Topic AK. The usual timescale for the accretion of planetesimals is millions of years or more. Consequently much of the heat generated by the collisions of planetesimals will be lost by conduction and radiation before the heated region is buried beneath later-arriving material. The total heating associated with accretion may thus be relatively small but the bodies will subsequently be heated by the decay of radioactive elements, especially by ^{238}U, ^{232}Th and ^{40}K.

Another possible mode of formation is by the gravitational collapse of a dusty cloud that would form a solid core and possibly retain an envelope of gas. This is what occurs if condensations form in a filament as described in section 12.7.2. Gravitational collapse when gas is present will not follow exactly the behaviour of free-fall because of the retarding effect of gas pressure, but the initial slow stage is similar. If the solid material coagulates and forms solid bodies that eventually become completely decoupled from the gas then the final stage, at least as far as the solid component is concerned, will be a very rapid accumulation of material to form the body. This may also be the pattern for the planetesimal model in the final stages of body formation when the growing mass of the core increases the gravitational cross section (section AK.2) leading to accelerated growth. For very rapid accretion there will not be sufficient time for much of the generated heat in a particular region to be lost before that region is buried by more material. This is the mode of formation with which we are primarily concerned here.

AE.2. ACCRETION WITHOUT MELTING

Consider a rapidly accreting body, of final radius R, consisting of material of density ρ and specific heat capacity C at the stage when a spherical core of radius x has formed. The mass of the core is

$$M_x = \tfrac{4}{3}\pi x^3 \rho. \tag{AE.1}$$

The material falling on to the core will not have come from an infinite distance starting at rest but, as long as it falls inwards from a distance large compared with x, it will be a good approximation to assume that it has done so. The gravitational energy released per unit mass heats the material from

411

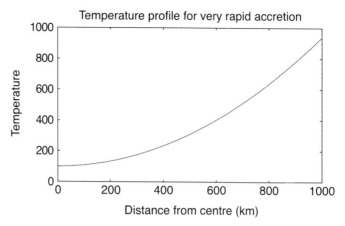

Figure AE.1. *The temperature profile for very rapid accretion.*

an initial temperature T_0 to a final temperature T_x coming from

$$\frac{GM_x}{x} = C(T_x - T_0) \quad \text{or} \quad T_x = \frac{4\pi G\rho x^2}{3C} + T_0. \tag{AE.2}$$

To see what this means we show in figure AE.1 the temperature profile within a body of final radius 1000 km consisting of material of intitial temperature 100 K, density $3 \times 10^3 \, \text{kg m}^{-3}$ and specific heat capacity 1000 J kg^{-1} K^{-1}. The melting point of most common minerals fall in the range approximately 1400 to 2100 K and pure iron has a melting point of 1806 K so it is clear that the body of this example would remain solid. However, if the radius of the body was 1740 km, that of the Moon, then the temperature at the surface, as indicated by equation (AE.2) would be 2640 K so that some material would be molten. In that case equation (AE.2) would not be valid. In the first place the equation takes no account of latent heat of fusion and secondly the specific heat capacities of the solid and liquid may be different. Even where the body remains as a solid the variation of specific heat with temperature should really be taken into account. However, in any practical situation the uncertainty in composition would be so great that taking a constant specific heat capacity is acceptable.

AE.3.   ACCRETION WITH MELTING

There are three possible types of thermal outcome from fast accretion. The first is that we have already seen, where the body remains solid throughout. The second is where at the surface the energy available gives partial melting and the third where at the surface the energy available is sufficient to completely melt the material.

Which of these three possibilities actually occurs can be found by calculating

$$Q_R = \frac{4\pi G\rho R^2}{3}, \tag{AE.3}$$

the intrinsic energy for heating material at the surface. For the first possible outcome, that has already been considered,

$$Q_R \leq C(T_m - T_0) \tag{AE.4}$$

where T_m is the melting temperature. The second outcome requires

$$C(T_m - T_0) \leq Q_R \leq C(T_m - T_0) + L \qquad \text{(AE.5)}$$

in which L is the latent heat of fusion of the material. The final outcome requires

$$Q_R > C(T_m - T_0) + L. \qquad \text{(AE.6)}$$

We now illustrate how to determine the thermal profile for the third case, where there is complete melting at the surface. We consider a body of radius 1740 km (the Moon's radius) consisting of material with specific heat capacity 1200 J kg^{-1} K^{-1} for both the solid and liquid phases, latent heat 4.8×10^5 J kg^{-1}, melting point 1500 K, density 3.4×10^3 kg m^{-3} and initial temperature 100 K. It is easily confirmed from equation (AE.6) that surface material is completely molten.

First we find the distance from the centre, x_{sol}, that marks the interface between solid and partially molten material. This comes from equation (AE.2) with $T_x = T_m$ and gives

$$x_{sol} = \left(\frac{3C(T_m - T_0)}{4\pi G \rho} \right)^{1/2} \qquad \text{(AE.7)}$$

and we find $x_{sol} = 1330$ km. Next we find x_{liq}, the interface between partially melted and completely molten material. This is given by

$$x_{liq} = \left(\frac{3C(T_m - T_0) + L}{4\pi G \rho} \right)^{1/2} \qquad \text{(AE.8)}$$

which gives $x_{liq} = 1392$ km.

For the region up to a radius x_{sol} the temperature profile is exactly as given in figure AE.1. For any value of x between x_{sol} and x_{liq} the temperature is T_m, and the proportion of the material that has melted is given by

$$\phi = \frac{Q_x - Q_{sol}}{Q_{liq} - Q_{sol}} \qquad \text{(AE.9)}$$

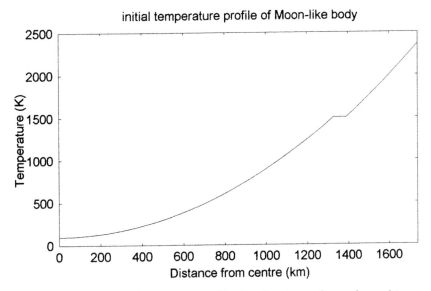

Figure AE.2. *The initial temperature profile when there is complete surface melting.*

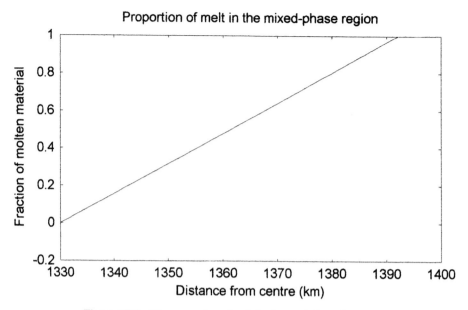

Figure AE.3. *The proportion of melt in the partially molten zone.*

where Q_x is the intrinsic energy for heating at distance x from the centre and the other values of Q are at the interfaces previously described. At $x = 1370$ km, $Q_x = 1.783 \times 10^6$ J kg^{-1}. We also find $Q_{\mathrm{sol}} = 1.680 \times 10^6$ J kg^{-1} and $Q_{\mathrm{liq}} = 1.841 \times 10^6$ J kg^{-1}. The proportion of molten material at $x = 1370$ km is 0.64.

Between $x = 1392$ km and the surface, the temperature may be found from

$$\frac{4\pi G\rho x^2}{3} = C(T_x - T_0) + L \qquad \text{or} \qquad T_x = \frac{4\pi G\rho x^2 - L}{3C} + T_0. \qquad \text{(AE.10)}$$

The final temperature profile is shown in figure AE.2 and the proportion of melt in the partially molten zone is shown in figure AE.3.

AE.4. A MORE REALISTIC INITIAL THERMAL PROFILE

So far we have considered that all the kinetic energy of colliding material is dumped at the point of impact and that no cooling takes place. A more realistic scenario was described by Toksöz and Solomon (1973). They found a modified thermal profile taking two extra factors into account. These were:

(i)   Colliding material sends shock waves into the body, the dissipation of which heats material interior to the point of contact.
(ii) Some cooling of the current surface will take place before it is buried and hence thermally insulated.

These two processes lead to a revised temperature profile that is higher on the inside but cooler on the outside with respect to that we have previously considered. The result of this modification on the temperature profile shown in figure AE.1 is shown in figure AE.4. The difference in the two profiles

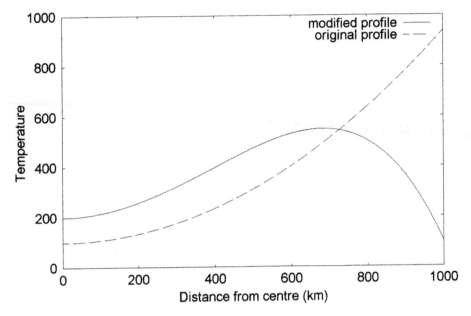

Figure AE.4 *The thermal profile with shock transfer of energy and cooling.*

is clearly greater the slower the accretion takes place and the illustration here is for a fairly slow accretion.

Problem AE

AE.1   A rapidly accreting body has a final radius of 1400 km. Its material has density $3200\,\mathrm{kg\,m^{-3}}$, specific heat capacity $1000\,\mathrm{J\,kg\,K^{-1}}$, latent heat of fusion $5\times10^5\,\mathrm{J\,kg^{-1}}$ and melting point $1450\,\mathrm{K}$. The initial temperature of the material is $150\,\mathrm{K}$.

(a)   Assuming that there is some melting then within what radius is it (i) solid, (ii) partially molten?

(b)   What is the surface temperature?

(c)   What is the average energy of infall per unit mass for the finally formed body? If this energy were equally shared by all the mass then what would be the final uniform temperature?

TOPIC AF

PERTURBATIONS OF THE OORT CLOUD

There are three major problems associated with the Oort cloud. The first is how it was formed in the first place, the second is how it survived the ravages of various influences for the lifetime of the Solar System and the third is how comets are perturbed from the Oort cloud to become visible 'new' comets. Here we shall consider the third problem in some detail but also touch on the other two problems.

AF.1. STELLAR PERTURBATIONS

For a comet in a very extended orbit such that $e \approx 1$ the intrinsic angular momentum

$$H = [GM_\odot a(1 - e^2)]^{1/2} \approx (2GM_\odot q)^{1/2} \tag{AF.1}$$

since $1 + e \approx 2$ and the perihelion distance $q = a(1 - e)$. A perturbation that changes a comet's orbit from being unobservable to observable must reduce the perihelion distance, and hence reduce the magnitude of the intrinsic angular momentum.

Calculations on stellar perturbation can be made by the *impulse approximation*. Figure AF.1 shows the passage of a star, T, of mass M_* moving in a straight line at speed v_* past a body S. This closely describes a star's motion with respect to the Sun at a large distance, of order 1 pc. The relative speeds of stars in the Sun's vicinity are usually in the range 20–30 km s^{-1} and at a distance of 1 pc there is little deviation from straight-line motion. With the star at distance x from the point of closest approach the magnitude of the acceleration of S towards the star is

$$a_S = \frac{GM_*}{D^2 + x^2} \tag{AF.2}$$

where D is the closest approach distance of the star. The acceleration has components parallel to, and perpendicular to, the motion of the star but, by symmetry, it can be seen that the net velocity imparted to S parallel to the star's motion will be zero. The component that is perpendicular to the star's motion gives a rate of change of speed in that direction

$$\frac{dv_\perp}{dt} = \frac{GM_* D}{(D^2 + x^2)^{3/2}}. \tag{AF.3}$$

We now write

$$\frac{dv_\perp}{dt} = \frac{dv_\perp}{dx}\frac{dx}{dt} = v_* \frac{dv_\perp}{dx}$$

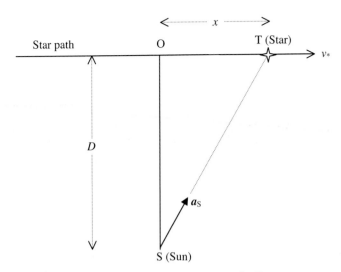

Figure AF.1. *Stellar passage past the Sun.*

so that

$$\frac{dv_\perp}{dx} = \frac{GM_*D}{v_*(D^2 + x^2)^{3/2}}.$$

(AF.4)

Using the symmetry about $x = 0$, the total change in velocity of S, which is along SO, is given by

$$\Delta v_\perp = \frac{2GM_*D}{v_*}\int_0^\infty \frac{dx}{(D^2 + x^2)^{3/2}}.$$

Substituting $x = D\tan\theta$ gives

$$\Delta v_\perp = \frac{2GM_*}{Dv_*}\int_0^{\pi/2}\cos\theta\,d\theta = \frac{2GM_*}{Dv_*}.$$

(AF.5)

In figure AF.2 the passage of the star is considered relative to the Sun at S and the comet at C. The closest approaches of the star to the two bodies are given by the vectors $\mathbf{D}_S$ and $\mathbf{D}_C$ that are not necessarily coplanar. Although the comet is moving relative to the Sun during the interaction this motion can be ignored. For a star passing at a distance 1 pc, strong interactions will occur for less than 10 pc of the star's path. At a speed of 20 km s^{-1} relative to the Sun this will take 5×10^5 years. For an Oort-cloud comet with $a = 40\,000$ AU the period is 8×10^6 years and the comet will not move very far in that time, especially near aphelion when the perturbation is strongest. Assuming that the comet is at rest during the perturbation is the basis of the impulse approximation.

The changes in velocity of the Sun and the comet, $\Delta\mathbf{v}_S$ and $\Delta\mathbf{v}_C$, are different so that the comet changes its velocity by $\Delta\mathbf{v}_{CS} = \Delta\mathbf{v}_C - \Delta\mathbf{v}_S$ relative to the Sun. The consequent change in the intrinsic angular momentum of the comet's orbit is

$$\Delta\mathbf{H} = \Delta\mathbf{v}_{CS}\times\mathbf{r}_C$$

(AF.6)

where $\mathbf{r}_C$ is the position of the comet relative to the Sun. If $\Delta\mathbf{H}$ is parallel to $\mathbf{H}$ then there is a maximum change of q found by differentiating equation (AF.1)

$$\Delta q = \sqrt{\frac{2q}{GM_\odot}}\Delta H.$$

(AF.7)

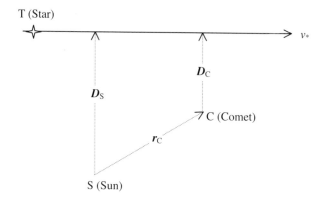

Figure AF.2. *The geometry of the impulse approximation applied to the Sun at S and the comet at C. In general the system is not planar.*

Assuming a coplanar system, if $\mathbf{r}_C$ is parallel to the star's motion then $D_C = D_S$ and there is no perturbation of the comet's orbit. At another extreme if $\mathbf{r}_C$ is perpendicular to the star's motion then $\Delta\mathbf{v}_{CS}$ is as large as possible but since it is parallel to $\mathbf{r}_C$ there is no change of $\mathbf{H}$ and hence of q. For the maximum change of q, $\mathbf{r}_C$ should be at about 45° to $\mathbf{v}_*$.

Table AF.1 shows the minimum possible perihelia for a number of initial values of q and r_C for the passage of a star of mass $M_\odot$ with $D_S = 0.5\,\mathrm{pc}$ and $v_* = 30\,\mathrm{km\,s^{-1}}$. The table shows that it is not really possible for a comet to have an initial perihelion distance outside the influence of the major planets and then to be perturbed to be easily visible with the parameters given in table AF.1. On the other hand if the initial perihelion was less than, say, 20 AU then planetary perturbation at the *previous* perihelion passage would almost certainly have removed it as a member of the Oort cloud (section 9.3). This suggests that impact parameters considerably less than 0.5 pc may be required. From the observed characteristics of the field stars in the solar neighbourhood it is possible to estimate the rate of stellar passages past the Sun within a distance D. For a stellar density n and a mean relative Sun–star speed v_*, the time between star passages within a distance D is

$$t_D = \frac{1}{\pi n D^2 v_*}, \tag{AF.8}$$

where gravitational focusing effects have been ignored. If it is assumed that new comets are produced at a steady rate, then this probably implies that stellar passages are required at intervals corresponding to the period of such comets before they enter the inner Solar System. This interval would be of order 5 million years for a comet with $a = 30\,000\,\mathrm{AU}$. Inserting that value in equation

Table AF.1. *Minimum final perihelia for a stellar impact parameter of 0.5 pc with various values of r and q and stellar mass $M_\odot$.*

	q_{init} (AU)			
r (AU)	40	30	20	10
80 000	23.47	15.68	8.31	1.73
60 000	33.25	24.15	15.23	6.62
40 000	37.59	27.92	18.30	8.80
20 000	39.49	29.56	19.64	9.75

(AF.8) gives $D = 0.14\,\mathrm{pc}$. For comparison, the interval between passages with $D \leq 0.5\,\mathrm{pc}$ is 3.9×10^5 years.

If $r_\mathrm{C} \ll D$ then v_CS can be expressed as

$$v_\mathrm{CS} = \frac{\mathrm{d}\Delta v_\perp}{\mathrm{d}D}\delta = -\frac{2GM_*}{v_* D^2}\delta \qquad (AF.9)$$

where δ is the projection of $\mathbf{r}_\mathrm{C}$ along the direction of $\mathbf{D}$. With $D = 0.14\,\mathrm{pc}$ this would give Δq more than 12 times that listed in table AF.1 since $\Delta q \propto \Delta H \propto 1/D^2$ and a steady stream of visible new comets will be produced with perihelia within $5\,\mathrm{AU}$. Actually the condition that $r \ll D$ will not be met for $D = 0.14\,\mathrm{pc}$ ($\sim 29\,000\,\mathrm{AU}$) with a comet near the aphelion of an orbit with $a = 30\,000\,\mathrm{AU}$. Such a star would plough through the Oort cloud and simple analysis does not indicate the outcome—although massive dispersion and loss of comets is likely. The statistics of Sun–star interactions seem to indicate many such events in the lifetime of the Solar System. There is no reason to suppose that the formation of new comets is a constant feature of the Solar System. They may be episodic events and we just happen to be living in one of the active periods. Equally there may be periods much more active than at present and a large flux of new comets may be associated with the catastrophic impacts that the Earth and other Solar-System bodies seem to have undergone from time to time.

AF.2. PERTURBATIONS BY GIANT MOLECULAR CLOUDS

Giant molecular clouds were first detected through CO emission indicating the presence of large clouds of molecular gas. A typical mass of a GMC is $5 \times 10^5 M_\odot$ with a radius of some tens of parsecs, typically $20\,\mathrm{pc}$. Since, from equation (AF.9), for a given V and δ the strength of the perturbation depends on the ratio M_*/D^2 a GMC at a distance of $350\,\mathrm{pc}$ gives the same perturbation as a solar-mass star at $0.5\,\mathrm{pc}$, the condition specified in table AF.1.

It is estimated that there are about 1500 ± 500 GMCs in the galaxy. They are star-forming regions (e.g. the Orion Nebula) and are concentrated in the galactic disk. The galactic disk can be crudely modelled as having a radius $50\,\mathrm{kpc}$ and a thickness of $500\,\mathrm{pc}$. This gives the average volume per GMC as $\sim 2.6 \times 10^9\,\mathrm{pc}^3$. The speeds of GMCs relative to the Sun are generally about $40\,\mathrm{km\,s}^{-1}$ and this enables us to find time for a GMC passage past the Sun within a distance D. This comes from

$$t_\mathrm{D} = \frac{V_\mathrm{GMC}}{\pi D^2 v_\mathrm{GMC}} \qquad (AF.10)$$

where V_GMC is the volume per GMC and v_GMC is the mean velocity of GMCs with respect to the Sun. With values for V_GMC and v_GMC given above the time (in years) for a GMC passage within a distance D (in parsecs) is

$$t_\mathrm{D} = \frac{2.1 \times 10^{12}}{D^2}. \qquad (AF.11)$$

The time between interactions with average GMCs equivalent in effect to that of a solar-mass star passing at $0.5\,\mathrm{pc}$ is found by inserting $D = 350\,\mathrm{pc}$ in equation (AF.11) and is 1.7×10^7 years. This is much longer than the interval for equivalent stellar interactions (3.9×10^5 years) so we can assume that GMC interactions of *this* type are not very important.

Biermann (1978) was the first to point out the significance of GMC perturbation of comets. If we insert $D = 20\,\mathrm{pc}$ in equation (AF.11) then we find the expected time for a Solar-System passage *through* a GMC as 8.4×10^8 years. This indicates that the Solar System has probably travelled through

~5 GMCs during its lifetime. For a grazing passage of a GMC and a comet–Sun vector of length 4×10^4 AU perpendicular to the relative motion of the GMC, $\Delta v_{CS} = \sim52$ m s^{-1}. If $a = 40\,000$ AU and $r_C = 60\,000$ AU, say, then the escape speed from the Sun at that distance is 172 m s^{-1} and the speed of the comet in its orbit is 61 m s^{-1}. The perturbation would not cause escape of the comet in this case but there are different parameters where it could do so.

The situation is different, however, if the Solar System moves through the GMC. Observations reveal that GMCs have a hierarchical substructure, containing clumps of masses typically $2 \times 10^4 M_\odot$ and radii 2 pc which in their turn contain smaller clumps of mass about $50 M_\odot$ and radius $16\,000$ AU. These clumps would greatly enhance the disruptive power of a GMC, given that the Solar System was passing through it. With five such passages it seems unlikely that many members of the original Oort cloud would remain. We have already seen that close stellar passages could have a similar ravaging effect and this raises the question of how the Oort cloud has survived for the lifetime of the Solar System. A very strong case has been made for an inner reservoir of comets (section 12.9.2) at distances of $\sim10^3$ AU that would also be perturbed by stellar and GMC passages but only to the extent that comets would move outwards to replenish the Oort cloud. Kuiper-belt objects could then be interpreted as the most visible innermost stragglers of this population that stretches from just beyond Neptune to the outer limits of the Oort cloud.

AF.3. PERTURBATIONS BY THE GALACTIC TIDAL FIELD

The Sun is situated in the disk of the Milky Way galaxy. The disk has a thickness of about 500 pc with a more-or-less uniform distribution of stars. The Solar System can be considered as existing within an infinite slab of material of uniform density. The uniform gravitational field external to an infinite plane of areal density σ is perpendicular to the plane and of magnitude

$$E = 2\pi G\sigma. \tag{AF.12}$$

Within a uniform slab there is a field gradient perpendicular to the slab. At the point P in the slab shown in figure AF.3, a distance z from the mean plane, the field is towards the mean plane and be due to a thickness $2z$ of the material since the gravitational fields of the two shaded regions will be equal and opposite. If the volume density of the slab is ρ then the field at P is

$$E_P = 4\pi G\rho z$$

that corresponds to a field gradient

$$\frac{dE_P}{dz} = 4\pi G\rho. \tag{AF.13}$$

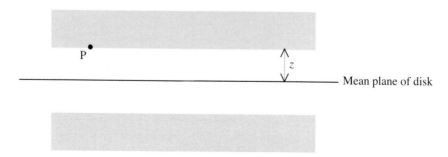

Figure AF.3. *The gravitational field at* P *is due to the unshaded portion of the disk.*

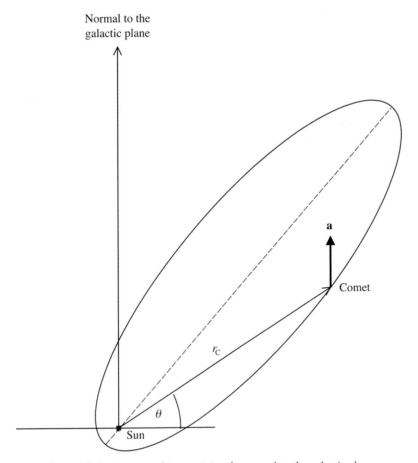

Figure AF.4. *A comet orbit containing the normal to the galactic plane.*

The difference in the z coordinates of a comet and the Sun then gives an acceleration of the comet relative to the Sun that constitutes a perturbation.

Figure AF.4 illustrates the simple case where the plane of the comet's orbit contains the z direction. The magnitude of the acceleration, $\mathbf{a}$, of the comet with respect to the Sun is

$$|\mathbf{a}| = 4\pi G \rho r_c \sin \theta \tag{AF.14}$$

giving a rate of change of intrinsic angular momentum

$$\frac{\mathrm{d}H}{\mathrm{d}t} = |\mathbf{a} \times \mathbf{r}_c| = 4\pi G \rho r_c^2 \sin \theta \cos \theta = 2\pi G \rho r_c^2 \sin 2\theta. \tag{AF.15}$$

Assuming that the comet is always at aphelion will give an upper bound for the change of angular momentum per period, δH. The estimate so obtained will probably be within a factor of two of the true value since Oort-cloud comets do spend most of the time near aphelion. With the approximation that the aphelion distance $r_a = 2a$ for an extended orbit this assumption gives

$$\delta H = 8\pi G \rho P a^2 \sin 2\theta_a \tag{AF.16}$$

where P is the period and θ_a the value of θ at aphelion. This has a maximum value when $\theta_a = \pi/4$.

To estimate ρ we take the mass density of stars in the solar region of the galaxy as $0.08 M_\odot \, \text{pc}^{-3}$. It is uncertain how one should allow for the so-called *missing mass*. It is not at all certain that this will be present in disk material and so we make no provision for it here. The estimate for the mean density of disk material is $5.4 \times 10^{-21} \, \text{kg m}^{-3}$. For a comet with $a = 20\,000 \, \text{AU}$ the period is $(20\,000)^{3/2} = 2.83 \times 10^6$ years. Substituting these values in equation (AF.16) gives a maximum change of intrinsic angular momentum $7.3 \times 10^{15} \, \text{m}^2 \, \text{s}^{-1}$. From equation (AF.1) and (AF.7) the change of perihelion

$$\Delta q = \frac{H}{GM_\odot} \delta H. \tag{AF.17}$$

An orbit with $q = 30 \, \text{AU}$, at the edge of the planetary region, has, from equation (AF.1), $H = 3.5 \times 10^{16} \, \text{m}^2 \, \text{s}^{-1}$. Inserting this in equation (AF.17) gives $\Delta q = 12.8 \, \text{AU}$. Since the galactic tidal field operates continuously this result indicates that a comet may be transferred by this effect from one with a perihelion well outside the planetary region to one with a perihelion well within the planetary region. However, unless the transfer is to a visible orbit in a single step this is not a good process for producing new comets since planetary perturbation will remove them from the Oort-cloud family before they can become visible.

Some new comets may be produced by the action of the galactic tidal field but probably not many. All the approximations in the analysis we have used have been in the direction of making the process more effective and the reality may be less efficient than the analysis indicates. This conclusion concerning the ineffectiveness of the galactic tidal field would be negated if the change of angular momentum per orbit was changed by a factor of 5—if, for example some enhancement of disk density due to missing mass was justifiable. It would then be possible to go from a perihelion outside the planetary region to one of a few AU in a single comet orbit.

If the galactic field was important in producing new comets then it might be expected that this would show itself statistically in that the directions of the perihelia should concentrate round an angle $\pi/4$ to the normal to the galactic plane. No such correlation has been found; on the contrary there is a slight tendency for perihelion directions to be aligned with the galactic plane. If in figure AF.4 the direction of the comets motion were reversed then the galactic tidal field would increase the perihelion of the orbit.

AF.4.   CONCLUSION

A model that seems consistent with observations and theory is that in the early Solar System there was an inner reservoir of comets, stretching out to one or two thousand AU, representing either icy planetesimals (section 12.6.2.2) or volatile debris from a planetary collision (section 12.9.2). All this material would have been in direct orbits. Massive perturbations, which occur frequently by the close passage of stars and less frequently by GMC passages, threw comets outwards, so establishing the Oort cloud. Due to large numbers of lesser perturbations of comet orbits at large distance from the Sun, velocities were randomized to the extent that Oort cloud comets have all possible orbital inclinations.

The main source of new comets is due to perturbations either by stars or, more rarely, by GMCs. It is likely that the frequency of new comets is highly variable and may be either more or less than the present value of somewhat less than one new comet per year. A violent episode of new comet formation could be a possible explanation for the large increase in bombardment rate of the Moon some 500 million years ago, as illustrated in figure 6.16.

Problem AF

AF.1 A comet with $(a, e) = (20\,000\,\text{AU}, 0.998)$ is at aphelion. A star of mass $1.3M_\odot$ moves with speed $20\,\text{km}\,\text{s}^{-1}$ relative to the Sun along a path in the plane of the comet's orbit at an angle of $45°$ to the line of apses of the orbit. If the closest approach of the star to the Sun is $60\,000\,\text{AU}$, estimate the new perihelion of the comet's orbit, assuming that it has been reduced.

TOPIC AG

RADIATION PRESSURE AND THE POYNTING–ROBERTSON EFFECT

AG.1. THE FORCE DUE TO RADIATION PRESSURE

Photons possess both energy and momentum. The energy flux (energy per unit area per unit time) at distance r from the Sun is

$$F_e = \frac{L_\odot}{4\pi r^2} \tag{AG.1}$$

where $L_\odot$ is the solar luminosity and the momentum flux is, correspondingly,

$$F_m = \frac{L_\odot}{4\pi r^2 c} \tag{AG.2}$$

where c is the speed of light. A spherical grain of radius a will gain momentum by absorption of the radiation and this will give rise to a radial force

$$F_{gr} = \pi a^2 F_m = \frac{L_\odot a^2}{4 r^2 c}. \tag{AG.3}$$

This force on the grain will oppose that due to gravity, which is

$$F_{gg} = \frac{GM_\odot \times \frac{4}{3}\pi a^3 \rho}{r^2}, \tag{AG.4}$$

where ρ is the density of the grain material. The ratio of F_{gr} to F_{gg} is independent of distance from the Sun and is

$$\frac{F_{gr}}{F_{gg}} = \frac{3 L_\odot}{16\pi G M_\odot c \rho a}. \tag{AG.5}$$

For a grain of radius $1\,\mu m$ and density $3 \times 10^3\,kg\,m^{-3}$ the ratio is 0.19 and for a grain of radius $0.2\,\mu m$ the ratio would be almost unity.

AG.2. THE POYNTING–ROBERTSON EFFECT

There is another effect of the absorption of radiation, the Poynting–Robertson effect, that is linked to the motion of the absorbing body around the Sun. The momentum vector corresponding to the absorbed radiation is in a radial direction. If the body is in thermal equilibrium then it must be re-radiating energy at the same rate. However, this energy radiation is isotropic *with respect to the frame of reference of the body* and hence, in the Sun's reference frame, the radiated energy has a net momentum component in a tangential direction, i.e. along the direction of motion of the body. From conservation principles the gain of momentum by the radiation in this direction must be balanced by an equal loss of momentum by the radiating body. We now find the magnitude of this effect for a body in a circular orbit.

The tangential rate of change of momentum (corresponding to a force) can be found by multiplying the rate of radiated energy (equal to the rate of absorbed energy), expressed in equivalent mass terms, by the speed of the body in its orbit, v. This gives

$$F_t = L_\odot \frac{a^2 v}{4r^2 c^2}.$$ (AG.6)

This force exerts a torque that changes the angular momentum of the body at a rate

$$\frac{dh}{dt} = -F_t r = -L_\odot \frac{a^2 v}{4rc^2}.$$ (AG.7)

If the mass of the body is m then its angular momentum is

$$h = m\sqrt{GM_\odot r}$$

so that

$$\frac{dh}{dt} = \frac{1}{2} m \sqrt{\frac{GM_\odot}{r}} \frac{dr}{dt} = \frac{1}{2} mv \frac{dr}{dt}.$$ (AG.8)

From equations (AG.7) and (AG.8) and expressing m in terms of a and ρ we find

$$\frac{dr}{dt} = -\frac{3L_\odot}{8\pi rac^2 \rho}.$$ (AG.9)

This can be integrated to give the total time for the particle to go from distance r_0 to r_f as

$$t = -\frac{8\pi ac^2 \rho}{3L_\odot} \int_{r_0}^{r_f} r\, dr = \frac{4\pi ac^2 \rho}{3L_\odot} (r_0^2 - r_f^2).$$ (AG.10)

For a spherical body of radius 1 cm and density $3 \times 10^3\ \mathrm{kg\,m^{-3}}$ in the vicinity of the Earth, the time to be absorbed by the Sun, i.e. going from distance 1 AU to zero, is 6.4×10^{14} s or 2.0×10^7 years. Since the time is proportional to a it seems that even bodies as large as 1 m, or greater, in radius would have been absorbed by the Sun from the vicinity of the Earth during the lifetime of the Solar System.

Problem AG

AG.1   A spherical body of density $2 \times 10^3\ \mathrm{kg\,m^{-3}}$ and radius 5 mm is in a circular orbit around the Sun at a distance of 2×10^{11} m. At the same time another spherical body made of similar material and with radius 1.5 cm is in a circular orbit of radius 1.5×10^{11} m. How long will it take for the spheres to be at the same distance from the Sun and what will then be their distance from the Sun?

TOPIC AH

ANALYSES ASSOCIATED WITH THE JEANS TIDAL THEORY

At the time that Jeans introduced his tidal theory there were no computers and so no possibility of making detailed computational models of any proposed system. Instead the approach was to carry out mathematical analyses related to the system under investigation. If the system was complicated then it was necessary to introduce simplifying approximations to make the analysis possible. Very often realistic analyses were difficult to achieve and theorists depended on rather hand-waving arguments to support their ideas.

Jeans' tidal theory, in a dynamical sense, depended on two main phenomena. The first was the disruption of the Sun by a passing star and the formation of a tidal filament, and the second was the break-up of the tidal filament into blobs that would form protoplanets. What made this theory quite different from all previous theories is that Jeans was an accomplished mathematical theorist and the important processes in his model were subjected to elegant analysis and shown to be valid. We shall now consider these analyses, all of which still have some relevance in the consideration of astronomical problems. To simplify matters our approach may sometimes differ in detail from that of Jeans himself.

AH.1. THE TIDAL DISTORTION AND DISRUPTION OF A STAR

In figure AH.1 the star S, which we may identify as the Sun, is approached by another star, T, of mass M_T. The material of S is highly centrally condensed and it is approximated as a *Roche model* (section AA.1) so that it has a fixed volume but, effectively, has all its mass concentrated at its centre. If T approaches slowly, so that there is no inertial lag in the reaction of S to its field, then S will be distorted in such a way that its boundary becomes an equipotential surface with respect to O, the centre of S. That is to say, the difference of potential between all points on the boundary of S and the point O is the same. Jeans gave this difference of potential for the point P in figure AH.1 as

$$\Omega_P = \frac{GM_\odot}{r} + \frac{GM_T}{r'} - \frac{GM_T x}{R^2} \qquad (AH.1)$$

where r, r', x and R are shown in the figure. This expression is the negative of that usually given. Its validity can be verified by finding $\partial\Omega_P/\partial x$ and $\partial\Omega_P/\partial y$ and seeing that they are the components of the acceleration of P with respect to O. The contours corresponding to a cross section of these surfaces is shown in figure AH.2 for the case $M_T/M_\odot = 3$.

The behaviour of S as T comes closer can now be described. When T is very far away it has virtually no effect on S and the volume of S can be contained within an approximately spherical

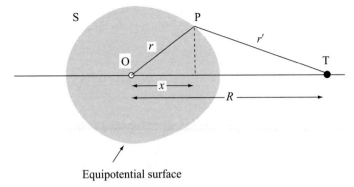

Figure AH.1. *An equipotential surface of a tidally distorted star.*

equipotential surface. As T approaches the system of surfaces shrinks in proportion to the distance ST ($= R$) and the volume of S occupies increasingly distorted surfaces. Eventually a value of R is reached where the volume of S occupies the largest closed equipotential surface, marked with the full line in figure AH.2. If T approaches any closer then S can no longer be contained within an equipotential surface and so must break up. Material then is emitted from the region of the tidal tip, Q, in the form of a stream or filament. This filament would move subsequently under the influence of the gravitational fields of both S and T but its behaviour would also be affected by self-gravitational effects, dealt with in the following section.

In Jeans' theory the relative motion of S and T was a hyperbola and the nature of the interaction meant that there was an inertial lag in the motion of S such that the direction of the tide, OQ, pointed

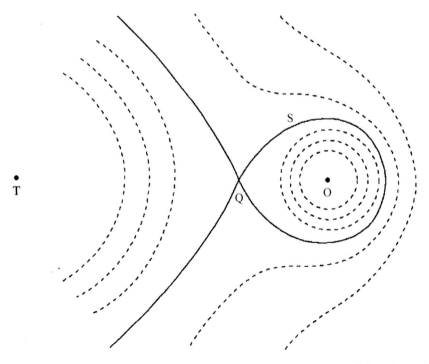

Figure AH.2. *Equipotential surfaces around a star* S *due to its own mass and that of star* T.

towards where T had been some time beforehand. Jeans seems to have been aware of this phenomenon, although a diagram in an article he wrote in 1943 shows the tide and filament always pointing towards the passing star.

AH.2.   THE BREAK-UP OF A FILAMENT AND THE FORMATION OF PROTOPLANETS

The filament streaming out of the Sun would have been mainly gas but with some solid component. The density and temperature would almost certainly have been different in different parts of the filament. However, it will be assumed that the stream had a uniform density and temperature, an assumption that will not affect the general conclusions drawn from the model.

Jeans showed that the filament would be gravitationally unstable and break up into a series of blobs. A simple argument, illustrated in figure AH.3, shows why this is so. Figure AH.3(a) shows a perturbation of the filament in the form of a small density excess in region A. Because of an unbalance of forces, material near A experiences an attraction towards A and this creates two lower density regions, B and B$'$, on either side of A (figure AH.3(b)). Material beyond B and B$'$, at C and C$'$ say, now experiences outward accelerations and produce high density regions at D and D$'$ (figure AH.3(c)). These high-density regions act like the original perturbation at A and so the wave-like disturbance of the filament travels outwards.

A formal analysis of this model in terms of the properties of the gas gives the distance, l, between the high-density condensations in the stream but the general form of the expression can be found just

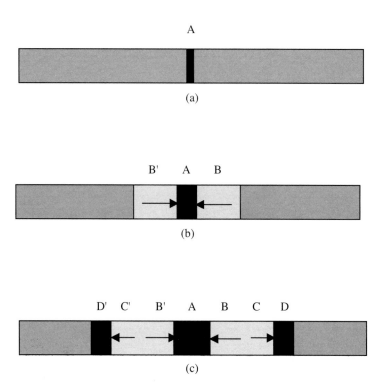

Figure AH.3. *(a) A filament with a density excess at* A. *(b) Material at* B *and* B$'$ *is attracted towards* A. *(c) Material at* C *and* C$'$ *moving away from the depleted regions at* B *and* B$'$ *so creating higher density regions at* D *and* D$'$.

from dimensional analysis. The rate at which a disturbance can move along the filament is related to the speed of sound in the gas given by

$$c = \sqrt{\frac{\gamma k T}{\mu}} \qquad (AH.2)$$

in which γ is the usual ratio of specific heats of the gas, k is the Boltzmann constant and T and μ are respectively the temperature and mean molecular weight of the gas. Other factors influencing l are the gravitational constant, G, and the density of the gas, ρ. The relationship found by Jeans, with the numerical constant not given by dimensional analysis, is

$$l = \left(\frac{\pi}{\gamma G \rho}\right)^{1/2} c = \left(\frac{\pi k T}{G \rho \mu}\right)^{1/2}. \qquad (AH.3)$$

In the Jeans analysis the stream is found to have a periodic variation of density and l is the wavelength of the density fluctuation. Something the analysis does not include is the line density of the filament, σ, i.e. the mass per unit length, a quantity that does not control the wavelength of the density wave. If σ is low then the blobs formed in the filament will disperse; if σ is sufficiently high then the blobs will collapse to form protoplanets.

Although the blobs are not necessarily spherical they would probably be roughly so and the fate of the blobs will depend on whether they are greater or less in mass than the Jeans critical mass (Topic D). If σl is greater than the critical mass corresponding to the T, ρ and μ of the filament material then a protoplanet will form. An additional factor that must be taken into account is the disrupting tidal influence of S and T. If σl only just satisfies the Jeans critical mass criterion then it is likely that tidal effects will prevent the blob from condensing. Essentially this is a Roche-limit problem (Topic AA).

Problem AH

AH.1   A stream of molecular hydrogen, drawn out of a star by tidal effects has temperature 30 K and density 10^{-8} kg m^{-3}. What is the length of the filament, l, that forms a blob?

Assuming that the filament has a circular cross section of diameter l then what is the mass contained in a blob? If the mass of this blob is considered as being in spherical form with the given density then what fraction of the Jeans critical mass (equation D.4) is it?

Actually, under the conditions for forming the blob as given, the ratio you find is independent of the type, density and temperature of the filament material. Why is this?

THE VISCOUS-DISK MECHANISM FOR THE TRANSFER OF ANGULAR MOMENTUM

If the collapsing nebula, in the form of a disk, had some kind of turbulence in it then a theory by Lynden-Bell and Pringle (1974) suggests that angular momentum transport could occur. Since turbulence quickly dissipates, without an input of energy the nebula disk would quickly settle down into quiet rotation with the only relative motions of material due to Keplerian shear. Suggested mechanisms for maintaining the turbulence have included material falling on the disk from outside or the effect of heating from the central condensing star. However, analysis of all these suggestions shows that either the effects they gave would be too weak or that they would only occur for a very short time in the early stages of formation of the nebula.

The basis of the Lynden-Bell and Pringle mechanism is that, for a rotating system in which energy is being lost but angular momentum must remain constant, inner material will move farther inward while outer material will move farther outward. This amounts to a transfer of angular momentum from inner material to outer. That this is so can be shown very simply. Consider two bodies in circular orbits, with radii r_1 and r_2 around a central body of much greater mass. The energy and angular momentum of the system are

$$E = -C\left(\frac{1}{r_1} + \frac{1}{r_2}\right)$$ (AI.1a)

and

$$H = K\left(\sqrt{r_1} + \sqrt{r_2}\right) \qquad \text{(see equation M.15a)}$$ (AI.1b)

where C and K are two positive constants. For small changes in r_1 and r_2 the changes in E and H are:

$$\delta E = C\left(\frac{1}{r_1^2}\delta r_1 + \frac{1}{r_2^2}\delta r_2\right).$$ (AI.2a)

and

$$\delta H = \frac{1}{2}K\left(\frac{1}{\sqrt{r_1}}\delta r_1 + \frac{1}{\sqrt{r_2}}\delta r_2\right).$$ (AI.2b)

If angular momentum remains constant then, from equation (AI.2b),

$$\delta r_1 = -\sqrt{\frac{r_1}{r_2}}\,\delta r_2$$ (AI.3)

and substituting this in equation (AI.2a) gives

$$\delta E = \frac{C}{\sqrt{r_2}} \left(\frac{1}{r_2^{3/2}} - \frac{1}{r_1^{3/2}} \right) \delta r_2. \tag{AI.4}$$

Given that δE is negative then it is clear that if $r_2 < r_1$ then δr_2 must be negative, that is to say that the inner body moves inwards and hence, from equation (AI.3), the outer body moves outwards.

Problem AI

AI.1   It has been suggested that in the early Solar System the three outer major planets were in circular orbits with the following orbital radii: Saturn 9.0 AU, Uranus 12.0 AU and Neptune 15 AU. Drag in the disk, in which the planets existed, pulled Jupiter inwards and this generated spiral waves in the disk that pushed the other three major planets outwards to their present positions. What was the original radius of Jupiter's orbit assuming that angular momentum was conserved in the major planet system? What was the total loss of energy in this process?

TOPIC AJ

MAGNETIC BRAKING OF THE SPINNING SUN

Theoretical models for the formation of the Sun usually predict that it would have formed spinning much more rapidly than at present, anything from just lower than the angular speed for disruption down to ten or so times its present rate. A plausible mechanism for slowing down the spin thereafter involves the coupling of ionized material moving out of the Sun with the solar magnetic field. Charged particles leaving the Sun, in the form of a solar wind, travel along field lines in the vicinity of the Sun where the field is strong. Since the magnetic field rotates with the Sun, so the escaping material will corotate with the Sun while moving outwards and hence gain angular momentum that is removed from the Sun.

AJ.1. COUPLING OF PARTICLES TO FIELD LINES

First we consider the mechanism by which the charged particles are initially coupled to field lines and later become decoupled from them. The condition that governs whether or not the charged particles remain coupled to field lines depends on the relative strengths of the magnetic pressure (energy density) given by

$$P_B = \frac{B^2}{2\mu_0},$$ (AJ.1)

in which B is the field and μ_0 the permeability of free space, and the total gas pressure

$$P_g = nkT + nmv^2,$$ (AJ.2)

in which T is the temperature and n the number density of particles of mean mass m and bulk flow velocity v. The first term in equation (AJ.2) is the normal gas pressure and the second term is the dynamic pressure due to the bulk flow (Topic AC). If the magnetic pressure exceeds the total gas pressure then the motion of the charged particles is controlled by the field. Equality of the two pressures gives the approximate conditions under which the particles decouple from the field.

AJ.1.1. The form of the magnetic field

The form of the solar magnetic field is quite complex because of the rapid flow of the solar wind. At larger distances from the Sun, where the field is weaker, the wind is more-or-less unconstrained by the field but, on the other hand, the magnetic field becomes frozen into the plasma and field lines take on

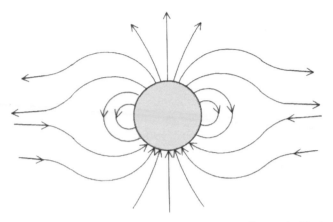

Figure AJ.1. *Magnetic field lines round a star with a strong stellar wind of ionized particles.*

the directions of the local flow (section W.4). The net effect on the field lines is shown schematically in figure AJ.1. The result is that close to the equatorial plane, and at large distances, r, from the Sun, the fall-off in field varies as r^{-1} rather than as r^{-3}, that is expected for a dipole field. For theoretical purposes Freeman (1978) suggested a form of field given by

$$B = D_\odot \left(\frac{1}{r^3} + \frac{1}{(30R_\odot)^2 r} \right) \tag{AJ.3}$$

in which $D_\odot$ is the magnetic dipole moment and $R_\odot$ the radius of the Sun.

AJ.1.2. The present rate of loss of angular momentum

With protons as the charged particles and with a typical solar-wind speed of $500\,\mathrm{km\,s^{-1}}$, $mv^2 > kT$ unless T is of order $10^7\,\mathrm{K}$ so that the first term on the right-hand side of equation (AJ.2) can be ignored. It also turns out that, up to distances where decoupling takes place, the field is effectively of dipole form. Equating the magnetic and gas pressures with these simplifications

$$\frac{D_\odot^2}{2\mu_0 r^6} = nmv^2. \tag{AJ.4}$$

The quantity nm is the local density of the ionized material and, in terms of the rate of mass loss from the Sun, $\mathrm{d}M/\mathrm{d}t$, assuming that all lost material is ionized,

$$nm = \frac{\mathrm{d}M/\mathrm{d}t}{4\pi r^2 v}. \tag{AJ.5}$$

Inserting equation (AJ.5) into (AJ.4) we find that corotation of ionized material will persist out to a distance

$$r_c = \left(\frac{2\pi D_\odot^2}{\mu_0 v (\mathrm{d}M/\mathrm{d}t)} \right)^{1/4}. \tag{AJ.6}$$

Present values for the Sun, $\mathrm{d}M/\mathrm{d}t = 2 \times 10^9\,\mathrm{kg\,s^{-1}}$, $D_\odot = 8 \times 10^{22}\,\mathrm{T\,m^3}$ and $v = 5 \times 10^5\,\mathrm{m\,s^{-1}}$, give $r_c = 3.4R_\odot$ which means that the lost mass takes from the Sun $(3.4)^2$ times the angular momentum it had when it was part of the Sun. At the present solar-wind rate the total mass loss of the Sun over its lifetime would have been about 1.4×10^{-4} of the initial mass and the loss of angular momentum about 0.16% of the original angular momentum.

AJ.2.   THE EARLY SUN

It is generally assumed that the early Sun was far more active than it is now, which would have given both a greater rate of mass loss and also a higher early magnetic field. Within the range of speculation, supported by theoretical and observational considerations, the magnetic field could have been up to one thousand times as strong as at present. It is also fairly certain that the rate of loss of mass was far higher than at present. The assumption is sometimes made that the Sun went through a T-Tauri stage when the loss of mass was of order $10^{-7} M_\odot$ yr^{-1} ($\sim 6 \times 10^{15}$ kg s^{-1}), sustained for a period of 10^6 years. However, such a rate of loss is at the upper end of expectations and we shall consider smaller rates.

Returning to the requirement of reducing the angular momentum of the Sun to a few percent of its original value, the magnetic braking effect is able to do this without any outlandish assumptions. We assume that the moment of inertia of the Sun is always of the form $\alpha M R_\odot^2$ where M is the changing mass but the radius is assumed not to change.

The rate of loss of angular momentum of the Sun is given by

$$\frac{dH}{dt} = \frac{dM}{dt} r_c^2 \Omega \tag{AJ.7}$$

where Ω is the spin angular speed. The expression for angular momentum is

$$H = \alpha M R_\odot^2 \Omega$$

which gives

$$\frac{dH}{dt} = \alpha R_\odot^2 \Omega \frac{dM}{dt} + \alpha R_\odot^2 M \frac{d\Omega}{dt}. \tag{AJ.8}$$

From equations (AJ.7) and (AJ.8)

$$\alpha R_\odot^2 M \frac{d\Omega}{dt} = (r_c^2 - \alpha R_\odot^2) \Omega \frac{dM}{dt} \tag{AJ.9}$$

since $\alpha R_\odot^2 \ll r_c^2$ we can write

$$\frac{d\Omega}{dM} = \frac{r_c^2 \Omega}{\alpha R_\odot^2 M}. \tag{AJ.10}$$

Integrating equation (AJ.10) gives

$$\Omega = \Omega_0 \left(\frac{M}{M_0} \right)^{r_c^2 / \alpha R_\odot^2} \tag{AJ.11}$$

where M and M_0 are the final and initial masses and Ω and Ω_0 are the final and initial spin angular speeds. For a particular rate of mass loss M is related to M_0 by

$$M = M_0 - \frac{dM}{dt} t$$

where t is the duration of the mass loss, taken here as 10^6 years.

The effect of different combinations of rate of loss of mass, between 10^4 and 10^6 the present rate, and magnetic dipole moment, between 10 and 1000 times the present value, is shown in table AJ.1 with

Table AJ.1. *The fraction of the initial angular momentum remaining after 10^6 years with different combinations of magnetic dipole moment and rates of mass loss.*

dM/dt (kg s^{-1})	D (T m^3)		
	8×10^{23}	8×10^{24}	8×10^{25}
2×10^{13}	0.9933	0.9350	0.5106
2×10^{14}	*	0.8083	0.1190
2×10^{15}	*	*	0.0011

* Indicates that equation (AJ.6) gives $r_c < R_\odot$.

α (angular momentum factor for the Sun) $= 0.055$. It is clear that combinations of rate of loss and dipole moment towards the upper ends of the ranges considered are capable of giving the required reduction of angular momentum

Problem AJ

AJ.1   A star rotating with a period of one day has a magnetic dipole moment 5×10^{24} T m^3 and emits ionized material at a rate 10^{12} kg s^{-1} and at speed 1000 km s^{-1}. What are the rates of energy loss due to (i) its loss of mass and (ii) the slowdown of its spin.

You may assume that $r_c^2 \gg \alpha_* R_*^2$ in equation (AJ.9). The kinetic energy of a spinning body is $\frac{1}{2}I\Omega^2$ where I is the moment of inertia and Ω the spin speed.

TOPIC AK

THE SAFRONOV THEORY OF PLANET FORMATION

AK.1. PLANETESIMAL FORMATION

The formation of a dust disk concentrates the essential material for the formation of terrestrial planets and the cores of the major planets. The process of forming these bodies consists of two stages. The first stage is the formation of planetesimals and the second is the accumulation of planetesimals to form planets or cores. Finally, for giant planets, the planetary cores capture gaseous material.

A strong argument in favour of planetesimal formation is that most solid bodies in the Solar System have been bombarded by projectiles of planetesimal size and that some planetesimals may still be visible as asteroids or even, perhaps, comets. There are two main ideas for planetesimal formation. The most favoured way is through gravitational instability within the dust disk. Material in the dust disk will have a tendency to form clumps through mutual gravitational attraction but disruptive solar tidal forces will oppose this tendency. This is a Roche-limit problem (Topic AA). Clumping will occur if the mean density of the distribution of solid material in the vicinity, ρ, satisfies

$$\rho > \frac{3M_\odot}{2\pi r^3} = \rho_{cr} \qquad (AK.1)$$

where r is the distance from the Sun and ρ_{cr} is the critical density for clumping. Safronov (1972) showed that a two-dimensional wave-like variation of density would develop in a uniform disk and that, with local density greater than the critical density, condensations would form. He also showed that, at the critical density, the wavelength would be about eight times the thickness of the disk.

For a nebula of mass $0.1M_\odot$ with a 2% solid component, in the form of a uniform disk of radius 40 AU, the mean surface density of solids, σ, is $35\,\mathrm{kg\,m^{-3}}$. Assuming critical density, the thickness of the dust disk, $h = \sigma/\rho_{cr}$, and hence the volume of material clumping together $\sim 60h^3$. The masses and dimensions of the resulting condensations, with solid material of mean material density, ρ_{sol}, of $2000\,\mathrm{kg\,m^{-3}}$, are given in table AK.1. Planetesimals of dimensions from a few kilometres up to, perhaps, 100 km in the outer Solar System are predicted but this conclusion was challenged by Goldreich and Ward (1973). From thermodynamics principles they showed that the condensations would be hundreds of metres in extent rather than the more than kilometre-size bodies predicted by Safronov. The escape velocity from a solid body of radius 500 m is about $0.5\,\mathrm{m\,s^{-1}}$ and the velocity dispersion of the initial Goldreich–Ward planetesimals is estimated as $\sim 0.1\,\mathrm{m\,s^{-1}}$ so that collisions with the largest planetesimals can give accretion to form larger bodies. Eventually planetesimals of the size predicted by Safronov would come about, albeit by a two-stage process.

Table AK.1. *The masses, m_p, and radii, r_p, of planetesimals, according to Safronov (1972) in the vicinity of the Earth and Jupiter.*

	Earth	Jupiter
ρ_{cr} (kg m^{-3})	2.83×10^{-4}	2.01×10^{-6}
$h = \sigma/\rho_{cr}$ (m)	1.24×10^{5}	1.74×10^{7}
$m_p = 60h^3\rho_{cr}$ (kg)	3.20×10^{13}	6.35×10^{17}
$r_p = (3M_p/4\pi\rho_{sol})^{1/3}$ (km)	1.6	42

Weidenschilling, Donn and Meakin (1989) argued for a different process because the presence of even a small amount of turbulence in the disk would inhibit gravitational instability. The free-fall time (Topic E) for the collapse of a planetesimal clump in the vicinity of Jupiter would be more than a year but if turbulence stirred up the material before the collapse was well under way then the condensation would simply not form. They suggest that the adhesion of fine-grained material to forming a dust disk (section 12.6.2.1) could also operate to form planetesimals. There is some justification for this criticism. A planetesimal clump in the Jupiter region before it began to collapse would have an escape speed $\sim 1.4 \, \mathrm{m \, s^{-1}}$ and turbulent speeds of this magnitude would disrupt the clump. There is no agreement about whether or not the nebula would be turbulent. For some purposes theorists postulate a quiet nebula, for example, to enable planetesimals to form, but then other theorists prefer a turbulent nebula, for example, as an aid to angular momentum transfer.

The general view of theorists in this area is that, whatever the uncertainties in the actual mechanism for forming planetesimals, kilometre-size bodies will form on a relatively short time scale. If this is so then almost all the lifetime of the dusty nebula ($\leq 10^7$ years) is available for the next stage of forming planets from planetesimals.

AK.2.   PLANETS FROM PLANETESIMALS

The basic theory for planet formation from planetesimals was developed by Safronov (1972) and most subsequent work has been developments, or variants, of it. He showed that if the random relative velocity between planetesimals is less than the escape speed from the largest of them then that body will grow and eventually accrete all other bodies which collide with it. Newly formed planetesimals move on elliptical orbits and gravitational interactions between them, equivalent to *elastic* collisions, will increase the random motions. Eventually, the consequential increase in the relative velocities and eccentricities of their orbits enhances the probability of *inelastic* collisions between planetesimals that then damp down the randomness in the motion. Safronov showed that a balance between the effects which increase and decrease random motions, and hence the relative speed of planetesimals, occurs when the mean random speed, v, is of the same order, but less than, v_e, the escape speed from the largest planetesimal. In general one could write

$$v^2 = \frac{Gm_L}{\beta r_L} \qquad \qquad (\text{AK.2})$$

where m_L and r_L are the mass and radius of the largest planetesimal and β is a factor in the range 2 to 5 in most situations.

In a simple case where all colliding bodies adhere, the rate of growth is proportional to the collision cross section, which takes into account the focusing effect of the mass of the accreting body. The rate of growth a spherical body of mass m and radius r is given by the Eddington accretion mechanism (Topic AL) as

$$\dot{m} = \pi r \left(r + \frac{2Gm}{v^2} \right) \rho v \tag{AK.3}$$

where ρ is the mean local density of the material being accreted.

From equation (AK.2) and the relationship

$$\frac{m}{m_L} = \frac{r^3}{r_L^3} \tag{AK.4}$$

equation (AK.3) becomes

$$\dot{m} = \pi r^2 \left[1 + 2\beta \left(\frac{r}{r_L} \right)^2 \right] \rho v. \tag{AK.5}$$

From equations (AK.4) and (AK.5) the ratio of the relative rate of growth of a general body to that of the largest body is

$$\frac{\dot{m}/m}{\dot{m}_L/m_L} = \frac{r_L}{r} \frac{1 + 2\beta(r/r_L)^2}{1 + 2\beta}. \tag{AK.6}$$

This ratio in equation (AK.6) is unity both when $r = r_L$ and $r = r_L/2\beta$. For values of r between those two values the ratio is less than unity and the relative size of the two bodies diverges. Eventually when $r = r_L/2\beta$ the ratio of masses will remain constant at r_L^3/r^3 or $8\beta^3$. For β between 2 and 5 this corresponds to the mass ratio of the largest forming body to the next largest of between 64 and 1000.

It is now possible to estimate the timescale for the formation of a terrestrial planet or the core of a major planet. For a particle in a circular orbit of radius r the speed in the orbit is $2\pi r/P$ where P is the period of the orbit. If the random speed perpendicular to the mean plane of the system is less than or equal to v then the orbital inclinations will vary up to $\phi = vP/2\pi r$. The material at distance r will be spread out perpendicular to the mean plane though a distance $h = 2r\phi = vP/\pi$ so that

$$\rho = \frac{\sigma}{h} = \frac{\pi\sigma}{vP}. \tag{AK.7}$$

For the largest body, from equation (AK.5),

$$\dot{m}_L = \frac{dm_L}{dt} = \pi r_L^2 (1 + 2\beta) \rho v. \tag{AK.8}$$

If ρ_s is the density of the material forming the body then $m_L = \frac{4}{3}\pi\rho_s r_L^3$ and we also have, from equation (AK.7), $\rho v = \pi\sigma/P$. Inserting these values into equation (AK.8) gives

$$\frac{dm_L}{dt} = A m_L^{2/3} \tag{AK.9}$$

where

$$A = \frac{\sigma(1 + 2\beta)}{P} \left(\frac{3\pi^2}{4\rho_s} \right)^{2/3}.$$

Integrating from $m_L = 0$ when $t = 0$ to the formation time, t_{form}, when the planet or core has its final mass, M_p gives the formation time as

$$t_{\text{form}} = \frac{3P}{\sigma(1 + 2\beta)} \left(\frac{4\rho_s}{3\pi^2} \right)^{2/3} M_p^{1/3}. \tag{AK.10}$$

Equation (AK.10) assumes that σ is constant although it will actually decrease with time. However, equation (AK.10) may be used to give an order-of magnitude lower bound to the formation time if the initial σ values are taken.

Many models for disk mass and the distribution of disk material have been proposed. Clearly, to reduce formation times the more massive the disk the better but this also introduces new problems, notably that of disposing of the surplus disk material. For illustration we consider a disk mass of $0.1 M_{\odot}$, as large as possible, with a 2% solid fraction. If the surface density varies as R^{-1}, where R is the distance from the Sun, then this gives a surface density of solids at 1 AU of $943 \, \mathrm{kg \, m^{-2}}$. Taking $(1 + 2\beta) = 8$, $\rho_s = 3 \times 10^3 \, \mathrm{kg \, m^{-3}}$ and $r_L = 6.4 \times 10^6 \, \mathrm{m}$ this gives a time for forming the Earth of 3.9×10^6 years. For the same disk model, at Jupiter's distance $\sigma = 181 \, \mathrm{kg \, m^{-2}}$ and the formation time for a $10 M_{\odot}$ core for Jupiter is 5.3×10^8 years. The formation time for Neptune ($\sigma = 31 \, \mathrm{kg \, m^{-2}}$) is 2.9×10^{10} years—which greatly exceeds the age of the Solar System.

Since the lifetimes of nebula disks are a few million years at most, modifications of the Safronov theory have been suggested drastically to reduce the planet-formation times. One way to do this is to have local enhancements of density in the regions of planetary formation that would not require the total mass of the disk to increase, which would introduce new problems. Another line has been to find ways of slowing down the relative speed of planetesimals since, from equation (AK.3) this will increase the capture cross section of the forming planets. The inclusion of viscous drag into the system makes a small improvement in this direction. Another suggestion by Stewart and Wetherill (1988) is that an energy equipartition law operates so that the larger masses move more slowly. This would increase the probability of large masses combining when they come together. An amalgam of these ideas gives what Stewart and Wetherill have called *runaway growth* with planet formation times from 3.9×10^5 years for Jupiter up to about 3×10^7 years for Neptune.

For giant planets, once the core has been formed, it is necessary to attract nebula gas to form the total planet as it appears today. Assuming that the nebula is still present this final stage should take of order 10^5 years and presents no tight constraint on theories.

Problem AK

AK.1   On the basis of the model described just below equation (AK.10), find the formation times for Venus and an $8 M_{\oplus}$ core for Saturn.

TOPIC AL

THE EDDINGTON ACCRETION MECHANISM

A spherical body in a uniform stream of matter moving at relative speed V may, in some circumstances, accrete all the matter that falls upon it. Inevitably the oncoming matter will arrive with greater than the escape speed from the body. If the excess over the escape speed is sufficiently small then the bombarding matter shares its energy with surface material of the body and all the involved material then has less than escape speed and so is retained, i.e. *accretion* is taking place. However, if V is very much greater than the escape speed then, after the sharing of energy, both the oncoming matter and some of the surface material may have enough energy to escape so that *abrasion* of the body occurs. Here we are concerned with the case of total accretion.

AL.1. THE ACCRETION CROSS SECTION

In figure AL.1 there is depicted a body of mass M and radius R situated in a uniform stream of material moving at speed V relative to the body. Shown in the figure are various streams of matter which are

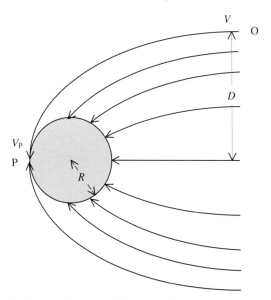

Figure AL.1. *Streams of matter falling onto a body. The accretion radius is D.*

focused by the gravitational attraction of the body. It is clear that the limiting stream is that marked OP where the matter arrives at the surface tangentially at P with speed V_P. The distance D is the *accretion radius* and the accretion cross-section, $A = \pi D^2$.

From conservation of angular momentum

$$VD = V_P R \qquad \text{or} \qquad V_P = \frac{D}{R} V. \tag{AL.1}$$

From conservation of energy

$$\frac{1}{2} V_P^2 = \frac{1}{2} V^2 + \frac{GM}{R} \tag{AL.2}$$

or, substituting for V_P from equation (AL.1) and rearranging,

$$D^2 = R\left(R + \frac{2GM}{V^2} \right). \tag{AL.3}$$

If V is small compared with the escape speed then the second term in the brackets on the right-hand side will dominate and this is the situation in the Safronov model (Topic AK).

Problem AL

AL.1   Find the ratio D/R for $V = kV_{esc}$ where V_{esc} is the escape speed from an accreting body and $k = 0.1, 0.5, 1.0, 2.0, 5.0$ and 10.0.

TOPIC AM

LIFE ON A HOSPITABLE PLANET

With planets having been discovered around other solar-type stars the inevitable question is whether animate material can be expected to be detected elsewhere as well, and if so what it might be like. Our aim here is very modest. We are not concerned with analysing the distinctions between animate and inanimate matter as such, nor how animate matter came about. Rather, we survey briefly the history and the general form of animate matter on Earth, its composition and requirements, and what is necessary to meet these. We will also ask what relevance these arguments might have for gaining an understanding of the form of possible extra-terrestrial life.

AM.1. WE ARE HERE

It is often asked why 'they' have not been 'here' if 'they' are 'there'. This question must be answered to accept the possible existence of life generally in the Universe as credible. It can be said that 'we' have not been 'there' and 'we' are most certainly 'here'. For us on Earth the distances would be in excess of four light years (beyond the nearest star) and with foreseeable power units the journey would take several hundred years. Everything must be taken—there are no breakdown centres or shops either on the way or at the destination!

There are at least four general capabilities that are required for 'animate' travel through space. One is a technology able to move a large group of individuals (to cover all the necessary skills) over the vast distances of space. This is not yet available—we can do no more than move three men to the Moon and back, over a period of about a week. The second is a social capability and commitment to achieve the allocation of the huge resources needed to support life over enormous time intervals—perhaps hundreds of years. The third is a life span for individuals sufficient to see a cosmic journey through, although there is the possibility of breeding en route so that several generations could be involved in the journey. However, might extra-terrestrials live longer? Oak trees do. The allocation of resources is a central problem and would use a very high proportion of the world's GDP during the preparatory period of the mission. The task of feeding a space crew for long periods of time presents severe scientific problems. A wider issue, the fourth, is physiological. There is, as yet, no quantitative knowledge of the effect of weightlessness on the mechanism of the body over lifetime periods. It appears that there can be a loss of strength of bone tissue and possibly other effects. It would be very difficult in practice to arrange an effective system of artificial gravity in the space ship. Extra-terrestrials would, of course, face the same problems that could well prevent them coming 'here' as we are prevented from going 'there'.

442

AM.2. EARLY LIFE ON EARTH

Life on Earth is the only life yet known and the nature and speed of the evolution of life on Earth is the best guide we have to that of all animate ecosystems.

The earliest evidence of life on Earth is a small fossil carbon ball embedded in rock, 3860 million years old and found in the oldest Greenland rock. This ball will not have been the initial development and the Greenland rock is not the earliest rock on Earth. It is probable that living material is older than this. The Earth was formed 4500 million years ago and every indication is that it formed in a molten state. Evidence throughout the Solar System attests to very substantial bombardment of the surface by large bodies at this stage, providing a hot, rough, plastic and changing surface for the Earth. Volcanic activity was also widespread over the surface. The surface probably began to solidify some 4000 million years ago, although volcanic activity persisted, holding a surface temperature much higher than it is at present. It was in this inferno that life seems to have begun with some vigour. Whether it formed spontaneously on Earth or whether it was 'seeded' from outside is not yet established. In any case it enjoyed heat—it seems that all early life was thermophilic.

The thermophiles divide into three main root sources, archaea, bacteria and eucarya (eukaryotes) that possibly branched out from an unknown initial life form. These roots have members as follows, very roughly in evolutionary order:

Archaea. The existence of these organisms was unknown until the 1970s. They are single-celled life forms, resembling bacteria in some ways but quite distinct from them in their DNA composition. They include members that inhabit some of the most extreme environments—near rift vents in the oceans at temperatures above 100°C and in extreme acid or alkali environments. Other places they inhabit include the digestive tracts of cows, termites and some marine creatures, where they generate methane. They also occur in oxygen-free environments such as marsh mud, ocean depths and even in petroleum deposits.

Bacteria. Although bacteria are best known for causing diseases, this aspect is almost a trivial component of their existence. Some require oxygen for their activity, others are anaerobic and cannot exist in the presence of oxygen while a third group thrive in oxygen but can manage without it. The Earth's ecosystem is heavily dependent on bacteria that break down rotting organic material, recycling CO_2 into the atmosphere, so enabling new plants to grow. Other bacteria, some within plant roots, fix nitrogen into nitrates and nitrites that make nitrogen available to plants.

Eucarya. These are bodies made of cells with a nucleus. Examples are slime moulds, algae and various types of fungi. These are the early life forms leading to plants and animals and, ultimately, *Homo sapiens*. The more complex of these creatures require a component of oxygen in the atmospheric gas.

This evolutionary pattern is associated with the way that the environment developed at the Earth's surface. At the beginning there was a reducing atmosphere, rich in hydrogen compounds and also at very high temperatures. Later oxygen became a component of the atmosphere, possibly due to the photo-dissociation of water and the loss of hydrogen, and later, as plants developed, through photosynthesis. At the same time there was an increase in oxygen so the overall temperature of the Earth's surface was reducing.

The distinctive nature of the DNA in archaea and bacteria may contain an important message. If the DNA in bacteria could have evolved from that of archaea, or if they both evolved from some common ancestor, then this simply indicates that as conditions changed so evolutionary changes also took place to adapt to the new conditions—a Darwinian explanation. On the other hand if there is no clear evolutionary pathway from archaean to bacterial DNA, or the possibility of a common source, then this may indicate two distinct origins. This would promote the suggestion

Time Ago (million years)	Living Forms
4000–2000	Single cell creatures RNA based
2000	First eukaryotes: cell structure with nucleus: DNA well established
1100	Sexual reproduction: multiplication of eukaryotes forms. Photosynthesis and the appearance of free oxygen
600	Multicelled animals: worms and sponges
550	Lobites, clams—many invertebrates; the first vertebrates
65	End of the dinosaurs; vertebrates get smaller; mammals
4	First ancestors of modern man
0.1	Homo sapiens

Figure AM.1. *A simplified chronology of life forms on Earth.*

that the generation of life is not an unlikely phenomenon and may arise in a variety of forms to thrive in whatever environment happens to be available.

Most of these elementary life forms still survive today. If there was just one initial common form that started everything then this has probably not survived. It appears that all the elementary life forms thrive in hot conditions. Geothermal hot springs (at temperatures of $60–70°C$) and rims of active volcanic cones all contain these elementary living forms.

The evolutionary sequence is becoming clear and is shown schematically in figure AM.1. The relative time-scales for the development of different life forms, shown in the figure, are just estimates since it is not possible to put precise dates on the important events. It will be most graphic to look at the various stages compressed in time to something within our experience. If the age of the Earth is taken to be 24 hours, the first 12 hours involve single celled creatures only. The first oxygen, later to benefit larger organisms, is seen at about 11 o'clock, just before noon. Eukaryotes appear at noon. Sexual reproduction, so important for the evolution of advanced creatures, appears much later, at about 6 o'clock in the evening. Multi-celled animals make an appearance at 21.00 hours. The first ancestors to modern man enter the picture about 22.40 and *Homo sapiens* appears about 1.2 seconds before midnight. It is seen that the evolutionary pattern was slow to emerge from the very elementary forms but eventually accelerated once the elementary forms were well established. It is astonishing how quickly mankind has developed and put a stamp on the Earth. Mankind's grasp of the Earth is increasing apace and it can only be wondered what the next 24 hours, or even 24 seconds, will bring.

Although life on Earth has been characterized by a mean evolution towards more advanced forms, evolution has occurred through a broadening process in which many elementary life forms have persisted during the development of the more advanced forms. The microbial life forms still account for perhaps half of the inventory of the life forms on Earth.

Two separate life systems exist on Earth. One is the familiar form living on the surface region of the Earth. The other has been discovered only over the last 30 years living over the deep high temperature ocean 'vents' of the mid-ocean ridge structure that circumvents the oceans (section 4.3.4). As far as is known, the chemical composition of the two forms is the same and the structures are similar. But each has developed quite independently with different heat sources (one solar and the other geothermal) to provide energy to their different environments. For instance, in the deep ocean there are blind shrimp-like creatures, eyes not having evolved because there is no sunlight at that depth in the ocean. The detailed study of the mid-ocean ridge animate systems, several kilometres below the surface of the sea, is still in its infancy.

The appearance of the mid-ocean ridge life on Earth has led to the possibility that sub-surface life might have developed below icy surfaces of satellites such as Europa (section 7.3.2), one of Jupiter's large satellites. The eventual exploration of that satellite will thus be of enormous interest.

AM.3. CHEMICAL COMPOSITION

The cosmic abundance of the chemical elements for inanimate matter applies also to the animate world. Animate matter is composed very largely of water (hydrogen and oxygen, the first and second most abundant chemically-active elements), carbon and nitrogen, supported by the remaining elements as minor constituents. This parallel composition suggests that living material is not of special origin but is a natural part of the cosmos. The inanimate components have been found to have a common form throughout the cosmos and the same might be expected for animate matter. The unity of the inanimate Universe suggests the elements would combine to construct animate matter of the same type everywhere that suitable conditions arise.

The basis of life on Earth now is deoxyribonucleic acid (DNA), although for the first 2000 million years it seems to have been the more primitive ribonucleic acid (RNA) that promoted life. DNA now provides the means for evolutionary changes through mutations caused in genetic material. DNA has the form of a double helix structure, the material of which contains a random admixture of components from the two parents. It is interesting to notice that the vast majority of the genetic material in the nucleus of a cell has an unknown function. It seems not to code for the production of any known molecule. In fact, genetic diversity between species is astonishingly small. Different species differ in genetic makeup by no more than a few percent.

AM.4. GENERAL PROPERTIES

We can summarize the general features of living materials on Earth which could give clues to life elsewhere, should it exist. At its roots, a living entity has some aspects of a thermodynamic engine. It takes in fuel and rejects waste in the same way that a heat engine accepts high quality heat and rejects lower quality heat. A heat engine wears out through the action of non-equilibrium processes, such as friction (wear and tear). Living material, however, is self-replicating although there are internal mechanisms that limit the number of replications per life.

It is necessary to upgrade the free energy continually, and this is achieved partly by the intake of food. As in an engine, the free energy absorbed by the creature cannot all be used and some must be rejected. There is a hierarchy of values. The energy required by one species may be of a higher order than that required by another. Energy rejected by one species may be accepted as valid free energy by another species. A simple example of this is garden manure, rejected by the horse but accepted as containing good food by plants.

The repair of components by a living system to prolong its life span is against the natural movement of thermodynamics. Living material is not, however, in thermodynamic equilibrium and living material can be detected this way, having a consistently higher temperature than its

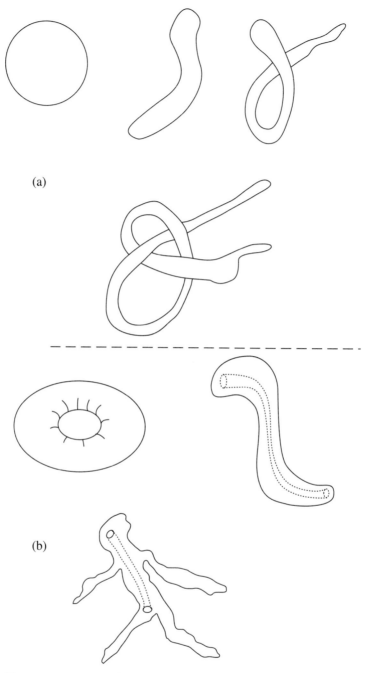

Figure AM.2. *Examples of the topology of (a) a sphere and (b) a doughnut.*

environment. It cannot defy thermodynamic requirements indefinitely and succumbs slowly to irreversible decay. The template for repair becomes impaired and repairs become progressively less precise. The breakdown of the system results partly from mismatched materials, which accumulate and ensure the demise of the body as a living entity.

The topology of the form of the creature is determined by the way it operates. Elementary life forms accept foods that can be absorbed through its surface. It has a simple spherical topology to maximize area for a given volume. Intrinsic mobility is not relevant, the material either staying in a chosen environment or being carried by fluid flows. Advanced forms require more. It is necessary to have a specific region where free energy can be extracted (a stomach) and this requires a track into, and a track out of, the region. This can be met by the topology of a doughnut. It is arranged to involve a head, limbs, wings and a region for reproduction in the more advanced life forms. Some topological examples of spheres and doughnuts are shown in figure AM.2.

Vertebrates need a framework of bone. This must be strong enough to withstand falls under gravity. The stronger the gravity, the stronger the frame of the structure must be. Size is also an important factor in determining the relative dimensions of various parts of a creature, particularly its limbs. If the dimensions of a creature are simply scaled up by a factor k then its mass will increase by a factor k^3. On the other hand the cross section of its legs, and hence its ability to support a tensional or compressive load, would increase by k^2 and the extra length of leg, increased by a factor k, would also give a greater tendency to bend and hence snap. This is why spiders can have long and slender legs in relation to their body sizes while elephants have relatively short thick legs. In addition a mammal such as a whale could not exist out of water since no legs of any practical type could support such a weight. Science fiction films that depict giant ants or spiders as scaled-up versions of the normal variety are making a mockery of mechanics.

The soft materials in creatures, such as muscle or fat, have more open atomic structure, the inter-atomic forces having energies in the general range 10^{-20}–10^{-21} J. The temperature of a solar type star is of the order 6000 K with a mean photon energy of about 8×10^{-20} J, precisely in the interparticle energy range. Photons of higher energy will be plentiful and the most energetic ones will be capable of disrupting atomic structures. Substantial breaking of atomic arrangements will disrupt the function of a living creature and lead to its death. There are two protective actions that avoid this—one involving the creature itself and the other involving its environment. The creature forms a skin to establish its body boundary and this is made sufficiently robust to absorb the least energetic of the harmful radiation wavelengths. The second protective mechanism is the atmosphere that shields out the most harmful stellar radiation. There cannot be a complete shielding from radiation, however, if photosynthesis is to occur.

AM.5. INSTABILITY DUE TO RADIATION: RÔLE OF AN ATMOSPHERE

An atmosphere can act as a shield against harmful radiation from outside. In section O.2.3.1 it was shown how an ozone layer is formed in the mesosphere and these molecules absorb high-energy radiation and so protect the surface. The Earth's atmosphere is transparent to visible light but blocks out much of the other radiation from the Sun, especially the high-energy harmful radiation. By comparison, the atmosphere of Venus is too dense to let very much radiation through to the surface while that of Mars is too thin to offer a substantial resistance to it. To secure an atmospheric column density appropriate to adequate screening of high-energy components, it turns out that the planetary mass should be greater than about 10^{24} kg. It would be possible for a body of smaller mass to possess a suitable atmosphere if the gravitation loss is countered precisely by some source.

For life forms of similar size to those on Earth there is also an upper limit to the planet mass dictated by the strength of gravity. If gravity is too strong then creatures would have to be small if their frames are made of normal bone. It would be possible to live in water, as does a whale, but it is unlikely that a highly evolved technologically-advanced society could develop in such an environment. The maximum mass for land living, based on gravity, is about 5×10^{25} kg, or a factor 8 greater than that of the Earth.

AM.6.   STABILITY OF THE SURFACE REGION

Elementary living forms seems able to survive substantial changes in the environment possibly because they can 'reseed'. These forms are subject to evolutionary pressures but, from experience on Earth, this is very slight. It took some 2000 million years for more advanced life forms to begin to appear while the elementary forms appear to have started very quickly. More advanced forms are more susceptible to environmental changes and it seems that there have been several occasions when natural events, such as volcanic eruptions or asteroid falls, have decimated large parts of life.

The surface region is subject to the effects of impacts by bodies of a large size range, the frequency depending on the mass of the body. The eruption of large volcanoes can also give rise to devastating effects. Large quantities of dust are thrown into the atmosphere and may even penetrate the stratosphere. There it can reside for a very long time, even years, cutting out heat and light from the central star. With the source of energy gone, living material will perish if it cannot hibernate. The result is a gap to be exploited by evolutionary forces but some lines of development may be irretrievably broken.

It is difficult to make precise estimates of the frequency of impact as a function of the impacting mass. From experience on Earth we know dust is continually falling and that there has not been an impact with a major body for some 65 million years. There was a near-miss encounter with a body in 1908 in the Tunguska region of Siberia. This laid waste several hundred square kilometres of forest but, fortunately, the area was uninhabited. There are families of asteroids in Earth-crossing orbits (section 8.2) but little of a quantitative nature is known about potential collisions.

It may be possible for an advanced life form, equipped with missiles and nuclear explosives, to deflect or destroy an oncoming projectile. Otherwise the only protection is a 'fly paper' that preferentially attracts the most dangerous of the wandering masses. The 'fly paper' is another planet of large mass outside the orbit of the smaller planet. Jupiter, and to a much lesser extent Saturn, play this rôle for the Solar System. Other systems will require such a protecting planet to have a semi-major axis greater than about 2 AU.

AM.7.   HOW MANY PLANETS MIGHT CARRY ADVANCED LIFE? THE DRAKE EQUATION

It is of interest to attempt to estimate how many stellar systems are likely to carry life, either elementary or advanced. The first steps were guided by a formula devised by Drake as the basis for discussion at the first SETI (Search for Extra-Terrestrial Intelligence) meeting at Green Bank, West Virginia in 1961. It involves recognizing a number of crucial factors and attempting to estimate numerical values for them. The factors are:

the rate of formation of suitable stars, f_S
the fraction of such stars that will have planets, p_S
the fraction of planets that will be suitable for life to develop, p_L
the fraction of planets where life will actually develop, p_{life}
the fraction where intelligent life will develop, p_{int}
the fraction where technological life will develop, p_{tech}
the mean lifetime of a technological civilisation, τ.

The number of advanced technological civilisations, N_{civ}, can be estimated by putting these factors together. First, the rate of formation of planets where any form of life develops is

$$N_{elem} = f_S \times p_S \times p_L \times p_{life}.$$

The rate of formation of those where intelligent life develops is

$$N_{int} = N_{elem} \times p_{int}.$$

The rate of formation of those in which technological life develops is

$$N_{tech} = N_{int} \times p_{tech}.$$

The number of technological civilizations that exist together at any given time is then

$$N = N_{tech} \times \tau.$$

Many factors in this expression are largely unknown. The rate of production of suitable stars can be estimated from observation in certain regions with a good degree of approximation. In this context 'suitable' implies that the stars remain on the main sequence for a sufficiently long time for intelligent life to develop—which rules out larger mass stars. A value for p_S can be estimated from recent observations on exoplanets and this could be translated into a value for p_L. Further values are not known. The importance of the formula is not to give an accurate numerical value now but to isolate for the future the variables whose magnitudes are needed to make such an estimate. It will be realized that, judging from the Earth, the mean lifetime for a technological civilization, τ, will be at least of the order of 1000 years. Such civilizations will, of course, be spread over the Galaxy so the number per unit volume may well be very small. It could be that the separation is too large for any effective communication to exist between the different members.

AM.8. CONCLUSION

Living materials on Earth are the only indication we have of life in the Universe. It is based on the chemical element carbon and the many molecules involved are controlled by the chemistry of carbon, which is determined by the electronic structure of matter. The evolutionary drive arises from random modifications due to mutation—DNA and RNA being the media for change.

The same situation most probably will develop wherever living materials are able to form. This could be over many areas for elementary life. Special conditions must apply for the development of advanced animate life and their occurrence in the Universe may be much more limited. Where they do occur, however, the same chemical and environmental constraints will apply and the probability is that advanced life forms elsewhere (should they exist) will have similarities to those on Earth. The forms on Earth are probably broadly representative samples, although there may be chance differences.

Observations of bodies orbiting other stars show many systems with very small semi-major axes, down to one-tenth of that of Mercury about the Sun. If appropriately sized planets also exist in these systems they will not have the 'gravitational protection' against impacts with large meteorites and comets that Jupiter affords the Earth. On the other hand table 2.1 shows that some stars, e.g. 16 CygB and 47 Uma, have above Jupiter mass planets that could give protection to a smaller, but undetectable, planet closer to the star.

Even if humankind is an exemplar for advanced life in the Universe then it does not follow that other examples will be precisely the same or that they will exist close together in space. Indeed, we cannot exclude the possibility that we are alone in the Universe although, human curiosity being what it is, we would always be wondering and searching. Detecting other examples of life would throw up profound philosophical questions although, even if it happened, it is unlikely that we would be able to meet or even to communicate in a meaningful way.

The considerations we have raised may allow an answer to the so-called Fermi paradox: if extra-terrestrials are there then why have they not been here? We have already noted that we are here but we have certainly not been there. Perhaps we shall never go. Perhaps we have already learned something about extra-terrestrials—their physiological and psychological constraints are like ours! These studies are the basis of astrobiology.

THE ROLE OF SPACE VEHICLES

Our view of the Universe from the surface of the Earth is very restricted. We look out through an atmosphere that is not completely transparent at any wavelength of the electromagnetic spectrum and is essentially highly opaque to most. Indeed, only the optical wavelengths penetrate the surface with any freedom, that is electromagnetic radiation with the general frequency range 10^{15} Hz. This, actually, is fortunate since the atmosphere protects living things from harmful radiation and, if this cover were withdrawn, we would not exist for long without it. Advanced life probably could not have evolved without it, at least in its present form. Nevertheless, from a scientific viewpoint the Earth's atmosphere has remained a nuisance and serious advances in astronomy and planetary science could only come about when this obstacle was overcome. Space vehicles have allowed us to do this by passing beyond the atmosphere to sample the full range of radiation outside.

The atmosphere is not the only problem. The Universe is very large and distances between even close objects are substantial. This means that the angles subtended at the Earth are small, even when magnified by a large telescope. Again, the level of illumination decreases with the square of the distance apart. Detailed study of an object requires a close approach to it. The large distances, however, act as a major constraint to travel in space, even for automated vehicles within the Solar System. Power sources, impressive on Earth, appear puny outside it. Carrying a sufficient reservoir of energy for a mission is also a constraint to exploration. It is possible to use solar energy in the vicinity of the Sun, out to about the orbit of Mars, but other sources must be found beyond that. At the moment this is usually the decay of a radioactive element and plutonium is perhaps the favourite.

The elementary requirements for placing a vehicle in a planetary orbit follow from Newton's law of gravitation and the Kepler laws of motion, and were known to Newton. The tangential speed is precisely arranged to move the body accelerating towards the body around the surface curvature to maintain a constant height from the surface. The requirement then is to place the body at the required altitude with the appropriate speed. This requires a rocket of sufficient power.

Russian workers, and particularly K E Tsiolkovsky, developed at the turn of the 20th century the theoretical requirements for such a rocket to reach the escape velocity. It was found to require a liquid fuel and three driving stages (no fewer) but the technology of those days was not sufficient to allow these ideas to be put to practical use. These requirements still stand although it remains the dream of rocket designers to develop a single stage solid fuel rocket. Liquid fuels are dangerous and require special handling but solid fuel could be stocked and carried safely from place to place.

The first practical development of these ideas was the German liquid-fuelled V2 rocket of World War II. This was a surface-to-surface single stage vehicle whose trajectory took it to the very edge of space. It could, of course, act as a first stage for a three-stage vehicle. Unused V2 rockets were developed for use in the exploration of the upper atmosphere of the Earth during the 1950s. Some were used at White Sands in the USA and others at Woomera in Australia allowing in situ investigations of the ionosphere but only at these two locations. These sub-orbital measurements, made near the top of the atmosphere, were still not outside it. They led to many important discoveries and acted as an appetizer for the future.

Russian scientists and engineers made the first move beyond the ionosphere in October 1957 with the launch of the automatic probe Sputnik. It orbited at a height of about 100 km in a time of approximately 90 minutes. The vehicle emitted a continuous radio signal but carried no scientific instruments. This historic achievement (which separates the pre-satellite and post-satellite worlds) was quickly followed first by an orbital vehicle containing a dog and later the single orbit by Major Yuri Gagarin. The special problems of carrying a human cargo became clear as was the importance of more powerful rockets. The United States soon joined the Soviet Union in this activity but went further in making a commitment to explore the Solar System. The National Aeronautics and Space Administration (NASA) was formed in 1958 and remains the central pivot of American space activities. Its achievements have been impressive and important. It began, in 1958, with the launch of Explorer 1 into Earth orbit to study the radiation and particle environments. The immediate result was the discovery, by van Allen and his collaborators, of the two radiation belts that now bear his name. It was realized that such belts would be associated with any planet with an atmosphere and an intrinsic magnetic field.

Satellite vehicles are now widely used to study the Earth. They have allowed the atmosphere and ionosphere to be explored in detail with essentially instantaneous world-wide coverage. The surface has been studied in detail to explore a range of properties, both industrial and geophysical—and also military. Orbiting satellites are now used for terrestrial communications and for surveillance. The external gravitational field was studied and the gravitational coefficients J_n (Topic T), the letter symbol chosen in honour of the geophysicist Sir Harold Jeffreys, have been found out to $n = 8$. A group of four satellites in orbit is allowing the solar radiation approaching the Earth to be studied in three space dimensions and over time. In these ways the space vehicles act as indispensable permanent sentinels.

It was inevitable that our nearest neighbour, the Moon, should come under scrutiny very early on. In September 1959 Russian space scientists, concentrating on accurate targeting, impacted the Moon's surface with the unmanned Luna 2 after a three-day mission. Hitting the Moon is not a simple task even now, but was an impressive achievement at that time. In October 1959 Luna 3 orbited the Moon and photographed the far side, never visible from the Earth. Although the photographs were of poor quality they showed a surface very different from the visible near side, heavily cratered but having only one substantial mare basin. The surface appeared quite solid and able to sustain heavy bodies. In July 1964 NASA's Ranger 7 photographed the lunar surface at close range and in February 1966 the Russian automatic Luna 9 returned photographs to Earth from the lunar surface itself. In April of that year Luna 10 orbited the Moon but in June NASA's Surveyor 1 made a soft landing on the surface. This was the first of several tests of the strength of the lunar surface prior to a manned landing. In August, Lunar Orbiter 1 orbited the Moon to return more photographs of the surface. In September 1968 the Russian Zond 5 circumnavigated the Moon (without landing) with various living creatures on board to gain information about the possibilities of life surviving in space.

Human involvement began in December 1968 when astronauts were sent by NASA in Apollo 8 to circumnavigate the Moon and return, but without landing. This followed the manned Apollo 7 launched in Earth orbit in October 1968 to provide a final test of the Apollo system. The first lunar landing finally took place in July 1969 in a mission by Apollo 11, which also saw the return of mineral samples for analysis by Earth laboratories. Other regions of the surface were explored in subsequent Apollo missions up to Apollo 17 of 1972. No manned landings have been attempted since. Twelve astronauts have sampled the lunar surface during the six landings. Disaster struck Apollo 13 in 1970 when an explosion aboard nearly wrecked the mission but the crew were returned safely. This gave invaluable experience of handling survival problems in space.

Russian scientists have not attempted to send a human cargo to the Moon but did send automatic probes to collect lunar surface samples. In September 1970 Luna 16 carried an automatic lander which returned samples to Earth. In November Luna 17 placed an automatic rover on the surface. The comparison of the Russian and American missions is interesting. The Russian missions were less complicated but their returns were smaller in size and hence rather less valuable. Human exploration allows greater flexibility and more thorough coverage than is possible with automatic missions, for the same aims.

While the Moon was studied in detail other members of the Solar System were not neglected. Venus, thought then to be the Earth's twin, came under study early on. A NASA flyby in 1962 was followed by a surface impact by Russia in 1966 (Venera 3). Venera 7 made a soft surface landing in 1967 and Venera 9 returned photographs from the surface in 1975 that give the first direct view. The surface pressure was found to be very high ($\sim$92 atmospheres) with a high surface temperature. The lander survived no more than 20 minutes on the surface, such are the alien conditions there. A crude chemical analysis of the surface was made, however, showing the surface composition to be very similar to that of the Earth. A long-life NASA orbiter was placed around Venus in 1978 to observe the surface by radar. The topological details of the surface are now well known though not entirely understood.

Detailed studies have been extended to Mars. NASA's Mariner 4 made an early flyby in 1965 but the surface coverage was relatively small in the southern hemisphere and the pictures showed an unexpected cratered surface. The photographs from the Mariner 9 orbiter of November 1971 showed this is not actually the case over all the globe and in December 1971 a soft landing was made on the surface, which was repeated in 1976 when photographs were returned to Earth. The missions Mars Pathfinder and Mars Global Surveyor have begun a detailed study of the surface with the special emphasis of attempting to identify life forms. None has been identified so far though it is now accepted that a final conclusion will not be reached until a manned landing ultimately becomes possible.

Mercury, the remaining terrestrial planet, is known only through three rather distant flybys in 1974 when about 45% of the surface was photographed. Rocket energy considerations make orbiting the planet difficult but an ESA mission is planned to do this.

Efforts to explore the major planets began in 1973 with the Pioneer 10 flyby of Jupiter. This was followed by the Jupiter encounter by Voyager 1 in 1979 which then went on to encounter Saturn in 1980. Voyager 2, launched in 1977, reached the Uranus environment in 1986 and the Neptune environment in 1989. The long mission times must be noted. The Pioneer and Voyager spacecraft have moved on towards the edge of the Solar System since then and are about to enter interstellar space. They are now about 10^9 km from the Sun, having been travelling for nearly 30 years. This is a mean speed of about 4000 km per hour over this period, involving gravitational accelerations and free movement. These craft will move through interstellar space indefinitely and might, long in the future, be found by an extraterrestrial. They contain a cartouche telling where they are from and something of the nature of humans, if an alien discoverer could interpret the message.

These missions were possible only by using other planets as appropriate 'slings' to increase the speed of the space probe from time to time. The gravitational attraction increases the speed of the craft which can then be directed on to its destination elsewhere. The effects allow the craft to fly close to intermediate objects. In this way asteroids have been photographed (for instance, Gaspra in 1991, Mathilde in 1997). The space craft ICE passed through the plasma tail of the comet 21P/Giacobini-Zinner in 1985. A great achievement occurred in 1986. The Russian Vega vehicle photographed the nucleus of Halley's comet during its appearance in 1986 while the ESA Giotto probe passed close to the nucleus, transmitting pictures to Earth all the time. The camera no longer works but, otherwise, a useful space vehicle remains in solar orbit for other encounters. This raises questions of the design strength of space vehicles—a problem associated with the cost.

These missions have involved a continuing development of spacecraft. The trend has been towards miniaturization leading to more compact, but more complex, systems. This has itself led to the development of multiple systems. Two examples are the Galileo Jupiter mission and the Cassini Saturn mission. In each case the main mission involves both an orbiter and an entry probe. The Galileo orbiter has been operating for several years. Changes of altitude have allowed unprecedented measurements of the Galilean satellites from altitudes as low as 100 km. The Galileo probe descended into the atmosphere and reported measurements of the Jovian upper atmosphere over a period of several hours. The Cassini spacecraft has been moving to Saturn since 1998 and is due to arrive in 2004. The high energy particles surrounding Jupiter (especially) and Saturn are particularly harmful to the electronic equipment. The space probes have been astonishingly reliable over long time periods and under very hostile physical conditions.

Missions to the Sun involve considerable propulsive energy and were not possible until recently. Two recent missions are the European built Ulysses craft and the joint NASA–ESA SOHO mission. A wide range of frequencies has been observed over a period of time, considerably increasing our knowledge of the solar environment.

There are other applications. One has been the 100-inch Hubble Space Telescope orbiting outside the atmosphere and controlled from the ground. Being outside the atmosphere it can receive a very wide range of frequencies, and particularly in the infrared and ultraviolet: its resolution is unsurpassed on Earth. A second, larger aperture, instrument is planned for deployment in the future.

Spacecraft missions have limited duration for various reasons but continuous measurements are often important for a full scientific programme. The first step is to move the base into Earth orbit, beyond the atmosphere. The Russian MIR orbiting space laboratory was the first attempt to provide continued human habitation in orbit. It was effective for over a decade. Its replacement is an International Space Laboratory which is being constructed in orbit about the Earth. This is meant to be a permanent laboratory with a permanent, though changing, human occupation. The cost in both resource and financial terms is reaching the stage of being too great for any one nation to carry alone. It is planned in the longer term to place such laboratories around the Moon and Mars.

The space laboratory could well be the first long-term step into space for humans but there are considerable problems. One is the need to protect astronauts against high-energy radiation that, without the protection of the atmosphere, can be fatal. A second is the effect of weightlessness on humans. Bones and muscles deteriorate without gravity, and the corpuscular contents of the blood changes making sustained weightlessness impossible without some artificial attractive force—perhaps by spinning the spacecraft—to simulate gravity. A third problem is the sustenance of the human crew. All food and support materials must be carried and humans require at least a minimum level of hygiene to remain healthy and so effective during a mission. Waste must be controlled, and perhaps recycled, if space is not to become unacceptably contaminated. Missions must surely always be restricted to be completed within a small proportion of the mean human life span.

There are certainly many places where a human observer cannot go but a suitably designed remote probe can. One might expect a relationship in space between a human crew and the support of remote vehicles as there is already on Earth, for instance in oceanographic work. It may be that interplanetary and especially interstellar distances and conditions will always prove too hostile for human conquest—or will they?

TOPIC AO

PLANETARY ATMOSPHERIC WARMING

In Topic X the equilibrium temperature of a body orbiting the Sun was given by equation (X.4). The assumptions built into this derivation were that the surface of the body was directly exposed to the solar radiation and that the radiation emitted by the body escaped into space without hindrance.

The presence of a planetary atmosphere changes this simple picture. A thick atmosphere with haze or clouds will reflect some of the oncoming radiation, this corresponding to a finite Bond albedo (Topic X). Thus the radiation absorbed by the planet is lessened. On the other hand the radiation emitted by the surface of the planet may be absorbed by the atmosphere and hence generate heat within it, so increasing the temperature of the planet and its environment. It turns out that the gases that constitute planetary atmospheres are virtually transparent to the higher radiation frequencies that come from the high-temperature Sun but heavily absorb the lower radiation frequencies emitted by the low-temperature planetary surface. It is this that constitutes the greenhouse effect (section 4.2.2). It should be noted that the atmosphere has two effects that balance one another to some extent. On the one hand the albedo effect reduces the energy received by the planet and so reduces the temperature. On the other hand the greenhouse effect enables the planet to retain more of that energy so increasing the temperature. If the atmosphere was stripped from the Earth then, by equation (X.4) with $A = 0$, the equilibrium temperature would be 279 K. Taking the albedo into account, but not the greenhouse effect, gives a temperature of 255 K but, in practice, *with* the greenhouse effect, the mean Earth temperature is ~288 K. For the Earth, the incident radiation is about 1375 W m^{-2}: the atmosphere absorbs about 20% (about 275 W m^{-2}), the surface 40% (about 550 W m^{-2}) and 30% (about 412 W m^{-2}) is reflected back into space.

Venus is another example which would have a mean surface temperature of about 300 K without the mainly CO_2 atmosphere, but is actually 500 K higher. However, the atmosphere is not the only factor controlling the surface temperature of a planet. Changes of emission from the Sun and changes of the planet's orbit over long periods of time can also have their effects although these are not always easy to assess. We do know that the temperature of the Earth's surface is rising by a small amount per year. The cause is complex and its details have still not been identified unambiguously. It is quite difficult to detect small systematic changes in the Earth's temperature against the background of large, apparently random, fluctuations.

Gases of geophysical interest that are particularly important in respect to the greenhouse effect are CO_2, CO, NH_4 and SO_2. Other gases are the artificially produced CF gases. The composition of the atmosphere can be expected to change over long periods of time. For instance, an increase of volcanic

456

eruptions during one geological period will increase the natural concentration of greenhouse gases in the atmosphere and so raise the surface temperature. This has happened a number of times throughout the geological history of the Earth, the mean temperature rising significantly at some periods but falling even below the freezing point of water at others. As recently as the Elizabethan age there were regular winter fairs held on the frozen River Thames, something that is unthinkable now. Other natural events can act to increase the mean surface temperature, especially those involving living material. Both flora and fauna emit methane and other greenhouse gases, increasing the greenhouse heating effect. The earliest composition of the Earth's atmosphere did not contain oxygen and probably led to high surface temperatures. This fell as the oxygen content increased but will have increased again as land animals became more plentiful. The effect of living materials on an atmosphere can be significant and this is a natural result of evolutionary processes.

Homo sapiens has introduced artificial pollutants into the atmosphere, and especially CO, CO_2 and H_2O involved with the burning of fossil fuels. These fuels—coal, oil and natural gas—have been buried in the Earth for billions of years, so storing large quantities of carbon and keeping it out of the atmosphere. When this fuel is burnt it releases carbon in the form of CO_2. In the biosphere carbon exists either as living matter or as CO_2 in the atmosphere and since the amount of life that can be generated is limited then, perforce, most of the output of fossil-fuel burning ends up in the atmosphere. It should be noted that burning vegetable products does not enhance CO_2 in a permanent way since this is really no different from the rotting of vegetation that is part of the normal process of recycling carbon.

There is an ongoing discussion about the importance of greenhouse gases for global warming. At one extreme we know that Venus, with an abundance of greenhouse gases, has been transformed from a temperature at which life would be possible to one in which metals such as tin actually melt. But there are those that argue that there are safety valves operating that would prevent large increases of temperature. For example, an enhanced temperature would increase the amount of water vapour in the atmosphere and also the cloud cover. Increased cloud cover would increase the albedo and hence *reduce* the radiation reaching the Earth. In the other direction, extra water vapour, a greenhouse gas, would *increase* the retained energy. What the balance of these two effects would be is difficult to assess; clearly temperature could not fall but the increase might be very small even for large factors of increase of greenhouse gases.

The effects of the heat content in the surface regions of a planet are not easy to assess. Only one hemisphere will receive heat from the Sun at any time leading to a disparity of temperatures between them. This could be a permanent difference if the planet corotates with the star but otherwise the heating of each hemisphere will fluctuate with a daily period. The effects on oceans will be rather different from those on the atmosphere due to the higher density of the liquid. Convection currents will be set up in each. Rotation will provide a Coriolis force which will also have an effect on the dynamical behaviour of the fluid. For the Earth, the water oceans act as a substantial thermal blanket helping to maintain a stable surface temperature. The motion of the fluids, gas or liquid, can be calculated in principle by solving the appropriate expressions of the conservation of mass, momentum and energy of the fluid. These expressions are strongly non-linear and have proved very difficult to solve, even numerically, for a realistic planetary model. Various features of the Earth's surface temperature remain unexplained, including the general feature that the south polar region has a lower temperature than the north polar region.

Concerns about the nature of planetary atmospheric warming, and its relationship with the composition of the atmosphere are for obvious reasons often restricted to the immediate past and future. Other considerations are relevant over geological timescales. Recent studies have shown that the whole surface of the Earth was covered with ice some 600 million years ago and, although the role of volcanism for ending this period is understood, there is no clear consensus of the reasons for such a catastrophic fall in temperature. Later geological periods have seen the temperature rise very substantially, again for reasons that are not entirely clear. Considerations of this kind are possibly relevant to the Martian water-channel landscape, although other explanations have been given for the formation of a temporary flood plain.

TOPIC AP

MIGRATION OF PLANETARY ORBITS

Current theories of the origin of planets do not produce them in orbits corresponding to their final positions in relation to the parent star. Some extra-solar planets have very small orbital radii, down to 0.04 AU (table 2.1), and although they are able to exist in such proximity to the star it is clear that they could not have formed there. They must have formed farther out and then migrated inwards. Similarly, according to the standard ideas about producing planets from planetesimals (section 12.6.2.3), the times of formation of Uranus and Neptune are too long so it is necessary for them to have formed farther in and then to have migrated outwards. For the capture-theory model the initial orbits are very extended and with high eccentricity (section 12.7.3) so it is necessary for them to both decay (reduce in scale) and to round off. For both these models planets are produced in a resisting medium, in the general form of a disk, and it is the interaction of the planet with the medium that modifies the orbits.

It is important to emphasize that the resistance is *not* due to viscosity effects where a body moving through a fluid produces velocity gradients within it and hence internal friction. For massive bodies this is a trivial component of the total resistance that is mainly due to gravitational effects. The planet, moving through the medium, exerts forces on it and the reaction to those forces is what influences the motion of the planet.

AP.1. DEFLECTION IN A HYPERBOLIC ORBIT

In order to understand the origins of resistance due to motion in a medium we first need to discover something about the nature of hyperbolic orbits. Such an orbit is shown in figure AP.1. The near *apsis* distance to the central body P (equivalent to perihelion for Sun-centred motion) is q and the approach direction line is tangential to the hyperbola at infinity. What we wish to determine is the angle β, the total deflection when the moving body travels from an infinite distance to an infinite distance. To find this we can modify equations we have already deduced for elliptical motion to a form appropriate to hyperbolae for which $e > 1$.

The equation for a hyperbola, equivalent to equation (3.2), is

$$r = \frac{q(1+e)}{1+e\cos\theta}. \tag{AP.1}$$

From this it is clear that the limiting angle, θ_L, corresponding to $r = \infty$, is given by

$$\cos(\theta_L) = -1/e. \tag{AP.2}$$

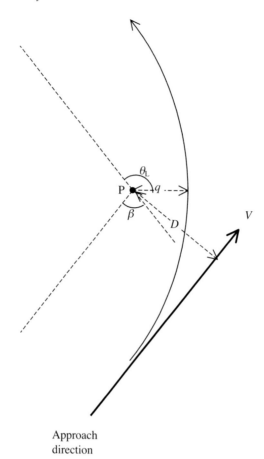

Figure AP.1. *Hyperbolic motion about a body* P.

From figure AP.1 it is found that $\theta_L = \frac{1}{2}\pi + \frac{1}{2}\beta$ so that

$$\sin(\beta/2) = 1/e \quad \text{or} \quad \text{cosec}(\beta/2) = e. \tag{AP.3}$$

The expression for the eccentricity of a hyperbola is exactly that given by equation (M.25),

$$e = \left[1 + \frac{2EH^2}{G^2M^2}\right]^{1/2}.$$

We thus deduce that

$$\cot^2(\beta/2) = \text{cosec}^2(\beta/2) - 1 = e^2 - 1 = \frac{2EH^2}{G^2M^2} \quad \text{or} \quad \tan^2(\beta/2) = \frac{G^2M^2}{2EH^2}. \tag{AP.4}$$

From the speed at infinite distance, V, and the distance, D (figure AP.1), the intrinsic energy and intrinsic angular momentum are given by

$$E = \frac{1}{2}V^2 \quad \text{and} \quad H = VD$$

which gives

$$\tan(\beta/2) = \frac{GM}{V^2 D}, \qquad \text{(AP.5)}$$

which is the required expression giving the deflection angle.

AP.2. MOTION IN AN INFINITE UNIFORM PLANAR MEDIUM

We consider a body P of mass M moving at speed V in the positive x direction in a uniform infinite planar medium of areal density σ. For convenience we consider the body at rest at the origin, so that it is the medium that is moving at speed V in the negative x direction. We now consider the motion of a small element of the medium, of mass m that is negligible compared with M. The element approaches from afar along the line at distance y from the x axis and is deflected through an angle β as shown in figure AP.2. There has been a change of momentum of the element and from the figure this is found to have magnitude

$$\Delta p = 2mV \sin(\beta/2). \qquad \text{(AP.5)}$$

The small element is just part of a stream of medium flowing past the body and, taking the stream between y and $y + dy$ the rate of flow of mass in the stream is

$$\frac{dm}{dt} = \sigma V \, dy. \qquad \text{(AP.6)}$$

Combining equations (AP.5) and (AP.6) the magnitude of the rate of transfer of momentum to the stream of medium is

$$\frac{d(\Delta p)}{dt} = 2\sigma V^2 \sin(\beta/2) \, dy. \qquad \text{(AP.7)}$$

This rate of change of momentum corresponds to a force acting on the stream of the medium along the direction BD in figure AP.2. We wish to find the net force on a section of medium from $y = D$ to $y = -D$ and it is clear from symmetry that this will be in the x direction. Since BD makes an angle $\beta/2$ with the y direction the x component of the force on the stream between y and $y + dy$ is

$$dF_x = \frac{d(\Delta p_x)}{dt} = 2\sigma V^2 \sin^2(\beta/2) \, dy. \qquad \text{(AP.8)}$$

From equation (AP.5) this becomes

$$dF_x = 2\sigma V^2 \frac{\tan^2(\beta/2)}{\sec^2(\beta/2)} \, dy = 2\sigma V^2 \frac{\tan^2(\beta/2)}{1 + \tan^2(\beta/2)} \, dy = 2\sigma \frac{G^2 M^2 V^2}{G^2 M^2 + V^4 y^2} \, dy. \qquad \text{(AP.9)}$$

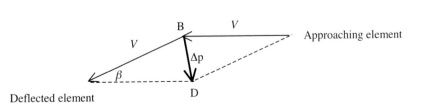

Figure AP.2. *The deflection of an element from passage along a hyperbolic path.*

The total force on the medium between $y = D$ and $y = -D$ in the positive x direction is thus

$$F_x = 2\sigma G^2 M^2 V^2 \int_{-D}^{D} \frac{1}{G^2 M^2 + V^4 y^2} \, \mathrm{d}y = 4\sigma G M \tan^{-1} \left(\frac{V^2 D}{GM} \right). \tag{AP.10}$$

This force is equal and opposite to that acting on the body so that the acceleration of the body is in the negative x direction (opposing the motion) and is

$$\frac{\mathrm{d}V}{\mathrm{d}t} = -4\sigma G \tan^{-1} \left(\frac{V^2 D}{GM} \right). \tag{AP.11}$$

If the argument of arctan is large then the deceleration is $2\pi G\sigma$ and is independent of the mass of the body. This is always true for an infinite medium. For a finite medium and a small argument, where the value of arctan approximately equals the argument, the deceleration is inversely proportional to the mass, independent of G and proportional to V^2.

This calculation does not correspond to any real situation in terms of orbital evolution but gives an idea of how the forces operate. In practice the medium is non-uniform and is in motion around the central star. An important aspect of this is that a stream of the medium being deflected by the planet will be changing its velocity and density as it flows. In addition since σ and V vary there will be no cancellation of the force perpendicular to the motion due to streams at $y = D$ and $y = -D$. The situation is complicated and some kind of numerical approach is required to give a realistic estimate of the resistive force.

AP.3.   RESISTANCE FOR A HIGHLY ELLIPTICAL ORBIT

Some models of planetary formation, including the capture-theory model, give planets in highly elliptical orbits. When such a condition prevails then for much of the orbit the relative speed of the planet with respect to the medium may be large, much larger than the variation in the speed of the medium within the region in which it is effective. A simple calculation will illustrate this. It uses the condition that the effective region of the medium is that for which D is less than the radius of the sphere of influence of the planet as given by equation (AA.12). Consider a planet of Jupiter mass orbiting the Sun in a highly elliptical orbit ($e \lesssim 1$) at perihelion with $q = 9$ au. Assuming that the medium is in Keplerian orbit its speed would be about $10 \, \mathrm{km \, s^{-1}}$ while that of the planet would be about $14 \, \mathrm{km \, s^{-1}}$, making $V = 4 \, \mathrm{km \, s^{-1}}$. From equation (AA.12) the radius of the sphere of influence is about $1.1 \times 10^{11} \, \mathrm{m}$ and this makes the argument of arctan in equation (AP.11) approximately 1300 so the arctan term can be replaced by $\pi/2$. The variation of the speed of the medium over the effective range of the medium is about $\pm 4\%$ so it is reasonable to assume in the application of equation (AP.11) that V is constant. Of course, there will be some error in this approximation but down to when the eccentricity is about 0.2 the errors will not be too severe. For lower eccentricities some more sophisticated approach is necessary.

AP.4.   RESISTANCE FOR A CIRCULAR ORBIT

For the solar–nebula theory planets are produced in circular orbits in a medium rotating in Keplerian fashion. Hence there is little if any relative speed of the planet with respect to the local medium but that does not mean that migration cannot take place. We have not dealt with the situation where both the planet and the medium are moving in circular orbits but we can use the general ideas behind equation (AP.10) to see how migration can occur in such a case. In figure AP.3 we show part of the orbit of the

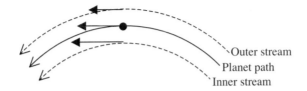

Figure AP.3. *The motions of the planet and neighbouring inner and outer streams.*

planet and streams of the orbiting medium, one outside the planet's orbit and the other inside. Although these orbits are curved we shall treat them as approximately linear so that we can apply equation (AP.8). Relative to the outer medium the planet is moving in a direct sense so the tangential force on it due to the outer stream will be in the retrograde sense. This provides a torque in a direction that reduces the angular momentum of the planet and so causes it to drift inwards. However, the torque on the outer stream is in the prograde direction so that its angular momentum is increased, causing it to drift outwards. We noted that there would also be a force normal to the motion and we cannot assume here that there is some cancellation of this force. We can effectively ignore this force since we are considering angular momentum which is not affected by forces along the radius vector. In fact the medium is a fluid with viscous and pressure forces operating so that any significant departures from circular motion are soon damped out. Thus we can consider the forces operating due to slight deflections of the medium while at the same time assuming that the medium affected by the planet is always in circular motion. This assumption may break down if the planet is very massive and the medium is so heavily disturbed that considerable turbulent motions are set up.

What we find for the inner stream is just the opposite of the effect of the outer stream. The angular momentum of the planet is increased, so that it drifts outwards, while that of the stream is reduced so that it drifts inwards. The net action on the planet depends on the difference of the effects of the outer and inner material which will be influenced by the radial distribution of the density of the medium. It could migrate either inwards or outwards.

The planet will accrete material with which it comes into contact so it opens up a circular gap in the medium around its orbit. The effect of the forces we have described above is to open up that gap since outer material is pushed outwards and inner material is pushed inwards. This mechanism is called Type II migration, which is important for orbital evolution in the solar–nebula theory. Modelling of the effect shows that the angular momentum is transported inwards and outwards in the form of spiral waves. Dissipation of the waves increases or reduces the local angular momentum where the dissipation takes place. If, for example, an outwardly moving spiral wave from an internal planet is dissipated by interaction with an outer planet then the outer planet can gain angular momentum and so move outwards. Thus spiral waves generated by an early Jupiter could have been the driving agent that pushed Uranus and Neptune out to their present orbits.

Problem AP

AP.1   A body is moving at a speed of $10\,\text{km}\,\text{s}^{-1}$ in an infinite planar medium of surface density $1\,\text{kg}\,\text{m}^{-2}$. How long will it take to come to rest and how far will it travel in doing so?

TOPIC AQ

INTERACTIONS IN AN EMBEDDED CLUSTER

In the core of an embedded cluster there are numbers of condensed stars which are either on the main sequence or approaching the main sequence along the Kelvin–Helmholtz evolutionary track (section F.2). The condition of the gas in the embedded cluster is such that the Jeans critical mass corresponds to smaller mass stars, say from one-half a solar mass or so down to brown dwarf masses. Within this gas, stars will be forming, initially in the form of large-radius protostars that will persist for a few thousand years as extended objects. If, during that period, they happen to interact with a condensed star in a suitable way then the conditions for a capture-theory interaction may occur.

A detailed model to investigate the probability of such events is quite complicated. The condensed stars should have a range of masses and velocities and they will be moving in a dense medium which, within the core, has several times the mass of the stars formed within it. Protostars that form will also be moving and, since they are not part of the existing stellar cluster, it is not possible to know with what speeds they will be moving when they form. Here we shall consider a simple model that will indicate the general level of the probabilities, almost certainly to within a factor of two or so one way or the other.

AQ.1. THE INITIAL CONDITIONS

We assume that in the core there are N condensed stars per unit volume, each of mass M_* and all moving with speed V. A protostar is formed from the gaseous material with the Jeans critical mass M_P given by equation (D.4). For such a protostar the radius is

$$R_P = \frac{G M_P \mu}{5kT} \tag{AQ.1}$$

in which μ is the mean molecular mass of the material and T its temperature. For molecular cloud material an appropriate value of μ is 4×10^{-27} kg and we shall take $T = 15$ K. From these values

$$R_P = 5.2 \times 10^{14} \mathcal{M}_P \, \text{m} \tag{AQ.2}$$

where $\mathcal{M}_P$ is the protostar mass in solar units. For example, a protostar with mass equal one half of the Sun would have a radius of about 1730 AU.

The root-mean-square velocities of stars in embedded clusters are between 0.5 and 2.0 km s^{-1} so we shall take $V = 1$ km s^{-1}. This will also be taken as the speed of the protostar when it forms.

AQ.2. CONDITIONS FOR AN INTERACTION

Numerical simulations of the type that led to figure 12.4 suggest that for a successful capture-theory interaction the extended protostar must have an orbit taking it to a distance between about $0.5R_P$ and $1.0R_P$ from the star. In the case that gave figure 12.4 the periastron was $0.75R_P$ but this does not lead to a collision as, well before the periastron position is reached, the protostar has been stretched into a long filament.

With both stars and the protostar having speed V the root-mean-square relative speed of the protostar to a star is $\sqrt{2}V$. Because of gravitational focusing the interaction cross section for approaching to within a distance r is, as given by equation (AL.3),

$$A_r = \pi r \left(r + \frac{G(M_* + M_P)}{V^2} \right). \tag{AQ.3}$$

The sum of the masses of the star and the protostar must be used in this equation since they are comparable. The effective interaction cross section to have an approach within $0.5R_P$ and R_P is therefore

$$A_e = A_{R_P} - A_{\frac{1}{2}R_P} = \pi R_P \left(\frac{3}{4} R_P + \frac{G(M_* + M_P)}{2V^2} \right). \tag{AQ.4}$$

The form of free-fall is that it is slow initially so that the protostar will remain as an extended object for a considerable fraction of the free-fall time (Topic E). For our simple model we shall assume that the protostar retains its original radius for an effective time $t_e = \frac{1}{2}t_{ff}$, where t_{ff} is the free fall time given by equation (E.7). From equation (D.4) and equation (E.7) we find

$$t_e = 0.05 \pi G M_P \sqrt{\frac{\mu^3}{10k^3 T^3}} = 5.63 \times 10^{11} \mathcal{M}_P \, \text{s}, \tag{AQ.5}$$

with the parameters that have been chosen. The protostar moves through the core with speed V and the total interaction volume during a period t_e is $A_e V t_e$. Thus the probability that the protostar makes a suitable interaction with a star is

$$P_{int} = N A_e V t_e. \tag{AQ.6}$$

AQ.3. NUMERICAL CALCULATIONS

To illustrate the range of probabilities protostars with masses $0.4M_\odot$, $0.5M_\odot$ and $0.6M_\odot$ together with embedded star densities of 2000, 8000 and 32 000 pc^{-3} will be taken. The order of carrying out the calculation is first to find R_P from (AQ.2), then A_e from (AQ.4) and t_e from (AQ.5). Finally the probability can be found from (AQ.6), remembering to express the stellar number densities in SI units. The results found are shown in table AQ.1

Actually the numbers in table AQ.1 are not quite what we want. They give the probability that a protostar will have a suitable interaction but we are really interested in the proportion of *stars* that have suitable interactions. Making the transformation requires a knowledge of the number distribution of stars as a function of mass. For lower mass stars this can be taken in the form

$$f(M) = CM^{-2} \tag{AQ.7}$$

where the *mass index* 2 is somewhat less than the Salpeter index, 2.35, usually taken for the complete mass range. If we take the number of condensed stars as 100 with their masses between $0.7M_\odot$ and

Table AQ.1. *Probability of an extended protostar approaching a star between distances $0.5R_P$ and R_P.*

	Protostar mass (Sun units)		
Star density (pc^{-3})	0.4	0.5	0.6
2000	0.002	0.005	0.008
8000	0.010	0.018	0.031
32 000	0.040	0.073	0.123

$1.4M_{\odot}$ then the number of stars with masses

$$0.35-0.45M_{\odot} = 89$$

$$0.45-0.55M_{\odot} = 56$$

$$0.55-0.65M_{\odot} = 39$$

With $N = 8000\,pc^{-3}$ in table AQ.1 the probability that a condensed star has a suitable interaction with a protostar is found as

$$P_{star}(8000) = \frac{39 \times 0.031 + 56 \times 0.018 + 89 \times 0.010}{100} = 0.031.$$

The result will be four times as big for $N = 32\,000\,pc^{-3}$ and four times smaller for $N = 2000\,pc^{-3}$.

   These results can only be regarded as indicative since there are approximations that influence the results both downwards and upwards. Computations indicate that if the interaction between the star and protostar is hyperbolic with $e \gtrsim 1.1$ then captured protoplanets are less likely to occur—although free protoplanets are still formed. This condition would reduce the probabilities we have found. In the other direction the condition that the protostar must reach periastron in an extended condition between the two limits of distance is too stringent. If the protostar is still extended when it is about two protostar radii from the star then stretching into a filament will take place and protoplanets will be produced. This relaxation will increase the probabilities we have found.

   The probabilities clearly depend linearly on the stellar number density within the embedded cluster. Stars observed with planetary companions will have come from different clusters, all with different characteristics. The very dense core of the Trapezium cluster has a stellar number density higher than the $32\,000\,pc^{-3}$ we have taken as the upper limit in the calculations, but presumably there will be other embedded states with less than the $2000\,pc^{-3}$ we have taken as the lower limit. All that can be said is that the calculations are not incompatible with an observed frequency of planetary companions for solar-type stars somewhere in the range 3–6%.

Problem AQ

AQ.1   In an embedded cluster there are 100 solar mass stars with stellar density, $N = 20\,000\,pc^{-3}$. During the embedded state 1000 brown dwarfs are produced with masses between $0.05M_{\odot}$ and $0.075M_{\odot}$. Using the parameters given in this topic, taking a mean brown-dwarf mass as $0.0625M_{\odot}$, what proportion of the stars will have a suitable interaction with a brown dwarf to possibly produce protoplanets?

APPENDIX I

```
C     PROGRAMME TROJANS
      PROGRAM NBODY
C     THIS IS A GENERAL N-BODY PROGRAMME WHERE INTER-BODY FORCES BETWEEN
C     BODIES i AND j ARE OF THE FORM CM(i)*CM(j)*F(Rij,Vij) WHERE F IS A
C     FUNCTION OF Rij, THE DISTANCE BETWEEN THE BODIES, AND Vij, THE
C     RELATIVE VELOCITIES OF THE TWO BODIES.
C     THE STRUCTURE OF THE PROGRAMME IS;
C
C     (I)    THE MAIN PROGRAMME ''NBODY'' IS WHICH INCLUDES THE RUNGE-KUTTA
C     ROUTINE WITH AUTOMATIC STEP CONTROL.
C     (II)   SUBROUTINE ''START'' WHICH ENABLES INPUT OF THE INITIAL BOUNDARY
C     CONDITIONS.
C     (III)  SUBROUTINE ''ACC'' WHICH GIVES THE ACCELERATION OF EACH BODY
C     DUE TO ITS INTERACTIONS WITH ALL OTHER BODIES.
C     (IV)   SUBROUTINE ''STORE'' WHICH STORES INTERMEDIATE COORDINATES AND
C     VELOCITY COMPONENTS AS THE COMPUTATION PROGRESSES.
C     (V)    SUBROUTINE ''OUT'' WHICH OUTPUTS THE RESULTS TO DATA FILES.
C
C     BY CHANGING THE SUBROUTINES DIFFERENT PROBLEMS MAY BE SOLVED.
C     THE CM'S CAN BE MASSES OR CHARGES OR BE MADE EQUAL TO UNITY WHILE
C     THE FORCE LAW CAN BE INVERSE-SQUARE OR ANYTHING ELSE -
C     e.g. LENNARD-JONES. SEE COMMENT AT THE BEGINNING OF SUBROUTINE
C     ''ACC'' FOR THE TYPES OF FORCES OPERATING.
C
C     THE FOUR-STEP RUNGE-KUTTA ALGORITHM IS USED. THE RESULTS OF TWO
C     STEPS WITH TIMESTEP H ARE CHECKED AGAINST TAKING ONE STEP WITH
C     TIMESTEP 2*H. IF THE DIFFERENCE IS WITHIN THE TOLERANCE THEN THE
C     TWO STEPS, EACH OF H, ARE ACCEPTED AND THE STEPLENGTH IS DOUBLED FOR
C     THE NEXT STEP. HOWEVER, IF THE TOLERANCE IS NOT SATISFIED THEN THE
C     STEP IS NOT ACCEPTED AND ONE TRIES AGAIN WITH A HALVED STEPLENGTH.
C     IT IS ADVISABLE, BUT NOT ESSENTIAL, TO START WITH A REASONABLE
C     STEPLENGTH; THE PROGRAMME QUICKLY FINDS A SUITABLE VALUE.
C
```

```
C     AS PROVIDED THE PROGRAMME HANDLES UP TO 20 BODIES BUT THIS CAN BE CHANGED FROM
C     20 TO WHATEVER IS REQUIRED IN THE DIMENSION STATEMENT.
C
      DIMENSION CM(20),X(20,3).V(20,3),DX(20,3,0:4),DV(20,3,0:4),WT(4),
     +XTEMP(2,20,3),VTEMP(2,20,3),XT(20,3),VT(20,3),DELV(20,3)
      COMMON/A/X,V,TOL,H,TOTIME,DELV,XT,VT,NB,IST,TIME,IG,XTEMP,
     +VTEMP,CM
      DATA WT/0.0,0.5,0.5,1.0/
      IST=0
      OPEN(UNIT=9,FILE='LPT1')
C
C     SETTING THE INITIAL BOUNDARY CONDITION CAN BE DONE EITHER
C     EXPLICITLY AS VALUES OF (X,Y,Z) AND (U,V,W) FOR EACH BODY OR
C     CAN BE COMPUTED. THIS IS CONTROLLED BY SUBROUTINE ''START''.
C     OTHER PARAMETERS ARE ALSO SET IN ''START'' WHICH ALSO INDICATES
C     THE SYSTEM OF UNITS BEING USED.
C
      CALL START
C
      TIME=0
C     INITIALIZE ARRAYS
      DO 57 I=1,20
      DO 57 J=1,3
      DO 57 K=0,4
      DX(I,J,K)=0
      DV(I,J,K)=0
   57 CONTINUE
C
C     WE NOW TAKE TWO STEPS WITH STEP LENGTH H FOLLOWED BY ONE STEP
C     WITH STEP LENGTH 2*H FROM THE SAME STARTING POINT BUT FIRST WE
C     STORE THE ORGINAL SPACE AND VELOCITY COORDINATES AND TIMESTEP.
C
   25 DO 7 IT=1,2
      DO 8 J=1,NB
      DO 8 K=1,3
      XTEMP(IT,J,K)=X(J,K)
      VTEMP(IT,J,K)=V(J,K)
    8 CONTINUE
      HTEMP=H
      DO 10 NOSTEP=1,3-IT
      DO 11 I=1,4
      DO 12 J=1,NB
      DO 12 K=1,3
      XT(J,K)=XTEMP(IT,J,K)+WT(I)*DX(J,K,I-1)
   12 VT(J,K)=VTEMP(IT,J,K)+WT(I)*DV(J,K,I-1)
C
      CALL ACC
C
```

```
      DO 13 J=1,NB
      DO 13 K=1,3
      DV(J,K,I)=IT*HTEMP*DELV(J,K)
      DX(J,K,I)=IT*HTEMP*VT(J,K)
   13 CONTINUE
   11 CONTINUE
      DO 14 J=1,NB
      DO 14 K=1,3
      XTEMP(IT,J,K)=XTEMP(IT,J,K)+(DX(J,K,1)+DX(J,K,4)+2*
     +(DX(J,K,2)+DX(J,K,3)))/6.0
      VTEMP(IT,J,K)=VTEMP(IT,J,K)+(DV(J,K,1)+DV(J,K,4)+2*
     +(DV(J,K,2)+DV(J,K,3)))/6.0
   14 CONTINUE
   10 CONTINUE
    7 CONTINUE
C
C  THE ABOVE HAS MADE TWO STEPS OF H AND, FROM THE SAME STARTING POINT,
C  A SINGLE STEP OF 2*H. THE RESULTS ARE NOW COMPARED
C
      DO 20 J=1,NB
      DO 20 K=1,3
      IF(ABS(XTEMP(1,J,K)-XTEMP(2,J,K)).GT.TOL)THEN
      H=0.5*H
      GOTO 25
      ENDIF
   20 CONTINUE
C
C  AT THIS STAGE THE DOUBLE STEP WITH H AGREES WITHIN TOLERANCE WITH
C  THE SINGLE STEP WITH 2*H. THE TIMESTEP WILL NOW BE TRIED WITH
C  TWICE THE VALUE FOR THE NEXT STEP. IF IT IS TOO BIG THEN IT WILL
C  BE REDUCED AGAIN.
C
      H=2*H
      DO 80 J=1,NB
      DO 80 K=1,3
      X(J,K)=XTEMP(1,J,K)
      V(J,K)=VTEMP(1,J,K)
   80 CONTINUE
      TIME=TIME+H
C
      CALL STORE
C
      IF(TIME.GE.TOTIME)GOTO 50
      IF(IG.GT.1000)THEN
      IG=1000
      GOTO 50
      ENDIF
      GOTO 25
```

```
C
   50 CALL OUT
C
      STOP
      END

      SUBROUTINE STORE
      DIMENSION CM(20),X(20,3),V(20,3),XSTORE(10000,20,3),
     +XTEMP(2,20,3),VTEMP(2,20,3),DELV(20,3),XT(20,3),VT(20,3)
      COMMON/A/X,V,TOL,H,TOTIME,DELV,XT,VT,NB,IST,TIME,IG,XTEMP,
     +VTEMP,CM
      COMMON/B/NORIG,XSTORE
      DO 21 J=1,NB
      DO 21 K=1,3
      X(J,K)=XTEMP(1,J,K)
      V(J,K)=VTEMP(1,J,K)
   21 CONTINUE
C
C  UP TO 1000 POSITIONS ARE STORED. THESE ARE TAKEN EVERY 5 STEPS.
C
      IST=IST+1
      IF((IST/5)*5.NE.IST)GOTO 50
      IG=IST/5
      IF(IG.GT.10000)GOTO 50
      DO 22 J=1,NB
      DO 22 K=1,3
      XSTORE(IG,J,K)=X(J,K)
   22 CONTINUE
   50 RETURN
      END

      SUBROUTINE START
      DIMENSION CM(20),X(20,3),V(20,3),XSTORE(10000,20,3),
     +XTEMP(2,20,3),VTEMP(2,20,3),DELV(20,3),XT(20,3),VT(20,3)
      COMMON/A/X,V,TOL,H,TOTIME,DELV,XT,VT,NB,IST,TIME,IG,XTEMP,
     +VTEMP,CM
      COMMON/B/NORIG,XSTORE
      OPEN(UNIT=21,FILE='LAST.DAT')
      OPEN(UNIT=22,FILE='FAST.DAT')
C
C  THE PROGRAMME AS PROVIDED IS FOR AN INVERSE-SQUARE LAW AND
C  USES UNITS FOR WHICH THE UNIT MASS IS THAT OF THE SUN, THE
C  THE UNIT OF DISTANCE IS THE ASTRONOMICAL UNIT (MEAN SUN-EARTH
C  DISTANCE),THE UNIT OF TIME IS THE YEAR AND THE GRAVITATIONAL
C  CONSTANT IS 4*PI**2
C
      WRITE(6,'(''INPUT THE NUMBER OF BODIES'')')
      READ(5,*)NB
```

```
      WRITE(6,'(''INPUT THE VALUES OF CM IN SOLAR-MASS UNITS.'')')
      WRITE(6,'(''THE TROJAN ASTEROID MASSES CAN BE PUT AS ZERO'')')
      DO 1 I=1,NB
      WRITE (6,500)I
  500 FORMAT(25H READ IN THE VALUE OF CM[,I3,1H])
      READ(5,*)CM(I)
    1 CONTINUE
      WRITE(6,'(''INPUT THE INITIAL TIMESTEP [years]'')')
      READ(5,*)H
      WRITE(6,'(''INPUT TOTAL TIME FOR THE SIMULATION [years]'')')
      READ(5,*)TOTIME
C
C     THE PROGRAMME ASKS THE USER TO SPECIFY A TOLERANCE, THE MAXIMUM
C     ABSOLUTE ERROR THAT CAN BE TOLERATED IN ANY POSITIONAL COORDINATE
C     (X, Y OR Z). IF THIS IS SET TOO LOW THEN THE PROGRAMME CAN BECOME
C     VERY SLOW. FOR COMPUTATIONS INVOLVING PLANETS A TOLERANCE OF 1.0E-6
C     (c. 150 KM) IS USUALLY SATISFACTORY.
C
      WRITE(6,'(''INPUT THE TOLERANCE'')')
      WRITE(6,'(''SEE COMMENT ABOVE THIS STATEMENT IN LISTING'')')
      READ(5,*)TOL
      WRITE(6,'(''THE CALCULATION CAN BE DONE RELATIVE TO AN'')')
      WRITE(6,'(''ARBITRARY ORIGIN OR WITH RESPECT TO ONE OF'')')
      WRITE(6,'(''THE BODIES AS ORIGIN. INPUT ZERO FOR AN'')')
      WRITE(6,'(''ARBITRARY ORIGIN OR THE NUMBER OF THE BODY'')')
      WRITE(6,'(''IF A BODY IS CHOSEN AS ORIGIN THEN ALL ITS'')')
      WRITE(6,'(''POSITIONAL AND VELOCITY VALUES ARE SET TO ZERO'')')
      READ(5,*)NORIG
      DO 31 J=1,NB
      WRITE(6,100)J
  100 FORMAT(23H INPUT [X,Y,Z] FOR BODY,I3)
      READ(5,*)X(J,1),X(J,2),X(J,3)
      WRITE(6,200)J
  200 FORMAT(32H INPUT [XDOT,YDOT,ZDOT] FOR BODY,I3)
      READ(5,*)V(J,1),V(J,2),V(J,3)
   31 CONTINUE
      RETURN
      END

      SUBROUTINE ACC
      DIMENSION CM(20),X(20,3),V(20,3),XSTORE(10000,20,3),R(3),
     +XTEMP(2,20,3),VTEMP(2,20,3),DELV(20,3),XT(20,3),VT(20,3),DD(3)
      COMMON/A/X,V,TOL,H,TOTIME,DELV,XT,VT,NB,IST,TIME,IG,XTEMP,
     +VTEMP,CM
      COMMON/B/NORIG,XSTORE
C
C     THE PROGRAMME AS PROVIDED IS FOR AN INVERSE-
C     SQUARE LAW AND USES UNITS FOR WHICH THE UNIT MASS IS THAT OF THE SUN,
```

```
C     THE UNIT OF DISTANCE IS THE ASTRONOMICAL UNIT (MEAN SUN-EARTH DISTANCE),
C     THE UNIT OF TIME IS THE YEAR AND THE GRAVITATIONAL CONSTANT IS 4*PI**2.
C     HOWEVER, THE USER MAY MODIFY THE SUBROUTINE ''ACC'' TO CHANGE TO ANY OTHER
C     FORCE LAW AND/OR ANY OTHER SYSTEM OF UNITS.
C
C     SET THE VALUE OF G IN ASTRONOMICAL UNITS
      PI=4.0*ATAN(1.0)
      G=4*PI*PI
      DO 1 J=1,NB
      DO 1 K=1,3
      DELV(J,K)=0
    1 CONTINUE
C     THE FOLLOWING PAIR OF DO LOOPS FINDS INTERACTIONS FOR ALL PAIRS
C     OF BODIES
      DO 2 J=1,NB-1
      DO 2 L=J+1,NB
      DO 3 K=1,3
      R(K)=XT(J,K)-XT(L,K)
    3 CONTINUE
      RRR=(R(1)**2+R(2)**2+R(3)**2)**1.5
      DO 4 K=1,3
C     THE NEXT TWO STATEMENTS GIVE THE CONTRIBUTIONS TO THE THREE
C     COMPONENTS OF ACCELERATION DUE TO BODY J ON BODY L AND DUE TO
C     BODY L ON BODY J.
      DELV(J,K)=DELV(J,K)-G*CM(L)*R(K)/RRR
      DELV(L,K)=DELV(L,K)+G*CM(J)*R(K)/RRR
    4 CONTINUE
    2 CONTINUE
C     IF ONE OF THE BODIES IS TO BE THE ORIGIN THEN ITS ACCELERATION IS
C     SUBTRACTED FROM THAT OF ALL OTHER BODIES.
      IF(NORIG.EQ.0)GOTO 10
      DD(1)=DELV(NORIG,1)
      DD(2)=DELV(NORIG,2)
      DD(3)=DELV(NORIG,3)
      DO 6 J=1,NB
      DO 6 K=1,3
      DELV(J,K)DELV(J,K)-DD(K)
    6 CONTINUE
   10 RETURN
      END

      SUBROUTINE OUT
      DIMENSION CM(20),X(20,3),V(20,3),XSTORE(10000,20,3),
     +XTEMP(2,20,3),VTEMP(2,20,3),DELV(20,3),XT(20,3),VT(20,3)
      COMMON/A/X,V,TOL,H,TOTIME,DELV,XT,VT,NB,IST,TIME,IG,XTEMP,
     +VTEMP,CM
      COMMON/B/NORIG,XSTORE
C       (X,Y) VALUES FOR THE LEADING ASTEROID ARE PLACED IN DATA FILE
```

```
C        LAST.DAT AND FOR THE FOLLOWING ASTEROID IN DATA FILE FAST.DAT.
C
C     FOR THE TROJAN ASTEROID PROBLEM THE JUPITER RADIUS VECTOR IS
C     ROTATED TO PUT IT ON THE Y AXIS. THE POSITIONS OF THE ASTEROIDS
C     RELATIVE TO JUPITER ARE PLOTTED.
         LAST=MIN(IG,10000)
         DO 60 I=1,LAST
C     THETA IS THE ANGLE BETWEEN THE X AXIS AND THE JUPITER RADIUS VECTOR
         THETA=ATAN2(XSTORE(I,2,2),XSTORE(I,2,1))
         XSTORE(I,2,2)=SQRT(XSTORE(I,2,2)**2+XSTORE(I,2,1)**2)
         XSTORE(I,2,1)=0
C     NOW THE ASTEROID RADIUS VECTORS ARE ROTATED BY PI/2-THETA
         DO 61 J=1,2
         AA=XSTORE(I,J+2,1)*SIN(THETA)-XSTORE(I,J+2,2)*COS(THETA)
         BB=XSTORE(I,J+2,1)*COS(THETA)+XSTORE(I,J+2,2)*SIN(THETA)
         XSTORE(I,J+2,1)=AA
         XSTORE(I,J+2,2)=BB
      61 CONTINUE
      60 CONTINUE
C     THE MODIFIED POSITIONS ARE NOW OUTPUT TO DATA FILES.
         DO 63 J=3,4
         N=18+J
         REWIND N
         DO 64 I=1,LAST
         WRITE(N,*)XSTORE(I,J,1),XSTORE(I,J,2)
      64 CONTINUE
      63 CONTINUE
         RETURN
         END
```

PHYSICAL CONSTANTS

The basic SI units of interest are as follows:

length	metre	m
mass	kilogram	kg
time	second	s
current	ampere	A
magnetic field	tesla	T
absolute temperature	kelvin	K

Some other units derived from these are:

force	newton	$N = kg\,m\,s^{-2}$
energy	joule	$J = N\,m$
power	watt	$W = J\,s^{-1}$
electric charge	coulomb	$C = A\,s$
electric potential	volt	$V = W\,A^{-1}$
capacitance	farad	$F = C\,V^{-1}$
inductance	henry	$H = V\,A^{-1}\,s$

Multiples or submultiples of the basic units, which differ by factors of 10^3 are acceptable. Thus the millimetre (mm) is an acceptable unit. In defining magnetic fields the gauss (10^{-4} T) is sometimes used but in modern work its use is declining. For very small fields the unit $\gamma = 10^{-9}$ T is useful.

Physical constants

Gravitational constant	$G = 6.673 \times 10^{-11}\,m^3\,kg^{-1}\,s^{-2}$
Electron charge	$e = 1.602 \times 10^{-19}\,C$
Electron mass	$m_e = 9.109 \times 10^{-31}\,kg$
Proton mass	$m_p = 1.673 \times 10^{-27}\,kg$
Boltzmann constant	$k = 1.381 \times 10^{-23}\,J\,K^{-1}$
Planck constant	$h = 6.626 \times 10^{-34}\,J\,s$
Electron volt	$eV = 1.602 \times 10^{-19}\,J$
Speed of light	$c = 2.998 \times 10^8\,m\,s^{-1}$
Permittivity of free space	$\varepsilon_0 = 8.854 \times 10^{-12}\,F\,m^{-1}$
Permeability of free space	$\mu_0 = 4\pi \times 10^{-7}\,H\,m^{-1}$

In some of the solutions to problems, where accuracy is inappropriate, approximate values are used for some solar-system quantities, e.g.

Mass of the Sun	2×10^{30} kg	$(1.989 \times 10^{30}$ kg$)$
Radius of the Sun	7×10^{8} m	$(6.963 \times 10^{8}$ m$)$
Mass of Jupiter	2×10^{27} kg	$(1.899 \times 10^{27}$ kg$)$
Radius of Jupiter	7×10^{7} m	$(7.14 \times 10^{7}$ m$)$
Mass of the Earth	6×10^{24} kg	$(5.974 \times 10^{24}$ kg$)$
Radius of the Earth	6.4×10^{6} m	$(6.378 \times 10^{6}$ m$)$
Astronomical unit	1.5×10^{-11} m^{3} kg^{-1} s^{-2}	$(1.496 \times 10^{-11}$ m^{3} kg^{-1} s$^{-2})$
Parsec	3×10^{16} m	$(3.086 \times 10^{16}$ m$)$

SOLUTIONS TO PROBLEMS

CHAPTER 1

1.1 (i) The total mass of Saturn is

$$M_S = \int_0^R 4\pi r^2 \rho_0 \left[1 - \left(\frac{r}{R} \right)^{1/4} \right] dr = 4\pi \rho_0 R^3 \left(\frac{1}{3} - \frac{4}{13} \right) = \frac{4}{39} \pi \rho_0 R^3.$$

(ii) The total mass of helium is

$$M_{He} = \int_0^R 4\pi r^2 \rho_0 \left[1 - \left(\frac{r}{R} \right)^{1/4} \right] p_0 (1 - \alpha r) \, dr.$$

Expanding the integrand in powers of r and integrating gives

$$M_{He} = 4\pi \rho_0 p_0 R^3 \left(\frac{1}{39} - \frac{\alpha R}{68} \right).$$

(iii) The proportion of helium in Saturn is

$$\frac{M_{He}}{M_S} = p_0 \left(1 - \frac{39}{68} \alpha R \right) = 0.15. \tag{1}$$

The proportion of helium at the surface is given by

$$p_0 (1 - \alpha R) = 0.033. \tag{2}$$

Eliminating p_0 from (1) and (2) gives $\alpha R = 0.893$. Hence, from (2), $p_0 = 0.31$.

CHAPTER 2

2.1 (i) Consider N representative bodies. The total number with masses between M_1 and M_2 is

$$Q = NC \int_{M_1}^{M_2} M^{-2.35} \, dM = \frac{NC}{1.35} \left(\frac{1}{M_1^{1.35}} - \frac{1}{M_2^{1.35}} \right). \tag{1}$$

With $M_2 = \infty$ and $M_1 = 0.075 M_\odot$ this gives the number of main sequence stars,

$$Q_{ms} = \frac{NC}{1.35(0.075 M_\odot)^{1.35}} = 24.5 \frac{NC}{M_\odot^{1.35}}.$$

476

The total mass of these stars is

$$M_{ms} = NC \int_{M_1}^{M_2} M \times M^{-2.35} \, dM = \frac{NC}{0.35} \left(\frac{1}{M_1^{0.35}} - \frac{1}{M_2^{0.35}} \right)$$

and with the same limits of mass this gives $M_{ms} = 7.07 NC/M_\odot^{0.35}$.
The average mass of a main sequence star is thus

$$\bar{M} = \frac{M_{ms}}{\varrho_{ms}} = 0.29 M_\odot.$$

(i) From (1) the number of brown dwarfs is

$$\varrho_{bd} = \frac{NC}{1.35} \left(\frac{1}{(0.013 M_\odot)^{1.35}} - \frac{1}{(0.075 M_\odot)^{1.35}} \right) = 236.1 \frac{NC}{M_\odot^{1.35}}.$$

The ratio of the number of brown dwarfs to main sequence stars is

$$\frac{\varrho_{bd}}{\varrho_{ms}} = 9.6.$$

CHAPTER 3

3.1 The relationship is equivalent to $\log r_N = n \log b + \log a$.

Planet	r_N		$\log r_N$	Formula
Mercury	0	0.387	−0.412	0.37
Venus	1	0.732	−0.141	0.63
Earth	2	1.000	0.000	1.09
Mars	3	1.524	0.188	1.86
Jupiter	5	5.203	0.716	5.42
Saturn	6	9.539	0.980	9.28
Uranus	7	19.19	1.283	15.86
Neptune	8	30.07	1.478	27.12
Pluto	9	39.46	1.596	46.38

Visual estimation of the best straight line will vary from one individual to another. For comparison an analytical least-squares line gives $\log b = 0.233$ and $\log c = -0.430$. This gives the best relationship of this type as

$$r_n = 0.371 \times 1.710^n.$$

that gives the values in the final column of the table.

3.2 (a) The proper motions of Uranus and Neptune are:

$$n_U = 2\pi/84.011 \text{ yr}^{-1} = 0.74790 \text{ yr}^{-1}$$

$$n_N = 2\pi/164.79 \text{ yr}^{-1} = 0.38128 \text{ yr}^{-1}.$$

This gives the relationship $2n_N - n_U = 0.001466 \text{ yr}^{-1}$.
(b) (i) $n_V = 2\pi/0.723 \text{ yr}^{-1} = 8.690 \text{ yr}^{-1}$ and $n_E = 2\pi/1.000 = 6.283 \text{ yr}^{-1}$.
(ii) $7n_E - 5n_V = 0.531$ is the best relationship that can be found.

CHAPTER 4

4.1   (i) Since the orbital sidereal period of Venus is less than that of the Earth from one conjunction to the next similar one it will orbit through $2\pi + \theta$ while the Earth orbits through angle θ. Thus

$$\frac{\theta}{2\pi + \theta} = \frac{0.6152}{1.0000} \quad \text{or} \quad \theta = 10.045 \, \text{radians}.$$

The orbital synodic period for Venus is the time for the Earth to orbit through angle θ and is 1.599 years.

(ii) Since the orbital sidereal period of Jupiter is greater than that of the Earth from one conjunction to the next similar one it will orbit through θ while the Earth orbits through angle $2\pi + \theta$. Thus

$$\frac{2\pi + \theta}{\theta} = \frac{11.862}{1.000} \quad \text{or} \quad \theta = \frac{2\pi}{10.862}.$$

The synodic orbital period for Jupiter is the time taken for the Earth to orbit through an angle $2\pi + \theta$ and is 1.092 years.

CHAPTER 5

5.1   (i) The rate of release of gravitational energy is

$$\frac{d\Omega}{dt} = \frac{d\Omega}{dR}\frac{dR}{dt} = \frac{GM_J^2}{R^2}\frac{dR}{dt} = 8 \times 10^{17} \, \text{W}.$$

Hence $dR/dt = 8 \times 10^{17}(R^2/GM_J^2) \, \text{m s}^{-1}$ using standard units.
With $R = 7.1 \times 10^7$ m and $M_J = 1.9 \times 10^{27}$ m, $dR/dt = 0.053$ m century^{-1}.
(ii) The total mass of helium is $0.15 \times 5.7 \times 10^{26}$ kg $= 8.6 \times 10^{25}$ kg.
If M_S is the mass of Saturn and m the mass of helium moved then the energy released in moving from distance r_0 to distance r_1 is $GM_S m(1/r_1 - 1/r_0)$. With $r_0 = 6.0 \times 10^7$ m and $r_1 = 10^7$ m this is $3.2 \times 10^9 m$ J. For this to equal 2.4×10^{34} J, $m = 7.5 \times 10^{24}$ kg or 8.7% of that present.

CHAPTER 6

6.1   The orbital period comes from $P^2 = 4\pi^2 a^3/GM_{\text{m}}$. With $a = 1.788 \times 10^6$ m and $M_{\text{m}} = 7.35 \times 10^{22}$ kg this gives $P = 6785$ s.

6.2   From equation (6.1) the semi-angle of the cone of light is

$$\alpha = \frac{1.22 \times 5.6 \times 10^{-7}}{0.02} = 3.42 \times 10^{-5} \, \text{radians}.$$

The radius of the patch of light on the Moon is thus

$$r = 3.84 \times 10^8 \times 3.42 \times 10^{-5} = 1.31 \times 10^4 \, \text{m}.$$

The number of photons, n, in the patch is given by

$$\frac{nhc}{\lambda} = 1 \, \text{J} \quad \text{or} \quad n = \frac{5.6 \times 10^{-7}}{6.63 \times 10^{-34} \times 3 \times 10^8} = 2.8 \times 10^{18}.$$

Hence the number density of photons is

$$\frac{2.8 \times 10^{18}}{\pi \times (1.31 \times 10^4)^2} = 5.2 \times 10^9 \, \text{m}^{-2}.$$

CHAPTER 7

7.1   Just including the masses given in tables 7.2 to 7.5 the ratios are:

Jupiter $\dfrac{3.95 \times 10^{23}}{1.90 \times 10^{27}} = 2.08 \times 10^{-4}$ Saturn $\dfrac{1.42 \times 10^{23}}{5.69 \times 10^{26}} = 2.50 \times 10^{-4}$

Uranus $\dfrac{9.18 \times 10^{21}}{8.66 \times 10^{25}} = 1.06 \times 10^{-4}$ Neptune $\dfrac{2.14 \times 10^{22}}{1.02 \times 10^{26}} = 2.09 \times 10^{-4}.$

The ratios for Jupiter, Saturn and Neptune are similar although, since Triton is clearly irregular, this may be just fortuitous. The ratio for Uranus is about one half that for the other planets but its large axial tilt may indicate that it suffered some violent event that could have disturbed an original satellite family.

CHAPTER 8

8.1   (i) Each kilogram of asteroid contains $10^{-2} \times 5 \times 10^{-6} \, \text{kg} = 5 \times 10^{-8} \, \text{kg}$ of ^{26}Al. The mass of one atom of ^{26}Al is $26 \times 1.67 \times 10^{-27} \, \text{kg}$. Hence the number of ^{26}Al atoms per kilogram of asteroid is

$$\frac{5 \times 10^{-8}}{26 \times 1.67 \times 10^{-27}} = 1.15 \times 10^{18}.$$

(ii) The energy released per kilogram is

$$1.15 \times 10^{18} \times 10^7 \times 1.6 \times 10^{-19} \, \text{J} = 1.84 \times 10^6 \, \text{J}.$$

(iii) The increase in temperature this gives is

$$\frac{1.84 \times 10^6}{1000} \, \text{K} = 1840 \, \text{K}.$$

For a very small asteroid most of this will be lost so the increase in temperature will be much less. For a very large asteroid almost all the heat generated would be retained on a timescale of a few half-lives of ^{26}Al.

CHAPTER 9

9.1   The values of a and e for the comet orbit are given by the perihelion, q, and aphelion, Q, as

$$a = \tfrac{1}{2}(q + Q) = 6.5 \, \text{AU} \qquad \text{and} \qquad e = \frac{Q - q}{Q + q} = 0.6154.$$

Inserting these values in equation (9.1) gives: $(a_p = 5.20 \, \text{AU})$ $T_{\text{Jup}} = 0.2464 \, \text{AU}^{-1}$ for Jupiter and $(a_p = 9.54 \, \text{AU})$ $T_{\text{Sat}} = 0.1451 \, \text{AU}^{-1}$ for Saturn.

These values of T can now be inserted into equation (9.2) to give the original perihelion of the Oort cloud comet. These are, for Jupiter and Saturn respectively, 4.27 AU and 9.14 AU.

9.2   The value of T can be deduced from the original perihelion, $6\,\text{AU}$, from equation (9.2).

$$T = \left(\frac{2q}{a_p^3}\right)^{1/2} = 0.1176\,\text{AU}^{-1}.$$

Since we know the new perihelion we transform equation (9.1) by $a = q/(1 - e)$ to

$$T = \frac{1-e}{2q} + \left(\frac{q(1+e)}{a_p^3}\right)^{1/2}.$$

With $q = 4\,\text{AU}$ this gives

$$0.1176 = 0.125(1 - e) + [0.004607(1 + e)]^{1/2}.$$

This can be transformed into a quadratic equation for e

$$15.625e^2 - 6.467e - 4.552 = 0$$

with a positive solution $e = 0.785$. Then $a = q/(1 - e) = 18.60\,\text{AU}$.

CHAPTER 10

10.1   From the SMOW values given in section 10.6, $[n(^{17}O)/n(^{16}O)]_{\text{SMOW}} = 0.007543$ and $[n(^{18}O)/n(^{16}O)]_{\text{SMOW}} = 0.04209$. Applying equation (10.1) gives the following δ values:

$\delta^{17}O$	−16.44	−11.93	−7.82	−11.14
$\delta^{18}O$	0.00	4.04	8.80	5.46

Plotting these points gives a best line with slope close to unity. In terms of its oxygen isotopic composition the grain material is similar to that giving a unit slope in figure 10.15 but it is more deficient in ^{17}O.

10.2   Let the mass of the grain be m. Then, since there is very little ^{15}N in normal nitrogen the total number of normal nitrogen atoms in the sample is

$$\frac{10^{-5}m}{14 \times 1.66 \times 10^{-27}} = 4.303 \times 10^{20}m.$$

Of these 4.287×10^{20} are ^{14}N and 1.588×10^{18} are ^{15}N.
  The mass of a SiC entity is $40\,\text{amu}$ and hence the number in the grain is

$$\frac{m}{40 \times 1.66 \times 10^{-27}} = 1.506 \times 10^{25}m.$$

This is also the number of carbon atoms and hence there were originally $1.506 \times 10^{21}\,^{14}C$ atoms which transform into ^{14}N. Hence the final number of ^{14}N atoms is $4.287 \times 10^{20}m + 1.506 \times 10^{21}m = 1.935 \times 10^{21}m$ and the final ratio

$$\frac{n(^{14}N)}{n(^{15}N)} = \frac{1.935 \times 10^{21}m}{1.588 \times 10^{18}m} = 1218.$$

This is below 2000 and hence a sensible result.

CHAPTER 11

11.1 Relative to the Earth the particles are moving at speed $(30^2 + 40^2)^{1/2}$ km s^{-1} = 50 km s^{-1}. The volume swept out by the cross section of the Earth per second is

$$V = \pi(6.4 \times 10^6)^2 \times 5 \times 10^4 \text{ m}^3 = 6.43 \times 10^{18} \text{ m}^3$$

and this will contain 643 particles. Hence the number per hour is $643 \times 3600 = 2.41 \times 10^6$. These will fall on an area $\pi \times 6400^2$ km^2 but from a given point only those falling in an area $\pi \times 400^2$ km^2 can be seen. Thus, assuming a uniform distribution, the number seen per hour is

$$2.41 \times 10^6 \times 400^2/6400^2 = 9400.$$

The Earth traverses the width of the stream in $10^9/3 \times 10^4$ s = 9 h 16 m.

CHAPTER 12

12.1 Let the radius of the core be a. Then the mass of Mars is the mass of a sphere of radius 3397 km and density 3.3×10^3 kg m^{-3} plus the mass of a sphere of radius a and density $(7.8 - 3.3) \times 10^3$ kg m^{-3}. Hence

$$\tfrac{4}{3}\pi(3.397 \times 10^6)^3 \times 3.3 \times 10^3 + \tfrac{4}{3}\pi a^3 \times 4.5 \times 10^3 = 6.42 \times 10^{23}$$

giving $a = 1745$ km.

The stripped planet now has a mass equivalent to the mass of a sphere of radius 2440 km and density 3.3×10^3 kg m^{-3} plus the mass of a sphere of radius 1745 km and density 4.5×10^3 kg m^{-3}. This is 3.01×10^{23} kg. The mass of Mercury, 3.30×10^{23} kg, is some 10% greater.

TOPIC A

A.1 Let the proportions of quartz, orthoclase, albite and olivine be a, b, c and d respectively. Then

	$a +$	$b +$	$c +$	$d = 1$	(i)
From SiO_2	$a +$	$0.65b +$	$0.54c +$	$0.40d = 0.6585$	(ii)
From Al_2O_3		$0.18b +$	$0.29c$	$= 0.1065$	(iii)
From $MgO + FeO$				$0.60d = 0.1500$	(iv)
From $Na_2O + K_2O$		$0.17b +$	$0.05c$	$= 0.0670$	(v)
From CaO			$0.12c$	$= 0.0180$	(vi)

There are more equations than unknowns so we find a solution from four of them and check for consistency with the other two.

From (vi)	$c = 0.15$
From (iv)	$d = 0.25$
From (iii)	$b = 0.35$
From (i)	$a = 0.25$

These values give consistency for equations (ii) and (v). If they did not then one would need to seek a least-squares solution of the six equations for the four unknowns.

A.2   The element that gives the chain by repetition is Si_4O_{11} within the rectangle.

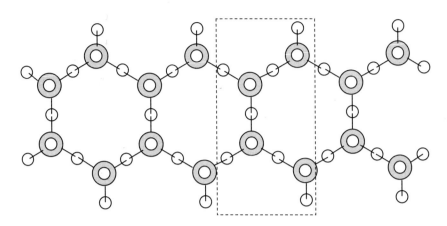

TOPIC B

B.1   (a) $T_{1/2} = 0.693/(5 \times 10^{-11})$ yr $= 1.39 \times 10^{11}$ yr.
      (b) $\lambda = 0.693/T_{1/2} = 0.0559\,h^{-1} = 489.9\,yr^{-1}$.
B.2   The following ratios are found:

| $^{87}Sr/^{86}Sr$ | 0.713 | 0.731 | 0.746 | 0.763 |
| $^{87}Rb/^{86}Sr$ | 0.200 | 0.540 | 0.787 | 1.093 |

When plotted these give a good straight line of slope ~ 0.05783. The slope can be taken as $\lambda\tau$ and this gives an age $0.05783/1.42 \times 10^{-11}$ yr $= 4.07 \times 10^{9}$ yr.

TOPIC C

C.1   We assume that the self-potential energy of the cluster is equivalent to that of a uniform distribution of matter with the same total mass and the same radius. Hence the equation giving the root-mean-square speed is

$$M\langle V^2 \rangle - \frac{3GM^2}{5R} = 0 \quad \text{or} \quad \langle V^2 \rangle^{1/2} = \left(\frac{3GM}{5R} \right)^{1/2}.$$

With $M = 300 \times 10^{30}$ kg and $R = 1.6 \times 3.09 \times 10^{16}$ m we find $\langle V^2 \rangle^{1/2} = 493\,m\,s^{-1}$.
C.2   The translational kinetic energy per particle is $1.5kT$ so that the total translational energy is

$$K = 1.5kT \times \frac{M}{\mu}$$

where M is the mass of the star and μ the mean particle mass. Hence the Virial Theorem gives

$$\frac{3kTM}{\mu} - \frac{3GM^2}{5R} = 0 \quad \text{or} \quad T = \frac{GM\mu}{5kR}.$$

With the values given $T = 4.60 \times 10^6$ K.
      With a more centrally-condensed distribution the potential energy would have a greater magnitude (be more negative). This would increase the estimate of the mean temperature.

TOPIC D

D.1 With $\mu = 1.67 \times 10^{-27}$ kg, $M_c = 7.5 \times 10^{21}(T^3/\rho)^{1/2}$. For the density and temperature given $M_c = 6.9 \times 10^{30}$ kg.

  The radius of the sphere is $R = (3M_c/4\pi\rho)^{1/3} = 8.95 \times 10^{14}$ m. Thus the potential energy is $\Omega = -3GM_c^2/5R = 2.13 \times 10^{36}$ J. The thermal kinetic energy is $K = \frac{3}{2}kT(M_c/\mu) = 1.07 \times 10^{36}$ J. Allowing for round-off error the values of Ω and K satisfy the Virial Theorem.

TOPIC E

E.1 By inserting the appropriate values of r/r_0 in equation (E.8) the following values of t/t_{ff} are found: (i) 0.396, (ii) 0.748, (iii) 0.923.

  An inspection of figure E.2 shows that for $t/t_{ff} = 0.5$ the value of r/r_0 is between 0.8 and 0.85. For 0.8 the value of $t/t_0 = 0.550$ while for 0.85 it is 0.481. Linear interpolation suggests $r/r_0 = 0.835$. The actual solution is $r/r_0 = 0.836$.

  This problem can be done by more sophisticated numerical methods but the trial-and-error approach here is fast and efficient.

TOPIC F

F.1 From equation (E.7) in terms of radius and mass

$$t_{ff} = \frac{\pi}{2}\left(\frac{(1.5 \times 10^{14})^3}{2 \times 6.67 \times 10^{-11} \times 2 \times 10^{30}}\right)^{1/2} = 1.77 \times 10^{11}\,\text{s} = 5590\,\text{years}.$$

From table F.1 the time on the K-H track is 5×10^7 years. Hence the ratio is ~ 9000.

F.2 The rate of change of total energy of the star is just due to its luminosity. Hence $dE/dt = -L_*$ and since, from equation (F.4), $E = -K$ we have $dK/dt = L_*$. The total thermal kinetic energy is given by $K = \frac{3}{2}kT(M_*/\mu)$ where M_* is the mass of the star and μ is the mean particle mass. Combining these results $dT/dt = 2L_*\mu/3kM_*$. With $L_* = 1.2 \times 10^{26}$ W, $M_* = 2 \times 10^{30}$ kg and $\mu = 10^{-27}$ kg we find $dT/dt = 2.9 \times 10^{-8}$ K s^{-1} = 0.92 K yr^{-1}.

TOPIC G

G.1 From equation (G.3)

$$P = \frac{10^4 \times 1.38 \times 10^{-23} \times 6 \times 10^6}{10^{-27}}\,\text{Pa} = 276 \times 10^{14}\,\text{Pa}.$$

For the small shift, $-\delta r$, towards the centre we assume that a finite-difference approach is justified so that, for example,

$$\delta P = \frac{GM_1\rho}{r^2}\,\delta r = \frac{6.67 \times 10^{-11} \times 1.3 \times 10^{30} \times 10^4}{(2.2 \times 10^8)^2} \times 10^7\,\text{Pa} = 5.4 \times 10^{13}\,\text{Pa}.$$

The positive sign indicates that pressure is higher and $P + \delta P = 3.30 \times 10^{14}$ Pa.

Similarly $\delta M_I = -4\pi(2.2 \times 10^8)^2 \times 10^4 \times 10^7$ kg $= -6.1 \times 10^{28}$ kg. Hence the included mass is lower and is 1.24×10^{30} kg.

$$\delta T = \frac{3 \times 2 \times 10^{26} \times 5 \times 10^4}{64 \times \pi \times 5.67 \times 10^{-8} \times (6 \times 10^6)^3 \times (2.2 \times 10^8)^2} \times 10^7 \text{ K} = 2.52 \times 10^6 \text{ K}.$$

The temperature is higher and is 8.52×10^6 K.

$$\delta L = -4\pi(2.2 \times 10^6)^2 \times 10^4 \times 1.5 \times 10^{-4} \times 10^7 \text{ W} = -9.1 \times 10^{24} \text{ W}.$$

The luminosity is lower and is 1.91×10^{26} W.

TOPIC H

H.1
$$\frac{\varepsilon_{CNO}}{\varepsilon_{pp}} = \frac{2.3 \times 10^{-8}}{9.5 \times 10^{-9}} \frac{X_C}{X_p} \left(\frac{T}{10^7}\right)^{17} \left(\frac{10^7}{T}\right)^4 = 1.04 \times 10^{-3} \left(\frac{T}{10^7}\right)^{13}.$$

If this ratio equals f then

$$T = 10^7 \left(\frac{f}{1.04 \times 10^{-3}}\right)^{1/13}.$$

This gives temperatures (i) 1.52×10^7 K, (ii) 1.70×10^7 K, (iii) 1.89×10^7 K. Energy generation at the two temperatures are as follows:

	1.50×10^7 K	1.51×10^7 K
p-p	$2.357 \times 10^{-8} \rho \,\text{W kg}^{-1}$	$2.420 \times 10^{-8} \rho \,\text{W kg}^{-1}$
CNO	$4.76 \times 10^{-9} \rho \,\text{W kg}^{-1}$	$5.33 \times 10^{-9} \rho \,\text{W kg}^{-1}$
Total	$2.833 \times 10^{-8} \rho \,\text{W kg}^{-1}$	$2.953 \times 10^{-8} \rho \,\text{W kg}^{-1}$

This corresponds to an increase of 4.2%.

TOPIC I

I.1   Inserting values into equations (I.1) give the following pressures in Pa:

	Gas	Radiation	Degeneracy
(i)	8.26×10^{10}	2.52	3.08×10^{13}
(ii)	8.26×10^{14}	2.52×10^{12}	1.43×10^{15}
(iii)	8.26×10^{15}	2.52×10^{16}	1.43×10^{15}
(iv)	8.26×10^{16}	2.52×10^{12}	3.08×10^{18}
(v)	8.26×10^{18}	2.52×10^{20}	3.08×10^{18}

TOPIC J

J.1   The radius is given by

$$r = \frac{9.05 \times 10^{16}}{M^{1/3}} = \frac{9.05 \times 10^{16}}{(1.6 \times 10^{30})^{1/3}} \text{ m} = 7740 \text{ km}.$$

$M \propto r^{-3}$ and $r^{-3} \propto \rho/M$ hence $\rho \propto M^2$.

J.2 Since the neutron star has the same density as a neutron its radius, r_*, is related to the radius of the neutron, r_N, by

$$\left(\frac{r_*}{r_N}\right)^3 = \frac{M_*}{m_N} \quad \text{or} \quad r_* = r_N \left(\frac{M_*}{m_N}\right)^{1/3}.$$

Hence

$$r_* = \left(\frac{4 \times 10^{30}}{1.67 \times 10^{-27}}\right)^{1/3} \times 10^{-15} \, \text{m} = 13.4 \, \text{km}.$$

For the period we have $r_* \omega^2 = 0.1(GM_*/r_*^2)$ and $P = 2\pi/\omega$. This gives

$$P = 2\pi \left(\frac{10 r_*^3}{GM_*}\right)^{1/2} = 2\pi \left(\frac{10 \times (1.34 \times 10^4)^3}{6.67 \times 10^{-11} \times 4 \times 10^{30}}\right)^{1/2} \, \text{s} = 1.9 \, \text{ms}.$$

TOPIC K

K.1 From equation (K.1)

$$R = \left(\frac{GM_* P^2}{4\pi^2}\right)^{1/3} = \left(\frac{6.67 \times 10^{-11} \times 1.15 \times 2 \times 10^{30} \times (2.06 \times 3.16 \times 10^7)^2}{4\pi^2}\right)^{1/3} \, \text{m}$$

$$= 2.54 \times 10^{11} \, \text{m} = 1.70 \, \text{AU}.$$

The speed of the planet, $v_p = 2\pi R/P = 2.45 \times 10^4 \, \text{ms}^{-1}$.
The minimum mass of the planet is given by

$$M_p > 1.15 \times 2 \times 10^{30} \times \frac{52}{2.45 \times 10^4} \, \text{kg} = 4.88 \times 10^{27} \, \text{kg} = 2.6 M_J.$$

TOPIC L

L.1 The Sun's rays make an angle $\tan^{-1} 1.508 = 0.9852$ radians with the vertical stick in Aberdeen and an angle $\tan^{-1} 1.190 = 0.8719$ radians with a vertical stick in Bournmouth. Hence the difference in latitude of the two locations is 0.1133 radians. The estimated radius of the Earth is thus

$$R = \frac{750}{0.1133} \, \text{km} = 6620 \, \text{km}.$$

L.2 The angular precision of measurement was $10^{-4}/2.1 \, \text{radians} = 4.8 \times 10^{-5} \, \text{radians}$. This is equivalent to $9.8''$.

L.3 The mean orbital radii for the Earth and Venus are $1.496 \times 10^{11} \, \text{m}$ and $1.083 \times 10^{11} \, \text{m}$ respectively. The diameter of Venus is $1.240 \times 10^7 \, \text{m}$.
    In full phase the angular diameter is

$$\frac{1.240 \times 10^7}{(1.496 + 1.083)^{11}} \, \text{radians} = 4.81 \times 10^{-5} \, \text{radians} = 9.9''.$$

In crescent phase the angular diameter is

$$\frac{1.240 \times 10^7}{(1.496 - 1.083)^{11}} \text{ radians} = 3.00 \times 10^{-4} \text{ radians} = 61.9''.$$

TOPIC M

M.1 The perihelion $a(1 - e) = 2.3\,\text{AU}$. With $\theta = 60°$

$$3.0 = \frac{a(1 - e^2)}{1 + e \cos 60°} = \frac{2.3(1 + e)}{1 + 0.5e}$$

This gives $e = 0.875$. We then find $a = q/(1 - e) = (2.3/0.125)\,\text{AU} = 18.4\,\text{AU}$.

M.2 The orbital radii of the Earth and Venus are 1.496×10^{11} m and 1.083×10^{11} m respectively. The velocities in their orbits are

$$V_{\text{E}} = \left(\frac{GM_\odot}{r_{\text{E}}}\right)^{1/2} = 29.86\,\text{km s}^{-1} \quad \text{and} \quad V_{\text{V}} = \left(\frac{GM_\odot}{r_{\text{V}}}\right)^{1/2} = 35.10\,\text{km s}^{-1}.$$

To go out of circular Earth orbit to an orbit with perihelion equal to the radius of the Venus orbit the speed must change to that given by equation (M.35) which gives $V_{\text{D}} = 27.37\,\text{km s}^{-1}$. This requires a change of velocity $2.49\,\text{km s}^{-1}$. The velocity at perihelion is given by equation (M.36) and is $V_{\text{F}} = 37.80\,\text{km s}^{-1}$. The speed must be changed at this point to that for a Venus orbit so again the spacecraft must be decelerated, this time by $2.70\,\text{km s}^{-1}$.

M.3 For the asteroid,

$$x = 1.2\,\text{AU} = 1.8 \times 10^{11}\,\text{m},$$

$$y = 0.8\,\text{AU} = 1.2 \times 10^{11}\,\text{m},$$

$$\dot{x} = 2.0 \times 10^4 \times \cos 55°\,\text{m s}^{-1} = 1.15 \times 10^4\,\text{m s}^{-1},$$

$$\dot{y} = 2.0 \times 10^4 \times \sin 55°\,\text{m s}^{-1} = 1.64 \times 10^4\,\text{m s}^{-1}.$$

The intrinsic energy of the orbit is

$$E = \tfrac{1}{2}(2.0 \times 10^4)^2 - \frac{6.67 \times 10^{-11} \times 2 \times 10^{30}}{(1.8^2 + 1.2^2)^{1/2} \times 10^{11}}\,\text{J kg}^{-1} = -4.17 \times 10^8\,\text{J kg}^{-1}.$$

Since the orbit is in the x–y plane the magnitude of the intrinsic angular momentum is

$$H = x\dot{y} - y\dot{x} = [(1.8 \times 1.64) - (1.2 \times 1.15)] \times 10^{15}\,\text{m}^2\,\text{s}^{-1} = 1.57 \times 10^{15}\,\text{m}^2\,\text{s}^{-1}.$$

From equation (M.24)

$$a = -\frac{6.67 \times 10^{-11} \times 2 \times 10^{30}}{-2 \times 4.17 \times 10^8}\,\text{m} = 1.60 \times 10^{11}\,\text{m} = 1.07\,\text{AU}.$$

From equation (M.25)

$$e = \left[1 - 2 \times 4.17 \times 10^8 \left(\frac{1.57 \times 10^{15}}{6.67 \times 10^{-11} \times 2 \times 10^{30}}\right)^2\right]^{1/2} = 0.940.$$

The asteroid is Earth-crossing.

TOPIC O

O.1 We apply equation (O.5b) with $P_0 = 10^5 \, \mathrm{N\,m^{-2}}$, $\mu = 28 \times 1.67 \times 10^{-27} \, \mathrm{kg}$, $g = 9.8 \, \mathrm{m\,s^{-2}}$, and $T = 260 \, \mathrm{K}$. For $h = 8 \times 10^3 \, \mathrm{m}$, $P = 3.6 \times 10^4 \, \mathrm{N\,m^{-2}}$. For $h = 2 \times 10^4 \, \mathrm{m}$, $P = 7.78 \times 10^3 \, \mathrm{N\,m^{-2}}$.

O.2 (i) The adiabatic lapse rate is

$$\frac{\mathrm{d}T}{\mathrm{d}x} = -(\gamma - 1)\frac{g\mu}{k} = -0.25 \times \frac{9.8 \times 28 \times 1.67 \times 10^{-27}}{1.38 \times 10^{-23}} \, \mathrm{K\,m^{-1}} = -8.3 \, \mathrm{K\,km^{-1}}.$$

At the top of Mount Everest the temperature is $[300 - (8 \times 8.3)] \, \mathrm{K} = 233.6 \, \mathrm{K}$. For an adiabatic change the relationship between pressure and temperature is $P = CT^{\gamma/(\gamma - 1)}$, where C is a constant. For this case $\gamma/(\gamma - 1) = 5$.

The pressure at the top of Mount Everest is

$$P = 1 \times 10^5 \left(\frac{233.6}{300}\right)^5 \, \mathrm{N\,m^{-2}} = 2.86 \times 10^4 \, \mathrm{N\,m^{-2}}.$$

(ii) From equation (O.23) with $\gamma = 1.28$, $\mu = 44 \times 1.67 \times 10^{-27} \, \mathrm{kg}$ and $g = 8.4 \, \mathrm{m\,s^{-2}}$ we find $\mathrm{d}T/\mathrm{d}x = -0.0125 \, \mathrm{K\,m^{-1}} = -12.5 \, \mathrm{K\,km^{-1}}$. This is somewhat higher than is indicated in figure 4.9. The temperature at a height of 50 km would be indicated as $[730 - (50 \times 12.5)] \, \mathrm{K} = 105 \, \mathrm{K}$. The value of $\gamma/(1 - \gamma) = 4.57$ in this instance so the pressure at a height of 50 km is given by

$$P = 1.0 \times 10^7 \left(\frac{105}{730}\right)^{4.57} \, \mathrm{N\,m^{-2}} = 1.42 \times 10^3 \, \mathrm{N\,m^{-2}}.$$

O.3 Inserting the given values, and assuming $A = 10^{-18} \, \mathrm{m^2}$ for all molecules, gives the following loss rates $(\mathrm{s^{-1}})$

Jupiter	H_2	0
Mars	CO_2	4.3×10^{-21}
Mars	H_2O	7.2×10^9
Moon	O_2	2.7×10^{25}
Io	SO_2	3.0×10^{11}
Titan	N_2	6.6×10^5

TOPIC P

P.1 Inserting the values $M_\mathrm{p} = 10^{28} \, \mathrm{kg}$, $A = 12$ and $Z = 6$ in equation (P.8) gives $R = 24\,260 \, \mathrm{km}$.

P.2 (i) With $K = K_0$ we can write, using ordinary differential notation

$$\int_{\rho_0}^{\rho} \frac{\mathrm{d}\rho}{\rho} = \frac{1}{K_0} \int_0^P \mathrm{d}p \quad \text{or} \quad \ln\left(\frac{\rho}{\rho_0}\right) = \frac{P}{K_0}.$$

This is transformed simply into the required result.

(ii) With $K = K_0 + BP$ we have

$$\int_{\rho_0}^{\rho} \frac{\mathrm{d}\rho}{\rho} = \int_0^P \frac{\mathrm{d}p}{K_0 + Bp} \quad \text{or} \quad \ln\left(\frac{\rho}{\rho_0}\right) = \frac{1}{B}\ln\left(\frac{K_0 + BP}{K_0}\right) = \ln\left(1 + B\frac{P}{K_0}\right)^{1/B}.$$

Taking the exponential of both sides

$$\rho = \rho_0\left(1 + B\frac{P}{K_0}\right)^{1/B} \quad \text{or} \quad \frac{K_0 + BP}{\rho^B} = \frac{K}{\rho^B} = \frac{K_0}{\rho_0^B} = \text{constant}.$$

(iii) Here the form of the relationship comes from

$$\ln\left(\frac{\rho}{\rho_0}\right) = \int_0^P \frac{dp}{K_0 + Bp + Cp^2}.$$

This can be solved by completing the square in the divisor of the right-hand side. We write the right-hand side as

$$\frac{1}{C}\int_0^P \frac{dp}{\frac{K_0}{C} + \frac{Bp}{C} + p^2} = \frac{1}{C}\int_0^P \frac{dp}{\left(p + \frac{B}{2C}\right)^2 + \left(\frac{K_0}{C} - \frac{B^2}{4C^2}\right)}.$$

Writing $B/2C = s$, $[(K_0/C) - (B^2/4C^2)] = t^2$ and $p + s = tx$, the integral becomes

$$\frac{1}{Ct}\int_{s/t}^{(P+s)/t} \frac{dx}{1 + x^2} = \frac{1}{Ct}\left[\tan^{-1}\left(\frac{P+s}{t}\right) - \tan^{-1}\left(\frac{s}{t}\right)\right].$$

Substituting for s and t gives the required solution.

TOPIC Q

Q.1 The thermometric conductivity for the asteroid material is

$$\alpha = \frac{80}{7850 \times 500} \, m^2\,s^{-1} = 2.0 \times 10^{-5}\,m^2\,s^{-1}.$$

Taking the dimension of the asteroid as its radius the cooling time to $1/e$ of its original temperature

$$\tau = \frac{L^2}{\alpha} = \frac{10^8}{2 \times 10^{-5}}\,s = 1.6 \times 10^5\,yr.$$

The cooling time to the environment temperature will be a few times this, of order one million years.

Q.2 We use the equation

$$\frac{\partial T}{\partial t} = \frac{\zeta}{\rho C_P}\frac{\partial^2 T}{\partial x^2} + \frac{h}{\rho C_P}.$$

We are given the values of $\partial T/\partial x$ at the surface and at a depth of 10 km and at 5 km we estimate

$$\frac{\partial^2 T}{\partial x^2} = \frac{-1.83 \times 10^{-3} + 1.71 \times 10^{-3}}{10^4}\,K\,m^{-2} = -1.2 \times 10^{-8}\,K\,m^{-2}.$$

Inserting other values we find

$$\frac{\partial T}{\partial t} = -\frac{20}{3 \times 10^3 \times 800} \times 1.2 \times 10^{-8} + \frac{3 \times 10^{-7}}{3 \times 10^3 \times 800}\,K\,s^{-1} = 2.5 \times 10^{-14}\,K\,s^{-1}.$$

This is equivalent to heating at 0.79 K per million years.

TOPIC R

R.1 (i) Equation (R.11) shows that when $m(r)$ is small then so is the gradient of the density. In fact for smoothness of the density variation through the centre the gradient must be zero there.

(ii) Assuming a constant density in the inner core

$$m(r) = \tfrac{4}{3}\pi(1 \times 10^6)^3 \times 1.29 \times 10^4 \,\mathrm{kg} = 5.40 \times 10^{22} \,\mathrm{kg}.$$

Also

$$V_P^2 - \tfrac{4}{3}V_S^2 = (1.12 \times 10^4)^2 - \tfrac{4}{3}(3.6 \times 10^3)^2 \,\mathrm{m^2\,s^{-2}} = 1.08 \times 10^8 \,\mathrm{m^2\,s}.$$

From equation (R.11)

$$\frac{\mathrm{d}\rho}{\mathrm{d}r} = -\frac{G \times 5.40 \times 10^{22} \times 1.29 \times 10^4}{1.08 \times 10^8 \times (1 \times 10^6)^2} \,\mathrm{kg\,m^{-4}} = -4.30 \times 10^{-4}\,\mathrm{kg\,m^{-4}}.$$

If this is the average gradient between the centre and the inner-core boundary then the density variation within the inner core is

$$\Delta\rho = 1.3 \times 10^6 \times 4.30 \times 10^{-4} \,\mathrm{kg\,m^{-3}} = 559 \,\mathrm{kg\,m^{-3}}.$$

The variation is just over 4% of the central density, that confirms the validity of the assumption that the density is constant within the inner core.

TOPIC S

S.1 Following the approach leading to equation (S.7) the moment of inertia is

$$I = \frac{8}{3}\pi\rho_0 \int_0^R \left(1 - \frac{r}{R}\right)^n r^4 \,\mathrm{d}r.$$

This can be simplified by the substitution $x = 1 - (r/R)$ to give

$$I = \frac{8}{3}\pi R^5 \rho_0 \int_0^1 x^n(1-x)^4 \,\mathrm{d}x = \frac{8}{3}\pi R^5 \rho_0 \left(\frac{1}{n+1} - \frac{4}{n+2} + \frac{6}{n+3} - \frac{4}{n+4} + \frac{1}{n+5}\right).$$

Similarly the mass is given by

$$M = 4\pi\rho_0 \int_0^R \left(1 - \frac{r}{R}\right)^n r^2 \,\mathrm{d}r = 4\pi R^3 \rho_0 \int_0^1 x^n(1-x)^2 \,\mathrm{d}x = 4\pi R^3 \rho_0 \left(\frac{1}{n+1} - \frac{2}{n+2} + \frac{1}{n+3}\right).$$

Substituting for the mass in the expression for I gives

$$I = \tfrac{2}{3} MR^2 \frac{\dfrac{1}{n+1} - \dfrac{4}{n+2} + \dfrac{6}{n+3} - \dfrac{4}{n+4} + \dfrac{1}{n+5}}{\dfrac{1}{n+1} - \dfrac{2}{n+2} + \dfrac{1}{n+3}}.$$

With $n = 7.57$ this gives $\alpha = 0.055$.

TOPIC T

T.1 Expressing the equatorial radius as a and the polar radius as c then from equation (S.12)

$$J_2 = \frac{C - A}{Ma^2} = \frac{1}{2}\alpha\left(1 - \frac{c^2}{a^2}\right).$$

The flattening $f = 1 - (c/a)$ or $c/a = 1 - f$. Hence

$$\alpha = \frac{2J_2}{1 - (1 - f)^2} = \frac{2 \times 1.082 \times 10^{-3}}{1 - 0.9966^2} = 0.319.$$

The accepted value is 0.331.

T.2   The gravitational potential is

$$\Phi = -\frac{GM}{r}\left[1 - J_2\left(\frac{r_e}{r}\right)^2 \frac{1}{2}(3\cos^2\theta - 1)\right] = -\frac{GM}{r}\left[1 + \frac{1}{2}J_2\left(\frac{r_e}{r}\right)^2\right] \quad \text{for } \theta = \pi/2.$$

It should be noted that the radial and tangential components of the field are given by

$$F_r = -\frac{\partial\Phi}{\partial r} \quad \text{and} \quad F_\theta = \frac{1}{r}\frac{\partial\Phi}{\partial\theta}.$$

Because of the form of the potential there is no tangential component either at the pole ($\theta = 0$) or at the equator ($\theta = \pi/2$).

At the equator

$$E_r = -\frac{GM}{r^2} - \frac{3}{2}J_2\frac{GMr_e^2}{r^4} = -\frac{GM}{r^2}\left[1 + \frac{3}{2}J_2\left(\frac{r_e}{r}\right)^2\right].$$

With $r_e/r = 6371/6471$ we find $E_r = -(GM/r^2)(1 + 0.0016)$. Hence the J_2 term enhances the field by 0.16%.

TOPIC U

U.1   From equations (U.8), (S.9) and (T.2), in the notation of Topic U,

$$\Omega = \frac{3GM\sin 2\theta}{2R^3\omega} \times \frac{C - A}{C} = \frac{3GM\sin 2\theta}{2R^3\omega}\frac{J_2 M_\oplus a^2}{\alpha M_\oplus a^2} = \frac{3GMJ_2\sin 2\theta}{2R^3\omega\alpha}.$$

Hence

$$\alpha = \frac{3GMJ_2\sin 2\theta}{2R^3\omega\Omega}.$$

We are given $\theta = 27°$ and $\omega = 7.29 \times 10^{-5}\,\text{s}^{-1}$, the spin angular speed of the Earth. The factor converting seconds of arc to radians is

$$\frac{1}{3600} \times \frac{\pi}{180} = 4.848 \times 10^{-6} \quad \text{so} \quad \Omega = \frac{0.204 \times 4.848 \times 10^{-6}}{50 \times 3600}\,\text{s}^{-1} = 5.49 \times 10^{-12}\,\text{s}^{-1}.$$

This gives

$$\alpha = \frac{3 \times 6.67 \times 10^{-11} \times 7.35 \times 10^{22} \times 1.08 \times 10^{-3} \times \sin 54°}{2 \times (3.65 \times 10^8)^3 \times 7.29 \times 10^{-5} \times 5.49 \times 10^{-12}} = 0.330.$$

TOPIC V

V.1   Using normal units, from equation (V.5a)

$$F(8 \times 10^7, 0) = 2C_1\left(\frac{1}{8 \times 10^7}\right)^3 + 4C_3\left(\frac{1}{8 \times 10^7}\right)^5 = 2.00 \times 10^{-4}$$

and from equation (V.5b)

$$F(1 \times 10^8, \pi/2) = C_1 \left(\frac{1}{10^8}\right)^3 - \frac{1}{4} C_3 \left(\frac{1}{10^8}\right)^5 = 4 \times 10^{-5}.$$

Solving these simultaneous equations gives $C_1 = 4.08 \times 10^{19}\,\mathrm{T\,m^3}$ and $C_3 = 3.32 \times 10^{34}\,\mathrm{T\,m^5}$. With these values at the polar surface

$$F_{dipole} = 2 \times 4.08 \times 10^{19} \left(\frac{1}{5 \times 10^7}\right)^3 = 6.53 \times 10^{-4}\,\mathrm{T}$$

and

$$F_{quad} = 4 \times 3.32 \times 10^{34} \left(\frac{1}{5 \times 10^7}\right)^5 = 4.25 \times 10^{-4}\,\mathrm{T}.$$

Hence the quadrupole component is a fraction $4.25/(4.24 + 6.53) = 0.39$ of the field.

TOPIC W

W.1 From equation (W.9)

$$r = \left(\frac{(1.6 \times 10^{20})^2}{2 \times 4\pi \times 10^{-7} \times 2 \times 10^5 \times 1.67 \times 10^{-27} \times (3 \times 10^5)^2}\right)^{1/6} = 2.6 \times 10^9\,\mathrm{m}.$$

This is about $36 R_{Jupiter}$.

TOPIC X

X.1 At quadrature the Venus–Earth line is tangential to the orbit of Venus. Hence the Earth–Venus distance comes from

$$r_{EB}^2 = r_{ES}^2 - r_{VS}^2 = (1.50 \times 10^{11})^2 - (1.08 \times 10^{11})^2\,\mathrm{m^2} = 1.08 \times 10^{22}\,\mathrm{m^2}.$$

From equation (X.2) and $A_{Venus} = 0.650$ (table X.1) we find

$$I_{Venus} = \frac{0.65 \times (6.05 \times 10^6)^2}{8\pi \times 1.08 \times 10^{22} \times (1.08 \times 10^{11})^2}\, L_{SV}\,\mathrm{W\,m^{-2}} = 7.51 \times 10^{-33} L_{SV}\,\mathrm{W\,m^{-2}}.$$

However, since Venus is only half-illuminated as seen from Earth the actual intensity will be one half that value, or $3.75 \times 10^{-33} L_{SV}\,\mathrm{W\,m^{-2}}$.

Jupiter is closest when the Earth is on the line between Jupiter and the Sun (Jupiter is at opposition). The distance

$$r_{EJ} = r_{JS} - r_{ES} = 7.78 \times 10^{11} - 1.50 \times 10^{11}\,\mathrm{m} = 6.28 \times 10^{11}\,\mathrm{m}.$$

From equation (X.2) with $A_{Jupiter} = 0.520$

$$I_{Jupiter} = \frac{0.52 \times (7.15 \times 10^7)^2}{8\pi \times (6.28 \times 10^{11})^2 \times (7.78 \times 10^{11})^2}\, L_{SV}\,\mathrm{W\,m^{-2}} = 4.43 \times 10^{-34}\,\mathrm{W\,m^{-2}}.$$

Hence Venus is brighter by a factor 8.5.

X.2 The equilibrium temperature is given by equation (X.4) with $A_B = 0$. For Icarus $a = 1.08$ AU and $e = 0.827$, giving a perihelion $q = 0.187$ AU $= 2.80 \times 10^{10}$ m and the aphelion $Q = 1.97$ AU $= 2.96 \times 10^{11}$ m. Substituting these values for r_{BS} gives

$$T_q = \frac{1}{2} \left(\frac{4 \times 10^{26}}{\pi \times 5.67 \times 10^{-8} \times (2.80 \times 10^{10})^2} \right)^{1/4} K = 650\,K$$

$$T_Q = \frac{1}{2} \left(\frac{4 \times 10^{26}}{\pi \times 5.67 \times 10^{-8} \times (2.96 \times 10^{11})^2} \right)^{1/4} K = 200\,K.$$

TOPIC Y

Y.1 The maximum tide is when the Sun, Earth and Moon are collinear, when the Earth is at perihelion and the Moon at perigee. The stretching tidal forces per unit mass are then additive and equal to

$$F_{max} = \frac{2GM_M r_\oplus}{a_M^3 (1 - e_M)^3} + \frac{2GM_\odot r_\oplus}{a_\oplus^3 (1 - e_\oplus)^3} = 4.33 \times 10^{-3} Gr_\oplus.$$

For the minimum tide the Moon must be at apogee and also at quadrature so that the Sun's compressive force opposes the stretching force of the Moon. In order that this reduces the effect of the Moon the Earth must be at *perihelion*. This gives the minimum tidal effect

$$F_{min} = \frac{2GM_M r_\oplus}{a_M^3 (1 + e_M)^3} - \frac{GM_\odot r_\oplus}{a_\oplus^3 (1 - e_\oplus)^3} = 9.57 \times 10^{-4} Gr_\oplus.$$

Thus the ratio is about 4.5.

Y.2 From equation (Y.5), ignoring the second term the angular momentum of the system is

$$a_\oplus M_\oplus r_\oplus^2 \omega_\oplus + (GM_\oplus R)^{1/2} M_M = 3.47 \times 10^{34} \,kg\,m^2\,s^{-1}.$$

When the day equals the month almost all the angular momentum will reside in the Moon. This is because then $M_M R^2 \gg M_\oplus r_\oplus^2$. Since for the lunar orbit $\omega^2 = GM_\oplus / R^3$ approximately then the angular momentum in the Moon's orbit is

$$M_M R^2 \omega = \left(\frac{G^2 M_\oplus^2}{\omega} \right)^{1/3} M_M.$$

For a 50-day orbit $\omega = 1.454 \times 10^{-6}\,s^{-1}$ and the angular momentum is $3.52 \times 10^{34} \,kg\,m^2\,s^{-1}$, close to the present angular momentum of the Earth–Moon system.

TOPIC AA

AA.1 The tensile strength does not influence disruption if, in equation (AA.10), the value of a_S is so large that the first term of the divisor within the bracket becomes insignificant. In that case

$$R_D = \left(\frac{9M_P}{4\pi\rho_S} \right)^{1/3} = \left(\frac{3\rho_P}{\rho_S} \right)^{1/3} a_P.$$

If the densities of satellite and primary are equal then $R_D = 3^{1/3} a_P$.

AA.2 From equation (AA.10)

$$a_S^2 = \frac{2T}{G\rho_S\left(\dfrac{3M_P}{R_D^3} - \dfrac{4}{3}\pi\rho_S\right)}.$$

Taking $R_D = 10^7$ m and the other values given in the question $a_S = 169$ km.

TOPIC AB

AB.1 Equation (AB.6) in its general form is

$$W = (GM_SM_P)^2 \frac{27\pi a_S e_S}{8R_S^6 Q P_S Y}$$

where subscript S refers to the satellite and P to the planet. For the Earth–Moon system:

$$M_{\text{Moon}} = 7.4 \times 10^{22}\,\text{kg}, \qquad M_{\text{Earth}} = 6.0 \times 10^{24}\,\text{kg}, \qquad a_{\text{Moon}} = 1.74 \times 10^6\,\text{m},$$

$$e_{\text{Moon}} = 0.056, \qquad R_{\text{Moon}} = 3.84 \times 10^8\,\text{m}, \qquad P_{\text{Moon}} = 27.3\,\text{d} = 2.36 \times 10^6\,\text{s}.$$

Inserting these values and other quantities as in section AB.2 gives $W = 1.2 \times 10^{10}$ W.

AB.2 For the Saturn–Mimas case

$$M_{\text{Mimas}} = 4.5 \times 10^{19}\,\text{kg}, \qquad M_{\text{Saturn}} = 5.7 \times 10^{26}\,\text{kg}, \qquad a_{\text{Mimas}} = 1.95 \times 10^5\,\text{m},$$

$$e_{\text{Mimas}} = 0.02, \qquad R_{\text{Mimas}} = 1.86 \times 10^8\,\text{m}, \qquad P_{\text{Mimas}} = 8.17 \times 10^4\,\text{s}.$$

Inserting these values and other quantities as in section AB.2 gives $W = 3.6 \times 10^9$ W.

TOPIC AC

AC.1 The solar wind force on the grain is the ram pressure of the solar wind times the cross-sectional area of the grain. This is

$$F_{\text{SW}} = n\mu u^2 \times \pi a^2.$$

The gravitational force is

$$F_g = \frac{GM_\odot}{r^2} \times \frac{4}{3}\pi a^3 \rho.$$

Equating the two forces

$$a = \frac{3n\mu u^2 r^2}{4GM_\odot \rho} = \frac{3 \times 5 \times 10^6 \times 1.67 \times 10^{-27} \times (4 \times 10^5)^2 \times (1.5 \times 10^{11})^2}{4 \times 6.67 \times 10^{-11} \times 2 \times 10^{30} \times 2 \times 10^3}\,\text{m} = 8.5 \times 10^{-11}\,\text{m}.$$

TOPIC AD

AD.1 (i)

$$V_{\text{Teth}} = \left(\frac{GM_{\text{sat}}}{R_{\text{Teth}}}\right)^{1/2} = \left(\frac{6.67 \times 10^{-11} \times 5.685 \times 10^{26}}{2.95 \times 10^8}\right)^{1/2}\,\text{m s}^{-1}$$

$$= 1.133 \times 10^4\,\text{m s}^{-1} = 2.392\,\text{AU yr}^{-1}.$$

(ii) For the trailing satellite

$$x = 2.95 \times 10^8 \times \cos 30° \, \text{m} = 2.555 \times 10^8 \, \text{m} = 1.703 \times 10^{-3} \, \text{AU}.$$
$$y = 2.95 \times 10^8 \times \sin 30° \, \text{m} = 1.475 \times 10^8 \, \text{m} = 9.833 \times 10^{-4} \, \text{AU}.$$

For the leading satellite

$$x = -1.703 \times 10^{-3} \, \text{AU}, \qquad y = 9.833 \times 10^{-4} \, \text{AU}.$$

(iii) For the trailing satellite

$$V_x = -2.392 \times \cos 60° = -1.196 \, \text{AU yr}^{-1}$$
$$V_y = 2.392 \times \sin 60° \, \text{AU yr}^{-1} = 2.071 \, \text{AU yr}^{-1}.$$

For the leading satellite

$$V_x = -1.196 \, \text{AU yr}^{-1}, \qquad V_y = 2.071 \, \text{AU yr}^{-1}.$$

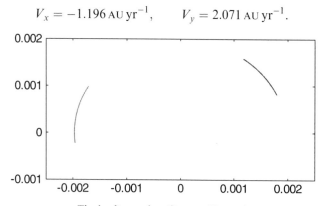

The leading and trailing satellite paths.

For this problem the table corresponding to table AD.1 is

(iv)

	Saturn	Tethys	Satellite 1	Satellite 2
Mass	2.843×10^{-4}	3.13×10^{-10}	0	0
x	0	0	-1.703×10^{-3}	1.703×10^{-3}
y	0	1.967×10^{-3}	9.833×10^{-4}	9.833×10^{-4}
z	0	0	0	0
V_x	0	-2.392	-1.196	-1.196
V_y	0	0	-2.071	2.071
V_z	0	0	0	0

The figures show the large-scale view of the satellite paths. Because of the small mass of Tethys relative to Saturn the satellites are very loosely bound and oscillate about the mean position with a very large amplitude. A larger scale view of part of the paths shows that they are in a state of oscillation about a mean position.

TOPIC AE

AE.1   (a) (i)
$$x_{\text{sol}} = \left(\frac{3C(T_m - T_0)}{4\pi G \rho} \right)^{1/2} = \left(\frac{3 \times 1000 \times 1300}{4\pi G \times 3200} \right)^{1/2} \, \text{m} = 1206 \, \text{km}$$

(ii)
$$x_{liq} = \left(\frac{3C(T_m - T_0) + L}{4\pi G\rho} \right)^{1/2} = 1281 \text{ km}$$

(b) From equation (AE.10), with body radius R, $T_x = (4\pi G\rho R^2 - L)/3C + T_0 = 1736 \text{ K}$.

(c) The potential energy per unit mass for a uniform sphere is

$$\Omega = -\frac{3GM}{5R} = -\frac{4\pi GR^2\rho}{5}$$

Equating $-\Omega$ with $C\Delta\theta$ gives

$$T_{av} = T_0 + \Delta\theta = T_0 + \frac{4\pi GR^2\rho}{5C} = 1201 \text{ K}.$$

TOPIC AF

AF.1 The distance to the Sun, $D_{Sun} = 60\,000 \text{ AU} = 9 \times 10^{15} \text{ m}$.
The distance to the comet, $D_{Com} = (60\,000 - 40\,000\cos 45°) \text{ AU} = 4.76 \times 10^{15} \text{ m}$.
Hence the impulsive velocity difference is

$$\Delta v_{CS} = \frac{2GM_*}{v_*} \left(\frac{1}{4.76 \times 10^{15}} - \frac{1}{9.00 \times 10^{15}} \right) \text{ m s}^{-1} = 1.72 \text{ m s}^{-1}.$$

The comet is at aphelion so its distance from the Sun is, closely, 40000 AU or $6.00 \times 10^{15} \text{ m}$.
Hence the impulsive change in angular momentum is

$$\Delta H = 6.00 \times 10^{15} \times \sin 45° \times 1.72 \text{ m}^2 \text{ s}^{-1} = 7.30 \times 10^{15} \text{ m}^2 \text{ s}^{-1}.$$

The original perihelion is 40 AU or $6.00 \times 10^{12} \text{ m}$. From equation (AF.7) the change in perihelion is

$$\Delta q = \sqrt{\frac{2 \times 6.00 \times 10^{12}}{G \times 2.6 \times 10^{30}}} \times 7.30 \times 10^{15} \text{ m} = 12.8 \text{ AU}.$$

Since the perihelion has reduced its final value is 27.2 AU.

TOPIC AG

AG.1 From equation (AG.10) we have

$$t = \frac{4\pi ac^2\rho}{3L_\odot} \left(r_0^2 - r_f^2 \right). \tag{i}$$

Designating the spheres by subscripts 1 and 2 then at the time they are at the same distance from the Sun

$$a_1(r_{0,1}^2 - r_f^2) = a_2(r_{0,2}^2 - r_f^2)$$

or

$$r_f^2 = \frac{a_2 r_{0,2}^2 - a_1 r_{0,1}^2}{a_2 - a_1} = \frac{0.015 \times (1.5 \times 10^{11})^2 - 0.005 \times (2 \times 10^{11})^2}{0.015 - 0.005} \text{ m}^2$$

giving $r_f = 1.17 \times 10^{11}$ m. To find the time we use (i) for one of the bodies and this gives $t = 2.54 \times 10^{14}$ s $= 8.04 \times 10^7$ yr.

TOPIC AH

AH.1   For this problem $T = 30$ K, $\rho = 10^{-8}$ kg m^{-3} and $\mu = 3.34 \times 10^{-27}$ kg. From equation (AH.3)

$$l = \left(\frac{\pi k T}{G \rho \mu} \right)^{1/2} = 7.64 \times 10^{11} \text{ m}.$$

Hence the volume of the blob $V = \pi(\tfrac{1}{2}l)^2 \times l = 3.50 \times 10^{35}$ m^3 and its mass $M_{\text{blob}} = V\rho = 3.50 \times 10^{27}$ kg.

From equation (D.4) the Jeans critical mass is

$$M_J = \left(\frac{375 k^3 T^3}{4 \pi \mu^3 G^3 \rho} \right)^{1/2} = 4.38 \times 10^{27} \text{ kg}.$$

The ratio of the blob mass to the Jeans critical mass is 0.80. If a condensation is to form then the filament must be about 12% thicker than that described—or perhaps somewhat more to allow for the fact that the blob would not be in an optimum spherical shape. The mass of the blob is

$$\frac{1}{4} \pi \rho l^3 = \frac{1}{4} \left(\frac{\pi^5 k^3 T^3}{G^3 \mu^3 \rho} \right)^{1/2}.$$

Comparing this with the expression for the Jeans critical mass it is seen that they differ only by a factor

$$\frac{\pi^{5/2}}{4} \sqrt{\frac{4\pi}{375}} = 0.80$$

which was what was found in the numerical example.

TOPIC AI

AI.1   The angular momentum of a planet in a circular orbit is $\sqrt{G M_\odot r_p} M_p$ where M_p is the mass of the planet and r_p is the radius of its circular orbit. This gives the following table of present and original angular momenta of the planets:

Planet	Original angular momentum (kg m^2 s^{-1})	Present angular momentum (kg m^2 s^{-1})
Jupiter		1.938×10^{43}
Saturn	7.629×10^{42}	7.854×10^{42}
Uranus	1.342×10^{42}	1.697×10^{42}
Neptune	1.774×10^{42}	2.512×10^{42}

The original angular momentum of Jupiter was 2.070×10^{43} kg m^2 s^{-1} for conservation corresponding to an orbital radius of 5.94 AU.

The energy of a planet in a circular orbit is $-(GM_\odot M_p)/2r_p$ and this gives the following table of original and present energies:

Planet	Original energy (J)	Present energy (J)
Jupiter	-1.422×10^{35}	-1.623×10^{35}
Saturn	-2.809×10^{34}	-2.650×10^{34}
Uranus	-3.209×10^{33}	-2.007×10^{34}
Neptune	-3.036×10^{33}	-1.514×10^{34}

This corresponds to a loss of energy 1.589×10^{34} J.

TOPIC AJ

AJ.1 The rate of loss of energy due to mass emission is $\frac{1}{2}v^2(dM/dt) = 5 \times 10^{23}$ J s^{-1}.
For spin $E_\Omega = \frac{1}{2}I\Omega^2 = \frac{1}{2}\alpha MR^2\Omega^2$ then

$$\frac{dE_\Omega}{dt} = \frac{1}{2}\alpha R^2\Omega^2 \frac{dM}{dt} + \alpha R^2 M\Omega \frac{d\Omega}{dt}.$$

From equation (AJ.9)

$$\frac{dE_\Omega}{dt} = \frac{1}{2}\alpha R^2\Omega^2 \frac{dM}{dt} + r_c^2\Omega^2 \frac{dM}{dt} \approx r_c^2\Omega^2 \frac{dM}{dt}.$$

From equation (AJ.6)

$$r_c = \left(\frac{2\pi \times (5 \times 10^{24})^2}{4\pi \times 10^{-7} \times 10^6 \times 10^{12}}\right)^{1/4} = 3.34 \times 10^9 \text{ m}.$$

The period of one day corresponds to $\Omega = 7.27 \times 10^{-5}$ s^{-1} giving $dE_\Omega/dt = 5.90 \times 10^{22}$ J s^{-1}.

TOPIC AK

AK.1 For Venus $\sigma = 1304$ kg m^{-2}, $P = 0.615$ years and $M_p = 4.87 \times 10^{24}$ kg. Inserting in equation (AK.10) gives

$$t_{form} = \frac{3 \times 0.615}{8 \times 1304}\left(\frac{4 \times 3 \times 10^3}{3\pi^2}\right)^{2/3} \times (4.87 \times 10^{24})^{1/3} \text{ yr} = 1.64 \times 10^6 \text{ yr}.$$

For the Saturn core $\sigma = 99$ kg m^{-2}, $P = 29.46$ years and $M_p = 4.8 \times 10^{25}$ kg. This gives $t_{form} = 2.2 \times 10^9$ years.

TOPIC AL

AL.1 Since $V_{esc}^2 = 2GM/R$ we have $D/R = \sqrt{1 + (V_{esc}^2/V^2)}$. This gives

V/V_{esc}	0.1	0.5	1.0	2.0	5.0	10.0
D/R	10.04	2.23	1.41	1.12	1.02	1.005

TOPIC AP

AP.1   For an infinite medium $dV/dt = -2\pi G\sigma$ so that the time to come to rest is $t = V_0/2\pi G\sigma$, where V_0 is the initial speed. This gives $t = 1.50 \times 10^{14}\,\text{s} = 4.74 \times 10^6\,\text{yr}$. Since the deceleration is uniform the distance travelled, d, is average speed $\times$ time or $d = 5 \times 10^3 \times 1.50 \times 10^{14}\,\text{m} = 7.5 \times 10^{17}\,\text{m} = 24.3\,\text{pc}$.

TOPIC AQ

AQ.1   From equation (AQ.2), $R_P = 5.2 \times 10^{14} \times 0.0625\,\text{m} = 3.25 \times 10^{13}\,\text{m}$.
From equation (AQ.4), $A_e = 9.72 \times 10^{27}\,\text{m}^2$.
From equation (AQ.5), $t_e = 5.63 \times 10^{11} \times 0.0625\,\text{s} = 3.52 \times 10^{10}\,\text{s}$.
Hence $P_{int} = 0.00023$.
However, since there are ten times as many brown dwarfs as stars the probability that a star has a suitable interaction with a brown dwarf is

$$P_{star}(20\,000) = 10 \times 0.00023 = 0.0023.$$

REFERENCES

Bailey M E 1983 *Mon Not R Astron Soc* **204** 603–33

Biermann L 1951 *Z Astrophys* **29** 274–86

Benz W, Slattery W L and Cameron A G W 1987 *Icarus* **71** 515–35

Burford R O 1966 *Proceedings of the Second International Symposium on Recent Crustal Movements* (ed V Auer and Kukkamaki) Helsinki

Butler R P and Marcy G W 1996 *Astrophys J* **464** L153–56

Charbonneau D, Brown T M, Latham D W and Mayor M 2000 *Astrophys J* **529** L45–48

Clayton R N, Grossman L and Mayeda T K 1973 *Science* **187** 485–8

Clube S V M and Napier W M 1984 *Mon Not R Astron Soc* **208** 575–88

Cole G H A 1984 *Physics of Planetary Interiors* (Bristol: Adam Hilger)

Cole G H A 1986 *Inside a Planet* (Hull: Hull University Press)

Connell A J and Woolfson M M 1983 *Mon Not R Astron Soc* **204** 1221–30

Darwin G H 1878 *Nature* **18** 580–2

Darwin G H 1906 *Phil Trans R Soc Lond* **A206** 161–248

Dewey F 1972 *Sci Amer* **226** 56–68

Dodd R T 1982 *Meteorites* (Cambridge: Cambridge University Press)

Dormand J R and Woolfson M M 1974 *Proc R Soc Lond* **A340** 307–31

Dormand J R and Woolfson M M 1977 *Mon Not R Astron Soc* **180** 243–279

Dormand J R and Woolfson M M 1989 *The Origin of the Solar System: the Capture Theory* (Chichester: Ellis Horwood)

Freeman F W 1978 *The Origin of the Solar System* (ed S F Dermott) (Chichester: Wiley) pp 635–40

Golanski Y and Woolfson M M 2001 *Mon Not R Astron Soc* **320** 1–12

Goldreich P and Ward W R 1973 *Astrophys J* **183** 1051–61

Gutenberg B and Richter C F 1954 *Seismicity of the Earth and Associated Phenomena* (Princeton: Princeton University Press)

Hayashi C 1961 *Publ Astron Soc Japan* **13** 450–2

Heezen B C and Menard H W 1963 In *The Sea* vol III (ed M N Hill) (New York: Interscience)

Holden P and Woolfson M M 1995 *Earth, Moon and Planets* **69** 201–36

Hurley P M and Rand J R 1969 *Science* **164** 1229–42

Iben I 1965 *Astrophys J* **141** 993–1018

Jeans J H 1917 *Mon Not R Astron Soc* **77** 186–199

Jeffreys H 1929 *Mon Not R Astron Soc* **89** 636–41

Jeffreys H 1930 *Mon Not R Astron Soc* **91** 169–73

Kuiper G P 1951 *Astrophysics: A Topical Symposium* (ed J A Hynek) (New York: McGraw-Hill) p 357

Laplace P S de 1796 *Exposition du Système du Monde* (Paris: Imprimerie Cercle-Social)

Lee T, Papanastassiou D A and Wasserberg G T 1976 *Geophys Res Lett* **3** 109–12

Lynden-Bell D and Pringle J E 1974 *Mon Not R Astron Soc* **168** 603–37

Marsden B G, Sekanina Z and Everhart E 1978 *Astron J* **83** 64–71

Mayor M and Queloz D 1995 *Nature* **378** 355–359

McCall G J 1973 *Meteorites and their Origins* (Newton Abbot: David & Charles)

Melita M D and Woolfson M M 1996 *Mon Not R Astron Soc* **280** 854–862

Michael D M 1990 *Evidence of a planetary collision in the early solar system and its implications for the origin of the solar system* DPhil thesis: University of York

Moore P and Hunt G 1983 The Atlas of the Solar System (London: Mitchell Beazley)

Mullis A M 1993 *Geophys J Int* **114** 196–208

Peale S J, Cassen P and Reynolds R T 1979 *Science* **203** 892–4

Ruskol E L 1955 Les particules solides dans les astres *Mém Soc Sci Liége* p 160

Ruskol E L 1960 *Sov Astron AJ* **4** 657–68

Russell H N 1935 *The Solar System and its Origin* (New York: Macmillan)

Rutherford Lord Ernest and Boltwood B B 1906 *Amer J Sci* 4 **22** 1–3

Safronov V S 1972 *Evolution of the Protoplanetary Cloud and Formation of the Earth and Planets* (Jerusalem: Israel Program for Scientific Translations)

Sclater J C, Fischer R L, Patriat P, Tapscott C and Parsons B 1981 *Geophys J R Astr Soc* **65** 587–604

Spitzer L 1939 *Astrophys J* **90** 675–88

Stewart G R and Wetherill G W 1988 *Icarus* **74** 542–53

Toksöz M N and Solomon S C 1973 *Moon* **7** 251–78

Watts A B and Daly S F 1981 *Ann Rev Earth Planet Sci* **9** 415–48

Weidenschilling S J 1980 *Icarus* **44** 172–89

Weidenschilling S J 1995 *Icarus* **116** 433

Weidenschilling S J, Donn B and Meakin P 1989 *The Formation and Evolution of Planetary Systems* (ed H A Weaver and L Danley) (Cambridge: Cambridge University Press) pp 131–50

Whitworth A P, Boffin H, Watkins S and Francis N 1998 *Astron Geophys* **290** 10–13

Whitworth A P, Chapman S J, Bhattal A S, Disney M J, Pongracic H and Turner J A 1995 *Mon Not R Astron Soc* **277** 727–46

Wierchert E and Zöppritz L 1907 *Nachr der König Gesellschaft der Wissenshcaft Göttingen Math-Phys* Kl 529

Williams I P and Cremin A W 1969 *Mon Not R Astr Soc* **144** 359–73

Woolfson M M 1964 *Proc R Soc Lond* **A282** 485–507

Woolfson M M 1979 *Phil Trans R Soc* **A291** 219–52

Woolfson M M 1999 *Mon Not R Astron Soc* **304** 195–8

Woolfson M M 2000 *The Origin and Evolution of the Solar System* (Bristol: Institute of Physics)

Yoder C F 1979 *Nature* **279** 767—70

INDEX